A-PLUS NOTES
FOR
ALGEBRA

(Algebra 2 and Pre-Calculus)

By
Rong Yang

Printed in the United States of America

A-Plus Notes Learning Center

Library of Congress Control Card Number: 2005910600
Washington, DC 20540-4320.

ISBN–13: 978-0-9654352-4-6
ISBN–10:   0-9654352-4-5
Printed in the United States of America.

A-Plus Notes Learning Center
A publisher
Redondo Beach, CA 90278
Tel: (310) 542-0029
Fax: (310) 542-6837

WHERE TO BUY:
Available at Barnes & Noble, Borders, and local educational bookstores
across the United States, such as Teacher Supply, School Supply,····, etc.

HOW TO ORDER:
Quantity discounts are available from the publisher. Please include information
on your letterhead concerning the number of books you wish to purchase.
See last page for the order form.

2006 Completely Revised and Updated,   10th printing

Preface

I am a math teacher in Los Angeles. For the past years, I have noticed that the students in the United States are different from the students I taught in Asia in many ways. They've come from different countries all over the world and have attained different academic levels. Many of the students have difficulty in comprehending math textbooks and lack basic math concepts and skills. Some of them are sometimes confused about the problem-solving steps in math, such as how to simplify a math expression or solve a math problem.

In order to provide a clear and easy way for my students to learn algebra in my classes, I designed notes which simplify and outline the concepts and skills in the course textbooks. The notes focus the right and easy steps in the problem-solving process. Each topic is organized in one or two pages for easy understanding. For the past few years, I have distributed the notes to my students and encouraged them to use the notes together with their course textbooks. My students have told me that my notes do help to facilitate their learning process in algebra. They have always asked for my notes whenever they have entered my classroom. Some teachers and students from other schools have also contacted me and requested copies of my notes. This is the reason why I have decided to print and publish my notes for use as a reference book to math students in the public.

This reference book provides a clear summary and outline of all topics in algebra. Each topic begins with concepts, definitions, formulas, theorems , and problem-solving steps, and is followed by well-designed examples which illustrate the summary. It covers the topics that most high school and college textbooks have, ranging from basic skills to advanced fields such as exponents, logarithms, series, matrices and determinants, parametric and polar equations, probabilities and statistics. It provides the students a useful guide for review and further study. There are 19,000 examples and exercises in this book-7,000 are real-life word problems modeled after typical questions on standardized tests that are given across the United States. In this book, I tried to cover all possible patterns of the test questions that a student needs to practice.

The Step-by-Step Graphing Calculator Approach shown in this book is specially designed for students who are studying for the new SAT Tests and college students who are studying courses in science and technology.

I wrote this book following the same pattern of the math books used in the schools in Asia. This pattern has been proved as excellent for student-learning.

Hopefully, this reference book can provide some assistance to math education in schools. If you are a math teacher, or a math professional, I would appreciate any comments or suggestions you may have to improve the contents for the next print. You may fax me at (310)542-6837.

Rong Yang
Los Angeles, California

HOW TO USE THIS BOOK

1. Understand the summary and outline in each section.
2. Study the examples and be familiar with the problem-solving steps.
3. Practice the exercises and check your answers at the end of each chapter.
4. Algebra 2 Students: If you have not studied trigonometry before, skip the sections involving trigonometry.
5. Pre-Calculus Students: If you have studied trigonometry before, you should study all chapters in this book.
6. If you are a top math student, read and study this book thoroughly to succeed in getting high scores in all kinds of achievement tests, and win nationwide brainpower math competition such as Super Quiz and Academic Decathlon.

If you do not understand any of the above steps, you should ask:
1. your classmates who can explain them to you;
2. your math teachers; or
3. write down your questions and fax them to the author for a free consultation. Fax number: Mr. Yang (310) 542-6837.

ATTENTION: MATH TEACHERS AND STUDENTS

This book will make the teachers' job and the students' learning process easier if the students have it on hand in the class.

It makes a teacher's job easier because the teacher need not " write notes " on the board in the class. It makes a student's learning process easier because the student need not spend time to " take notes " in the class. This book is a complete set of lecture " notes " for the students. They need only to read it.

Most schools in the United States take back the textbook from the student after the course is completed. The student has no material on hand for future review. This book is simple and easy like a dictionary for the student to review when needed in the future.

This book is strongly recommended for math teachers and students. You may use it as a reference book together with the textbook, or use it as the textbook in your class.

For more information, visit this book and readers' comments, or write your comments at :

www.amazon.com and **www.barnes & noble.com**
www.borders.com

The Publisher

Chapter 5: Polynomial Equations and Functions

Graphing Calculator: (TI-83 and TI-84)

Chapter 6: Rational Equations and Functions

Graphing Calculator: Graphing a rational function   229
   (TI-83 and TI-84)

> **Now, you have completed this book. Are you ready to study Calculus ?**
> **See you soon in next book !!**

CHAPTER ONE

Equations and Problem Solving

1-1 Variables and Expressions

Variable: A letter that can be used to represent one or more numbers.

If $x = 1$, then $2x = 2(1) = 2$. If $x = 2$, then $2x = 2(2) = 4$. x is a variable.

We use variables to form algebraic expressions and equations.

Numerical expression: An expression that includes only numbers.

Algebraic expression: An expression that combines numbers and variables by four operations (add, subtract, multiply or divide).

$xy + x + 2y - 3$ is an algebraic expression.

Algebraic equation: A statement by placing an equal sign "=" between two expressions.

$2x + 5 = 7$ is an equation. $2x + 5 = 9$ is an equation.

Term: A single number or a product of numbers and variables in an expression.

A monomial: An expression with only one term. $5x$ is a monomial.

A polynomial: An expression formed by two or more terms: $xy + x + 2y - 3$.

A polynomial or binomial: An expression formed by only two terms: $3x^2 + 6$.

A polynomial or trinomial: An expression formed by three terms: $3x^2 - 5x - 2$.

Constant: It is a number only. In the expression $x + 3$, 3 is a constant.

Coefficient: A number that is multiplied by a variable.

In the term $2y$, 2 is the coefficient. The coefficient of the term x is 1.

Degree of an expression: The greatest of the degrees of its terms after it has been simplified and its like terms are combined.

The degree of $2x + 1$ is 1, a linear form. The degree of $3x^2 + x - 1$ is 2, a quadratic form.

The degree of $4x^3 + 5x$ is 3, a cubic form. The degree of $x^4 - 1$ is 4, a quartic form.

The degree of $4a^2b + 2a^3b^4 - 2$ is 7.

Evaluating an expression: The process of replacing variables with numbers in an algebraic expression and finding its value.

Examples

1. Evaluate $2x + 15$ when $x = 4$.
Solution:
$2x + 15 = 2(4) + 15 = 23$. Ans.

2. Evaluate $a^2 + b - 2$ when $a = 3$, $b = 1$.
Solution:
$a^2 + b - 2 = 3^2 + 1 - 2 = 9 + 1 - 2 = 8$.
Ans.

3. Evaluate $\frac{1}{2}x + 3$ if $x = 3$.
Solution:
$\frac{1}{2}x + 3 = \frac{1}{2}(3) + 3 = 4\frac{1}{2}$. Ans.

4. Evaluate $\frac{1}{2}x - 3$ if $x = 3$.
Solution:
$\frac{1}{2}x - 3 = \frac{1}{2}(3) - 3 = -1\frac{1}{2}$. Ans.

EXERCISES

Evaluate each expression.

1. $2x+1$ if $x=1$.

2. $2x+1$ if $x=2$.

3. $2x+1$ if $x=3$.

4. $2x-1$ if $x=1$.

5. $2x-1$ if $x=2$.

6. $2x-1$ if $x=3$.

7. $2x+4$ if $x=4$.

8. $2x+4$ if $x=5$.

9. $2x+4$ if $x=6$.

10. $2x-6$ if $x=0$.

11. $2x-6$ if $x=3$.

12. $2x-6$ if $x=5$.

13. $4x-2$ if $x=-1$.

14. $4x-2$ if $x=-2$.

15. $4x-2$ if $x=-3$.

16. $4x+2$ if $x=-1$.

17. $4x+2$ if $x=-2$.

18. $4x+2$ if $x=-3$.

19. $-3x+5$ if $x=1$.

20. $-3x+5$ if $x=-1$.

21. $-3x+5$ if $x=3$.

22. $-2x-4$ if $x=3$.

23. $-2x-4$ if $x=-2$.

24. $-2x-4$ if $x=-3$.

25. $\frac{1}{2}x+5$ if $x=0$.

26. $\frac{1}{2}x+5$ if $x=3$.

27. $\frac{1}{2}x+5$ if $x=4$.

28. $\frac{1}{2}x-2$ if $x=4$.

29. $\frac{1}{2}x-2$ if $x=-3$.

30. $\frac{1}{2}x-2$ if $x=1$.

31. $\frac{3}{2}x+1$ if $x=2$.

32. $\frac{3}{2}x-3$ if $x=3$.

33. $\frac{3}{2}x-2$ if $x=-3$.

34. $2.5x+3$ if $x=4$.

35. $2.5x-3$ if $x=4$.

36. $2.5x-1.5$ if $x=-2$.

37. $2x^2-3x+5$ if $x=3$.

38. $2x^2-3x+5$ if $x=-2$.

39. x^3-x-3 if $x=-1$.

40. $2x^2+y-1$ if $x=1$, $y=2$.   **41.** a^2-2b+3 if $a=3$, $b=-2$.   **42.** $3a^2b$ if $a=2$, $b=3$.

43. The cost to buy a suit is given by the equation $c=p-0.20p$, where p is the regular price, and $0.20p$ is a 20% discount on p. Find the cost to buy the suit if the suit is regularly $50.

44. The cost to rent a car is $20 per day plus 20 cents per mile. Find the cost per day to rent the car if you drive 120 miles. (Hint: $c=20+0.20d$, where d is the mileage.)

45. The area of a square is given by the formula $A=s^2$, where s is the length of one side. Find the area of a square with one side 15 meters.

46. The distance in feet that an object falls in t seconds is given by the formula $d=16t^2$. Find the distance of an object falls in 8 seconds.

1-2 Adding and Subtracting Expressions

Like terms (Similar terms): Terms that contain the same form of variables in an algebraic expression.

$5x$ and $9x$ are like terms. $4a$ and $10a$ are like terms. $3x^2y$ and $7x^2y$ are like terms.

To simplify an expression, we combine the like terms and follow the **order of operations**.

Order of Operations:

   1) Do all operations within the grouping symbols (parentheses, brackets, and braces), start with the innermost grouping symbols.
   2) Do all operations with exponents.
   3) Do all multiplications and divisions in order from left to right.
   4) Do all additions and subtractions in order from left to right.

If there is only a negative sign in front of the parenthesis, we remove the parenthesis by writing **the opposite** of each term inside the parenthesis. Or, consider the negative sign as "-1" and use the **Distributive Property** to remove the parentheses.

Distribution Property: $a(b+c) = ab + ac$ **and** $a(b-c) = ab - ac$

If there are like terms inside the parentheses, we combine them first.

Examples: $-(x-4) = -x+4$ or: $-(x-4) = -1(x)+(-1)(-4) = -x+4$.
$$-(-x-4) = -1(-x)+(-1)(-4) = x+4.$$
$$(-x-4)(-1) = -x(-1)+(-4)(-1) = x+4.$$
$$-(6x-5-2x) = -(4x-5) = -4x+5$$

To add or subtract expressions, we remove the parentheses using the Distributive Property and then regroup like terms.

When we write an expression (a polynomial), we always write it in **standard form**.

In writing standard form, the terms are ordered from the greatest exponent to the least.

Examples

1. $2x+3x = 5x$.   2. $2x-3x = -x$.   3. $-(2x-3x) = -(-x) = x$.

4. $-(2x+5) = -2x-5$.   5. $-(2x-5) = -2x+5$.   6. $-(-2x-5) = 2x+5$.

7. $(3x-5)+(5x-3) = 3x-5+5x-3 = (3x+5x)+(-5-3) = 8x+(-8) = 8x-8$.

8. $(3x-5)-(5x-3) = 3x-5-5x+3 = (3x-5x)+(-5+3) = -2x+(-2) = -2x-2$.

9. $(4x+3y)+(6x-2y) = 4x+3y+6x-2y = (4x+6x)+(3y-2y) = 10x+y$.

10. $(4x+3y)-(6x-2y) = 4x+3y-6x+2y = (4x-6x)+(3y+2y) = -2x+5y$.

11. $3a^2-2a+4-(2a^2+3a) = 3a^2-2a+4-2a^2-3a = (3a^2-2a^2)+(-2a-3a)+4$.
$$= a^2+(-5a)+4 = a^2-5a+4.$$

12. $x^3-2x^2+x-4+2x^3+5x^2+7 = (x^3+2x^3)+(-2x^2+5x^2)+x+(-4+7)$
$$= 3x^3+3x^2+x+3.$$

EXERCISES

Simplify each expression.

1. $3x + 6x$

2. $3x - 6x$

3. $-3x - 6x$

4. $-3x + 6x$

5. $-(3x - 5 + 6x)$

6. $-(3x + 5 - 6x)$

7. $3x + (5 + 6x)$

8. $3x - (5 - 6x)$

9. $-3x + (5 + 6x)$

10. $3x - (-5 - 6x)$

11. $9 - (4x - 6)$

12. $7x - 4y + 2x + 8y$

13. $7x + 4y + (2x + 8y)$

14. $(7x + 4y) - (2x + 8y)$

15. $(7x - 4y) - (2x - 8y)$

16. $(2a + 9) - (6a - 10)$

17. $(4a - 6b) + (-3a - 2b)$

18. $(3u - 4w) + (6u - 5w)$

19. $(5w - 8u) - (2w - 4u)$

20. $(5w + 8u) + (4w - 9u)$

21. $(5m + 6) - (-4m - 7)$

22. $6 + x - (-x - 7)$

23. $-(2 - 3y) - (2y + 12)$

24. $n - (2n + 7)$

25. $-(5 - x) - (x - 12)$

26. $\frac{1}{2}x + \frac{1}{3}y + \frac{1}{3}x + \frac{1}{3}y$

27. $\frac{1}{2}x - \frac{1}{3}y - \frac{1}{3}x + \frac{1}{2}y$

28. $\frac{2}{3}x + \frac{3}{4}y - \frac{1}{4}x - \frac{1}{5}y$

29. $(\frac{1}{5}x + \frac{2}{3}y) + (\frac{3}{5}x - \frac{1}{3}y)$

30. $(\frac{1}{5}x + \frac{2}{3}y) - (\frac{3}{5}x - \frac{1}{3}y)$

31. $(1.2x - 3.5y) + (3.5x + 1.2y)$

32. $(1.2x - 3.5y) - (3.5x + 1.2y)$

33. $4.5x - (6.8x - 4)$

34. $2x^2 + 5 + 4x^2 - 6$

35. $4y^2 - 7 - 6y^2 + 9$

36. $3x^2 - (-x^2 + 7)$

37. $(5x^2 - x) + (4x^2 - 7x)$

38. $(-6x^2 - 4) - (2x^2 + 6)$

39. $-6n^2 + (3n^2 - 4)$

40. $x - 2x^2 - (3x^2 - 4)$

41. $(-a^2 - 2) + (3a^2 - 7)$

42. $(-k^2 + 4) - (k^2 - 4)$

43. $(x + 2y) + (2x - 3y) - (3x + 4y)$

44. $(3a - 4b) - (4a - b) + (9 - 2b)$

45. $(m - 2n + 3) - (4 + 3n) + (2m - 7)$

46. $(3p - 4q) - (p - 2q) - (3p + 4)$

47. $(3x^2 - 2x + 4) + (2x^2 + 4x - 6) - 7x^2$

48. $(3x^2 - 2x + 4) - (2x^2 + 4x - 6) + 7x^2$

49. $a - \{-a - a - [-a + (-a) - (-a) - a]\}$

50. $-(x - y) + \{-x + y - [x + y - (x - y)]\}$

51. Write an expression for the perimeter.

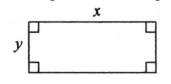

52. Write an expression for the perimeter.

1-3 Multiplying and Dividing Expressions

To multiply any two expressions, we use the **distributive property** and **rule of exponents**.

$$\textbf{Distributive Property: } a(b+c) = ab + ac$$
$$a(b-c) = ab - ac$$

To divide an expression by a monomial (a divisor), each term in the expression is divided by the divisor.

1) $\dfrac{a+b}{c} = \dfrac{a}{c} + \dfrac{b}{c},\ c \neq 0$ **2)** $\dfrac{a-b}{c} = \dfrac{a}{c} - \dfrac{b}{c},\ c \neq 0$

Rules of Exponents (Powers): To multiply two powers having the same base, we add the exponents.

To divide two powers having the same base, we subtract the exponents.

1) $a^m \cdot a^n = a^{m+n}$ **2)** $\dfrac{a^m}{a^n} = a^{m-n}$

Examples:

1. $x^5 \cdot x^2 = x^{5+2} = x^7$.   **2.** $\dfrac{x^5}{x^2} = x^{5-2} = x^3$.

3. $4(3x-5) = 4(3x) + 4(-5) = 12x - 20$.   **4.** $5(-2a+1) = 5(-2a) + 5(1) = -10a + 5$.

5. $3x(2x^2 - 4x - 6) = 3x(2x^2) + 3x(-4x) + 3x(-6) = 6x^3 - 12x^2 - 18x$.

6. $\dfrac{4x+8}{2} = \dfrac{4x}{2} + \dfrac{8}{2} = 2x + 4$.   **7.** $\dfrac{12a-6}{3} = \dfrac{12a}{3} - \dfrac{6}{3} = 4a - 2$.

8. $\dfrac{4x^2 - 8x}{2x} = \dfrac{4x^2}{2x} - \dfrac{8x}{2x} = 2x - 4$.   **9.** $\dfrac{-28x^3 + 7x^2}{7x} = -\dfrac{28x^3}{7x} + \dfrac{7x^2}{7x} = -4x^2 + x$.

10. $\dfrac{20n^3 - 16n^2 - 12n}{4n} = \dfrac{20n^3}{4n} - \dfrac{16n^2}{4n} - \dfrac{12n}{4n} = 5n^2 - 4n - 3$.

11. Write an expression for the area.

$x + 4$

x

Solution:

Area $= x(x+4) = x^2 + 4x$. Ans.

12. Write an expression for the area.

$2n + 5$

$2n$

Solution:

Area $= 2n(2n+5) = 4n^2 + 10n$. Ans.

EXERCISES

Simplify each expression.

1. $x \cdot x$ 2. $2x \cdot 3x$ 3. $-4x \cdot 6x^3$ 4. $-4x^5 \cdot x^7$ 5. $2.4n^6 \cdot 3n^4$

6. $\dfrac{x}{x}$ 7. $-\dfrac{2x}{3x}$ 8. $\dfrac{6x^3}{4x}$ 9. $\dfrac{x^7}{4x^5}$ 10. $\dfrac{2.4n^6}{3n^4}$

11. $3x \cdot 2 + 4x \cdot 3$ 12. $5a \cdot 3 - 2a \cdot 6$ 13. $-4n \cdot 4 - 3n \cdot 3$

14. $-(4x - 5)$ 15. $-(-5 + 4x)$ 16. $-(-2x^2 - x)$

17. $-(3a^2 - 4a + 6)$ 18. $3x^2 - (2x^2 + 6)$ 19. $x^2 - (4x^2 - 3)$

20. $2(3x + 5)$ 21. $-2(5 - 3x)$ 22. $-4(3a + 5)$

23. $5x(3x - 6)$ 24. $6a(3a - 7)$ 25. $4a(2a + 5)$

26. $k(k - 1)$ 27. $k(2k - 4)$ 28. $3x(y - 9z)$

29. $4a(2b - c)$ 30. $5a(2b - 7c)$ 31. $2x(-y + x)$

32. $3(x^2 - 4x - 6)$ 33. $4(y^2 + 3y - 2)$ 34. $-5(2a^2 + 3a - 7)$

35. $-2(3n^2 - 3n + 5)$ 36. $3y(2y^2 - y + 6)$ 37. $2x(x^2 - 3x - 9)$

38. $-4x(2x + y - z)$ 39. $-3x(-5x - 2y + 7z)$ 40. $-a(a - b + c)$

41. $5x^2 - 4(2 - 2x^2)$ 42. $8x^2 - 4(x^2 + 2)$ 43. $7a^2 + 3(2 - 4a^2)$

44. $\dfrac{12x - 4}{4}$ 45. $\dfrac{-12x + 3}{3}$ 46. $\dfrac{-12x^2 + 4x}{4}$ 47. $\dfrac{-8x^2 + 6x}{-2}$

48. $\dfrac{-15a^2 + 6a}{-3}$ 49. $\dfrac{a^3 - a^2}{a}$ 50. $\dfrac{4x^3 - 2x^2}{x}$ 51. $\dfrac{6a^4 - 3a^2}{3a}$

52. $\dfrac{12x^3 + 15x^2 + 9x}{3x}$ 53. $\dfrac{4a^3 - 8a^2 - 4a}{4a}$ 54. $\dfrac{9a^2b - 3ab^2}{3ab}$ 55. $\dfrac{8x^2y^2 - 4xy}{xy}$

56. $\dfrac{4x + 6}{3}$ 57. $\dfrac{4x - 3}{5}$ 58. $\dfrac{12a - 8b}{6}$ 59. $\dfrac{4x - 2x^2}{6x}$

60. Write an expression for the area.

    a) $2n + 2$ b)

1-4 Solving Equations in One Variable

Equation: A statement formed by placing an equals sign " = " (read " equals" or "is equal to") between two numerical or variable expressions.

Solution (root): Any value of the variable that makes the equation true.

To write the solutions , we may say either "the solutions are 2 and 5" or "the solutions set is {2, 5}".

To find the solution of an equation, we transfer the given equation into a simpler and equivalent equation that has the same solution. The last statement of the given equation after we transfer it should isolate the variable to one side of the equation.

An easier way to simplify an equation is to transfer (isolate) all variables to one side of the equation and transfer (isolate) all numbers to the other side. Reverse (change) the signs (+, −) in the process. (See Example 5 and Example 6.)

Examples

1. Solve $2x + 6 = 4x$.

Solution:
$$2x + 6 = 4x$$
$$2x + 6 - 4x = 4x - 4x$$
$$-2x + 6 = 0$$
$$-2x + 6 - 6 = -6$$
$$-2x = -6$$
$$\frac{-2x}{-2} = \frac{-6}{-2}$$
$$\therefore x = 3. \quad \text{Ans.}$$

2. Solve $2x - 5 = 7 - 6x$.

Solution:
$$2x - 5 = 7 - 6x$$
$$2x - 5 + 6x = 7 - 6x + 6x$$
$$8x - 5 = 7$$
$$8x - 5 + 5 = 7 + 5$$
$$8x = 12$$
$$\frac{8x}{8} = \frac{12}{8}$$
$$\therefore x = 1\tfrac{1}{2}. \quad \text{Ans.}$$

3. Solve $3(a - 5 + 2a) = 6$.

Solution:
$$3(a - 5 + 2a) = 6$$
$$3(3a - 5) = 6$$
$$9a - 15 = 6$$
$$9a - 15 + 15 = 6 + 15$$
$$9a = 21$$
$$\frac{9a}{9} = \frac{21}{9}$$
$$\therefore a = 2\tfrac{1}{3}. \quad \text{Ans.}$$

4. Solve $3(2y + 3) = 4y - 7$.

Solution:
$$3(2y + 3) = 4y - 7$$
$$6y + 9 = 4y - 7$$
$$6y + 9 - 4y = 4y - 7 - 4y$$
$$2y + 9 = -7$$
$$2y + 9 - 9 = -7 - 9$$
$$2y = -16, \quad \frac{2y}{2} = \frac{-16}{2}$$
$$\therefore y = -8. \quad \text{Ans.}$$

5. Solve $2x + 6 = 4x$.

Solution:
$$2x + 6 = 4x$$
$$2x - 4x = -6$$
$$-2x = -6 \quad \therefore x = \tfrac{-6}{-2} = 3. \text{ Ans.}$$

6. Solve $2x - 5 = 7 - 6x$.

Solution:
$$2x - 5 = 7 - 6x$$
$$2x + 6x = 7 + 5$$
$$8x = 12 \quad \therefore x = \tfrac{12}{8} = 1\tfrac{1}{2}. \text{ Ans.}$$

EXERCISES

Solve each equation. Check your solution.

1. $3x - 16 = x$ **2.** $2x + 15 = -3x$ **3.** $5x + 12 = 3x$

4. $4x - 9 = x$ **5.** $-4y - 7 = 3y$ **6.** $-6y + 8 = 2y$

7. $2(x - 5 + 3x) = 14$ **8.** $5(6x + 8 - 2x) = -10$ **9.** $4(-8 + y + 6) = 4$

10. $2(5 - 2y - 7) = 8$ **11.** $3(m - 5) = -30$ **12.** $4(2m + 3) = 28$

13. $-6(a + 3) = 9$ **14.** $-2(4 - 2x) = -16$ **15.** $-3(5 + 3x) = 4$

16. $-4(2 + 5a) = 32$ **17.** $-(-5 + 2x) = -13$ **18.** $5(2n - 3) = -30$

19. $3x - 6 = 4x - 8$ **20.** $3x - 6 = 4x + 8$ **21.** $3x + 6 = -4x - 8$

22. $y + 4 = 5y - 7$ **23.** $y - 4 = 5y - 7$ **24.** $y - 4 = -5y + 8$

25. $6m - 3 = m + 11$ **26.** $6m + 3 = m + 11$ **27.** $-6m - 3 = m - 11$

28. $-9 + 4n = 10n - 3$ **29.** $9 - 4n = 10n - 5$ **30.** $9 - 4n = -10n + 15$

31. $-8 - x = x + 2$ **32.** $-8 + y = -y - 2$ **33.** $8 + z = -z + 2$

34. $20c + 5 = 10c + 25$ **35.** $20c - 5 = 10c - 25$ **36.** $20c - 5 = -10c + 25$

37. $10 + k = 4k - 20$ **38.** $10 - k = 4k - 30$ **39.** $10 - k = -4k - 14$

40. $-12 + b = 3b + 30$ **41.** $-12 + b = -3b + 32$ **42.** $12 - b = 3b - 16$

43. $15 + 2p = 39 - p$ **44.** $15 - 2p = p - 21$ **45.** $-15 + 2p = p + 40$

46. $100x - 200 = 50x + 50$ **47.** $100x + 200 = 50x - 50$ **48.** $10x + 20 = 50x - 60$

49. $40y - 40 = -40y + 40$ **50.** $40y + 40 = 50y + 50$ **51.** $40y + 40 = -50y - 50$

52. $7(x + 1) = 5(x - 3)$ **53.** $7(x - 1) = 5(x + 3)$ **54.** $-7(x - 1) = 5(x + 3)$

55. $10(y - 2) = -2(2y + 3)$ **56.** $10(y + 2) = -2(2y - 3)$ **57.** $-10(-y + 2) = 2(-2y)$

58. $2(x + 3) - 5 = 3x - 6$ **59.** $2(x - 3) - 5 = 3x + 6$ **60.** $2(x - 3) + 5 = 3x + 6$

61. $5x - 3(x + 1) = x - 10$ **62.** $5y - 3(y - 1) = 4y + 1$ **63.** $6a - 4(2a + 3) = 7a$

64. $3x - 2(5x - 4) = 4x - (2x - 26)$ **65.** $7y - 3(2y + 6) = 5y - (3y + 10)$

1-5 Solving Word Problems in One Variable

To solve a word problem using an equation in algebra, we use the following steps.

Steps for solving a word problem:
      1. Read the problem carefully.
      2. Choose a variable and assign it to represent the unknown (answer).
      3. Write an equation based on the given fact in the problem.
      4. Solve the equation and find the answer.
      5. Check your answer with the problem (usually on a scratch paper).

Examples

1. Eight less than a number is 18. Find the number.
    Solution:

Let $n =$ the number
The equation is

$$n - 8 = 18$$
$$n - 8 + 8 = 18 + 8$$
$$\therefore n = 26 . \text{ Ans.}$$

Or: $n - 8 = 18$
$$n = 18 + 8$$
$$n = 26 . \text{ Ans.}$$

Check: $n - 8 = 18$
$$26 - 8 = 18 \checkmark$$

2. A number increased by 15 is 32. Find the number.
    Solution:

Let $n =$ the number
The equation is

$$n + 15 = 32$$
$$n + 15 - 15 = 32 - 15$$
$$\therefore n = 17 . \text{ Ans.}$$

Or: $n + 15 = 32$
$$n = 32 - 15$$
$$n = 17 . \text{ Ans.}$$

Check: $n + 15 = 32$
$$17 + 15 = 32 \checkmark$$

3. Find two consecutive integers whose sum is 37.
    Solution:

Let $n = 1^{st}$ integer, $n + 1 = 2^{nd}$ integer
The equation is

$$n + (n + 1) = 37$$
$$2n + 1 = 37$$
$$2n = 37 - 1$$
$$2n = 36$$

$\therefore n = 18$ 1st integer.
$n + 1 = 19$ 2nd integer.

Ans: 18 and 19

Check: $18 + 19 = 37 \checkmark$

Examples

4. Twice the sum of a number and 3 is 22. Find the number.
   Solution:

   Let $n =$ the number
   The equation is

   $$2(n+3) = 22$$
   $$2n + 6 = 22$$
   $$2n = 22 - 6$$
   $$2n = 16 \qquad\qquad \text{Check: } 2(n+3) = 22$$
   $$\therefore n = 8 . \quad \text{Ans.} \qquad\qquad 2(8+3) = 22 \checkmark$$

5. Find three consecutive even integers whose sum is 120.
   Solution:

   Let $\quad n = 1^{st}$ even integer
   $n + 2 = 2^{nd}$ even integer
   $n + 4 = 3^{rd}$ even integer
   The equation is

   $$n + (n+2) + (n+4) = 120$$
   $$3n + 6 = 120$$
   $$3n = 114$$
   $$\therefore n = 38 \quad 1^{st} \text{ integer} \qquad \text{Ans: } 38, 40, 42$$
   $$n + 2 = 40 \quad 2^{nd} \text{ integer}$$
   $$n + 4 = 42 \quad 3^{rd} \text{ integer} \qquad \text{Check: } 38 + 40 + 42 = 120 \checkmark$$

6. John has twice as much money as Carol. Together they have \$36. How much money does each have ?
   Solution: Let $c =$ Carol's money
   $2c =$ John's money
   The equation is

   $$c + 2c = 36$$
   $$3c = 36$$
   $$\therefore c = 12 \rightarrow \text{Carol's money.}$$
   $$2c = 24 \rightarrow \text{John's money.} \quad \text{Ans.} \qquad \text{Check: } 12 + 24 = 36 \checkmark$$

7. A number divided by 2 is equal to the number increased by 2. Find the number.
   Solution: Let $n =$ the number

   The equation is $\quad \dfrac{n}{2} = n + 2$

   Multiply each side by 2: $\quad \not{2} \cdot \dfrac{n}{\not{2}} = 2(n+2), \quad n = 2n + 4, \quad n - 2n = 4, \quad -n = 4$

   $$\therefore n = -4 . \quad \text{Ans.} \qquad \text{Check: } \tfrac{-4}{2} = -4 + 2 \checkmark$$

Examples

8. John picked up two-thirds of the books in the shelf. There are 12 books left in the shelf. How many books were in the shelf to begin with ?

Solution:

Let n = the number of books to begin with

There are $(1 - \frac{2}{3})\, n$ books left in the shelf after John picked up $\frac{2}{3}$ of the books.

The equation is

$$(1 - \tfrac{2}{3})n = 12$$
$$\tfrac{1}{3}n = 12$$
$$\therefore n = 36 \text{ books. Ans.}$$

Check: $\frac{2}{3}(36) = 24$

$36 - 24 = 12$ ✔

9. Roger picked up two-fifths of the books in the shelf. Maria picked up one-half of the remaining books. Jack picked up 6 books that were left. How many books were in the shelf to begin with ?

Solution:

Let n = the number of books to begin with

There are $(1 - \frac{2}{5})\, n$ remaining books in the shelf after Roger picket up $\frac{2}{5}$ of the books.

Then, Maria picked up $\frac{1}{2}(1 - \frac{2}{5})n$ books after Roger picked up.

The equation is

$$(1 - \tfrac{2}{5})n - \tfrac{1}{2}(1 - \tfrac{2}{5})n = 6$$
$$\tfrac{3}{5}n - \tfrac{1}{2} \cdot \tfrac{3}{5}n = 6$$
$$\tfrac{3}{5}n - \tfrac{3}{10}n = 6$$
$$\tfrac{6n-3n}{10} = 6$$
$$\tfrac{3n}{10} = 6$$
$$3n = 60$$
$$n = 20 \text{ books. Ans.}$$

Check: $\frac{2}{5}(20) = 8$

$20 - 8 = 12$

$\frac{1}{2}(12) = 6$

$12 - 6 = 6$ ✔

10. There are one-dollar bills and five-dollar bills in the piggy bank. It has 60 bills with a total value of $228. How many ones and how many fives in the piggy bank ?

Solution:

Let $\quad x$ = number of one-dollar bills

$60 - x$ = number of five-dollar bills

The equation is

$$1x + 5(60 - x) = 228$$
$$x + 300 - 5x = 228$$
$$-4x = -72$$
$$\therefore x = 18 \rightarrow \text{one-dollar bills}$$
$$60 - x = 42 \rightarrow \text{five-dollar bills}$$

Check:

$\$18 + 5(42) = \$228$ ✔

Examples

11. The length of a rectangle is $14\,cm$ longer than the width. The perimeter is $76\,cm$.
Find its length and width.
Solution:

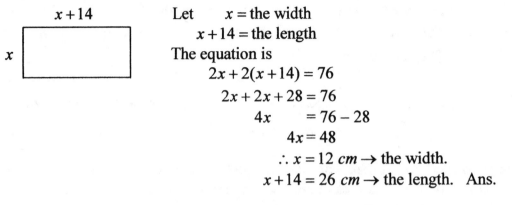

Let $x =$ the width
$x + 14 =$ the length
The equation is
$$2x + 2(x + 14) = 76$$
$$2x + 2x + 28 = 76$$
$$4x \qquad = 76 - 28$$
$$4x = 48$$
$$\therefore x = 12 \; cm \rightarrow \text{the width.}$$
$$x + 14 = 26 \; cm \rightarrow \text{the length.} \quad \text{Ans.}$$

Check: $(12 + 26) \times 2 = 76\; \checkmark$

12. A suit is on sale for 20% discount. Roger paid \$8.82, including a 5% sales tax.
Find the original price of the suit.
Solution:

Let $x =$ the original price
The sales price = $80\% \cdot x = 0.80x$
The sales tax = $0.80x(5\%) = 0.80x(0.05)$
The equation is
$$0.80x + 0.80x(0.05) = 8.82$$
$$0.80x + 0.04x = 8.82$$
$$0.84x = 8.82$$
$$\therefore x = \tfrac{8.82}{0.84} = \$10.50 . \;\text{Ans.}$$

Check: $\$10.50 \times 0.80 = \$8.40$
$\quad\;\; \$ \; 8.40 \times 0.05 = \$0.42$
$\quad \$8.40 + \$0.42 = \$8.82\; \checkmark$

13. A 16-gallon salt-water solution contains 25% pure salt. How much water should be
added to produce the solution to 20% salt ?
Solution:

$16 \times 25\% = 16 \times 0.25 = 4$ gallons of pure salt in the original solution.
Let $x =$ gallons of water added
The equation is
$$\frac{4}{16+x} = 20\%, \quad \frac{4}{16+x} = 0.20,$$
$$4 = 0.20(16 + x)$$
$$4 = 3.2 + 0.20x$$
$$4 - 3.2 = 0.20x$$
$$0.8 = 0.20x$$
$$\therefore x = \tfrac{0.8}{0.20} = 4 \text{ gallons of water added. Ans.}$$

Check:
$\tfrac{4}{16+4} = \tfrac{4}{20} = 0.2 = 20\%\; \checkmark$

EXERCISES

Solve each equation. Check your solution.

1. $x + 6 = 15$

2. $x - 6 = -15$

3. $\dfrac{x}{5} = 7$

4. $\dfrac{y}{-5} = 7$

5. $7n = -42$

6. $-7n = 42$

7. $15 = x + 3$

8. $15 = -x - 3$

9. $12 = \dfrac{x}{3}$

10. $-12 = \dfrac{x}{-3}$

11. $-12 = 3x$

12. $12 = -3x$

13. $x + \dfrac{1}{2} = -3$

14. $\dfrac{k}{12} = 1.5$

15. $-\dfrac{2}{9}x = \dfrac{4}{3}$

16. $5x = 2x + 6$

17. $5x = -2x - 6$

18. $4x - 5 = 19$

19. $-5y + 7 = -18$

20. $4x - 6x = 18$

21. $-12 = 3a - 5a$

22. $2.5p = 0.5p - 10$

23. $2.5p = -0.5p + 14$

24. $3.2x = 1.8x - 4.2$

25. $\dfrac{x}{2} = \dfrac{x}{4} - 7$

26. $\dfrac{3}{4}x + \dfrac{1}{4} = 2$

27. $\dfrac{x - 4}{5} = -10$

28. $3(2x - 4 + x) = 14$

29. $-2(3y + 4 - 5y) = 4$

30. $5(2n - 5) = 5$

31. $-5(5 - 3n) = 5$

32. $3x + 4 = -5x - 20$

33. $3a + 5 = a - 4$

34. $7(x + 2) = 3(x - 4)$

35. $5y - 2(y + 3) = 4y$

36. A number increased by 21 is 28. Find the number.
37. 8 less than a number is 15. Find the number.
38. A number decreased by 12 is equal to 30. Find the number.
39. A number less than 8 is 15. Find the number.
40. 5 more than a number is 25. Find the number.
41. 5 times a number is 45. Find the number.
42. 5 more than twice a number is 13. Find the number.
43. Twice the sum of a number and 5 is −35. Find the number.
44. One-third of a number is 12. Find the number.
45. 3 times a number decreased by 4 is 18. Find the number.
46. Six times the sum of a number and two is fourteen. Find the number.
47. Half the quotient of a number and three is ten. Find the number.
48. 12 more than 4 times a number is twice the number. Find the number.
49. Nine times a number is twice the difference between the number and seven. Find the number.
50. Five times the number which is two less than x is fifteen. Find x.   -----Continued-----

51. Find two consecutive integers whose sum is 55.
52. Find two consecutive even integers whose sum is 114.
53. Find two consecutive odd integers whose sum is 116.
54. Find three consecutive integers whose sum is 69.
55. Find three consecutive even integers whose sum is 222.
56. Find four consecutive even integers whose sum is 172.
57. Find four consecutive odd integers whose sum is 152.
58. Find four consecutive multiples of 5, whose sum is 230.
59. Twice the sum of a number and 4 is 32. Find the number.
60. Roger has twice as much money as David. Together they have $63. How much does each have ?
61. A number divided by 3 is equal to the number decreased by 14. Find the number.
62. A number divided by 3 is equal to the number decreased by 3. Find the number.
63. John picked up one-third of the books in the shelf. There are 18 books left in the shelf. How many books were in the shelf to begin with ?
64. Roger picked up one-third of the books in the shelf. Maria picked up two-thirds of the remaining books. There are 8 books left in the shelf. How many books were in the shelf to begin with ?
65. The width of a rectangle is 6 inches shorter than the length. The perimeter is 48 inches. Find its length and width.
66. In an isosceles triangle, the length of each leg exceeds the base by 4 inches. The perimeter is 92 inches. Find the length of each leg.
67. A product is on sale for 30% discount. Maria paid $15.12, including a 8% sales tax. Find the original price of the product.
68. A 15-gallon salt-water solution contains 20 % pure salt. How much water should be added to produce the solution to 15% salt ?
69. A 200-gram chlorine-water solution contains 40% chlorine. How much water should be added to make a new solution that is only 25% chlorine ?
70. A rope 27 *cm* long is cut into two pieces. One piece is 8 *cm* longer than the other piece. Find the lengths of the pieces.
71. A board 39 inches long is cut into three pieces. The largest piece is 5 inches longer than the median piece. The shortest piece is 5 inches shorter than the median piece. Find the length of each piece.
72. Roger has twice as much money as Carol. Maria has $\frac{1}{2}$ as much money as Carol. Together they have $77. How much money does each have ?
73. There are one-dollar bills and five-dollar bills in the piggy bank. It has 54 bills with a total value of $174. How many ones and how many fives are there in the piggy bank ?
74. John gave three-fifths of his money to Joe and gave $40 to Janet. John had $8 left. How much money did John have to start with ?
75. Tom ate one-third of the candies in a box. Roger ate two-fifths of the remaining candies. There were 18 candies left. How many candies were in the box to start with ?

1-6 Literal Equations and Formulas

A literal equation is an equation that contains two or more variables.
In science and math, every formula is a literal equation.
In a literal equation, we can solve it for a given variable by **isolating** the variable to one side of the equation. We follow the same steps as we solve an equation.

The area of a rectangle is given by the formula: $A = l \cdot w$

Solve it for l, we divide each side by w: $l = \dfrac{A}{w}$.

Solve it for w, we divide each side by l: $w = \dfrac{A}{l}$.

The perimeter of a rectangle is given by the formula: $p = 2(l + w)$

Solve it for w, we have $p = 2(l + w)$

$$p = 2l + 2w$$

$$p - 2l = 2w$$

$$\frac{p - 2l}{2} = w \qquad \therefore w = \frac{p - 2l}{2} \ \text{ or } \ \frac{p}{2} - l . \ \text{ Ans.}$$

Examples

1. Solve $x = y + 5$ for y.
Solution:

$$x = y + 5$$

$$x - 5 = y$$

$$\therefore y = x - 5. \ \text{ Ans.}$$

2. Solve $x = 2y - 5$ for y.
Solution:

$$x = 2y - 5$$

$$x + 5 = 2y$$

$$\frac{x + 5}{2} = y \quad \therefore y = \frac{x + 5}{2}. \ \text{ Ans.}$$

3. The formula for the area of a triangle is $A = \frac{1}{2}bh$, where b is the base, and h is the height. Solve the formula for h.
Solution:

$$A = \tfrac{1}{2}bh$$

Multiply each side by 2, we have $\quad 2A = bh \qquad \therefore h = \dfrac{2A}{b}. \ \text{ Ans.}$

4. The formula to find the Celsius temperature for each Fahrenheit temperature is given by the formula $C = \dfrac{5}{9}(F - 32)$. Solve the formula for F.

Solution: $\quad C = \dfrac{5}{9}(F - 32)$

Multiply each side by $\dfrac{9}{5}$, we have

$$\frac{9}{5}C = F - 32 \quad \therefore F = \frac{9}{5}C + 32. \ \text{ Ans.}$$

EXERCISES

Solve each equation for the indicated variable.

1. $x + y = 5$, for y

2. $x + y = 5$, for x

3. $x - y = 5$, for x

4. $x - y = 5$, for y

5. $2x + y = 10$, for y

6. $2x + y = 10$, for x

7. $-2x + y = 10$, for y

8. $2x - y = 10$, for x

9. $2x - 4y = -5$, for y

10. $2a + b = c$, for b

11. $2a + b = c$, for a

12. $2a + 2b = c$, for a

13. $a - b = 3a + b$, for a

14. $ax = a - 2$, for x

15. $ax = a + 2$, for a

16. If $\dfrac{a}{b} = 1$, then $a - b = ?$

17. The formula for the circumference of a circle is $C = 2\pi r$, where $\pi \approx 3.14$, and r is the radius.
   a. Solve the formula for r.
   b. Find the radius of a circle with a circumference of 4 meters.

18. The formula for the area of a triangle is $A = \frac{1}{2}bh$, where b is the base, and h is the height.
   a. Solve the formula for h.
   b. Find the height of a triangle if $A = 20\,ft^2$, $b = 5\,ft$.

19. The formula for the perimeter of a rectangle is $p = 2(l + w)$, where l is the length, and w is the width.
   a. Solve the formula for l.
   b. Find the length of a rectangle if $p = 88\,cm$, $w = 16\,cm$.

20. The formula for the distance traveled in a single direction at a constant speed is $d = rt$, where r is the speed (rate), and t is the time traveled.
   a. Solve the formula for r.
   b. If a plane left an airport at a constant speed to a distance of 1,950 miles in three hours, what is its speed ?

21. The formula for the volume of a pyramid is $V = \frac{1}{3}bh$, where b is the area of the base, and h is the height. Solve the formula for h.

22. The formula for the total surface area of a right circular cylinder is $A = 2\pi r h + 2\pi r^2$, where r is the radius of the circular base and top, and h is the height of the cylinder. Solve the formula for h.

CHAPTER 1   EXERCISES

Evaluate each expression.

1. $3x - 5$ if $x = 4$

2. $3x - 5$ if $x = -4$

3. $-3x + 5$ if $x = -4$

4. $-3x - 5$ if $x = 4$

5. $\frac{1}{2}x - 8$ if $x = 6$

6. $\frac{1}{2}x - 8$ if $x = -6$

7. $\frac{3}{4}y + 9$ if $y = 16$

8. $\frac{3}{4}y - 9$ if $y = -16$

9. $\frac{3}{4}y - 9$ if $y = 0$

10. $4 - 2.5a$ if $a = 3$

11. $4 - 2.5a$ if $a = 1.4$

12. $4 - 2.5a$ if $a = -1.4$

13. $3x^2 - 2x + 1$ if $x = 5$

14. $x^2 + y - 2$ if $x = -5$, $y = 2$

15. $3a^2b^3$ if $a = 2$, $b = -3$

Simplify each expression.

16. $x + 5x$

17. $5x - 2x$

18. $-5x + 2x$

19. $6x + (4x - 2)$

20. $6x + (2 - 4x)$

21. $-6x + (4x + 2)$

22. $7 + (2x + 5)$

23. $7 - (2x + 5)$

24. $7 - (2x - 5)$

25. $3x + 7 + (5x - 8)$

26. $3x + 7 + (8 - 5x)$

27. $3x - 7 + (-5x + 8)$

28. $3x + 7 - (5x - 8)$

29. $3x + 7 - (8 - 5x)$

30. $3x - 7 - (-5x + 8)$

31. $-(2 + x) - (x - 2)$

32. $x - 2y - (3y - 4x)$

33. $x - 2y + (3y - 4x)$

34. $\frac{1}{4} + \frac{2}{3}y - (1 - \frac{1}{3}y)$

35. $\frac{2}{5}x + \frac{1}{3}y + (\frac{1}{5}x - \frac{2}{3}y)$

36. $\frac{2}{5}x + \frac{1}{3}y - (\frac{1}{5}x - \frac{2}{3}y)$

37. $\frac{2}{5}x + \frac{1}{5}y - (\frac{1}{3}x - \frac{2}{3}y)$

38. $1.2x + 3.5y - (2y - 4.5x)$

39. $5.6n - (4 - 2.5n)$

40. $2x^2 - 3 - 4x^2 - 6$

41. $2x^2 - 3 - (4x^2 - 6)$

42. $3x^2 - (-4x^2 + 6)$

43. $x + (2x - 2) - (4y + 3)$

44. $3p + (p - q) - (4p - 3q)$

45. $2x^2 - (4x^2 - 3x - 5)$

46. $2x \cdot 4x$

47. $-2x \cdot 4x^2$

48. $-1.5a^3 \cdot 3a^4$

49. $4x(2x - 8)$

50. $-5(3x - 4y + 5)$

51. $-2a(a - 2b + 5)$

52. $4x - 6(2x - 3)$

53. $4x^2 + 5x - 6(x^2 + x - 2)$

54. $4p^2 - (2p^2 - 3p + 1)$

55. $\dfrac{5x + 10}{5}$

56. $\dfrac{5x^2 - 10x}{5x}$

57. $\dfrac{3a^3 - 2a^2}{a^2}$

58. $\dfrac{10x^3 + 5x^2 + 15x}{5x}$

59. $\dfrac{4a^3 - 12a^2 - 4a}{4a}$

60. $\dfrac{9x^3y - 6xy^2}{3xy}$

-----Continued-----

Solve each equation. Check your solution.

61. $x - 9 = 30$ **62.** $25 = y + 15$ **63.** $\dfrac{a}{15} = 6$

64. $32 = -5p$ **65.** $x + \dfrac{3}{4} = 8$ **66.** $y - \dfrac{3}{4} = 8$

67. $-\dfrac{4}{5}p = \dfrac{8}{25}$ **68.** $3x = 6x - 8$ **69.** $\dfrac{1}{3}x = \dfrac{3}{4}x - 9$

70. $-5x + 8 = 4x - 10$ **71.** $3y - 7 = -4y + 7$ **72.** $4(2p - 3) = 6$

73. $2(x - 1) = 4(2x + 5)$ **74.** $6y - 3(3y - 1) = 5y$ **75.** $\dfrac{2x + 15}{3} = -31$

76. $2x - y = 6$, for y **77.** $2x - y = 6$, for x **78.** $2x - 4y = 8$, for y

79. $4x + 5y = -25$, for y **80.** $2a + 2b = 3c$, for b

81. 25 less than a number is 40. Find the number.

82. A number less than 25 is 40. Find the number.

83. Five times a number is 95. Find the number.

84. The quotient of a number and 1.5 is 9. Find the number.

85. Eight times the difference between a number and 15 is 40. Find the number.

86. 12 less than 4 times a number is twice the number. Find the number.

87. A number divided by 6 is equal to the number increased by 6. Find the number.

88. Find two consecutive integers whose sum is 135.

89. Find two consecutive even integers whose sum is 174.

90. Find three consecutive multiples of 5, whose sum is 255.

91. A board is 75 inches. The carpenter cut it into two pieces. One piece is 11 inches less than the other piece. Find the lengths of pieces.

92. A board is 75 inches. The carpenter cut it into three pieces. The median piece is 5 inches longer than the shortest piece. The longest piece is 5 inches longer than the median piece. Find the lengths of the pieces.

93. The length of a rectangle is 7 feet longer than the width. The perimeter is 38 feet. Find its length and width.

94. A 20-gallon salt-water solution contains 15% pure salt. How much water should be added to make a new solution that is only 12% salt ?

95. A 20-gallon alcohol-water solution contains 15% pure alcohol. How much alcohol should be added to make a new solution that is 20% alcohol ?

96. In an isosceles triangle, the length of each leg exceeds the base by $14\,cm$. The perimeter is $73\,cm$. Find the length of the leg and the length of the base.

97. A suit is on sale for 25% discount. You paid $93.15, including a 8% sales tax. Find the original price of the suit.

98. There are one-dollar bills and five-dollar bills in the piggy bank. It has 80 bills with a total value of $184. How many ones and how many fives in the piggy bank ?

99. John's land is $500\,ft^2$ less than Roger's land. Use the symbol J for John's land, and R for Roger's land. Write an equation to represent John's land.

ANSWERS

CHAPTER 1 – 1   Variable and Expressions   page 10

1. 3   **2.** 5   **3.** 7   **4.** 1   **5.** 3   **6.** 5   **7.** 12   **8.** 14   **9.** 16   **10.** –6   **11.** 0   **12.** 4

13. –6   **14.** –10   **15.** –14   **16.** –2   **17.** –6   **18.** –10   **19.** 2   **20.** 8   **21.** –4   **22.** –10

23. 0   **24.** 2   **25.** 5   **26.** $6\frac{1}{2}$ or 6.5   **27.** 7   **28.** 0   **29.** $-3\frac{1}{2}$ or –3.5   **30.** $-1\frac{1}{2}$ or –1.5

31. 4   **32.** $1\frac{1}{2}$ or 1.5   **33.** $-6\frac{1}{2}$ or –6.5   **34.** 13   **35.** 7   **36.** –6.5   **37.** 14   **38.** 19

39. –3   **40.** 3   **41.** 16   **42.** 36   **43.** $40   **44.** $44   **45.** 225 square meters   **46.** 1024 feet

CHAPTER 1 – 2   Adding and Subtracting Expressions   page 12

1. $9x$ **2.** $-3x$ **3.** $-9x$ **4.** $3x$ **5.** $-9x+5$ **6.** $3x-5$ **7.** $9x+5$ **8.** $9x-5$

9. $3x+5$ **10.** $9x+5$ **11.** $15-4x$ **12.** $9x+4y$ **13.** $9x+12y$ **14.** $5x-4y$

15. $5x+4y$ **16.** $-4a+19$ **17.** $a-8b$ **18.** $9u-9w$ **19.** $3w-4u$ **20.** $9w-u$

21. $9m+13$ **22.** $13+2x$ **23.** $-14+y$ **24.** $-n-7$ **25.** 7   **26.** $\frac{5}{6}x+\frac{2}{3}y$ **27.** $\frac{1}{6}x+\frac{1}{6}y$

28. $\frac{5}{12}x+\frac{11}{20}y$ **29.** $\frac{4}{5}x+\frac{1}{3}y$ **30.** $-\frac{2}{5}x+y$ **31.** $4.7x-2.3y$ **32.** $-2.3x-4.7y$ **33.** $-2.3x+4$

34. $6x^2-1$ **35.** $-2y^2+2$ **36.** $4x^2-7$ **37.** $9x^2-8x$ **38.** $-8x^2-10$ **39.** $-3n^2-4$

40. $-5x^2+x+4$ **41.** $2a^2-9$ **42.** $-2k^2+8$ **43.** $-5y$ **44.** $-a-5b+9$ **45.** $3m-5n-8$

46. $-p-2q-4$ **47.** $-2x^2+2x-2$ **48.** $8x^2-6x+10$ **49.** a **50.** $-2x$

51. $2x+2y$ or $2(x+y)$ **52.** $4x+4y$ or $4(x+y)$

CHAPTER 1 – 3   Multiplying and Dividing Expressions   page 14

1. x^2 **2.** $6x^2$ **3.** $-24x^4$ **4.** $-4x^{12}$ **5.** $7.2n^{10}$ **6.** 1   **7.** $-\frac{2}{3}$ **8.** $\frac{3}{2}x^2$ **9.** $\frac{1}{4}x^2$ **10.** $0.8n^2$

11. $18x$ **12.** $3a$ **13.** $-25n$ **14.** $-4x+5$ **15.** $5-4x$ **16.** $2x^2+x$ **17.** $-3a^2+4a-6$

18. x^2-6 **19.** $-3x^2+3$ **20.** $6x+10$ **21.** $-10+6x$ **22.** $-12a-20$ **23.** $15x^2-30x$

24. $18a^2-42a$ **25.** $8a^2+20a$ **26.** k^2-k **27.** $2k^2-4k$ **28.** $3xy-27xz$ **29.** $8ab-4ac$

30. $10ab-35ac$ **31.** $-2xy+2x^2$ **32.** $3x^2-12x-18$ **33.** $4y^2+12y-8$

34. $-10a^2-15a+35$ **35.** $-6n^2+6n-10$ **36.** $6y^3-3y^2+18y$ **37.** $2x^3-6x^2-18x$

38. $-8x^2-4xy+4xz$ **39.** $15x^2+6xy-21xz$ **40.** $-a^2+ab-ac$ **41.** $13x^2-8$

42. $4x^2-8$ **43.** $-5a^2+6$

44. $3x-1$ **45.** $-4x+1$ **46.** $-3x^2+x$ **47.** $4x^2-3x$ **48.** $5a^2-2a$ **49.** a^2-a

50. $4x^2-2x$ **51.** $2a^3-a$ **52.** $4x^2+5x+3$ **53.** a^2-2a-1 **54.** $3a-b$ **55.** $8xy-4$

56. $\frac{4}{3}x+2$ **57.** $\frac{4}{3}x-\frac{3}{5}$ **58.** $2a-\frac{4}{3}b$ **59.** $\frac{2}{3}-\frac{1}{3}x$ **60.** a) $2n^2+2n$ b) x^2+2xy

CHAPTER 1 – 4   Solving Equations in One Variable   page 16

1. $x=8$ **2.** $x=-3$ **3.** $x=-6$ **4.** $x=3$ **5.** $y=-1$ **6.** $y=1$ **7.** $x=3$ **8.** $x=-2.5$

9. $y=3$ **10.** $y=-3$ **11.** $m=-5$ **12.** $m=2$ **13.** $a=-4\frac{1}{2}$ **14.** $x=-2$ **15.** $x=-2\frac{1}{9}$

16. $a=-2$ **17.** $x=9$ **18.** $n=-1\frac{1}{2}$ **19.** $x=2$ **20.** $x=-14$ **21.** $x=-2$ **22.** $y=2\frac{3}{4}$

23. $y=\frac{3}{4}$ **24.** $y=2$ **25.** $m=2\frac{4}{5}$ **26.** $m=1\frac{3}{5}$ **27.** $m=1\frac{1}{7}$ **28.** $n=-1$ **29.** $n=1$

30. $n=1$ **31.** $x=-5$ **32.** $y=3$ **33.** $z=-3$ **34.** $c=2$ **35.** $c=-2$ **36.** $c=1$

37. $k=10$ **38.** $k=8$ **39.** $k=-8$ **40.** $b=-21$ **41.** $b=11$ **42.** $b=7$ **43.** $p=8$

44. $p=12$ **45.** $p=55$ **46.** $x=5$ **47.** $x=-5$ **48.** $x=2$ **49.** $y=1$ **50.** $y=-1$

51. $y=-1$ **52.** $x=-11$ **53.** $x=11$ **54.** $x=-\frac{2}{3}$ **55.** $y=1$ **56.** $y=-1$ **57.** $y=1\frac{3}{7}$

58. $x=7$ **59.** $x=-17$ **60.** $x=-7$ **61.** $x=-7$ **62.** $y=1$ **63.** $a=-1\frac{1}{3}$ **64.** $x=-2$

65. $y=-8$

CHAPTER 1 – 5   Solving Word Problems in One Variable   page 21 ~ 22

1. $x = 9$ **2.** $x = -9$ **3.** $x = 35$ **4.** $x = -35$ **5.** $n = -6$ **6.** $n = -6$ **7.** $x = 12$

8. $x = -18$ **9.** $x = 36$ **10.** $x = 36$ **11.** $x = -4$ **12.** $x = -4$ **13.** $x = -3\frac{1}{2}$ **14.** $k = 18$

15. $x = -6$ **16.** $x = 2$ **17.** $x = -\frac{6}{7}$ **18.** $x = 6$ **19.** $y = 5$ **20.** $x = -9$ **21.** $a = 6$

22. $p = -5$ **23.** $p = 4\frac{2}{3}$ **24.** $x = -3$ **25.** $x = -28$ **26.** $x = 2\frac{1}{3}$ **27.** $x = -46$ **28.** $x = 2\frac{8}{9}$

29. $y = 3$ **30.** $n = 3$ **31.** $n = 2$ **32.** $x = -3$ **33.** $a = -4\frac{1}{2}$ **34.** $x = -6\frac{1}{2}$ **35.** $y = -6$

36. 7   **37.** 23   **38.** 42   **39.** –7   **40.** 20   **41.** 9   **42.** 4   **43.** –22.5   **44.** 36   **45.** $7\frac{1}{3}$ **46.** $\frac{1}{3}$

47. 60   **48.** –6   **49.** –2   **50.** $x = 5$ **51.** 27 and 28   **52.** 56 and 58   **53.** 57 and 59

54. 22, 23, and 24   **55.** 72, 74, and 76   **56.** 40, 42, 44, and 46   **57.** 35, 37, 39, and 41

58. 50, 55, 60, and 65   **59.** 12   **60.** Roger has $42, David has $21   **61.** 21   **62.** 4.5

63. 27 books   **64.** 36 books   **65.** length = 15 inches, width = 9 inches   **66.** length = 32 in.

67. $20   **68.** 5 gallons of water should be added.   **69.** 120 grams of water should be added.

70. $9.5\,cm$ and $17.5\,cm$

71. the largest piece = 18 in., the median piece = 13 in., the shortest piece = 8 in.

72. Roger has $44, Carol has $22, Maria has $11

73. 24 one-dollar bills and 30 five-dollar bills   **74.** $120   **75.** 45 candies

CHAPTER 1 – 6   Literal Equations and Formulas   page 24

1. $y = -x + 5$ **2.** $x = -y + 5$ **3.** $x = y + 5$ **4.** $y = x - 5$ **5.** $y = -2x + 10$

6. $x = -\frac{1}{2}y + 5$ **7.** $y = 2x + 10$ **8.** $x = \frac{1}{2}y + 5$ **9.** $y = \frac{1}{2}x + \frac{5}{4}$ **10.** $b = c - 2a$

11. $a = \frac{c-b}{2}$ **12.** $a = \frac{c}{2} - b$ **13.** $a = -b$ **14.** $x = 1 - \frac{2}{a}$ **15.** $a = \frac{2}{x-1}$

16. $a - b = 0$ **17. a.** $r = \frac{C}{2\pi}$,   **b.** $r \approx 0.64$ meters   **18. a.** $h = \frac{2A}{b}$,   **b.** $h = 8$ feet

19. a. $l = \frac{p}{2} - w$,   **b.** $l = 28\,cm$ **20. a.** $r = \frac{d}{t}$ **b.** $r = 650$ miles per hour

21. $h = \frac{3V}{b}$ **22.** $h = \frac{A}{2\pi r} - r$

CHAPTER 1 EXERCISES   page 25 ~ 26

1. 7   **2.** –17   **3.** 17   **4.** –17   **5.** –5   **6.** –11   **7.** 21   **8.** –21   **9.** –9   **10.** –3.5

11. 0.5   **12.** 7.5   **13.** 66   **14.** 25   **15.** –324

16. $6x$ **17.** $3x$ **18.** $-3x$ **19.** $10x - 2$ **20.** $2x + 2$ **21.** $-2x + 2$ **22.** $2x + 12$

23. $-2x + 2$ **24.** $-2x + 12$ **25.** $8x - 1$ **26.** $-2x + 15$ **27.** $-2x + 1$ **28.** $-2x + 15$

29. $8x - 1$ **30.** $8x - 15$ **31.** $-2x$ **32.** $5x - 5y$ **33.** $-3x + y$ **34.** $y - \frac{3}{4}$ **35.** $\frac{3}{5}x - \frac{1}{3}y$

36. $\frac{1}{5}x + y$ **37.** $\frac{1}{15}x + \frac{13}{15}y$ **38.** $5.7x + 1.5y$ **39.** $8.1n - 4$ **40.** $-2x^2 - 9$ **41.** $-2x^2 + 3$

42. $7x^2 - 6$ **43.** $3x - 4y - 5$ **44.** $2q$ **45.** $-2x^2 + 3x + 5$ **46.** $8x^2$ **47.** $-8x^3$

48. $-4.5a^7$ **49.** $8x^2 - 32x$ **50.** $-15x + 20y - 25$ **51.** $-2a^2 + 4ab - 10a$ **52.** $-8x + 18$

53. $-2x^2 - x + 12$ **54.** $2p^2 + 3p - 1$ **55.** $x + 2$ **56.** $x - 2$ **57.** $3a - 2$ **58.** $2x^2 + x + 3$

59. $a^2 - 3a - 1$ **60.** $3x^2 - 2y$

61. $x = 39$ **62.** $y = 10$ **63.** $a = 90$ **64.** $p = -6\frac{2}{5}$ **65.** $x = 7\frac{1}{4}$ **66.** $y = 8\frac{3}{4}$ **67.** $p = -\frac{2}{5}$

68. $x = 2\frac{2}{3}$ **69.** $x = 21\frac{3}{5}$ **70.** $x = 2$ **71.** $y = 2$ **72.** $p = 2\frac{1}{4}$ **73.** $x = -3\frac{2}{3}$ **74.** $y = \frac{3}{8}$

75. $x = -54$ **76.** $y = 2x - 6$ **77.** $x = \frac{1}{2}y + 3$ **78.** $y = \frac{1}{2}x - 2$ **79.** $y = -\frac{4}{5}x - 5$

80. $b = -a + \frac{3}{2}c$

81. 65   **82.** –15   **83.** 19   **84.** 13.5   **85.** 20   **86.** 6   **87.** –7.2   **88.** 67 and 68   **89.** 86 and 88

90. 80, 85, and 90   **91.** 32 inches and 43 inches   **92.** 20, 25, and 30 inches

93. length = 13 feet, width = 6 feet   **94.** 5 gallons of water added

95. 1.25 gallons of alcohol added   **96.** leg = $29\,cm$, base = $15\,cm$ **97.** $115

98. 54 one-dollar bills and 26 five-dollar bills   **99.** $J = R - 500$

Evaluating an algebraic expression by a graphing calculator

Hints: 1. The graphing calculator follows the order of operations.
2. (–) is the negation key. – is the subtraction key.
3. Enter 5^2 as **5 → ∧ → 2** or as **5 → x^2**. Enter 3^4 as **3 → ∧ → 4**.
4. Enter parentheses for all grouping symbols.
5. Go to home screen, press: **2nd → QUIT**.
6. To lighten or darken the screen, press and release **2nd** and then press the cursors ▲ or ▼.
7. To ensure the illustrated results, reset all memory (RAM) to clear RAM memory. Press **2nd → MEM → 7↓ Reset → ENTER → 1: ALL RAM → ENTER → 2:Reset → ENTER → RAM cleared → ENTER**.

Example 1: Evaluate $-3 + 5^2 \cdot 4 + \dfrac{3^4 - 1}{20} - 10$.

Solution: Press: **(–) → 3 → + → 5 → ∧ → 2 → × → 4 → + (→ 3 → ∧ → 4 → – → 1 →)**
→ ÷ → 20 → – → 10 → ENTER
The screen shows: **91**

Example 2: Evaluate $x^2 + 1$ if $x = 3$.

Solution: **Method 1:**

First, store 3 as x:
Press: **3 → STO → $\boxed{X,T,\theta,n}$**
Use the colon **:** to begin the expression:
Press: **ALPHA → : → $\boxed{X,T,\theta,n}$ → x^2**
→ + → 1 → ENTER
The screen shows: **10**

Method 2:

To begin the expression:
Press: **Y= → $\boxed{X,T,\theta,n}$ → x^2 → +**
→ 1 → MATH → 2nd → CALC
The screen shows: **1:Value**
Press: **ENTER**
The screen shows: the graph and **X=**
Type: **X= 3**   Press: **ENTER**
The screen shows: **Y= 10**

Example 3: Evaluate $4x^3 + 2(5 - 2x)^2 + 3x$ if $x = 2$, 3, and -1.5.

Solution: First, store 2 as x. Press: **2 → STO → $\boxed{X,T,\theta,n}$**
Use the colon **:** to begin the expression:
Press: **ALPHA → : → 4 → $\boxed{X,T,\theta,n}$ → ∧ → 3 → + → 2 → (→ 5 → – → 2**
→ $\boxed{X,T,\theta,n}$ →) → x^2 → + → 3 → $\boxed{X,T,\theta,n}$ → ENTER
The screen shows: **40**
To evaluate the value of $x = 3$, press: **2nd → ENTRY**
Move the cursor to the 2 and change it to 3:
Press: **ENTER**   The screen shows: **119**
To evaluate the value of $x = -1.5$:
Press: **2nd → ENTRY** and move the cursor to the 3 and press **DEL**:
Press: **2nd → INS → (–) → 1.5 → ENTER**.   The screen shows: **110**

Making a table or a spreadsheet to solve a linear equation by a graphing calculator

Examples

1: Solve the equation $5x + 4 = 14$.
Solution:
Method 1: Rewrite the equation:
$$5x - 10 = 0$$
Let $y = 5x - 10$ and enter:
Press: **Y=** $\rightarrow$ **5** $\rightarrow$ $\boxed{\text{X,T,}\theta\text{,}n}$ $\rightarrow - \rightarrow$ **10**
Press: **2nd** $\rightarrow$ **TBLSET**
Set up the starting x – value
with $x = 0$ and an increment
for x – values to 1:

    TblStart = 0, $\triangle$**Tbl = 1**
    Indpnt: Auto
    Depend: Auto

Press: **2nd** $\rightarrow$ **TABLE**
The screen shows that $x = 2$
when $y = 0$:

x	y_1
0	–10
1	–5
2	**0**
3	5
4	10
5	15
6	20

$x = 2$. Ans.

Method 2:
Press: **Y1= 5** $\rightarrow$ $\boxed{\text{X,T,}\theta\text{,}n}$ $\rightarrow - \rightarrow$ **10**
Press: **GRAPH** $\rightarrow$ **ZOOM** $\rightarrow$ **2:Zoom In**
    (or **6: ZStandard**) $\rightarrow$ **ENTER**
Move the cursor $\boxed{\triangleleft}$ and $\boxed{\triangleright}$ to estimate
the x-intercept.

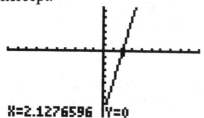

X=2.1276596   Y=0

2: Solve the equation $15x + 4 = 11$.
Solution:
Rewrite the equation:
$$15x - 7 = 0$$
Let $y = 15x - 7$ and enter:
Press: **Y=** $\rightarrow$ **15** $\rightarrow$ $\boxed{\text{X,T,}\theta\text{, }n}$ $\rightarrow - \rightarrow$ **7**
Press: **2nd** $\rightarrow$ **TBLSET**
Set up the starting x – value
with $x = 0$ and an increment for
x – values to 0.1:

    TblStart = 0, $\triangle$**Tbl = 0.1**
    Indpnt: Auto
    Depend: Auto

Press: **2nd** $\rightarrow$ **TABLE**
Scroll down the table to locate
where $y_1 = 0$. The screen shows:

x	y_1
0.4	–1
0.5	5

The solution is between 0.4 and 0.5.
Continue to set up the value of x:
Press: **2nd** $\rightarrow$ **TBLSET**
    TblStart = 0, $\triangle$**Tbl = 0.01**
Press: **2nd** $\rightarrow$ **TABLE**
Scroll down the table to locate
where $y_1 = 0$. The screen shows:

x	y_1
0.46	–0.1
0.47	0.05

The solution is between 0.46 and
0.47.
Repeating the process, the solution
is between 0.466 and 0.477.

We have the solution to the nearest
hundredth, $x \approx 0.47$.

Functions and Graphs

2-1 Number Lines and Coordinate Planes

We can graph all of the real numbers on the **number line**. Each number is assigned a point on the line.

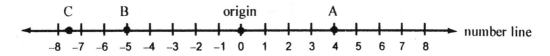

The arrowheads on the number line mean that the line continues in both directions. The number is called the **coordinate** of the point. The coordinate of the point tells the distance and direction from the origin of the line. The dot mark on the point is the graph of the point. The coordinate of point A is 4. The coordinate of point B is -5. The coordinate of point C is -7.5.

We can graph the solution of an equation on the number line.

Example 1: Solve $x - 5 = -8$.

Solution:
$$x - 5 = -8$$
$$x = -8 + 5$$
$$x = -3. \text{ Ans.}$$

Example 2: Solve $2x + 6 = 3$.

Solution:
$$2x + 6 = 3$$
$$2x = 3 - 6$$
$$2x = -3$$
$$x = -1\tfrac{1}{2} \text{ or } -1.5. \text{ Ans.}$$

31

The graph of a point on a number line shows only the horizontal location of the point.
We can graph a point showed both the horizontal and vertical locations on a **coordinate plane**.

Cartesian Coordinate Plane:

It is a rectangular plane used to locate, or plot points (x, y), lines, and graphs.
The horizontal line is called the x – axis.
The vertical line is called the y – axis.
Two number lines perpendicular to each other at the **origin** (O).
The coordinates of origin is (0, 0).
The axes separate the plane into four quadrants.
A point (x, y), called **ordered pair**, in the plane, such as (4, 2), is the coordinates of the point.
x is the x – coordinate (or **abscissa**). y is the y – coordinate (or **ordinate**).

Ordered pair (x, y) : Ordered pair is used to indicate the location of a point on a plane.

x is the number of horizontal units moved from 0.
y is the number of vertical units moved from 0.
A **grid** is often used to find the ordered pair (x, y) for each point.

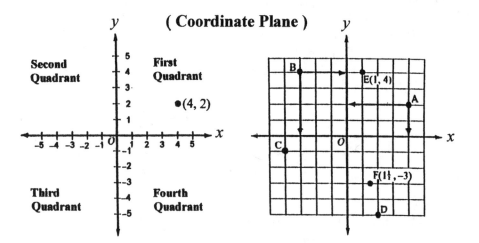

(**Coordinate Plane**)

To find the coordinates of a point in the coordinate plane, first draw a vertical line from the point to the x – axis, next draw a horizontal line from the point to the y – axis. Together, we have the coordinates of the point.

Example 3: Give the coordinates of each of the points A, B, C, and D on the above grid.
Solution:

The coordinates of A is (4, 2).   The coordinates of B is (–3, 4).
The coordinates of C is (–4, –1).   The coordinates of D is (2, –5).

Example 4: Graph each of the ordered pairs on a coordinate plane: E(1, 4), F($1\frac{1}{2}$, –3).

Solution: The answers are shown on the above grid.

EXERCISES

Identify the coordinate of each point described on the number lines.

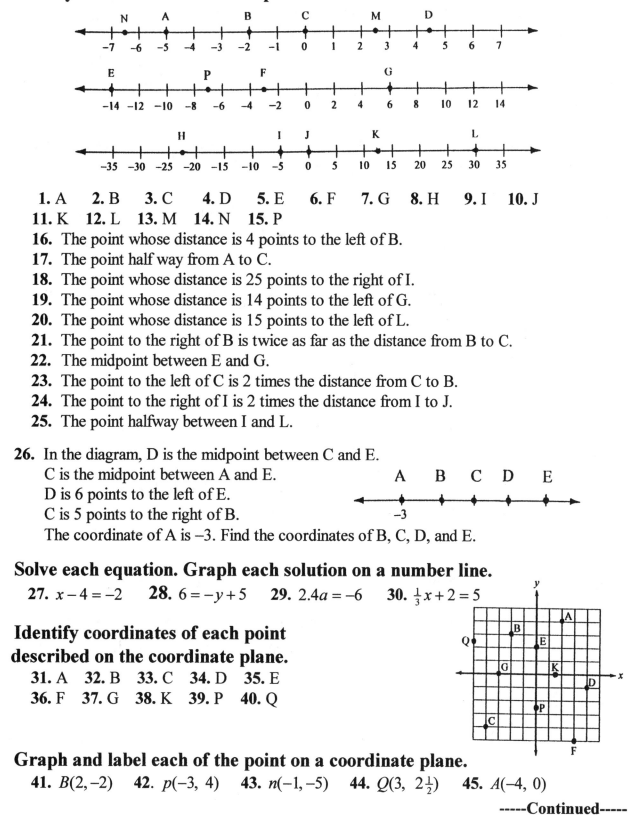

1. A   **2.** B   **3.** C   **4.** D   **5.** E   **6.** F   **7.** G   **8.** H   **9.** I   **10.** J
11. K   **12.** L   **13.** M   **14.** N   **15.** P

16. The point whose distance is 4 points to the left of B.
17. The point half way from A to C.
18. The point whose distance is 25 points to the right of I.
19. The point whose distance is 14 points to the left of G.
20. The point whose distance is 15 points to the left of L.
21. The point to the right of B is twice as far as the distance from B to C.
22. The midpoint between E and G.
23. The point to the left of C is 2 times the distance from C to B.
24. The point to the right of I is 2 times the distance from I to J.
25. The point halfway between I and L.

26. In the diagram, D is the midpoint between C and E.
C is the midpoint between A and E.
D is 6 points to the left of E.
C is 5 points to the right of B.
The coordinate of A is −3. Find the coordinates of B, C, D, and E.

Solve each equation. Graph each solution on a number line.

27. $x - 4 = -2$ **28.** $6 = -y + 5$ **29.** $2.4a = -6$ **30.** $\frac{1}{3}x + 2 = 5$

Identify coordinates of each point described on the coordinate plane.

31. A   **32.** B   **33.** C   **34.** D   **35.** E
36. F   **37.** G   **38.** K   **39.** P   **40.** Q

Graph and label each of the point on a coordinate plane.

41. $B(2, -2)$ **42.** $p(-3, 4)$ **43.** $n(-1, -5)$ **44.** $Q(3, 2\frac{1}{2})$ **45.** $A(-4, 0)$

-----Continued-----

Name the quadrant containing the point.

46. (–3, 5)   **47.** (–4, –1)   **48.** (7, 9)   **49.** (5, –5)   **50.** (–2, –2)

Graph each point and find the length of the segment between them.

51. A(–2, –3) and B(–2, 2)   **52.** M(3, 0) and N(3, –4)   **53.** P(–4, 5) and T(2, 5)
54. C(6, 3) and D(1, 3)   **55.** E(4, 5) and F(4, –6)   **56.** G(5, 4) and H(–6, 4)
57. J(–4, –3) and K(–4, –7)   **58.** S(3, 0) and T(–5, 0)   **59.** U(0, –2) and V(0, 7)
60. W(–8, 0) and Z(7, 0)

Graph the given points on a coordinate plane. Draw line segments to connect the points in the order given and to connect the first and last points. Name the closed figure.

61. (4, 3), (–4, 3), (–4, –3), (4, –3)   **62.** (1, 3), (–1, –3), (1, –3)
63. (–2, 0), (–3, –2), (2, –2), (3, 0)   **64.** (–3, 0), (1, 0), (1, 3), (–1, 3)
65. (2, 1), (4, 1), (4, –4), (2, –4)

Determine whether the line connecting the two points is horizontal, vertical, or neither.

66. (4, 1), (4, –5)   **67.** (2, –3), (2, –4)   **68.** (5, 0), (–4, 0)   **69.** (2, 5), (5, 2)
70. (–2, –6), (–2, –3)   **71.** (0, 4), (–5, 0)   **72.** (3, –6), (–1, –6)   **73.** (0, 5), (0, 0)
74. (5, 5), (–5, –5)   **75.** (5, –5), (–5, –5)

Graph the triangle connecting the given points on a coordinate plane. Change the values of the coordinates as directed. Identify the change of the graph.

(1, 1), (5, 1), (3, 3)

76. Increase all x – coordinates by 2.   **77.** Decrease all x – coordinates by 2.
78. Increase all y – coordinates by 2.   **79.** Decrease all y – coordinates by 2.
80. Multiply all x – coordinates by –1.   **81.** Multiply all y – coordinates by –1.

Graph the given points on a coordinate plane. State whether they lie on a straight line.

82. (1, 1), (2, 3), (3, 5), (0, –1)   **83.** (1, –1), (–1, 1), (–3, 3)
84. (–2, –2), (0, 1), (2, 3)   **85.** (–1, –5), (0, –3), (2, 1), (3, 3)

86. The vertices of a rectangle on a coordinate plane are (3, –1), (3, 3), (–2, 3), and (–2, –1). What are the dimensions of the rectangle ?

87. The vertices of a triangle on a coordinate plane are (1, 1), (1, –3.5), and (–2, –3.5). What is the area of the triangle ?

2-2 Linear Equations

In Chapter 1, we have learned to solve equations that have only one variable.
In this chapter, we will learn to solve equations that have two variables of degree 1.
Here is an example in different forms:

$$2x - y = 3, \quad \text{or } 2x = y + 3, \quad \text{or } y = 2x - 3$$

The solutions to equations in one variable are numbers (see Chapter 1).
The solutions to equations in two variables are ordered pairs of numbers (x, y).

> **Example:** Identify whether the ordered pair $(0, -3)$ is a solution of $2x - y = 3$.
>
> Solution:
>
> We substitute $x = 0$ and $y = -3$ in the equation:
>
> $$2x - y = 2(0) - (-3) = 0 + 3 = 3 \checkmark$$
>
> Yes. It is a solution.

Therefore, $x = 0$ and $y = -3$ is a solution of $2x - y = 3$ and can be written as $(0, -3)$.
The equation $2x - y = 3$ has infinitely many solutions. Here are some of its solutions:

$$(-2, -7), (-1, -5), (0, -3), (1, -1), (2, 1), \cdots\cdots.$$

Each solution of the equation is a point (x, y) on the coordinate plane.

The graph of connecting all the solutions (points) of the equation $2x - y = 3$ forms a straight line. Therefore, we call the equation $2x - y = 3$ a **linear equation**.

We can write any linear equation into the form $ax + by = c$, where a, b, and c are real numbers with a and b not both zero.

The equation $ax + by = c$ is called the **standard form** of a linear equation. If a is negative, we change it to positive by multiplying each side of the equation by -1 (see example 3).

Examples

1. Write $2x = y + 3$ in standard form.

    Solution:

$$2x = y + 3$$

    Move y to the left side(change its sign):

$$2x - y = 3. \text{ Ans.}$$

2. Write $-y = -2x + 3$ in standard form.

    Solution:

$$-y = -2x + 3$$

    Move $-2x$ to the left side(change its sign):

$$2x - y = 3. \text{ Ans.}$$

3. Write $y = 2x - 3$ in standard form.

    Solution:

$$y = 2x - 3$$

    Move $2x$ to the left side(change its sign):

$$-2x + y = -3$$

    Multiply each side by -1:

$$2x - y = 3. \text{ Ans.}$$

4. Write $y + 3 = 2x$ in standard form.

    Solution:

$$y + 3 = 2x$$

    Move y to the right side(change its sign):

$$3 = 2x - y$$

$$2x - y = 3. \text{ Ans.}$$

EXERCISES

Identify whether each ordered pair is a solution of $x + 2y = 6$.

   1. $(0, 3)$ **2.** $(2, 1)$ **3.** $(-2, 4)$ **4.** $(6, 0)$ **5.** $(-1, 3)$

Identify whether each ordered pair of number is a solution of $2x - y = 3$.

   6. $(2, 1)$ **7.** $(3, 3)$ **8.** $(4, 2)$ **9.** $(-1, -5)$ **10.** $(-2, 5)$

Identify whether each ordered pair of number is a solution of the equation.

   11. $x + y = 3$, $(1, 2)$ **12.** $x - y = 3$, $(3, 1)$ **13.** $2x - 3y = 2$, $(4, 2)$

   14. $3x - 2y = 5$, $(1, -1)$ **15.** $5a - 2b = 5$, $(2, 2)$ **16.** $2m - 4n = 5$, $(\frac{1}{2}, -1)$

   17. $6s - t = 2$, $(\frac{1}{3}, 1)$ **18.** $3a + 4b = 3$, $(-2, 3)$ **19.** $x + 3y = 12$, $(2, 3)$

   20. $8x - y = -15$, $(-\frac{3}{4}, 9)$ **21.** $2x + y = 7$, $(0, 5)$ **22.** $2x + y = 7$, $(0, 7)$

   23. $2x - y = 7$, $(0, -7)$ **24.** $-2x - 3y = 1$, $(1, -1)$ **25.** $-2x + 3y = -5$, $(1, -1)$

   26. $y = 2x - 1$, $(1, 1)$ **27.** $y = 3x + 2$, $(2, 9)$ **28.** $y = \frac{1}{2}x - 5$, $(4, -3)$

   29. $y = \frac{2}{3}x$, $(-6, -4)$ **30.** $y = -\frac{2}{3}x + 1$, $(-3, 3)$

Write each of the linear equations in standard form.

   31. $3x = y - 5$ **32.** $y = -4x + 7$ **33.** $3x = -4y + 6$ **34.** $-4y = -5x - 9$

   35. $2x - 8 = -4y$ **36.** $-3y + 20 = 5x$ **37.** $-4x = 7y - 1$ **38.** $-2x + 4y = 7$

   39. $-2x = -5y - 4$ **40.** $4y - 6x = -5$

Find the values for y **by substituting** -2, 0, **and** 2 **for** x.

   41. $y = 2x$ **42.** $y = 2x + 1$ **43.** $y = 2x - 1$ **44.** $y = -2x + 1$

   45. $y = -2x - 1$ **46.** $y = -x + 5$ **47.** $y = -x - 5$ **48.** $y = 4 - x$

   49. $y = 4 - 2x$ **50.** $y = -4 + 2x$

51. The cost to buy a card is $1.50. Write an equation which expresses the relationship between the total cost (c) and the number (n) of cards bought.

52. A book regularly priced at p dollars is on sale at a 30% discount. Write an equation which expresses the relationship between the cost (c) of the book after the discount.

53. The price of a suit is regularly at p dollars. The sales-tax rate is 7.5%. What is the equation of the cost (c) including tax ?

54. The cost to rent a car is $20 per day plus 20 cents per mile. What is the equation of the cost (c) to rent the car for one day with a mileage (m) ? What is the equation of the cost to rent the car for n days ?

55. A-Plus Company charges $11.50 per copy of a book plus $5 handling fee per order. Write an equation which expresses the relationship between the total charge (c) and the number (n) of books per order. What is the total charge if Barnes and Noble Bookstore places an order of 34 copies ?

2-3 Equations and Functions

If we rewrite a linear equation by solving for y in terms of x, we have a **linear function**.
Example: Solve the equation $3x + 2y = 6$ for y in terms of x.

Solution: $3x + 2y = 6$, $2y = -3x + 6$

Divide each side by 2: $y = -\frac{3}{2}x + 3$. Ans.

In the equation $y = -\frac{3}{2}x + 3$, x is called the **independent variable**, and y is called the **dependent variable**. Each value of y depends on the value chosen for x.
To find the value of y in the equation, we could substitute any value (any number) chosen
for x into the equation: $x = 2$, $y = -\frac{3}{2}(2) + 3 = 0$. The solution is $(2, 0)$.
The equation $y = -\frac{3}{2}x + 3$ has infinitely many solutions. The x-values are called the
domain of the function, and the y-values are called the **range** of the function.
In the equation $y = -\frac{3}{2}x + 3$, each value of x is assigned exactly one value of y. We say
that "y **varies directly with** x", or "y **is a function of** x".
"y is a function of x" is written by **functional notation** $y = f(x)$.
Therefore, the equation $y = -\frac{3}{2}x + 3$ can be written by **functional notation**:

$$f(x) = -\frac{3}{2}x + 3 \quad ; \quad f(x) = -\frac{3}{2}x + 3 \text{ is equivalent to } y = -\frac{3}{2}x + 3.$$

The other way to write a function is to use **arrow notation**: $f : x \to -\frac{3}{2}x + 3$

Definition of a relation: A relation is any set of ordered pairs (x, y).

Each value of x is paired with one or more values of y.

Definition of a function: A function is a relation in which each value of x is paired
with exactly one value of y.

A vertical line intersects the graph of a function in only one point.

Examples:

1. Given $f(x) = 2x - 4$, find the value for each function.

    a) $f(3)$ **b)** $f(-2)$ **c)** $f(0)$ **d)** $f(\frac{1}{2})$

Solution: **a)** $f(3) = 2(3) - 4 = 2$.   **b)** $f(-2) = 2(-2) - 4 = -8$.
    c) $f(0) = 2(0) - 4 = -4$.   **d)** $f(\frac{1}{2}) = 2(\frac{1}{2}) - 4 = -3$. Ans.

2. Given $f : x \to 4 - 3x$, find the value for each function.

    a) $f(-4)$ **b)** $f(1)$

Solution: Write the function: $f(x) = 4 - 3x$

    a) $f(-4) = 4 - 3(-4) = 4 + 12 = 16$.   **b)** $f(1) = 4 - 3(1) = 1$.

3. If the domain of $f(x) = 2x - 3$ is D = $\{-1, 0, 2, 3\}$, find the range of $f(x)$.

Solution: $f(-1) = 2(-1) - 3 = -5$, $f(0) = 2(0) - 3 = -3$

    $f(2) = 2(2) - 3 = 1$, $f(3) = 2(3) - 3 = 3$

The range of $f(x)$ is R = $\{-5, -3, 1, 3\}$. Ans.

EXERCISES

Solve each linear equation for y in terms of x.

1. $x + y = 5$ **2.** $x - y = 5$ **3.** $-x + y = -5$ **4.** $-x - y = 5$

5. $2x + y = 6$ **6.** $2x - y = -6$ **7.** $-2x + 2y = 7$ **8.** $y - 2x = -5$

9. $6x - 3y = 8$ **10.** $4x - 5y = 20$ **11.** $x - 5y = 10$ **12.** $3x - y = 1$

13. $-x + \frac{1}{2}y = 12$ **14.** $2x - \frac{2}{3}y = 10$ **15.** $\frac{1}{2}x - \frac{3}{2}y = 6$

Find the solutions for the given values of x.

16. $y = 2x$ for $x = -2, 0, 2$ **17.** $y = 2x + 3$ for $x = -1, 0, 2$

18. $y = 2x - 3$ for $x = -2, 1, 3$ **19.** $y = \frac{1}{2}x + 1$ for $x = -2, 0, 1$

20. $y = \frac{2}{3}x - 1$ for $x = -3, 0, 1$

Write each of the equation in functional notation $f(x)$.

21. $y = 2x - 4$ **22.** $y = -2x + 5$ **23.** $y = 7 - 8x$ **24.** $y = -5 + 8x$

25. $x + y = 4$ **26.** $x - y = 4$ **27.** $2x + y = 4$ **28.** $-2x - y = 4$

29. $2x + 2y = 7$ **30.** $4x - 2y = 3$

Find the value for each function.

31. $f(x) = 3x - 6$ **a)** $f(-1)$ **b)** $f(0)$ **c)** $f(1)$ **d)** $f(2)$

32. $f(x) = 2 - 5x$ **a)** $f(-2)$ **b)** $f(\frac{1}{2})$ **c)** $f(3)$ **d)** $f(5)$

33. $p(a) = a^2 - 2a$ **a)** $p(-\frac{1}{2})$ **b)** $p(0)$ **c)** $p(3)$ **d)** $p(6)$

34. $q(c) = 2c^2 + c - 2$ **a)** $q(-2)$ **b)** $q(0)$ **c)** $q(1)$ **d)** $q(\frac{1}{2})$

35. $f : x \rightarrow 4x - 5$ **a)** $f(-3)$ **b)** $f(-\frac{1}{2})$ **c)** $f(1)$ **d)** $f(1\frac{1}{4})$

36. Given $f(x) = 4x - 16$, solve $f(x) = 0$.   **37.** Given $g(x) = 3 - 4x$, solve $g(x) = 0$.

38. Given $f(x) = 2x - 1$ and $g(x) = x^2$, find $f(1) + g(2)$.

39. Given $f(x) = 3 - 2x$ and $g(x) = 2x - 1$, find $f(2) \cdot g(3)$.

40. Given $f(x) = x^2$ and $g(x) = 3 - 2x$, find $f[g(2)]$ and $g[f(2)]$.

41. If the domain of $f(x) = 3x - 1$, D $= \{-4, -2, 3, 5, 6\}$, find the range of $f(x)$.

42. Find the domain and range of the function $f(x) = 2x - 1$.

43. Find the domain and range of the function $|f(x)| = x - 2$.

44. Does the equation $y = 2x - 2$ determine y is a function of x? Explain.

45. Does the equation $y^2 = 2x - 2$ determine y is a function of x? Explain.

46. John drives at a speed of 60 miles per hour. Write a function for distance (d) as a function of the number of hours (h).

2-4 Graphing Linear Functions

An equation in two variables can be linear or not linear. In this section, we discuss how to graph a linear equation. We will discuss how to graph an equation which is not linear in the later chapters.

An equation in the form $ax + by = c$ is called a linear equation, because all its ordered pairs (points) lie on a straight line.
The graph of every linear equation in two variables is a straight line in a coordinate plane. Therefore, the equation $2x - y = 3$ is a linear equation.

If we solve the equation $2x - y = 3$ for y, we have a **linear function**:
$$y = 2x - 3 \quad \text{or} \quad f(x) = 2x - 3 \quad \text{where } y = f(x) \text{ and read as } "y \text{ is a function of } x".$$
A linear equation or a linear function can have infinitely many solutions (points).
Each solution represents a point (x, y) on its graph. To graph a linear equation or a linear function, we select a few points and graph the points on the coordinate plane, draw a line that connects the points.

Examples

1. Graph $f(x) = 2x$.

Solution:

x	$y = 2x$	(x, y)
0	$y = 2(0) = 0$	$(0, 0)$
1	$y = 2(1) = 2$	$(1, 2)$
2	$y = 2(2) = 4$	$(2, 4)$

y **or** $f(x)$

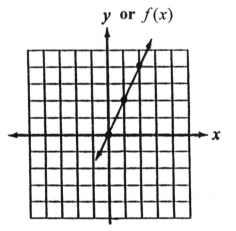

2. Graph $f(x) = 2x - 3$.

Solution:

x	$y = 2x - 3$	(x, y)
0	$y = 2(0) - 3 = -3$	$(0, -3)$
1	$y = 2(1) - 3 = -1$	$(1, -1)$
2	$y = 2(2) - 3 = 1$	$(2, 1)$
3	$y = 2(3) - 3 = 2$	$(3, 3)$
4	$y = 2(4) - 3 = 5$	$(4, 5)$

y **or** $f(x)$

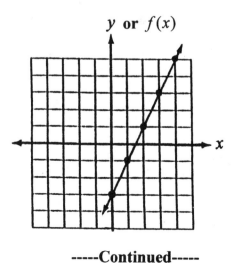

-----Continued-----

Examples

3. Graph $y = \frac{1}{2}x - 1$.

Solution:

y	$y = \frac{1}{2}x - 1$	(x, y)
0	$y = \frac{1}{2}(0) - 1 = -1$	$(0, -1)$
1	$y = \frac{1}{2}(1) - 1 = -\frac{1}{2}$	$(1, -\frac{1}{2})$
2	$y = \frac{1}{2}(2) - 1 = 0$	$(2, 0)$
3	$y = \frac{1}{2}(3) - 1 = \frac{1}{2}$	$(3, \frac{1}{2})$
4	$y = \frac{1}{2}(4) - 1 = 1$	$(4, 1)$

Since two points determine a straight line, we need to find only two points to graph a linear equation.

The points where the line crosses the $x-axis$ (let $y = 0$) and the $y-axis$ (let $x = 0$) are the easiest two points to find. We call them the $x-$intercept and the $y-$intercept.

4. Graph $2x - y = 4$.

Solution:

Let $y = 0$	Let $x = 0$
$2x - 0 = 4$	$2(0) - y = 4$
$2x = 4$	$-y = 4$
$x = 2$	$y = -4$
Point $(2, 0)$	Point $(0, -4)$
$x-$intercept $= 2$	$y-$intercept $= -4$

On the number line, the graph of an equation in one variable is a point (see Chapter 6).

On the coordinate plane, an equation in one variable is a linear equation. Its graph is a vertical line or a horizontal line.

A vertical line has no restriction on y. A horizontal line has no restriction on x.

5. a. Graph $x = 4$.   **b.** Graph $y = 3$.   **c.** Graph $x = -2$.   **d.** Graph $y = -3$.

Solution:

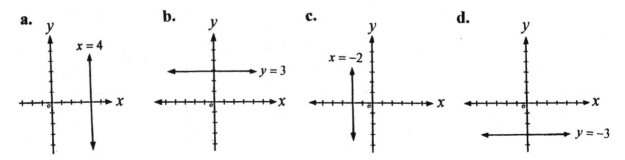

EXERCISES

First Complete the values in each table. Then graph the equation on a coordinate plane. (Hint: They are on a straight line.)

1. $f(x) = 2x$

x	$y = 2x$	(x, y)
0	$y = 2(0) = 0$	$(0, 0)$
1		
2		
3		
4		

2. $f(x) = -2x$

x	$y = -2x$	(x, y)
-1	$y = -2(-1) = 2$	$(-1, 2)$
0		
1		
2		
3		

3. $f(x) = 2x - 1$

x	$y = 2x - 1$	(x, y)
-2	$y = 2(-2) - 1 = -5$	$(-2, -5)$
-1		
0		
1		
2		

4. $f(x) = 3x + 2$

x	$y = 3x + 2$	(x, y)
-4		
-2		
0		
2		
3		

5. $f(x) = \frac{1}{2}x - 1$

x	$y = \frac{1}{2}x - 1$	(x, y)
-2		
-1		
0		
2		
5		

Use the x - intercept **and** y - intercept **to graph each function or equation.**

(**Hint:** $y = x + 1$ **and** $f(x) = x + 1$ **are equivalent.**)

6. $f(x) = x + 1$ **7.** $f(x) = x - 2$ **8.** $f(x) = 2x - 4$

9. $f(x) = 4 - x$ **10.** $f(x) = 4 - 2x$ **11.** $f(x) = 3x - 6$

12. $y = -3x + 6$ **13.** $y = -4x + 2$ **14.** $y = x + \frac{1}{2}$

15. $y = x - \frac{1}{2}$ **16.** $y = \frac{1}{2}x + 4$ **17.** $y = -\frac{1}{3}x - 3$

18. $y = 0.5x - 2$ **19.** $y = -0.5x + 2$ **20.** $x - y = 5$

21. $x + y = 5$ **22.** $2x + y = 4$ **23.** $2x - y = 4$

24. $x + 3y = 9$ **25.** $x - 3y = 9$ **26.** $4x - y = 2$

27. $-4x + y = 2$ **28.** $-2x + y = 1$ **29.** $-2x + 5y = 10$

30. $y = x$ **31.** $y = 3x$ **32.** $\frac{1}{2}x - \frac{1}{3}y = 3$

33. $\frac{1}{3}x - \frac{1}{2}y = 3$ **34.** $\frac{2}{3}x + \frac{1}{2}y = 1$ **35.** $\frac{1}{2}x + \frac{2}{3}y = 3$

-----Continued-----

Graph each equation in a coordinate plane.

36. $y = 3$ **37.** $x = 5$ **38.** $y = -1$ **39.** $x = -4$ **40.** $y = 6$

41. $y = -5$ **42.** $x = 7$ **43.** $x = -7$ **44.** $x = 0$ **45.** $y = 0$

46. Describe the graph of the equation $x = 0$ in a coordinate plane.

47. Describe the graph of the equation $y = 0$ in a coordinate plane.

48. For what value of k is the point $(2, -3)$ in the graph of the equation $6x + ky = 3$?

49. For what value of k is the point $(-2, 3)$ in the graph of the equation $kx - 5y = 1$?

50. For what value of k is the point $(6, k)$ so that it is a solution of the equation $3x + 5y = 3$?

51. For what value of k is the point $(k, -4)$ so that it is a solution of the equation $2x + y = 1$?

52. Find an equation for the group of ordered pairs $(1, 6)$, $(2, 7)$, $(3, 8)$.

53. Find an equation for the group of ordered pairs $(1, \frac{1}{2})$, $(2, 1)$, $(4, 2)$.

54. Find an equation for the group of ordered pairs $(2, 3)$, $(4, 7)$, $(6, 11)$.

55. Write an equation for the data in the table below. Graph the equation.

x	0	1	2	3	4	5	6	7
y	4	6	8	10	12	14	16	18

56. The table shows the distance that a bus travels over time. Let the horizontal axis represent time (t) in hours. Let the vertical axis represent distance (d) in miles. Find and graph the equation for the data.

Time (hours)	0	1	2	3	4	5
Distance (miles)	0	60	120	180	240	300

57. The equation $d = 50\,t$ represents distance in miles and time in hours for a car traveling at a constant speed of 50 miles per hour. Graph the equation for the data.

t	0	2	4	6
d	0	100	200	300

In an English class, the teacher asked the kids to write a letter to his (her) mother regarding what he (she) did today in school. The teacher noticed that John was writing very slowly.

Teacher: John, why do you write so slowly ?

   John: My mom could not read fast.

2-5 Slope of a line

Slope is used to describe the **steepness** of a straight line. To find the slope of a line, we choose any two points on the line and form a right triangle which has the line as its hypotenuse. A line that rises more steeply has a greater slope. A line that runs more horizontally has a smaller slope. Therefore, the slope of a line is defined as the ratio of its vertical rise to its horizontal run. The letter m is commonly used to represent the slope.

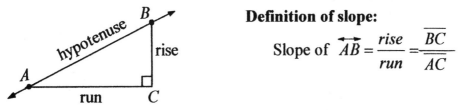

Definition of slope:

$$\text{Slope of } \overleftrightarrow{AB} = \frac{rise}{run} = \frac{\overline{BC}}{\overline{AC}}$$

In the coordinate plane, the slope of a line between two points (x_1, y_1) and (x_2, y_2) is given in the following slope formula:

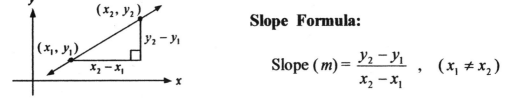

Slope Formula:

$$\text{Slope } (m) = \frac{y_2 - y_1}{x_2 - x_1} \ , \quad (x_1 \neq x_2)$$

When we use the slope formula to find the slope of a line, we can choose any point as the first point and the other as the second point. The result will be the same.

Examples

1. Find the slope of the line passing through the points (2, 3) and (6, 5).
    Solution:

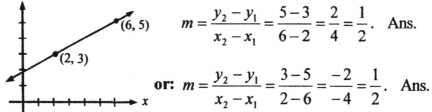

$$m = \frac{y_2 - y_1}{x_2 - x_1} = \frac{5-3}{6-2} = \frac{2}{4} = \frac{1}{2}. \ \text{Ans.}$$

$$\text{or: } m = \frac{y_2 - y_1}{x_2 - x_1} = \frac{3-5}{2-6} = \frac{-2}{-4} = \frac{1}{2}. \ \text{Ans.}$$

2. Find the slope of the line containing the points (–2, 4) and (5, 2).
    Solution:

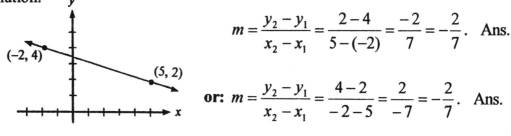

$$m = \frac{y_2 - y_1}{x_2 - x_1} = \frac{2-4}{5-(-2)} = \frac{-2}{7} = -\frac{2}{7}. \ \text{Ans.}$$

$$\text{or: } m = \frac{y_2 - y_1}{x_2 - x_1} = \frac{4-2}{-2-5} = \frac{2}{-7} = -\frac{2}{7}. \ \text{Ans.}$$

(See Examples 1 and 2): 1) **If a line slants up from left to right, the slope is positive.**
                         2) **If a line slants down from left to right, the slope is negative.**

Examples

3. Find the slope of the line passing the points $(-7, 4)$ and $(3, 4)$.

Solution:

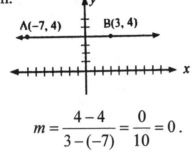

$$m = \frac{4-4}{3-(-7)} = \frac{0}{10} = 0.$$

Ans: The slope is 0.

4. Find the slope of the line passing through the points $(7, 5)$ and $(7, -3)$.

Solution:

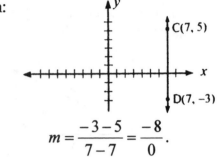

$$m = \frac{-3-5}{7-7} = \frac{-8}{0}.$$

Ans: The slope is undefined.

(See examples 3 and 4): 1) **The slope of a horizontal line is 0.**
                         2) **The slope of a vertical line is undefined.**
                            It means that the slope of a vertical line does not exist.

Slope-intercept form of a straight line

We can choose any two points on a straight line and find its slope by the slope formula (see examples 1 and 2).

However, an easier way to find the slope of a straight line is to rewrite the equation in the slope- intercept form:

$$y = mx + b$$

Then, the slope is m and $y-$ intercept is b.

5. Find the slope of the line $y + 2x = 3$.

Solution:

$$y + 2x = 3$$
$$y = -2x + 3$$
$$\therefore m = -2. \text{ Ans.}$$

6. Find the slope of the line $y - 2x = 3$

Solution:

$$y - 2x = 3$$
$$y = 2x + 3$$
$$\therefore m = 2. \text{ Ans.}$$

7. Find the slope and $y-$ intercept of the line whose equation is $2x + 3y - 6 = 0$.

Solution:

$$2x + 3y - 6 = 0$$
$$3y = -2x + 6$$
$$y = -\tfrac{2}{3}x + 2$$

Ans: slope $= -\tfrac{2}{3}$.

$y-$ intercept $= 2$.

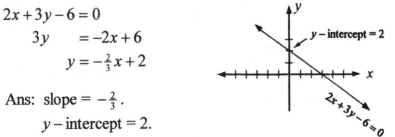

EXERCISES

Find the slope of each line using the given two points on the line.
(Hint: The rise of a line is negative if it falls from left to right.)

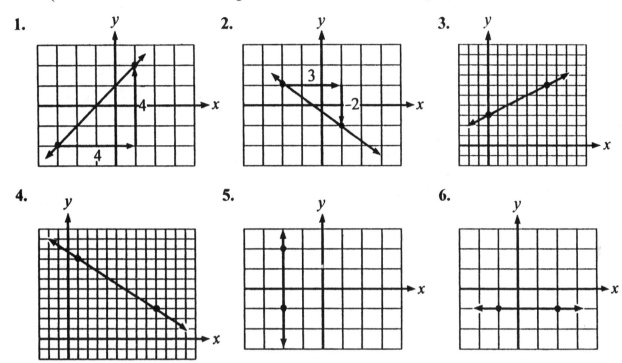

Find the slope of each line with its rise and run.

7. rise 3, run 1   8. rise 1, run 3   9. rise −4, run 2   10. rise 2, run 6
11. rise −2, run 6   12. rise 5, run 2   13. rise 2, run 5   14. rise 8, run 0
15. rise 0, run 8   16. rise $-\frac{1}{2}$, run 3   17. rise $\frac{1}{2}$, run $\frac{1}{3}$ 18. rise $-1\frac{1}{2}$, run $\frac{3}{4}$
19. rise 1.5, run 3   20. rise 4.5, run 1.5

Find the slope of each line containing the given two points.

21. (4, 2) and (2, 4)   22. (−4, 2) and (2, 4)   23. (4, −2) and (2, 4)
24. (4, −2) and (6, −2)   25. (−6, 0) and (0, 6)   26. (0, −6) and (6, 0)
27. (−4, 0) and (−4, 8)   28. (0, 0) and (−2, 4)   29. (−3, −7) and (−7, −3)
30. (1.2, 5) and (0.6, 6.2)   31. (−3.4, 1.4) and (0.4, 9)   32. $(\frac{1}{2}, 1)$ and $(1, \frac{1}{2})$
33. $(\frac{1}{2}, \frac{1}{2})$ and $(\frac{1}{4}, \frac{1}{5})$ 34. $(-\frac{1}{4}, 2)$ and $(2, \frac{1}{4})$ 35. $(\frac{1}{4}, 4)$ and $(4, \frac{1}{2})$
36. (1, 4) and (−3, 0)   37. (1, 5) and (−6, −2)   38. (−2, −4) and (2, 4)
39. (−1, −3) and (4, −2)   40. (−7, −8) and (−3, −4)

Draw a line passing through the given point with the given slope.

41. (1, 2), slope $\frac{1}{2}$ 42. (−2, 4), slope $-\frac{1}{2}$ 43. (−2, 3), slope 3   44. (−3, 2), slope $\frac{2}{3}$
45. (3, −1), slope 0   46. (3, −2), slope $-\frac{3}{5}$ 47. (0, 3), slope $-\frac{2}{5}$ 48. (−5, 2), slope 4
49. (−5, −2), slope $\frac{4}{5}$ 50. (−3, 4), slope undefined   -----Continued-----

Find the slope and y –intercept **of each line.**

51. $y - 2x - 5$ **52.** $y = -2x + 5$ **53.** $y = -7x - 8$ **54.** $y = 7x - 8$

55. $y = \frac{2}{3}x + 4$ **56.** $y = -\frac{2}{3}x + 4$ **57.** $y = -\frac{1}{2}x - \frac{3}{4}$ **58.** $y = \frac{1}{2}x + \frac{3}{4}$

59. $y = -4x$ **60.** $y = -x + 5$ **61.** $y = x - 10$ **62.** $y = 7x$

63. $y = x$ **64.** $y = -x$ **65.** $y = 5$ **66.** $y = -2$

67. $y = 7 - 8x$ **68.** $y = 1 + 2x$ **69.** $y = -2 - \frac{1}{3}x$ **70.** $y = 8 - \frac{4}{5}x$

71. $y - 2x = 15$ **72.** $y + 2x = 15$ **73.** $2x + y = 6$ **74.** $2x - y = 6$

75. $2y - 2x = 3$ **76.** $2y + 4x = 1$ **77.** $4y + 2x = 5$ **78.** $4y - 2x = 3$

79. $3x - 2y = 6$ **80.** $3x + 2y = 7$

Graph each equation using the slope and y –intercept **.**

81. $y = 2x + 3$ **82.** $y = -2x + 1$ **83.** $y = 2x - 3$ **84.** $y = -2x - 1$

85. $y = 3x - 2$ **86.** $y = -3x + 1$ **87.** $y = -3x - 1$ **88.** $y = 3x - 4$

89. $2y - 4x = 5$ **90.** $2x - 4y = 5$

91. What is the slope of a vertical line ?   **92.** What is the slope of a vertical line ?

93. What is the value of $\dfrac{y_2 - y_1}{x_2 - x_1}$ if $x_1 = x_2$?

94. In the equation $ax + 4y = 6$, for what value of a that the equation has a slope 2 ?

95. In the equation $6x - ay = 3$, for what value of a that the equation has a slope –2 ?

96. Determine whether the given points (1, 3), (2, 5), and (4, 9) lie on the same line.

97. Determine whether the given points (2, –6), (0, –2), and (–2, 3) lie on the same line.

98. The points $(2, -3)$, $(4, 1)$, and $(5, y)$ lie on the same line. Find the value of y.

99. The points (2, –3), (0, –2), and (x, 3) lie on the same line. Find the value of x.

100. Find the formula for the slope (m) and the formula for the y – intercept of the linear equation $ax + by = c$ in terms of a, b, and c.

101. A ladder hit the wall at a height of 39 feet with its base 6 feet from the wall. Find the slope of the ladder.

102. If the temperature in the City of Hawthorne drops from $96°F$ at 2 P.M. to $78°F$ at 8 P.M., what is the average rate of change in the temperature ?
Estimate the temperature at 12 P.M..
(Hint: The average rate of change is the slope connecting the points (2, 96) and (8, 78).

Roger was absent yesterday.
Teacher: Roger, why were you absent ?
Roger: My tooth was aching.
    I went to see my dentist.
Teacher: Is your tooth still aching ?
Roger: I don't know. My dentist has it.

2-6 Finding the Equation of a Line

We have learned how to find its points, slope, and intercepts of a given line.
Now we will learn how to find the equation of a line under certain given information.
Such as, find the equation of a line:

> **1)** having its slope and y – intercept.
>
> **2)** having its slope and a point on the line.
>
> **3)** having its two points on the line.

The following examples show how to find the equation of a line.
The equation of a line can be written in two different ways:

> **1) slope-intercept form** $y = mx + b$.   **2) standard form** $ax + by = c$.

Examples

1. Find the equation of the line having slope –5 and y – intercept 2.

Solution:

$$m = -5, \quad b = 2$$
$$y = mx + b$$
$$\therefore y = -5x + 2. \quad \text{Ans. (slope-intercept form)}$$
$$\text{or: } 5x + y = 2. \quad \text{Ans. (standard form)}$$

2. Find the equation of the line having slope $\frac{3}{4}$ and y – intercept $\frac{5}{2}$.

Solution:

$$m = \tfrac{3}{4}, \quad b = \tfrac{5}{2}$$
$$y = mx + b$$
$$y = \tfrac{3}{4}x + \tfrac{5}{2}. \quad \text{Ans. (slope-intercept form)}$$

or: Multiply each side by 4:

$$4y = 3x + 10$$
$$-3x + 4y = 10 \quad \therefore 3x - 4y = -10. \quad \text{Ans. (standard form)}$$

3. Find the equation of the line having slope 4 and passing through the point (2, –3).

Solution:

$$m = 4$$
$$y = mx + b$$
$$y = 4x + b$$

To find b, substitute the point (2, –3) into the above equation.

$$-3 = 4(2) + b$$
$$-11 = b$$
$$\therefore y = 4x - 11. \quad \text{Ans. (slope-intercept form)}$$
$$\text{or: } 4x - y = 11. \quad \text{Ans. (standard form)}$$

Examples

4. Find the equation of the line passing through the points $(-2, 3)$ and $(4, 6)$.
Solution:

$$\text{Find the slope:} \quad m = \frac{y_2 - y_1}{x_2 - x_1} = \frac{6-3}{4-(-2)} = \frac{3}{6} = \frac{1}{2}$$

$$y = mx + b$$
$$y = \tfrac{1}{2}x + b$$

To find b, substitute the point $(-2, 3)$ into the above equation.
$$3 = \tfrac{1}{2}(-2) + b$$
$$3 = -1 + b$$
$$4 = b$$
$$\therefore y = \tfrac{1}{2}x + 4. \quad \text{Ans. (slope-intercept form)}$$
$$\text{or: } x - 2y = -8. \quad \text{Ans. (standard form)}$$

Point-Slope form of a line: If m is the slope of a line and (x_1, y_1) is one of its points, we can write the equation of the line in **point-slope form**.

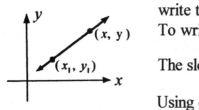

To write the equation, we choose any other point (x, y) on the line.

$$\text{The slope is:} \quad m = \frac{y - y_1}{x - x_1}, \quad x \neq x_1$$

Using cross multiplication, we have:
$$y - y_1 = m(x - x_1) \quad \rightarrow \textbf{Point-Slope form of a line}$$

Examples :

5. Find the equation in point-slope form of a line having slope 4 and passing through the point $(2, -3)$.
Solution:

$$y - y_1 = m(x - x_1)$$
$$y - (-3) = 4(x - 2)$$
$$\therefore y + 3 = 4(x - 2). \quad \text{Ans. (point-slope form)}$$

6: Find the equation in point-slope form of a line passing through the points $(-2, 3)$ and $(4, 6)$. (Hint: There are two answers.)

Solution: $\text{Find the slope:} \quad m = \dfrac{y_2 - y_1}{x_2 - x_1} = \dfrac{6-3}{4-(-2)} = \dfrac{3}{6} = \dfrac{1}{2}$

Substitute one of the points, say $(-2, 3)$, and the slope into the equation.
$$y - y_1 = m(x - x_1)$$
$$y - 3 = \tfrac{1}{2}[x - (-2)]$$
$$\therefore y - 3 = \tfrac{1}{2}(x + 2). \quad \text{Ans. (point-slope form)}$$

If we choose $(4, 6)$, the other answer is: $y - 6 = \tfrac{1}{2}(x - 4)$.

Both answers have the same standard form: $x - 2y = -8$.

EXERCISES

Find the equation of the line having the slope and y–intercept. Express each equation in standard form $ax + by = c$.

1. slope –3, y–intercept 5
2. slope 3, y–intercept –5
3. slope –3, y–intercept –5
4. slope 6, y–intercept –9
5. slope $\frac{1}{2}$, y–intercept 7
6. slope $-\frac{1}{2}$, y–intercept 7
7. slope $-\frac{2}{3}$, y–intercept 8
8. slope $\frac{3}{4}$, y–intercept 8
9. slope $\frac{2}{3}$, y–intercept –4
10. slope $-\frac{3}{4}$, y–intercept – 4

Find the equation of the line having the slope and passing through the given point. Express each equation in slope-intercept form $y = mx + b$.

11. slope 2, (2, 5)
12. slope 2, (5, 2)
13. slope –2, (3, 4)
14. slope –2, (–3, 4)
15. slope –2, (3, –4)
16. slope 2, (–3, –4)
17. slope –2, (–3, –4)
18. slope 7, (1, 6)
19. slope 6, (–5, 8)
20. slope –1, (4, 9)
21. slope 1, (5, –1)
22. slope 0, (–7, 5)
23. slope 0, (3, 0)
24. slope undefined , (5, 0)
25. slope undefined, (0, 5)
26. slope $\frac{1}{2}$, (4, –2)
27. slope $-\frac{3}{5}$, (10, 3)
28. slope $-\frac{2}{3}$, (2, –1)
29. slope $\frac{4}{5}$, (0, 9)
30. slope $\frac{5}{4}$, (9, 0)

Find the equation of the line passing through the given two points. Express each equation in standard form $ax + by = c$.

31. (1, 2), (3, 4)
32. (2, 1), (4, 3)
33. (–2, 3), (0, 4)
34. (2, –3), (4, 0)
35. (0, 5), (2, 6)
36. (0, 0), (–2, 4)
37. (5, 0), (6, 3)
38. (–3, –4), (0, 7)
39. (–3, –6), (–4, 7)
40. (–6, –2), (7, 0)
41. (2.5, 3), (4, 7.5)
42. (3, 4.5), (2.5, 3.5)
43. (1, $\frac{1}{2}$), (2, $\frac{2}{3}$)
44. (2, 3), ($\frac{1}{2}$, $\frac{1}{2}$)
45. ($\frac{3}{2}$, $\frac{1}{2}$), ($-\frac{1}{2}$, $\frac{1}{3}$)

Find the equation of each line described. Express each equation in point-slope form $y - y_1 = m(x - x_1)$.

46. slope 3, passes through (2, 4)
47. slope –3, passes through (–2, 4)
48. slope 4, passes through (–1, –2)
49. slope –5, passes through (7, –5)
50. slope $\frac{3}{4}$, passes through (–1, 5)
51. slope $-\frac{1}{5}$, passes through (4, –7)
52. passes through (4, 5) and (8, –3)
53. passes through (5, –2) and (0, 2)
54. passes through (0, 5) and (5, 0)
55. passes through (0, 0) and (–3, –1)

-----Continued-----

56. Find the equation in standard form of a line passing through (4, 6) and parallel to the line $20x - 2y = 5$.

57. Find the equation in standard form of a line passing through (4, −6) and parallel to the line $20x - 2y = 5$

58. Find the equation in standard form of a line having slope 5 and x – intercept 4.

59. Find the equation in standard form of a line having slope −5 and x – intercept 4.

60. Find the equation in standard form of a line having x – intercept −6 and y – intercept 4.

61. Find the equation in standard form of a line having x – intercept 6 and y – intercept −4.

62. Find the equation of a horizontal line passing through (3, −2).

63. Find the equation of a vertical line passing through (3, −2).

64. Find the equation of a vertical line passing through (−2, 3).

65. Find the equation of a horizontal line passing through (−2, 3).

66. Find a if two points (3, 2) and (−2, a) are on a line having slope 2.

67. Find a if two points (3, 2) and (a, −2) are on a line having slope 2

68. Find a if two points (3, 2) and (−2, $a+1$) are on a line having slope −2.

69. Find a if three points (−1, 3), (−3, 2), and (−4, $a-5$) are on a line.

70. Find a

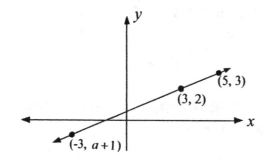

In a biology class, a student said to his teacher.
Student: I just killed five mosquitoes, three were males and two were females.
Teacher: How did you know which were males and which were females ?
Student: Three were killed on a beer bottle. Two were killed on a mirror.

2-7 Parallel and Perpendicular Lines

Now we will learn how to identify parallel lines and perpendicular lines by comparing their slopes.

Two different lines that have the same slope are **parallel**. Two different lines that intersect to form a right angle (90° angle) are **perpendicular**.

Rules: 1. **Two lines are parallel if and only if their slopes are equal:** $m_1 = m_2$
     2. **Two lines are perpendicular if and only if the product of their slopes is -1:**
      (They are negative reciprocals of each other.)

$$m_1 \cdot m_2 = -1, \quad \text{or} \quad m_1 = -\frac{1}{m_2}, \quad m_2 = -\frac{1}{m_1}$$

Examples

1. Find the slope of a line that is perpendicular to the line $8x - 4y = 3$.

 Solution:
$$8x - 4y = 3$$
  Rewrite it in slope-intercept form: $-4y = -8x + 3$

  Divide each side by -4, we have $y = 2x - \frac{3}{4}$. slope $m = 2$

 The slope of the perpendicular line is the negative reciprocal of $2 = -\frac{1}{2}$. Ans.

2. Tell whether or not the following two lines are parallel, perpendicular, or neither.
  $3x - 4y = -5$ and $6x - 8y = 7$

 Solution:

    To find their slopes, we rewrite each equation in slope-intercept form.
    $3x - 4y = -5$, $-4y = -3x - 5$, $y = \frac{3}{4}x + \frac{5}{4}$ $\therefore m_1 = \frac{3}{4}$
    $6x - 8y = 7$, $-8y = -6x + 7$, $y = \frac{3}{4}x - \frac{7}{8}$ $\therefore m_2 = \frac{3}{4}$
    $m_1 = m_2$ Ans: They are parallel.

3. Tell whether or not the following two lines are parallel, perpendicular, or neither.
  $x - 2y = 4$ and $2x + y = 5$

  Solution:

    $x - 2y = 4$, $-2y = -x + 4$, $y = \frac{1}{2}x - 2$ $\therefore m_1 = \frac{1}{2}$
    $2x + y = 5$, $y = -2x + 5$ $\therefore m_2 = -2$
    $m_1 \cdot m_2 = \frac{1}{2} \cdot (-2) = -1$ Ans: They are perpendicular.

4. Tell whether or not the following two lines are parallel, perpendicular, or neither.
  $x + 3y = 4$ and $6y = 4x - 5$

  Solution: $x + 3y = 4$, $3y = -x + 4$, $y = -\frac{1}{3}x + \frac{4}{3}$ $\therefore m_1 = -\frac{1}{3}$
      $6y = 4x - 5$, $y = \frac{2}{3}x - \frac{5}{6}$ $\therefore m_2 = \frac{2}{3}$
    $m_1 \neq m_2$ and $m_1 \cdot m_2 \neq -1$ Ans: Neither.

EXERCISES

Rewrite each equation in slope-intercept form $y = mx + b$.

1. $x + y = 5$ **2.** $x - y = 5$ **3.** $x + y = -5$ **4.** $-x + y = 5$

5. $2x + y = 6$ **6.** $2x - y = -6$ **7.** $-2x + 2y = 7$ **8.** $2y + 3x = -5$

Find the slope of a line that is perpendicular to the given line.

9. $6x - 3y = 8$ **10.** $4x - 5y = 20$ **11.** $x + 5y = 10$ **12.** $3x + y = 1$

13. $x + y = 5$ **14.** $x - y = 5$ **15.** $y - 2x = -5$ **16.** $2x + y = 9$

17. $y = 5$ **18.** $x = -3$ **19.** $2x - \frac{1}{3}y = 10$ **20.** $\frac{1}{2}x - \frac{3}{2}y = 6$

Tell whether or not each pair of equations are parallel, perpendicular, or neither.

21. $2x - 3y = 8$
$-2x + 3y = 14$

22. $2x - 3y = 8$
$2x + 3y = 14$

23. $2x - 3y = 8$
$3x + 2y = 14$

24. $y - 4x = -3$
$2y - 8x = 5$

25. $y - 4x = 8$
$2y + 8x = 3$

26. $5y - 3x = 25$
$3y + 5x = 15$

27. $x + y = 5$
$x - y = 5$

28. $x + y = 5$
$x + y = 7$

29. $x - y = 10$
$x - y = 9$

30. $5y - 4x = 10$
$5x + 4y = 1$

31. $4x + 8y = 7$
$2x + 4y = 9$

32. $4x - 2y = 13$
$2x - 4y = 11$

33. $4x - 2y = 13$
$2x + 4y = 11$

34. $\frac{2}{3}x + y = 15$
$y = \frac{3}{2}x - 10$

35. $2x - \frac{1}{2}y = 3$
$y = -\frac{1}{2}x - 3$

36. What is the slope of a line that is parallel to the line $5x - 4y = 7$?
37. What is the slope of a line that is perpendicular to the line $y = 2$?
38. What is the slope of a line that is parallel to the line $x = 2$?
39. What is the slope of a line that is perpendicular to a vertical line ?
40. What is the slope of a line that is parallel to a horizontal line ?
41. Write an equation in slope-intercept form for a line perpendicular to the line with slope 7 and y-intercept 15.
42. Find the equation in standard form of a line that passes the point (10, −2) and is parallel to the line $2x + 5y = 3$.
43. Find the equation in standard form of a line that passes the point (10, −2) and is perpendicular to the line $2x + 5y = 3$.

2-8 Direct Variations

A linear equation (or function) with the form $y = kx$ is called an equation having a **direct variation**. k is a nonzero constant. k is called the **constant of variation**.

In the equation $y = 3x$, 3 is a constant. If x is increased, then y is increased.

We say that y varies directly as x. Therefore, $y = 3x$ is an example of a direct variation.

The graph of $y = 3x$ is a straight line with slope 3 and passes through the origin $(0, 0)$.

If (x_1, y_1) and (x_2, y_2) are two ordered pairs of an equation having a direct variation defined by $y = kx$ and that neither x_1 nor x_2 is zero, we have: $y_1 = kx_1$ and $y_2 = kx_2$,

$$k = \frac{y_1}{x_1} \text{ and } k = \frac{y_2}{x_2}. \quad \text{Therefore: } \frac{y_1}{x_1} = \frac{y_2}{x_2}. \quad \textbf{The ratios are equal.}$$

If $y = kx$, we say that y varies directly as x, or y is directly proportional to x.

In science, there are many word problems which involve the concept of direct variations.

Examples

1. If y varies directly as x, and if $y = 6$ when $x = 2$, find the constant of variation.

   Solution: Let $y = kx$, $6 = k \cdot 2$ $\therefore k = \frac{6}{2} = 3$. Ans.

2. If y varies directly as x, and if $y = 6$ when $x = 2$, find y when $x = 3$.

   Solution: Let $y = kx$, $6 = k \cdot 2$ $\therefore k = \frac{6}{2} = 3$, $y = 3x$ is the equation.

       When $x = 3$, $y = 3x = 3 \cdot 3 = 9$. Ans.

3. If n varies directly as m, and if $n = 64$ when $m = 16$, find n when $m = 5$.

   Solution: Let $n = km$, $64 = k \cdot 16$ $\therefore k = \frac{64}{16} = 4$, $n = 4m$ is the equation.

       When $m = 5$, $n = 4m = 4(5) = 20$. Ans.

4. If (x_1, y_1) and (x_2, y_2) are ordered pairs of the same direct variation, find y_1.

   $$x_1 = 3, \; y_1 = ?, \; x_2 = 12, \; y_2 = 8$$

   Solution: The ratios are equal: $\frac{y_1}{x_1} = \frac{y_2}{x_2}$, $\frac{y_1}{3} = \frac{8}{12}$, $12y_1 = 24$ $\therefore y_1 = 2$. Ans.

       Or: $y_2 = kx_2$, $8 = k \cdot 12$, $k = \frac{8}{12} = \frac{2}{3}$ $\therefore y_1 = kx_1 = \frac{2}{3}x_1 = \frac{2}{3} \cdot 3 = 2$. Ans.

5. The total length of the wire spring is directly proportional to the mass of the weight attached. A 25 grams weight causes a wire spring to a total length of $10\,cm$. What will be the total length of the wire spring if a 55 grams weight is attached ?

   Solution: Let $x =$ the weight in grams, $y =$ the total length of the spring in cm

       Let $y = kx$, $10 = k \cdot 25$, $k = \frac{10}{25} = 0.4$ $\therefore y = 0.4x$ is the formula.

       When $x = 55$ grams, $y = 0.4x = 0.4(55) = 22\,cm$. Ans.

   (Hint: It is called **Hooke's Law**. k is called the coefficient of elasticity for a wire spring.)

EXERCISES

Find the constant of variation and write an equation of direct variation.

(Hint: "varies directly as" and "is directly proportional to" have the same meaning.

1. If y varies directly as x, and if $y = 10$ when $x = 2$.
2. If y is directly proportional to x, and if $y = 10$ when $x = 5$.
3. If y varies directly as x, and if $y = 5$ when $x = 2$.
4. If y varies directly as x, and if $y = 6.5$ when $x = 13$.
5. If n is directly proportional to m, and if $n = 200$ when $m = 20$.
6. If p varies directly as n, and if $p = 36$ when $n = 30$.
7. If s varies directly as t, and if $s = 15$ when $t = 30$.
8. If m varies directly as n, and if $m = 30$ when $n = 90$.
9. If y varies directly as $(x - 4)$, and if $y = 15$ when $x = 9$.
10. If y is directly proportional to $(x + 8)$, and if $y = 150$ when $x = 2$.

For each direct variation described, find each missing value.

11. If y varies directly as x, and if $y = 10$ when $x = 2$, find y when $x = 12$.
12. If y varies directly as x, and if $y = 10$ when $x = 2$, find x when $y = 12$.
13. If y is directly proportional to x, and if $y = 2$ when $x = 10$, find y when $x = 12$.
14. If n varies directly as m, and if $n = 150$ when $m = 6$, find n when $m = 20$.
15. If n varies directly as m, and if $n = 150$ when $m = 6$, find m when $n = 20$.

If (x_1, y_1) and (x_2, y_2) are ordered pairs of the same direct variation, find each missing value.

16. $x_1 = 1$, $y_1 = ?$, $x_2 = 3$, $y_2 = 12$
17. $x_1 = 3$, $y_1 = ?$, $x_2 = 4$, $y_2 = 80$
18. $x_1 = 6$, $y_1 = ?$, $x_2 = 8$, $y_2 = 9.6$
19. $x_1 = 6$, $y_1 = 4$, $x_2 = 9$, $y_2 = ?$
20. $x_1 = ?$, $y_1 = 1.5$, $x_2 = 15$, $y_2 = 9$
21. $x_1 = 1$, $y_1 = 7$, $x_2 = ?$, $y_2 = 17.5$
22. $x_1 = \frac{4}{5}$, $y_1 = ?$, $x_2 = \frac{2}{5}$, $y_2 = \frac{1}{10}$
23. $x_1 = \frac{1}{4}$, $y_1 = \frac{1}{5}$, $x_2 = \frac{5}{3}$, $y_2 = ?$
24. $x_1 = \frac{7}{8}$, $y_1 = ?$, $x_2 = \frac{1}{2}$, $y_2 = \frac{1}{7}$
25. $x_1 = \frac{1}{5}$, $y_1 = 5$, $x_2 = 5$, $y_2 = ?$

26. Does the equation $xy = 5$ define a direct variation ? Explain.
27. Are the ordered pairs (6, 4), (9, 6), (27, 18) in the same direct variation ? Explain.
28. The actual distance D is directly proportional to the length L on the map. If 0.5 cm represents 5 miles, write a formula for computing the actual distance.
29. The total length of the wire spring is directly proportional to the mass of the weight attached. A 30 grams weight causes a wire spring to a total length of 7.5 cm. What will be the total length of the wire spring if a 58 grams weight is attached ?
30. The simple interest you pay is directly proportional to the amount of money you have borrowed. If you paid $245 interest on $3,500, how much interest would you pay on $6,300 in the same period of time ?

CHAPTER 2   EXERCISES

Find the length of the segment between two points.

1. (5, 7) and (5, –3) **2.** (–5, 4) and (3, 4) **3.** (–2, 7) and (–2, 1)
4. (–6, –1) and (–3, –1) **5.** (6, –8) and (6, –7) **6.** (1, 1) and (1, 9)

Identify whether each ordered pair of number is a solution of the equation.

7. $x + 2y = 3$, (1, 1)   **8.** $x + 2y = 3$, (–1, 2)   **9.** $x + 2y = 3$, (3, 1)

10. $2x - y = 5$, (2, 1)   **11.** $2x - y = 5$, (2, –1)   **12.** $2x - y = 5$, $(\frac{1}{2}, -4)$

13. $2a - 4b = 1$, $(3, \frac{5}{4})$ **14.** $2a - 4b = 1$, $(-2, \frac{5}{3})$ **15.** $y = \frac{2}{3}x - 4$, (3, –2)

Write each of the linear equations in standard form $ax + by = c$.

16. $2x = y - 5$ **17.** $y = -2x + 5$ **18.** $-y = 2x - 5$ **19.** $-y = -2x + 5$
20. $4x = 5y + 7$ **21.** $3x - 9 = 8y$ **22.** $6y + 7 = 2x$ **23.** $\frac{2}{3}x = -2y + 1$

Solve each linear equation for y in terms of x.

24. $2x + y = 4$ **25.** $-2x + y = 4$ **26.** $2x - y = 4$ **27.** $y - 4 = 2x$
28. $4x + 2y = 5$ **29.** $4x - 2y = 5$ **30.** $5x - 3y = 6$ **31.** $3x + 5y = 6$

Find the value for each function.

$f(x) = 2x - 7$	**32)** $f(1)$	**33)** $f(0)$	**34)** $f(-1)$	**35)** $f(2)$
$f(x) = 7 - 2x$	**36)** $f(1)$	**37)** $f(0)$	**38)** $f(2)$	**39)** $f(-\frac{5}{4})$
$p(n) = n^2 - n + 1$	**40)** $p(2)$	**41)** $p(0)$	**42)** $p(5)$	**43)** $p(\frac{1}{2})$
$f : x \rightarrow 5x - 4$	**44)** $f(2)$	**45)** $f(-2)$	**46)** $f(\frac{1}{5})$	**47)** $f(-1\frac{1}{5})$

Given the domain of each function, find the range.

48. $f(x) = 3x - 7$, D = {0, 1, 2, 3, 4}   **49.** $f(x) = 4x - 2$, D = {–2, –1, 0, 1, 2}
50. $f(x) = x^2 - 2$, D = {–3, –1, 0, 1, 3}   **51.** $f(x) = 2x^2 - 2$, D = {–2, 0, 1, 2}
52. $f(x) = |x| - 4$, D = {–5, –1, 0, 3}   **53.** $f(x) = 4 - |x|$, D = {–5, –1, 0, 3}

Find the domain and range of each relation (ordered pairs).
Determine whether or not it is a function.

54. {(0, –2), (1, –1), (2, 1), (3, 4)}   **55.** {(–2, 1), (–2, 3), (0, 2), (2, 1)}

56. {(–2, –2), (–1, 1), (1, 0), (1, 2)}   **57.** {(–3, 1), (–1, 3), (2, –3), (3, –2)}

58. {(0, 2), (1, 1), (2, 0), (3, –1), (4, –2)}   **59.** {(1, 2), (2, 3), (8, 9)}

60. {(1, 2), (–4, 7), (1, –2)}   -----Continued-----

Find the x – **intercept and** y – **intercept. Use the** x – **intercept and** y – **intercept to graph each equation.**

61. $y = x - 3$ **62.** $y = 3 - x$ **63.** $x + y = -3$ **64.** $y = 2x + 2$

65. $x + 2y = 6$ **66.** $x - 2y = 6$ **67.** $2x - 3y = 12$ **68.** $2x + 3y = 12$

69. $y = -2$ **70.** $x = 1$ **71.** $y = 2x$ **72.** $y = -2x$

73. $\frac{2}{3}x - y = 4$ **74.** $\frac{1}{2}x + \frac{1}{3}y = 4$ **75.** $\frac{1}{2}x - \frac{2}{3}y = 4$

Find the slope of each line with its rise and run.

76. rise 5, run 2   **77.** rise 2, run −5   **78.** rise 0, run 9   **79.** rise 9, run 0

80. rise 6, run $\frac{1}{2}$ **81.** rise $-\frac{2}{3}$, run $\frac{1}{6}$ **82.** rise $1\frac{1}{2}$, run $-\frac{5}{4}$ **83.** rise 2.5, run 0.5

Find the slope of each line containing the given two points.

84. (3, 5) and (5, 3)   **85.** (−3, 5) and (3, −5)   **86.** (−3, −5) and (3, 5)

87. (0, −4) and (0, 5)   **88.** (−4, 0) and (5, 0)   **89.** (−8, 7) and (−8, −7)

90. $(2, \frac{2}{3})$ and (4, 2)   **91.** $(\frac{1}{2}, 2)$ and $(2, -4)$ **92.** $(\frac{2}{5}, -\frac{3}{5})$ and $(\frac{1}{5}, \frac{1}{5})$

93. (−1.2, −1.5) and (1.3, 3)   **94.** (0, 0) and (4.5, 7.2)   **95.** $(\frac{1}{2}, \frac{1}{2})$ and $(\frac{2}{3}, -\frac{1}{3})$

Find the slope (m) **and** y – **intercept** (b) **of each line.**

96. $y = -4x + 9$ **97.** $y = 9 + 4x$ **98.** $y = \frac{4}{5}x - 15$ **99.** $y = -\frac{4}{5}x + 15$

100. $8x + y = 10$ **101.** $8x - y = 10$ **102.** $y = 8x$ **103.** $y = -8$

104. $3x + 6y = 8$ **105.** $3x - 6y = -8$ **106.** $\frac{1}{2}y - 4x = 7$ **107.** $-5y + 10x = 7$

Find the equation in standard form $ax + by = c$ **of each line described.**

108. slope 5, y – intercept 7   **109.** slope -4, y – intercept –2

110. slope $\frac{3}{2}$, y – intercept 12   **111.** slope $-\frac{4}{5}$, y – intercept $-\frac{2}{3}$

112. slope 7, passes through (2, 3)   **113.** slope $-\frac{1}{2}$, passes through (4, −5)

114. slope $\frac{3}{4}$, passes through (−2, −1)   **115.** slope 0, passes through (−7, 9)

116. slope undefined, (5, 1)   **117.** passes through (2, −3) and (6, 5)

118. passes through (1, 7) and (−6, −7)   **119.** passes through $(\frac{2}{3}, 4)$ and (1, −5)

120. passes through (3, −1) and $(\frac{1}{3}, 3)$

Tell whether or not they are parallel, perpendicular, or neither.

121. $5x - 6y = 4$ **122.** $2x - 4y = 11$ **123.** $2x - 4y = 13$ **124.** $4x - 8y = 21$

   $6x + 5y = 7$ $4x + 8y = 15$ $4x - 8y = 19$ $4x + 2y = 25$

125. If y varies directly as x, and $y = 48$ when $x = 4$. Find y when $x = \frac{2}{3}$.

126. If n varies directly as m, and $n = 210$ when $m = 21$. Find n when $m = 5\frac{1}{2}$.

127. If (x_1, y_1) and (x_2, y_2) are ordered pairs of the same variation, find y_2.

$$x_1 = \tfrac{5}{8}, \ y_1 = \tfrac{1}{4}, \ x_2 = \tfrac{2}{3}, \ y_2 = ?$$

ANSWERS

CHAPTER 2 –1   **Number line and Coordinate Plane**   **page 33 ~ 34**

1. –5   **2.** –2   **3.** 0   **4.** 4.5 or $4\frac{1}{2}$ **5.** –14   **6.** –3   **7.** 6   **8.** –22.5 or $-22\frac{1}{2}$ **9.** – 5   **10.** 0

11. 12.5 or $12\frac{1}{2}$ **12.** 30   **13.** 2.5 or $2\frac{1}{2}$ **14.** –6.5 or $-6\frac{1}{2}$ **15.** –7

16. –6   **17.** –2.5 or $-2\frac{1}{2}$ **18.** 20   **19.** –8   **20.** 15   **21.** 2   **22.** –4   **23.** –4   **24.** 5

25. 12.5 or $12\frac{1}{2}$ **26.** B = 4,   C = 9,   D = 15,   E = 21

27. $x = 2$

28. $y = -1$

29. $a = -2.5$

30. $x = 9$

31. $A(2,4)$ **32.** $B(-2,3)$

33. $C(-4,-4)$ **34.** $D(4,-1)$

35. $E(0,2)$ **36.** $F(3,-5)$

37. $G(-3,0)$ **38.** $K(1\frac{1}{2},0)$

39. $P(0, -2\frac{1}{2})$ **40.** $Q(-5, 2\frac{1}{2})$

41 ~ 45

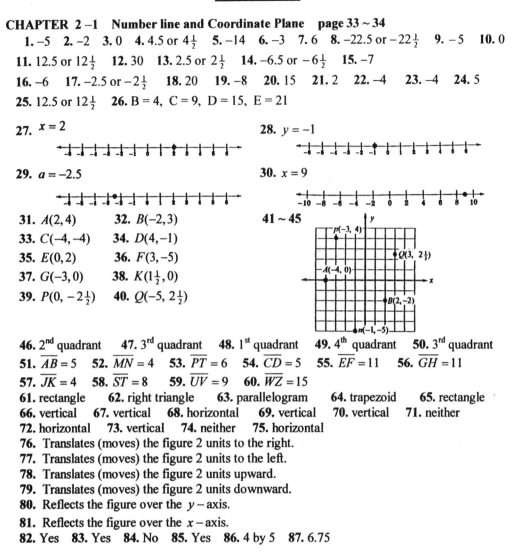

46. 2nd quadrant   **47.** 3rd quadrant   **48.** 1st quadrant   **49.** 4th quadrant   **50.** 3rd quadrant

51. $\overline{AB} = 5$ **52.** $\overline{MN} = 4$ **53.** $\overline{PT} = 6$ **54.** $\overline{CD} = 5$ **55.** $\overline{EF} = 11$ **56.** $\overline{GH} = 11$

57. $\overline{JK} = 4$ **58.** $\overline{ST} = 8$ **59.** $\overline{UV} = 9$ **60.** $\overline{WZ} = 15$

61. rectangle   **62.** right triangle   **63.** parallelogram   **64.** trapezoid   **65.** rectangle

66. vertical   **67.** vertical   **68.** horizontal   **69.** vertical   **70.** vertical   **71.** neither

72. horizontal   **73.** vertical   **74.** neither   **75.** horizontal

76. Translates (moves) the figure 2 units to the right.

77. Translates (moves) the figure 2 units to the left.

78. Translates (moves) the figure 2 units upward.

79. Translates (moves) the figure 2 units downward.

80. Reflects the figure over the y – axis.

81. Reflects the figure over the x – axis.

82. Yes   **83.** Yes   **84.** No   **85.** Yes   **86.** 4 by 5   **87.** 6.75

CHAPTER 2 –2   **Linear Equations**   **page 36**

1. Yes   **2.** No   **3.** Yes   **4.** Yes   **5.** No   **6.** Yes   **7.** Yes   **8.** No   **9.** Yes   **10.** No

11. Yes   **12.** No   **13.** Yes   **14.** Yes   **15.** No   **16.** Yes   **17.** No   **18.** No   **19.** No   **20.** Yes

21. N0   **22.** Yes   **23.** Yes   **24.** Yes   **25.** Yes   **26.** Yes   **27.** No   **28.** Yes   **29.** Yes   **30.** Yes

31. $3x - y = -5$ **32.** $4x + y = 7$ **33.** $3x + 4y = 6$ **34.** $5x - 4y = -9$ **35.** $2x + 4y = 8$

36. $5x + 3y = 20$ **37.** $4x + 7y = 1$ **38.** $2x - 4y = -7$ **39.** $2x - 5y = 4$ **40.** $6x - 4y = 5$

41. (–2, –4), (0, 0), (2, 4)   **42.** (–2, –3), (0, 1), (2, 5)   **43.** (–2, –5), (0, –1), (2, 3)

44. (–2, 5), (0, 1), (2, –3)   **45.** (–2, 3), (0, –1), (2, –5)   **46.** (–2, 7), (0, 5), (2, 3)

47. (–2, –3), (0, –5), (2, –7)   **48.** (–2, 6), (0, 4), (2, 2)   **49.** (–2, 8), (0, 4), (2, 0)

50. (–2, –8), (0, –4), (2, 0)

51. $c = 1.50n$ **52.** $c = p - 0.30p$, or $c = 0.70p$ **53.** $c = p + 0.075p$, or $c = 1.075p$

54. for one day: $c = 20 + 0.20m$ **55.** $c = 11.50n + 5$

     for n days: $c = 20n + 0.20m$ $c = \$396$ for 34 copies

CHAPTER 2 – 3 Equations and Functions   Page 38

  1. $y = -x + 5$ **2.** $y = x - 5$ **3.** $y = x - 5$ **4.** $y = -x - 5$ **5.** $y = -2x + 6$

  6. $y = 2x + 6$ **7.** $y = x + \frac{7}{2}$ **8.** $y = 2x - 5$ **9.** $y = 2x - \frac{8}{3}$ **10.** $y = \frac{4}{5}x - 4$

  11. $y = \frac{1}{3}x - 2$ **12.** $y = 3x - 1$ **13.** $y = 2x + 24$ **14.** $y = 3x - 15$ **15.** $y = \frac{1}{3}x - 4$

  16. (–2, –4), (0, 0), (2, 4)   **17.** (–1, 1), (0, 3), (2, 7)   **18.** (–2, –7), (1, –1), (3, 3)

  19. (–2, 0), (0, 1), (1, $1\frac{1}{2}$)   **20.** (–3, –3), (0, –1), (1, $-\frac{1}{3}$)

  21. $f(x) = 2x - 4$ **22.** $f(x) = -2x + 5$ **23.** $f(x) = -8x + 7$ **24.** $f(x) = 8x - 5$

  25. $f(x) = -x + 4$ **26.** $f(x) = x - 4$ **27.** $f(x) = -2x + 4$ **28.** $f(x) = -2x - 4$

  29. $f(x) = -x + \frac{7}{2}$ **30.** $f(x) = 2x - \frac{3}{2}$

  31. a) $f(-1) = -9$ **b)** $f(0) = -6$ **c)** $f(1) = -3$ **d)** $f(2) = 0$

  32. a) $f(-2) = 12$ **b)** $f(\frac{1}{2}) = -\frac{1}{2}$ **c)** $f(3) = -13$ **d)** $f(5) = -23$

  33. a) $p(-\frac{1}{2}) = 1\frac{1}{4}$ **b)** $p(0) = 0$ **c)** $p(3) = 3$ **d)** $p(6) = 24$

  34. a) $q(-2) = 4$ **b)** $q(0) = -2$ **c)** $q(1) = 1$ **d)** $q(\frac{1}{2}) = -1$

  35. a) $f(-3) = -17$ **b)** $f(-\frac{1}{2}) = -7$ **c)** $f(1) = -1$ **d)** $f(1\frac{1}{4}) = 0$

  36. $x = 4$ **37.** $x = \frac{3}{4}$ **38.** $f(1) + g(2) = 5$ **39.** $f(2) \cdot g(3) = -5$

  40. $f[g(2)] = 1$, $g[f(2)] = -5$. (Hint: we call it "function of function")

  41. R = {–13, –7, 8, 14, 17}   **42.** D = all real numbers, R = all real numbers.

  43. D = $x \geq 2$, R = all real numbers.

  44. Yes, y is a function of x because each value of x can find exactly one value of y.

  45. No, y is not a function of x because each value of x can find two values of y.

  46. $d(h) = 60\, h$

CHAPTER 2 – 4 Graphing Linear Equations   page 41 ~ 42

  1~5 graphs are left to the students (see examples 1, 2, 3 on page 39 and 40).

  1. (0, 0), (1, 2), (2, 4), (3, 6), (4, 8)   **2.** (–1, 2), (0, 0), (1, –2), (2, –4), (3, –6)

  3. (–2, –5), (–1, –3), (0, –1), (1, 1), (2, 3)   **4.** (–4, –10), (–2, –4), (0, 2), (2, 8), (3, 11)

  5. (–2, –2), (–1, $-1\frac{1}{2}$), (0, –1), (2, 0), (5, $1\frac{1}{2}$)

  6~35 graphs are left to the students (see example 4 on page 40).

6. x-intercept = –1 y-intercept = 1	**7.** x-intercept = 2 y-intercept = –2	**8.** x-intercept = 2 y-intercept = –4	**9.** x-intercept = 4 y-intercept = 4
10. x-intercept = 2 y-intercept = 4	**11.** x-intercept = 2 y-intercept = –6	**12.** x-intercept = 2 y-intercept = 6	**13.** x-intercept = $\frac{1}{2}$ y-intercept = 2
14. x-intercept = $-\frac{1}{2}$ y-intercept = $\frac{1}{2}$	**15.** x-intercept = $\frac{1}{2}$ x-intercept = $-\frac{1}{2}$	**16.** x-intercept = –8 y-intercept = 4	**17.** x-intercept = –9 y-intercept = –3
18. x-intercept = 4 y-intercept = –2	**19.** x-intercept = 4 y-intercept = 2	**20.** x-intercept = 5 y-intercept = –5	**21.** x-intercept = 5 y-intercept = 5
22. x-intercept = 2 y-intercept = 4	**23.** x-intercept = 2 y-intercept = –4	**24.** x-intercept = 9 y-intercept = 3	**25.** x-intercept = 9 y-intercept = –3
26. x-intercept = $\frac{1}{2}$ y-intercept = –2	**27.** x-intercept = $-\frac{1}{2}$ y-intercept = 2	**28.** x-intercept = $-\frac{1}{2}$ y-intercept = 1	**29.** x-intercept = –5 y-intercept = 2
30. x-intercept = 0 y-intercept = 0	**31.** x-intercept = 0 y-intercept = 0	**32.** x-intercept = 6 y-intercept = –9	**33.** x-intercept = 9 y-intercept = –6
34. x-intercept = $1\frac{1}{2}$ y-intercept = 2	**35.** x-intercept = 6 y-intercept = $4\frac{1}{2}$		

  36~45 graphs are left to the students (see example 5).

  46. The graph is the $y-axis$.   **47.** The graph is the $x-axis$.   **48.** $k = 3$ **49.** $k = -8$

  50. $k = -3$ **51.** $k = 2\frac{1}{2}$ **52.** $y = x + 5$ **53.** $y = \frac{1}{2}x$ **54.** $y = 2x - 1$

  55. $y = 2x + 4$ **56.** $d = 60\, t$ **57.** $d = 50\, t$

CHAPTER 2 – 5   Slope of a Line   page 45 ~ 46

1. slope $= \frac{4}{4} = 1$ **2.** slope $= \frac{-2}{3} = -\frac{2}{3}$ **3.** slope $= \frac{1}{2}$ **4.** slope $= -\frac{5}{8}$ **5.** Slope is undefined.

6. slope $= 0$

7. 3   **8.** $\frac{1}{3}$ **9.** -2 **10.** $\frac{1}{3}$ **11.** $-\frac{1}{3}$ **12.** $\frac{5}{2}$ **13.** $\frac{2}{5}$ **14.** undefined   **15.** 0   **16.** $-\frac{1}{6}$

17. $\frac{3}{2}$ **18.** -2 **19.** 0.5   **20.** 3

21. -1 **22.** $\frac{1}{3}$ **23.** -3 **24.** 0   **25.** 1   **26.** 1   **27.** undefined   **28.** -2 **29.** -1 **30.** -2

31. 2   **32.** -1 **33.** $\frac{6}{5}$ **34.** $-\frac{7}{9}$ **35.** $-\frac{14}{15}$ **36.** 1   **37.** 1   **38.** 2   **39.** $\frac{1}{5}$ **40.** 1

41.   **42.**

43 ~ 50 graphs are left to the students.

51. $2 ; -5$ **52.** $-2 ; 5$ **53.** $-7 ; -8$ **54.** $7 ; -8$ **55.** $\frac{2}{3} ; 4$ **56.** $-\frac{2}{3} ; 4$ **57.** $-\frac{1}{2} ; -\frac{3}{4}$

58. $\frac{1}{2} ; \frac{3}{4}$ **59.** $-4 ; 0$ **60.** $-1 ; 5$ **61.** $1 ; -10$ **62.** $7 ; 0$ **63.** $1 ; 0$ **64.** $-1 ; 0$ **65.** $0 ; 5$

66. $0 ; -2$ **67.** $-8 ; 7$ **68.** $2 ; 1$ **69.** $-\frac{1}{3} ; -2$ **70.** $-\frac{4}{5} ; 8$ **71.** $2 ; 15$ **72.** $-2 ; 15$

73. $-2 ; 6$ **74.** $2 ; -6$ **75.** $1 ; \frac{3}{2}$ **76.** $-2 ; \frac{1}{2}$ **77.** $-\frac{1}{2} ; \frac{5}{4}$ **78.** $\frac{1}{2} ; \frac{3}{4}$ **79.** $\frac{3}{2} ; -3$

80. $-\frac{3}{2} ; \frac{7}{2}$ **81.**   **82.**

83 ~ 90 graphs are left to the students.

91. undefined (The slope of a vertical line does not exit.)   **92.** 0

93. undefined (The value does not exit.)   **94.** $a = -8$ **95.** $a = -3$

96. Yes. The slope between $(1, 3)$ and $(2, 5)$ equals the slope between $(2, 5)$ and $(4, 9)$.

97. No.   **98.** $y = 3$ **99.** $x = -10$ **100.** $m = -\frac{a}{b}$; y – intercept $= \frac{c}{b}$ **101.** 6.5

102. $-3°F$ per hour, $66°F$ at 12 P.M.

CHAPTER 2 – 6   Finding the Equation of a Line   page 49 ~ 50

1. $3x + y = 5$ **2.** $3x - y = 5$ **3.** $3x + y = -5$ **4.** $6x - y = 9$ **5.** $x - 2y = -14$

6. $x + 2y = 14$ **7.** $2x + 3y = 24$ **8.** $3x - 4y = -32$ **9.** $2x - 3y = 12$ **10.** $3x + 4y = -16$

11. $y = 2x + 1$ **12.** $y = 2x - 8$ **13.** $y = -2x + 10$ **14.** $y = -2x - 2$ **15.** $y = -2x + 2$

16. $y = 2x + 2$ **17.** $y = -2x - 10$ **18.** $y = 7x - 1$ **19.** $y = 6x + 38$ **20.** $y = -x + 13$

21. $y = x - 6$ **22.** $y = 5$ **23.** $y = 0$ **24.** $x = 5$ **25.** $x = 0$ **26.** $y = \frac{1}{2}x - 4$

27. $y = -\frac{3}{5}x + 9$ **28.** $y = -\frac{2}{3}x + \frac{1}{3}$ **29.** $y = \frac{4}{5}x + 9$ **30.** $y = \frac{5}{4}x - \frac{45}{4}$ **31.** $x - y = -1$

32. $x - y = 1$ **33.** $x - 2y = -8$ **34.** $3x - 2y = 12$ **35.** $x - 2y = -10$ **36.** $2x + y = 0$

37. $3x - y = 15$ **38.** $11x - 3y = -21$ **39.** $13x + y = -45$ **40.** $2x - 13y = 14$

41. $3x - y = 4.5$ **42.** $2x - y = 1.5$ **43.** $y = \frac{1}{6}x + \frac{1}{3}$ **44.** $5x - 3y = 1$ **45.** $2x - 24y = -9$

46. $y - 4 = 3(x - 2)$ **47.** $y - 4 = -3(x + 2)$ **48.** $y + 2 = 4(x + 1)$ **49.** $y + 5 = -5(x - 7)$

50. $y - 5 = \frac{3}{4}(x + 1)$ **51.** $y + 7 = -\frac{1}{5}(x - 4)$ **52.** $y - 5 = -2(x - 4)$ or $y + 3 = -2(x - 8)$

53. $y + 2 = -\frac{4}{5}(x - 5)$ or $y - 2 = -\frac{4}{5}x$ **54.** $y - 5 = -x$ or $y = -(x - 5)$

55. $y = \frac{1}{3}x$ or $y + 1 = \frac{1}{3}(x + 3)$

56. $10x - y = 34$ **57.** $10x - y = 46$ **58.** $5x - y = 20$ **59.** $5x + y = 20$ **60.** $2x - 3y = -12$

61. $2x - 3y = 12$ **62.** $y = -2$ **63.** $x = 3$ **64.** $x = -2$ **65.** $y = 3$ **66.** $a = -8$

67. $a = 1$ **68.** $a = 11$ **69.** $a = 6.5$ **70.** $a = -2$

CHAPTER 2 – 7 Parallel and Perpendicular Lines   page 52

1. $y = -x + 5$ **2.** $y = x - 5$ **3.** $y = -x - 5$ **4.** $y = x + 5$ **5.** $y = -2x + 6$

6. $y = 2x + 6$ **7.** $y = x + \frac{7}{2}$ **8.** $y = -\frac{3}{2}x - \frac{5}{2}$

9. $-\frac{1}{2}$ **10.** $-\frac{5}{4}$ **11.** 5   **12.** $\frac{1}{3}$ **13.** 1   **14.** -1 **15.** $-\frac{1}{2}$ **16.** $\frac{1}{2}$ **17.** undefined

18. 0   **19.** $-\frac{1}{6}$ **20.** -3

21. parallel   **22.** neither   **23.** perpendicular   **24.** parallel   **25.** neither   **26.** perpendicular
27. perpendicular   **28.** parallel   **29.** parallel   **30.** perpendicular   **31.** parallel   **32.** neither
33. perpendicular   **34.** perpendicular   **35.** neither

36. $\frac{5}{4}$ **37.** undefined   **38.** undefined   **39.** 0   **40.** 0   **41.** $y = -\frac{1}{7}x + 15$

42. $2x + 5y = 10$ **43.** $5x - 2y = 54$

CHAPTER 2 – 8 Direct Variations   page 54

1. $k = 5$; $y = 5x$ **2.** $k = 2$; $y = 2x$ **3.** $k = \frac{5}{2}$; $y = \frac{5}{2}x$ **4.** $k = 0.5$; $y = 0.5x$

5. $k = 10$; $n = 10m$ **6.** $k = \frac{6}{5}$; $p = \frac{6}{5}n$ **7.** $k = \frac{1}{2}$; $s = \frac{1}{2}t$ **8.** $k = \frac{1}{3}$; $m = \frac{1}{3}n$

9. $k = 3$; $y = 3(x - 4)$ **10.** $k = 15$; $y = 15(x + 8)$ **11.** $y = 60$ **12.** $x = \frac{12}{5}$ or $2\frac{2}{5}$

13. $y = \frac{12}{5}$ or $2\frac{2}{5}$ **14.** $n = 500$ **15.** $m = \frac{4}{5}$ **16.** $y_1 = 4$ **17.** $y_1 = 60$ **18.** $y_1 = 7.2$ **19.** $y_2 = 6$

20. $x_1 = 2.5$ **21.** $x_2 = 2.5$ **22.** $y_1 = \frac{1}{5}$ **23.** $y_2 = \frac{4}{3}$ or $1\frac{1}{3}$ **24.** $y_1 = \frac{1}{4}$ **25.** $y_2 = 125$

26. No. If $xy = 5$, then $y = \frac{5}{x}$. y is decreased when x is increases. It is an inverse variation.

27. Yes. The ratios are equal. $\frac{4}{6} = \frac{2}{3}$, $\frac{6}{9} = \frac{2}{3}$, $\frac{18}{27} = \frac{2}{3}$ **28.** $D = 10L$ **29.** 14.5 cm **30.** \$441

CHAPTER 2 EXERCISES   page 55 ~ 56

1. 10   **2.** 8   **3.** 6   **4.** 3   **5.** 1   **6.** 8
7. yes   **8.** yes   **9.** no   **10.** no   **11.** yes   **12.** yes   **13.** yes   **14.** no   **15.** Yes
16. $2x - y = -5$ **17.** $2x + y = 5$ **18.** $2x + y = 5$ **19.** $2x - y = 5$ **20.** $4x - 5y = 7$
21. $3x - 8y = 9$ **22.** $2x - 6y = 7$ **23.** $2x + 6y = 3$
24. $y = -2x + 4$ **25.** $y = 2x + 4$ **26.** $y = 2x - 4$ **27.** $y = 2x + 4$ **28.** $y = -2x + \frac{5}{2}$
29. $y = 2x - \frac{1}{2}$ **30.** $y = \frac{1}{3}x - 2$ **31.** $y = -\frac{3}{5}x + \frac{6}{5}$
32. $f(1) = -5$ **33.** $f(0) = -7$ **34.** $f(-1) = -9$ **35.** $f(2) = -3$ **36.** $f(1) = 5$ **37.** $f(0) = 7$
38. $f(2) = 3$ **39.** $f(-\frac{1}{4}) = 9\frac{1}{2}$ **40.** $p(2) = 3$ **41.** $p(0) = 1$ **42.** $p(5) = 21$ **43.** $p(\frac{1}{2}) = \frac{3}{4}$
44. $f(2) = 6$ **45.** $f(-2) = -14$ **46.** $f(\frac{1}{3}) = -3$ **47.** $f(-1\frac{1}{3}) = -10$
48. R = {−7, −4, −1, 2, 5}   **49.** R = {−10, −6, −2, 2, 6}   **50.** R = { −2, −1, 7}   **51.** R = {−2, 0, 6}
52. R = {−4, −3, −1, 1}   **53.** R = {−1, 1, 3, 4}
54. D = {0, 1, 2, 3}, R = {−2, −1, 1, 4}, a function
55. D = {−2, 0, 2}, R = {1, 2, 3}, not a function
56. D = {−2, −1, 1}, R = {−2, 0, 1, 2}, not a function
57. D = {−3, −1, 2, 3}, R = {−3, −2, 1, 3}, a function
58. D = {0, 1, 2, 3, 4}, R = {−2, −1, 0, 1, 2}, a function
59. D = {1, 2, 8}, R = {2, 3, 9}, a function
60. D = {−4, 1}, R = {−2, 2, 7}, not a function

61 ~ 75 graphs are left to the students (see Section 2 – 4, example 4).
61. x – intercept 3, y – intercept −3   **62.** x – intercept 3, y – intercept 3
63. x – intercept −3, y – intercept −3   **64.** x – intercept −1, y – intercept 2
65. x – intercept 6, y – intercept 3   **66.** x – intercept 6, y – intercept −3
67. x – intercept 6, y – intercept −4   **68.** x – intercept 6, y – intercept 4
69. x – intercept "none", y – intercept −2   **70.** x – intercept 1, y – intercept "none"
71. x – intercept 0, y – intercept 0   **72.** x – intercept 0, y – intercept 0
Need a third point to graph it, say (1, 2).   Need a third point to graph it, say (1, −2).
73. x – intercept 6, y – intercept −4   **74.** x – intercept 8, y – intercept 12
75. x – intercept 8, y – intercept −6

76. $\frac{4}{5}$ **77.** $-\frac{2}{3}$ **78.** 0   **79.** undefined   **80.** 12   **81.** −4   **82.** $-\frac{6}{5}$ **83.** 5   **84.** −1   **85.** $-\frac{5}{3}$
86. $\frac{5}{3}$ **87.** undefined   **88.** 0   **89.** undefined   **90.** $\frac{2}{3}$ **91.** −4   **92.** −4   **93.** 1.8   **94.** 1.6   **95.** −5
96. −4 ; 9   **97.** 4 ; 9   **98.** $\frac{4}{5}$; −15   **99.** $-\frac{4}{5}$; 15   **100.** −8 ; 10   **101.** 8 ; −10   **102.** 8 ; 0
103. 0 ; −8   **104.** $-\frac{1}{2}$; $\frac{4}{3}$ **105.** $\frac{1}{2}$; $\frac{4}{3}$ **106.** 8 ; 14   **107.** 2 ; $-\frac{7}{5}$
108. $5x - y = -7$ **109.** $4x + y = -2$ **110.** $3x - 2y = -24$ **111.** $12x + 15y = -10$
112. $7x - y = 11$ **113.** $x + 2y = -6$ **114.** $3x - 4y = -2$ **115.** $y = 9$ **116.** $x = 5$
117. $2x - y = 7$ **118.** $2x - y = -5$ **119.** $27x + y = 22$ **120.** $3x + 2y = 7$
121. perpendicular   **122.** neither   **123.** parallel   **124.** perpendicular
125. $y = 8$ **126.** $n = 55$ **127.** $y_2 = \frac{4}{15}$

Graphing a linear function by a graphing Calculator

Hints: 1. To define a function, press **Y=**. We can store up to 10 functions.

2. When we enter each function, the = is highlighted.

3. To graph the function, the = must be highlighted, indicating the function is selected. If the = is not highlighted, move the cursor to the = and press **ENTER**.

4. To change the on/off status of a stat plot, move the cursor on **Plot1, Plot2,** or **Plot3** and press **ENTER**. When a plot is on, its name is highlighted.

5. To graph a function, we must select **FUNC** and **Connected** from mode settings before we enter the function.
   Press: **MODE → Func → Connected → ENTER**

Examples:

1. Graph $y - 3x = 2$.

    Solution: Rewrite the equation:
$$y = 3x + 2$$
Press: **Y=** $\to$ **3** $\to$ $\boxed{\text{X,T},\theta,n}$ $\to$ **+** $\to$ **2**
$\to$ **GRAPH**
The screen shows:

2. Graph $3x + y = 2$.

    Solution: Rewrite the equation:
$$y = -3x + 2$$
Press **Y=** $\to$ **(–)** $\to$ **3** $\to$ $\boxed{\text{X,T},\theta,n}$
$\to$ **+** $\to$ **2** $\to$ **GRAPH**
The screen shows:

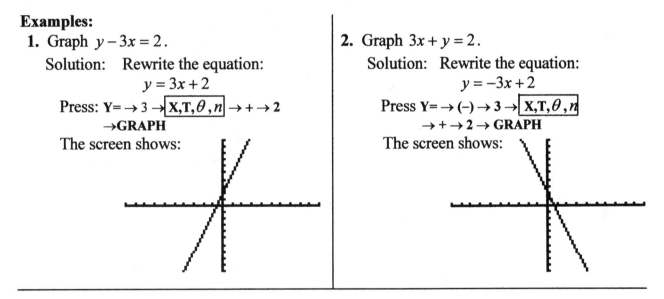

3. Graph $y = 20x + 5$.

    Solution: Press: **Y=** $\to$ **20** $\to$ $\boxed{\text{X,T},\theta,n}$ $\to$ **+** $\to$ **5**
       $\to$ **GRAPH**

If the screen does not show a complete graph, change the viewing window:

Press: **WINDOR**

Change the viewing window to:
$$x_{\min} = -10 , \quad x_{\max} = 10$$
$$y_{\min} = -40 , \quad y_{\max} = 40$$

Change the scale factors to:
$$x_{scl} = 1 , \quad y_{scl} = 5$$

Press: **GRAPH**

The screen shows:

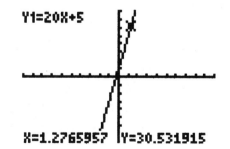

Hint: Press: **TRACE**

    A flashing trace cursor appears. Move right and left cursor to get estimate of y at the bottom of the screen.

Graphing a scatter plot and best-fitting line
by a graphing calculator

Example: Use a graphing calculator to make a scatter plot of the data.

x	1	2	3	4	5	6	7
y	–4	–2	1	2	5	8	9

Solution:

1. To enter the data:
Press: **STAT → 1:EDIT → ENTER**
Move the cursor and enter the data
on the screen:

```
L1    L2
1     –4
2     –2
3      1
4      2
5      5
6      8
7      9
```

3. To graph the scatter plot, press: **GRAPH**
The screen shows:

4. To graph the best-fitting line:
Press: **STAT → CALC →LinReg($ax+b$) → ENTER**
Select L1 as the x list and L2 as the y list:

Press: **2nd → L1 → , → 2nd → L2 → ,**
Press: **VARS → Y-VARS → ENTER**
        → FUNCTION → ENTER
The screen shows: **LinReg L1, L2, Y1 → ENTER**
The screen shows: **LinReg**
$$y = ax+b, \quad a = 2.25, \quad b = -6.28571486$$
Press: **GRAPH** (the screen shows the best-fitting line.)

5. To evaluate the y – value at $x = 8$:

Press: **2nd → CALC → 1:Value → ENTER**
Enter: $x = 8$ → **ENTER**

2. To adjust the size of the graph:
Press: **WINDOR**
Change the viewing window to:
$$x_{min} = -5 , \quad x_{max} = 10$$
$$y_{min} = -5 , \quad y_{max} = 10$$
Set the scale factors to:
$$x_{scl} = 1 , \quad y_{scl} = 1$$
Press: **2nd → STAT PLOT → ENTER**
Draw the scatter plot on the screen:
Move cursor on it and press **ENTER**
to highlight it.

Plot 1   Plot 2   Plot 3
On Off   Type: ⊡
x List: L_1 y List: L_2 Mark: •

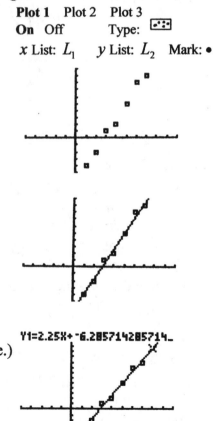

Finding a linear function (a regression line) of best-fit by a graphing calculator

In real life situation, we want to organize and analyze numerical data and make prediction. Enter the data in a graphing calculator having regression feature and make a scatter diagram, we can obtain a linear regression (line) of Best Fit.

In advanced statistics, we can find a number called **correlation coefficient**, r, $0 \le |r| \le 1$, to measure how good (accurate) is the line of best fit to describe the linear relationship between two variables. The closer $|r|$ is to 1, the better the linear relationship is. A negative value of r, $r < 1$, indicates that y decreases as x increases. If r is close to 0, there is little or no linear relationship between the variables.

Example: Use a graphing calculator to find the linear function of best-fit to the following data.

x	-2	-1	0	1	2	3	4	5	6
$f(x)$	-7	-5	-3	-1	1	3	5	7	9

Solution:

1. To enter the data:
   Press: **STAT → 1:EDIT → ENTER**
   Move the cursor and enter the data on the screen:

L₁	L₂
-2	-7
-1	-5
0	-3
1	-1
2	1
3	3
4	5
5	7
6	9

2. Find the linear function:
   Press: **2nd → QUIT**
   We have the home screen.
   Press: **STAT → CALC → LinReg** $(ax + b)$
   → ENTER
   Screen shows **LinReg** $(ax + b)$
   Select L₁ as the x list and L₂ as the y list:
   Press: **2nd → L₁ →, → 2nd → L₂ →,**
   Screen shows: **LinReg(ax+b) L₁, L₂,**
   Press: **VARS → Y-VARS → ENTER**
   Screen shows: **1: Function → ENTER**
   Screen shows: **1: Y1 → ENTER**
   Screen shows: **LinReg L₁, L₂, Y1 → ENTER**
   Screen shows: $y = ax + b$, $a = 2$, $b = -3$
   $y = 2x - 3$ is the answer.

3. Graph the function:
   Press: **Y= → GRAPH**
   Screen shows:

4. Check the answer:
   Press: **VARS**
   Move cursor to **Y-VARS → ENTER**
   Screen shows: **1:Y1 → ENTER**
   Screen shows the function and **Y1**
   Type: Y1(–2) → **ENTER**

   Screen shows: Y1(–2) = –7
   Press: **VARS**
   Move cursor to **Y-VARS → ENTER → ENTER**
   Type: **Y1(–1) → ENTER**
   Screen shows: Y1(–2) = – 7
   Y1(–1) = –5

Printing Graphs from Computer
Using TI Connect Solftware
(TI-83, TI-84 and TI-89)

The TI Connect Software and the TI-GRAPH LINK cable can connect the graphing calculator and PC. The TI Connectivity Kit USB can be found at most computer stores or Texas Instruments.
1. Download the TI Connect Software for Windows.
2. Connect the TI-GRAPH LINK cable between PC USB port and the calculator.
3. Display the desired graph on the calculator screen.
4. Open the TI Connect software and click on TI Screen Capture. The graph on the calculator is displayed on the TI screen.
5. To print the graph on TI screen, click on the printer icon in the toolbar above the graph.

Copy and Paste the graph on TI Screen Capture to Windows Word screen and adjust the location and size of the graph:
1. Click on the copy icon (the third icon with red arrow from the second row).
2. Click on the Windows Word at the bottom to open the Word screen.
3. Click on the right side of the mouse and click Paste. The graph is shown on the Window Word screen.
4. To adjust the location, click on the graph and drag it to the desired location.
5. To adjust the size, move the mouse pointer on outer line until an arrow "↔" shows, then drag it to the desired size.
6. To print the graph on Word screen, click on the printer icon.

Abraham Lincoln was the 16th President of the United States. One day, as he was giving a speech before the crowd, a pro-slave audience wrote " foolish " on a piece of paper and gave it to Lincoln.
Lincoln: Yesterday, I received a letter from a friend who forgot to sign his name.
Today, I just received a letter from a friend who signed his name only.

Inequalities and Absolute Values

3-1 Inequalities on the Number Line

An **inequality** is formed by placing an inequality sign or symbol ($>, <, \leq, \geq$) between numerical or variable expressions.

$$5 > 3 \quad \text{is read " 5 is greater than 3."}$$
$$3 < 5 \quad \text{is read " 3 is less than 5."}$$
$$x > -4 \text{ is read " } x \text{ is greater than } -4."$$

An inequality containing a variable is called **an open sentence**.

The inequality $x > -4$ indicates that all numbers greater than -4 are solutions of x.

$$x \geq -3 \text{ is read "} x \text{ is greater than or equal to } -3."$$
$$x \leq 2 \quad \text{is read "} x \text{ is less than or equal to 2."}$$
$$-3 < x < 3 \quad \text{is read "} x \text{ is greater than } -3 \text{ and less than 3."}$$
$$-3 \leq x < 2 \quad \text{is read "} x \text{ is greater than or equal to } -3 \text{ and less than 2."}$$

We can graph an inequality on a number line.

On a number line, the number on the right is greater than the number on the left.

To graph an inequality on a number line, we use a **close circle** "•" to show "included", and use an **open circle** "○" to show "not included".

Properties for inequalities:

1. Adding or subtracting the same number to or from each side of an inequality does not change the direction (order) of its inequality sign.

2. Multiplying or dividing each side of an inequality by the same positive number does not change the direction (order) of its inequality sign.

3. Multiplying or dividing each side of an inequality by the same negative number reverses (changes) the direction (order) of its inequality sign.

Examples

1. Add 6 to each side of $5 > 2$.
Then write a true inequality.
Solution:
$$5 > 2$$
$$5 + 6 > 2 + 6$$
$$11 > 8 \text{ (true)}.$$

2. Multiply each side of $5 > 2$ by -6.
Then write a true inequality.
Solution:
$$5 > 2$$
$$5(-6) < 2(-6)$$
$$-30 < -12 \text{ (true)}.$$

EXERCISES

Compare each pair of numbers, using the symbols <, >, =.
1. 6 and 4   2. 3 and 8   3. −3 and 5   4. −3 and −5
5. −7 and −4   6. 4.5 and 6   7. −12 and −12   8. −4.2 and −4.5
9. −18 and −24   10. −32 and −15

Classify each statement as true or false.
11. $7 < 5$ 12. $9 > 5$ 13. $-5 > -4$ 14. $-3 > -8$
15. $-5 \geq -5$ 16. $-2 < 3 < 9$ 17. $-3 < -6 < 8$ 18. $\left| -0.36 \right| < 0.24$
19. $\left| 5 - 8 \right| > \left| 8 - 6 \right|$ 20. $\left| 5 - 8 \right| \leq \left| 8 - 5 \right|$

Perform the indicated operation on each side of the inequality. Then write a true inequality.
21. $6 < 9$, add 4   22. $6 < 9$, subtract 4   23. $6 < 9$, multiply by 4
24. $6 < 9$, divide by 4   25. $6 < 9$, multiply by −4   26. $6 < 9$, divide by −4
27. $-5 < 6$, multiply by −2   28. $2 > -3$, multiply by −5   29. $-4 < 6$, divide by 2
30. $-6 > -9$, divide by −3   31. $-6 > -9$, multiply by −3   32. $-6 > -9$, divide by 3
33. $2x > 4$, divide by 2   34. $-2x > 4$, divide by −2   35. $\frac{1}{2}x \leq 4$, multiply by 2
36. $-\frac{1}{2}x \leq 4$, multiply by −2   37. $\frac{2}{3}x > 6$, multiply by $\frac{3}{2}$ 38. $-\frac{2}{3}x \leq 6$, multiply by $-\frac{3}{2}$
39. $-4x \geq 6$, divide by −4   40. $-4x \leq -6$, divide by −4

Translate each statement into symbols.
41. 8 is less than 15.   42. −8 is greater than −15.
43. −7 is less than −2.   44. −7 is between −9 and 2.
45. −2 is greater than −7 and less than 3.   46. −7 is greater than −10 and less than −2.
47. x is greater than 15.   48. x is less than or equal to 15.
49. n is greater than −8 and less than 2.   50. n is greater than 4 and less than or equal to 9.
51. The absolute value of n is less than 5.
52. The absolute value of x is greater than 7.
53. The absolute value of x is greater than or equal to 7.
54. The absolute value of n is less than or equal to 5.

Graph each inequality on a number line.
55. $x < -1$ 56. $x \geq 4$ 57. $x \geq -3$ 58. $-5 < n < 3$
59. $-4 < n \leq 2$ 60. $x < -3$ or $x \geq 3$ 61. $-3 < x < 3$ 62. $-2 \leq x \leq 5$

3-2 Solving Inequalities in One Variable

To solve an inequality, we use similar methods which are used to solve equations. Simply transfer all variables to one side of the inequality (usually to the left side), and transfer the others to the right side. Reverses (changes) the signs (+, −) in the process. Reverses the direction (order) of the inequality sign if we **multiply** (or **divide**) each side by **the same negative number**.

$$-\tfrac{1}{2}x > 4$$

$$(-2)\cdot -\tfrac{1}{2}x < 4\cdot(-2) \text{ (That is, } x < -8 \text{)}.$$

If the final statement is a "true statement", the inequality has solutions for all real numbers. If the final statement is a "false statement", the inequality has no solution.(See Ex. 5 and 6)

Examples

1. Solve $4x-1<11$.

Solution:

$$4x-1<11$$
$$4x \quad <11+1$$
$$4x \quad <12$$
$$\therefore x<3. \text{ Ans.}$$

2. Solve $4x-1\geq11$.

Solution:

$$4x-1\geq11$$
$$4x \quad \geq11+1$$
$$4x \quad \geq12$$
$$\therefore x\geq3. \text{ Ans.}$$

3. Solve $-4x-1<11$.

Solution:

$$-4x-1<11$$
$$-4x \quad <12$$

Divide each side by −4:
$$\therefore x>-3. \text{ Ans.}$$

4. Solve $-\tfrac{1}{4}x-1\geq11$.

Solution:

$$-\tfrac{1}{4}x-1\geq11$$
$$-\tfrac{1}{4}x \quad \geq12$$

Multiply each side by −4:
$$\therefore x\leq-48. \text{ Ans.}$$

5. Solve $3x<3(x+4)$.

Solution:
$$3x<3(x+4)$$
$$3x<3x+12$$
$$3x-3x<12$$
$$0<12 \text{ (true)}$$
The inequality is true for all real numbers. Ans.

6. Solve $3x>3(x+4)$.

Solution:
$$3x>3(x+4)$$
$$3x>3x+12$$
$$3x-3x>12$$
$$0>12 \text{ (false)}$$
The inequality has no solution.   Ans.
The solution set is ϕ (empty).

EXERCISES

Solve each inequality. Graph each solution on a number line.

1. $x - 3 < 8$
2. $x + 3 > 8$
3. $x + 4 \geq -5$
4. $x - 5 \leq -2$
5. $a - 4 > 7$
6. $n - 3 \geq -9$
7. $y + 7 < -3$
8. $y - 7 \leq -3$
9. $p - 5 \leq -8$
10. $k - 4 > 0$
11. $x + 5 \geq 0$
12. $v - 4 < -1$
13. $2x < 8$
14. $-2x < 8$
15. $-2x \geq 8$
16. $2x > -8$
17. $-2x \leq -8$
18. $\frac{1}{2}x \geq 8$
19. $-\frac{1}{2}x \geq 8$
20. $-\frac{1}{3}n \leq -1$
21. $-\frac{1}{5}n > -2$
22. $8p < -24$
23. $7a > -21$
24. $-8a < -24$
25. $-21a > 7$
26. $-24p \leq 8$
27. $-15y \geq 15$
28. $\frac{1}{5}x < -2$
29. $\frac{1}{4}x > 5$
30. $\frac{1}{6}x \leq -2$
31. $-\frac{2}{3}x \geq 4$
32. $-\frac{1}{4}x > 5$
33. $-\frac{2}{3}x \leq 6$
34. $-\frac{4}{5}p < 16$
35. $\frac{5}{6}p > 15$
36. $\frac{3}{10}a < -12$
37. $-\frac{3}{7}k \geq -9$
38. $-\frac{2}{5}k \leq -1$
39. $\frac{3}{4}x > \frac{3}{2}$
40. $-\frac{3}{8}x < \frac{3}{2}$
41. $-\frac{3}{5}x \geq \frac{1}{5}$
42. $-\frac{5}{6}x \leq \frac{5}{9}$
43. $2x - 5 > 3$
44. $2x + 5 > 3$
45. $-2x + 5 > 3$
46. $4x - 6 < 10$
47. $-4x - 6 \leq 10$
48. $-4x + 6 \geq -10$
49. $\frac{1}{3}x + 2 > -4$
50. $\frac{1}{4}x + 5 < -3$
51. $\frac{1}{5}x - 8 \leq 7$
52. $4x < 4x + 5$
53. $4x > 4x - 5$
54. $4x < 4(x - 3)$
55. $6x > 6(x + 1)$
56. $-5x + 8 > 14$
57. $-6x - 7 < 8$
58. $-\frac{1}{2}n + 12 < 10$
59. $-\frac{2}{3}n - 6 \geq 8$
60. $-\frac{3}{4}p + 4 \leq 4$
61. $4x - 2 < 3x$
62. $5x + 4 > -3x$
63. $2x - 3 \leq -3x + 1$
64. $x + 8 \geq 4x - 1$
65. $3x - 4 < 6x + 2$
66. $\frac{1}{2}x - 8 > \frac{3}{4}x + 4$
67. $\frac{4}{3}y - 9 > -1$
68. $3(x - 2) \leq 5x + 6$
69. $4(2x - 3) \geq 6x + 3$
70. $3x - 2 \leq -2(x + 1) + 3$
71. $2(2x + 3) < 3(x - 1) - 1$
72. $6x - 3(3x + 5) > -1$
73. $2 - (6 - 2n) > 4n - (n - 1)$
74. $4p + 3(p - 2) \geq 4(p + 1)$
75. $2.5y + 2 < 3y - 5$

76. If $a > b$, identify the values of c for which the statement is true: $a + c > b + c$.
77. If $a > b$, identity the values of c for which the statement is true: $ac > bc$.
78. If $a > b$, identity the values of c for which the statement is true: $ac < bc$.
79. If $a > b$, identify whether the statement is true or false: $a^2 > b^2$.
80. If $a > b$, identify whether the statement is true or false: $a^3 > b^3$.
81. If $a > b > 0$, identify whether the statement is true or false: $a^2 > b^2$.
82. Identify the values of y for which the statement is true: $3^y > 2^y$.

3-3 Solving Combined Inequalities

A **combined inequality** (or **compound inequality**) is an inequality formed by joining two inequalities with "**and**" or "**or**".

There are two forms of combined inequalities, **conjunction** and **disjunction**.

1. Conjunction of Inequality

It is an inequality formed by two inequalities with the word "**and**".

A conjunction is true when **both inequalities are true**.

Here is an example:

$$-5 < n \text{ and } n < 3$$
$$(\text{or } -5 < n < 3)$$

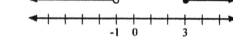

To solve a conjunction of inequality, we isolate the variable between the two inequality signs. Or, solve each part separately.

2. Disjunction of Inequality

It is an inequality forms by two inequalities with the word "**or**".

A disjunction is true when **at least one of the inequalities is true**.

Here is an example:

$$n < -1 \text{ or } n \geq 3$$

To solve a disjunction of inequality, we solve each part separately.

Examples

1. Solve $2 < \frac{1}{3}x + 4 \leq 6$.

Solution:

Method 1:	**Method 2:**
(Solve each part separately.)	(Solve between the inequality signs.)

Method 1:

$2 < \frac{1}{3}x + 4$ and $\frac{1}{3}x + 4 \leq 6$

$-2 < \frac{1}{3}x$ $\bigg|$ $\frac{1}{3}x \quad \leq 2$

$-6 < x$ $\bigg|$ $x \leq 6$

$\therefore -6 < x \leq 6$. Ans.

Method 2:

$2 < \frac{1}{3}x + 4 \leq 6$

Subtract 4 to each expression:

$-2 < \frac{1}{3}x \leq 2$

Multiply each expression by 3:

$\therefore -6 < x \leq 6$. Ans.

2. Solve $1 - 2x < -3$ or $3x + 14 \leq 2 - x$.

Solution:

Solve each part separately:

$1 - 2x < -3$ or $3x + 14 \leq 2 - x$ Ans: $x \leq -3$ or $x > 2$.

$-2x < -4$ $\bigg|$ $4x \leq -12$

$x > 2$ $\bigg|$ $x \leq -3$

Examples

3. Solve $n - 2 \le 2n \le 3n - 1$.

Solution:

$$n - 2 \le 2n \quad \text{and} \quad 2n \le 3n - 1$$
$$-2 \le n \qquad\qquad 1 \le n$$

It is a conjunction "and" inequality.
Both must be true.

Ans: $n \ge 1$.

4. Solve $3x - 1 \le 2x$ or $3x + 10 \ge -2x$.

Solution:

$$3x - 1 \le 2x \quad \text{or} \quad 3x + 10 \ge -2x$$
$$x \le 1 \qquad\qquad 5x \ge -10$$
$$\qquad\qquad\qquad x \ge -2$$

It is a disjunction "or" inequality.
At least one is true.

Ans: All real numbers of x.

5. Solve $3a + 5 < a + 7$ and $3 - a < 1$.

Solution:

$$3a + 5 < a + 7 \quad \text{and} \quad 3 - a < 1$$
$$2a < 2 \qquad\qquad -a < -2$$
$$a < 1 \qquad\qquad a > 2$$

It is a conjunction "and" inequality.
Both must be true.

Ans: No solution.

ϕ (The solution set is empty.)

6. Solve $x - 1 \le 0$ and $x + 2 \ge 0$.

Solution:

$$x - 1 \le 0 \quad \text{and} \quad x + 2 \ge 0$$
$$x \le 1 \qquad\qquad x \ge -2$$

It is a conjunction "and" inequality.
Both must be true.

Ans: $-2 \le x \le 1$.

7. Solve $x - 1 \le 0$ or $x + 2 \ge 0$

Solution:

$$x - 1 \le 0 \quad \text{or} \quad x + 2 \ge 0$$
$$x \le 1 \qquad\qquad x \ge -2$$

It is a disjunction "or" inequality.
At least one is true.

Ans: All real numbers of x.

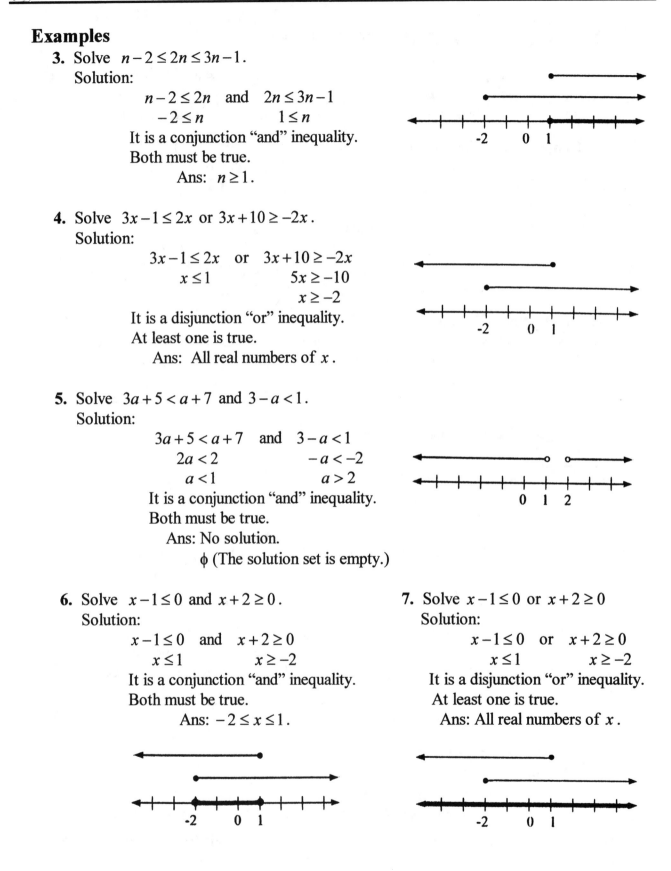

EXERCISES

Match each inequality with its graph.

1. $x \geq -1$ **2.** $x < 1$ **3.** $x \leq 1$

4. $-1 < x \leq 5$ **5.** $x < -2$ or $x \geq 1$ **6.** $x > -3$ or $x < 4$

7. $-3 < x < 3$ **8.** $-4 \leq x < 2$ **9.** $x \geq -4$ and $x < 2$

10. $x \geq -4$ or $x < 2$ **11.** $x \geq -1$ or $x \geq 1$ **12.** $x < -2$ and $x \geq 1$

A.

B.

C.

D.

E.

F.

G.

H.

Write each inequality with an equivalent inequality without "and" or "or".

13. $x < 5$ and $x > -2$ **14.** $x = 8$ or $x > 8$ **15.** $x > 1$ and $x \leq 4$

16. $x > -3$ and $x \leq 7$ **17.** $x \geq -5$ and $x \leq -1$ **18.** $x < 9$ or $x = 9$

19. $n \leq 4$ and $n \geq -6$ **20.** $a \leq -4$ and $a > -12$

Solve each inequality and graph its solution set.

21. $x + 3 < -2$ **22.** $x - 3 < -2$ **23.** $3x + 6 \leq 12$

24. $3x - 6 \geq 12$ **25.** $-3x + 6 > -12$ **26.** $-3x - 6 > 12$

27. $3x + 5 > -\frac{3}{2}$ **28.** $-\frac{2}{3}x \leq x - 4$ **29.** $-4 \geq 5 + \dfrac{x}{2}$

30. $4(2 - a) > -(a - 5)$

-----Continued-----

Graph each inequality on a number line.

31. $-1 \le x \le 5$ **32.** $x < -2$ or $x \ge 1$ **33.** $x \le -2$ or $x > 1$

34. $-4 < x < 2$ **35.** $0 \le a \le 5$ **36.** $-1 < a \le 6$

37. $n \le 0$ or $n \ge 5$ **38.** $n < -3$ or $n \ge 4$ **39.** $x > 1$ or $x = 1$

40. $x > 1$ and $x = 1$

Solve each combined (compound) inequality. Graph its solution set.

41. $2 < x - 5 < 9$ **42.** $-2 \le x + 5 < 9$

43. $0 < a + 4 \le 7$ **44.** $-1 < a - 4 \le 2$

45. $-5 < -2 + n \le 3$ **46.** $1 \le 4 + n < 6$

47. $-4 < -x < 8$ **48.** $4 < -x < -8$

49. $-4 < -x + 5 < 7$ **50.** $-2 \le 2y + 6 \le 8$

51. $-4 \le 3y - 1 < 5$ **52.** $5 < 2b - 1 < 9$

53. $4 < 4p - 4 < 12$ **54.** $-6 < -2x + 2 \le 8$

55. $-7 \le -3x - 1 < 5$ **56.** $2x - 1 > -5$ and $3x < 9$

57. $4x + 2 > -10$ and $3x - 2 < 4$ **58.** $4x + 2 > 10$ and $2x < -10$

59. $4x + 2 > -10$ and $-3x - 2 < 4$ **60.** $4x + 2 > 10$ or $2x < -10$

61. $4x + 2 > -10$ or $3x - 2 < 4$ **62.** $4x + 2 > 10$ or $-2x < 10$

63. $4x + 2 > -10$ or $-3x - 2 < 4$ **64.** $4x + 2 \le -10$ or $2x \ge 10$

65. $4x + 2 \le -10$ and $-3x + 2 < -4$ **66.** $-4x + 2 \le -10$ and $2x \le 10$

67. $n - 4 \le 3n \le 2n + 4$ **68.** $n - 4 < 3n < 4n + 4$

69. $3p + 5 \ge -1$ and $p - 4 < -3p + 4$ **70.** $3p + 5 > -1$ and $-2p + 4 < -p + 1$

71. $3p + 5 \ge -1$ or $p - 4 < -3p + 4$ **72.** $3p + 5 \ge -1$ or $-2p + 4 < -p + 1$

73. $1.5 < 3.2x + 7.9 \le 12.7$ **74.** $-8 \ge \frac{5}{3}x - 3$ or $3x \ge 5$

75. $0.2x + 2.4 < 1.5$ or $0.5x > 1.7$ **76.** $-11 \le \frac{2}{3}x + 1 \le 9$

3-4 Solving Word Problems involving Inequalities

Steps for Solving a word problem involving inequality:

1. Read the problem.
2. Assign a variable to represent the unknown.
3. Determine which inequality sign ($<$, $>$, $\leq$, $\geq$) needed.
4. Write an inequality based on the given facts.
5. Solve the inequality to find the answer.
6. Check your answer (usually on a scratch paper).

Examples

1. You want to rent a car for a rental cost of $50 plus $30 per day, or a van for $80 plus $25 per day. For how many days at most is it cheaper to rent a car ?
   Soluion:

   Let $d =$ days needed

   The inequality is:

   $$50 + 30d \leq 80 + 25d$$
   $$5d \leq 30$$
   $$\therefore d \leq 6$$

   Ans: At most 6 days (or less).

   Check: $50 + 30d = 50 + 30(6) = \$230$ ✓
   $\qquad\qquad 80 + 25d = 80 + 25(6) = \$230$ ✓

   The rental costs are the same for 6 days. It is cheaper to rent the car for less than 6 days. It is cheaper to rent the van for more than 6 days.

2. Your scores on your 4 math tests were 71, 74, 75 and 86. What is the lowest score you need in your next test in order to have an average score of at least 80 ?
   Solution:

   Let $x =$ score for next test

   The inequality is:

   $$\frac{71 + 74 + 75 + 86 + x}{5} \geq 80, \quad \frac{306 + x}{5} \geq 80, \quad 306 + x \geq 400 \quad \therefore x \geq 94$$

   Ans: 94 (or more).

3. John wants to rent a car and pay at most $250. The car rental costs $150 plus $0.25 a mile. How far can he travel ?
   Solution:

   Let $x =$ mileage he can travel

   The inequality is:

   $$150 + 0.25x \leq 250$$
   $$0.25x \leq 100 \quad \therefore x \leq 400 \qquad \text{Ans: 400 miles or less.}$$

4. The sum of two consecutive integers is less than 79. Find the integers with the greatest sum.
   Solution: Let $n = 1^{st}$ integer. $\quad n + 1 = 2^{nd}$ integer.

   The inequality is:

   $$n + (n + 1) < 79, \quad 2n < 78, \quad \therefore n < 39$$

   The greatest integer in n is 38. $\qquad$ Ans. 38 and 39.

EXERCISES

Solve each inequality. Graph each solution on a number line.

1. $x - 4 < 9$

2. $x + 4 < 9$

3. $2x + 5 \geq 97$

4. $2n - 8 \leq 78$

5. $40 + 10n \leq 90 + 5n$

6. $20 + 15n \geq 80 + 10n$

7. $45 + 0.25m > 150$

8. $115 + 0.15m < 220$

9. $2w + 2(w - 5) > 70$

10. $2w + 2(2w + 10) > 41$

11. $\dfrac{235 + x}{4} \geq 80$

12. $\dfrac{335 + x}{5} \geq 85$

13. You want to rent a car and pay at most \$125. The car rental costs \$80 plus \$0.15 a mile. How far can you travel?

14. You want to rent a car for a rental cost of \$40 plus \$25 per day, or a truck for \$75 plus \$20 per day. For how many days at most is it cheaper to rent a car?

15. John wants to rent a car for one day. Company A charges \$65 plus \$0.25 a mile. Company B charges \$110 with no mileage fee. How many miles John would need to travel to rent the car from Company A with the less expensive charges?

16. Your scores on your 2 tests were 75 and 74. What is the lowest score you need in your next test in order to have an average score of at least 80?

17. Your scores on your 4 tests were 92, 86, 87 and 90. What is the lowest score you need in next test in order to have an average score of at least 90?

18. The sum of two consecutive integers is less than 97. Find the integers with the greatest sum.

19. The sum of two consecutive integers is larger than 97. Find the integers with the smallest sum.

20. The sum of two consecutive integers is no more than 97. Find the integers with the greatest sum. (Hint: "no more than" or "at most" is "$\leq$".)

21. The sum of three consecutive integers is no less than 129. Find the integers with the smallest sum. (Hint: "no less than" or "at least" is "$\geq$".)

22. Find all the sets of three consecutive, positive and odd integers whose sum is less than 23.

23. The width of a rectangle is $12\,cm$ shorter than the length. The perimeter is at least $52\,cm$. What is the smallest possible dimensions of the rectangle?

24. The length of a rectangle is $6\,cm$ longer than the width. The perimeter is at least $48\,cm$. What is the smallest possible dimensions of the rectangle?

25. The length of a rectangle is $6\,cm$ longer than the width. The perimeter in no more than $48\,cm$. What is the greatest possible dimensions of the rectangle?

26. The width of a rectangle is 15 in. shorter than twice the length. The perimeter is at least 45 in.. What is the smallest possible dimensions of the rectangle if each dimension is an integer?

27. There are quarters and dimes in a box. Their total value is at most \$5.40. Quarters are five times as many as dimes. How many quarters and how many dimes are at most in the box?

28. At most how many grams of pure salt must be added with 54 grams of water to produce an salt-water solution that is no more than 20% pure salt?

3-5 Operations and Graphs with Absolute Values

Absolute Value: The positive number of any real number a. It is written by $|a|$.

$$\text{If } a > 0, \; |a| = a. \qquad \text{Example: } |5| = 5.$$

$$\text{If } a = 0, \; |a| = |0| = 0$$

$$\text{If } a < 0, \; |a| = -a. \qquad \text{Example: } |-5| = -(-5) = 5.$$

The absolute value of a negative number is the opposite of the number.

Absolute value is commonly used as the distance between two points on a number line. The distance between two points is always a positive number. A distance can never be a negative number. Therefore, we use the absolute value to represent the distance.

$$\overline{AB} = |3 - (-4)| = |3 + 4| = |7| = 7. \quad \text{Or: } \overline{AB} = |-4 - 3| = |-7| = 7.$$

The absolute value of a number is its distance from zero on a number line.

$$\overline{BC} = |3 - 0| = |3| = 3. \qquad \overline{AC} = |0 - 4| = |-4| = 4.$$

The domain and range of the absolute-value function

$y = |x|$ is the most basic absolute-value function. Its domain is all real numbers of x.

Its range is all real numbers of $y \geq 0$, or all non-negative real numbers of y.

The graph of the absolute-value function

To graph an absolute-value function, we can plot its ordered pairs.

Or, we can plot its separated equations without absolute-value sign.

Examples

1. Find the domain and range of $y = |x - 2|$.

    Solution: D = all real numbers.   R = all non-negative real numbers.

2. Find the domain and range of $y = |x| - 2$.

    Solution: D = all real numbers.   R = all real numbers of $y \geq -2$.

3. Find the domain and range of $y = -|x - 2|$.

    Solution: D = all real numbers.   R = all non-positive real numbers.

4. Graph $y = |x|$.

    Solution:

Method 1:	Method 2:
x: −3 −2 −1 0 1 2 3	For $x > 0$, $y = x$
y: 3 2 1 0 1 2 3	For $x < 0$, $y = -x$
	For $x = 0$, $y = 0$

EXERCISES

Evaluate each absolute-value.

1. $|15|$ **2.** $|-15|$ **3.** $|-7.2|$ **4.** $\left|-\frac{4}{5}\right|$

5. $|4-5|$ **6.** $|5-4|$ **7.** $|-4-4|$ **8.** $|6-6|$

9. $|12-15|$ **10.** $|15-12|$ **11.** $|1-10|$ **12.** $|0-(-8)|$

13. $|-9+9|$ **14.** $|-9-9|$ **15.** $|-12-(-4)|$

Identify each statement as true or false.

16. $|3|>2$ **17.** $|-3|>2$ **18.** $|-3|<-2$

19. $|-5|<4$ **20.** $|-5|>-4$ **21.** $|0.24|<0.21$

22. $|-0.24|>0.21$ **23.** $|-0.27|<0.19$ **24.** $|-0.35|>0.25$

25. $|-0.29|<0.19$ **26.** $|5-2|\le|2-5|$ **27.** $|5-2|<|2-5|$

28. $|1.2-3|>|2-1.5|$ **29.** $|3.5-5.8|\ge|6.5-5.5|$ **30.** $\left|\frac{1}{2}-\frac{2}{3}\right|<\left|\frac{3}{4}-\frac{1}{4}\right|$

Find the domain and range of each function.

31. $y=|x+8|$ **32.** $y=|x|+8$ **33.** $y=|x-8|$

34. $y=|x|-8$ **35.** $y=-|x+8|$ **36.** $y=-|x-8|$

37. $y=-|x|+8$ **38.** $y=-|x|-8$ **39.** $y=8|x|$

40. $y=-8|x|$ **41.** $y=-\frac{1}{8}|x|$ **42.** $y=8|x-1|$

43. $y=8|x|-1$ **44.** $y=|x+8|+1$ **45.** $y=-|x+8|+1$

Graph each absolute-value function.

46. $|x|=3$ **47.** $y=-|x|$ **48.** $y=|x-3|$

49. $y=|x+3|$ **50.** $y=-|x-3|$ **51.** $y=|x|+3$

52. $y=|x|-3$ **53.** $y=-|x|+3$ **54.** $y=|x|+x$

55. $y=|x|+3x$

56. Identity the statement as true or false: $|a-b|=|b-a|$.

57. Identify the statement as true or false: $|x+5|=|x|+5$.

58. Identify the value of x for which the statement is true: $|x+5|=|x|+5$.

59. Identify the values of a and b for which the statement is true: $|a|+|b|=|a+b|$.

60. Identify the following statement as true or false:
"The absolute value of every real number is a positive number"

3-6 Absolute-Value Equations and Inequalities

To solve equations and inequalities involving absolute values, we start with the following most basic patterns. Following these patterns, we can solve all inequalities easily.

   Rules: On the number line:

$$|x| = 5 \text{ means “ the distance from } x \text{ to } 0 \text{ is 5 ”.}$$

$$|x - c| = 5 \text{ means “ the distance from } x \text{ to } c \text{ is 5 ”.}$$

$$|x + c| = 5 \text{ means “ the distance from } x \text{ to } -c \text{ is 5 ”.}$$

$$|2x - c| = 5 \text{ means “ the distance from } (2x) \text{ to } c \text{ is 5 ”. Divide } 2x \text{ by 2 later.}$$

   These rules are also valid for most absolute-value inequalities ($<, \leq, >, \geq$).

1. Solve $|x| = 5$.

   Solution: $|x| = 5$

   Ans: $x = 5$ or -5.
   (or $x = \pm 5$)

2. Solve $|x| > 5$.

   Solution: $|x| > 5$

   Ans: $x < -5$ or $x > 5$.

3. Solve $|x| < 5$.

   Solution: $|x| < 5$

   Ans: $-5 < x < 5$.

Examples

1. Solve $|x - 2| = 8$.

   Solution: $x - 2 = \pm 8$

   $x = \pm 8 + 2$

   Ans: $x = 10$ or -6.

2. Solve $|x - 2| > 8$.

   Solution: $x - 2 < -8$ or $x - 2 > 8$

   $x < -6$ | $x > 10$

   Ans: $x < -6$ or $x > 10$.

3. Solve $|x + 2| < 8$.

   Solution: $-8 < x + 2 < 8$

   $-10 < x < 6$. Ans.

4. Solve $|4x - 7| = 9$.

   Solution: $4x - 7 = \pm 9$, $4x = \pm 9 + 7$

   $4x = 16$ or -2

   Ans. $x = 4$ or $-\frac{1}{2}$.

5. Solve $|3n + 5| - 3 \geq 7$.

   Solution: $|3n + 5| \geq 10$

   $3n + 5 \leq -10$ or $3n + 5 \geq 10$

   $3n \leq -15$ | $3n \geq 5$

   $\therefore n \leq -5$ or $n \geq \frac{5}{3}$. Ans.

6. Solve $|20 - 5y| \leq 25$.

   Solution: $-25 \leq 20 - 5y \leq 25$

   $-45 \leq -5y \leq 5$

   $9 \geq y \geq -1$

   Ans: $-1 \leq y \leq 9$.

EXERCISES

Solve each equation or inequality involving absolute value.

1. $|x| = 10$ **2.** $|x| > 10$ **3.** $|x| < 10$

4. $|x| \geq 10$ **5.** $|x| \leq 10$ **6.** $|x - 4| = 10$

7. $|x + 4| = 10$ **8.** $|x - 4| \geq 10$ **9.** $|x + 4| \leq 10$

10. $|5x - 7| = 13$ **11.** $|5x + 7| < 13$ **12.** $|6x + 5| < 7$

13. $|6x - 5| > 7$ **14.** $|4y + 12| + 5 > 9$ **15.** $|4y + 12| - 5 \leq 7$

16. $2|x| = 12$ **17.** $3|x + 2| = 15$ **18.** $4|x - 2| > 16$

19. $5|3a - 4| + 1 < 6$ **20.** $5|2a + 6| - 3 = 7$ **21.** $|2n - \frac{1}{3}| = \frac{2}{3}$

22. $|p - \frac{2}{3}| \geq \frac{5}{3}$ **23.** $|\frac{1}{2}x - 2| \geq 2$ **24.** $|\frac{1}{4}y - 5| \leq 7$

25. $|0.5x + 0.25| > 0.75$ **26.** $|0.5x - 0.25| < 0.75$ **27.** $|0.25 - 0.5x| \leq 0.75$

28. $|-3x| > 6$ **29.** $|-4x| < 12$ **30.** $|4 - 2x| < 16$

31. $|5 - 4x| > 7$ **32.** $|6 - 3x| \geq 9$ **33.** $|2 - x| \geq 7$

34. $|4 - x| \leq 12$ **35.** $|x| > 0$ **36.** $|x| < 0$

37. $|x| \geq 0$ **38.** $|x| \leq 0$ **39.** $|x - 4| > 0$

40. $|7x + 14| > 0$ **41.** $|5n - 20| \leq 0$ **42.** $|5n - 20| < 0$

43. $1 \leq |x| \leq 5$ **44.** $2 < |x + 1| \leq 3$ **45.** $0 < |2 - x| < 4$

46. In a random survey in a city election that 46% of the voters will choose candidate A for mayor. The result of the survey is considered to be accurate within a range of $\pm 5\%$. Write an absolute-value inequality that represents the upper and lower boundaries of the percent (p) of voters who will choose candidate A.

47. In question **46**, find the boundaries (the upper and the lower) of the percents.

48. The normal human body temperature is $98.6° F$ within a range of $\pm 1° F$. Write an inequality that represents the range of normal body temperatures (t).

49. Solve $|x - a| \leq b$ for x. Assume a and b are positive numbers.

50. Solve $|x + a| \geq b$ for x. Assume a and b are positive numbers.

CHAPTER 3   EXERCISES

Classify each statement as true or false.

1. $-5 > -3$　　　　**2.** $-4 < -2$　　　　**3.** $-2 < 0 < 1$　　　　**4.** $-7 < -5 > -4$

5. $-1 \le -1$　　　　**6.** $-1 \ge -1$　　　　**7.** $|-4-7| = |-4| + |-7|$　　**8.** $|-4+7| = |-4| + |7|$

9. $|3-9| < |9-4|$　　**10.** $|-17+5| \ge |17-5|$

Translate each statement into symbols.

11. 9 is greater than 1.

12. -9 is less than -1.

13. 2 is between -3 and 7.

14. -4 is greater than -8 and less than 1.

15. x is less than -1.

16. n is greater than or equal to 15.

17. x is greater than 0 and less than 12.

18. n is less than -3 or greater than 3.

19. The absolute value of x is greater than or equal to 25.

20. The absolute value of $x - a$ is less than or equal to b.

Solve each inequality.

21. $x + 5 \ge 26$　　　　　　**22.** $x - 5 \le 26$　　　　　　**23.** $-x + 5 \ge 26$

24. $-x - 5 \le 26$　　　　　**25.** $3x - 6 > 9$　　　　　　**26.** $3x + 6 < -9$

27. $5 - 4x > 17$　　　　　　**28.** $8 - 5x < 3$　　　　　　**29.** $\frac{4}{5}x \ge 24$

30. $-\frac{4}{5}x \ge 24$　　　　　**31.** $0.2a > 10$　　　　　　**32.** $-0.2a < 20$

33. $3x - 6 \le x + 4$　　　　**34.** $4x + 8 \le 7x - 1$　　　**35.** $6x + 8 > 6x + 2$

36. $6(x + 2) < 6x$　　　　　**37.** $\frac{4}{5}x + 10 < 35$　　　　**38.** $-\frac{4}{5}x - 10 > 35$

39. $5x + 4 \ge 2(x - 3) - 2$　　**40.** $3(x + 4) \le 6(2x - 1) - 9$　**41.** $4 - (n - 3) > 7(n + 2)$

42. $3 < a + 4 < 10$　　　　**43.** $-3 \le a - 7 < 8$　　　　**44.** $0 < 2x - 6 < 10$

45. $-6 < -2x < 8$　　　　　**46.** $-1 < -2n - 3 \le 7$

47. $7x + 6 > -8$ and $4x - 3 < 9$　　　　**48.** $6x - 5 > 13$ or $5x < -25$

49. $7x + 6 > -8$ or $4x - 3 < 9$　　　　**50.** $6x - 5 > 13$ and $5x < -25$

-----Continued-----

Graph each absolute-value function.

51. $|x| = 2$ **52.** $|y| = 2$ **53.** $y = |x + 2|$

54. $y = |x - 2|$ **55.** $y = -|x - 2|$ **56.** $y = -|x + 2|$

57. $y = |x| - 5$ **58.** $y = -|x| + 5$ **59.** $y = |x| - 2x$

60. $y = |2x| - x$

Solve each equation or inequality involving absolute value.

61. $|x| = -5$ **62.** $|x| = 15$ **63.** $|x + 7| = 12$

64. $|x + 7| < 12$ **65.** $|x + 7| > 12$ **66.** $|2x - 8| = 16$

67. $|2x - 8| < 16$ **68.** $|2x - 8| > 16$ **69.** $6|3x + 6| = 18$

70. $|7x - 6| \geq 1$ **71.** $|7x - 6| \leq 1$ **72.** $4|5x - 2| \leq 32$

73. $|k - \frac{4}{5}| \geq \frac{1}{5}$ **74.** $|a + 0.5| > 5.5$ **75.** $|1.5x - 3| < 2.5$

76. $|-5x| > 15$ **77.** $|-8x| \leq 16$ **78.** $|3 - 5x| < 8$

79. $|4 - x| \geq 20$ **80.** $1 < |x - 1| < 5$

81. If $a < b$, identify the values of c for which the statement is true: $ac > bc$.

82. If $a < b < 0$, identify whether the state is true or false: $a^2 < b^2$.

83. Find the domain and range of $y = |x + 5|$.

84. Find the domain and range of $y = |x| + 5$.

85. Find the domain and range of $y = -|x + 5|$.

86. Solve $|x - a| = b$ for x. Assume a and b are positive numbers.

87. Solve $|x - a| < b$ for x. Assume a and b are positive numbers.

88. Your scores on your 4 test were 74, 82, 77 and 78. What is the lowest score you need in your next test in order to have an average score of at least 80 ?

89. You want to rent vehicle for one day. The rental cost for a car is $45 plus $0.25 per mile. The rental cost for a truck is $60 plus $0.15 per mile. How many miles you would need to travel to rent the car with least expensive cost ?

90. The sum of three consecutive even integers is no less than 218. Find the integers with the smallest sum.

91. The length of a rectangle field is 7 feet longer than the width. The perimeter is no more than 350 feet. What is the greatest possible dimensions of the field ?

92. At most how many grams of pure salt must be added with 84 grams of water to produce an salt-water solution that is no more than 5% pure salt ?

93. A alcohol-water solution is produced with 5% pure alcohol. The acceptable error is $\pm 0.2\%$. Write an absolute-value inequality that represents the acceptable values of percent (p).

Answers

CHAPTER 3 – 1   Inequalities on the Number Line   page 66

1. $6 > 4$ **2.** $3 < 8$ **3.** $-3 < 5$ **4.** $-3 > -5$ **5.** $-7 < -4$ **6.** $4.5 < 6$ **7.** $-12 = -12$

8. $-4.2 > -4.5$ **9.** $-18 > -24$ **10.** $-32 < -15$

11. False   **12.** True   **13.** False   **14.** True   **15.** True (Hint: It is true since $-5 = -5$ is true.)

16. True   **17.** False   **18.** False   **19.** True   **20.** True (Hint: It is true since $3 = 3$ is true.)

21. $10 < 13$ **22.** $2 < 5$ **23.** $24 < 36$ **24.** $1.5 < 2.25$ **25.** $-24 > -36$ **26.** $-1.5 > -2.25$

27. $10 > -12$ **28.** $-10 < 15$ **29.** $-2 < 3$ **30.** $2 < 3$ **31.** $18 < 27$ **32.** $-2 > -3$ **33.** $x > 2$

34. $x < -2$ **35.** $x \le 8$ **36.** $x \ge -8$ **37.** $x > 9$ **38.** $x \ge -9$ **39.** $x \le -\frac{3}{2}$ **40.** $x \ge \frac{3}{2}$

41. $8 < 15$ **42.** $-8 > -15$ **43.** $-7 < -2$ **44.** $-9 < -7 < 2$ **45.** $-7 < -2 < 3$

46. $-10 < -7 < -2$ **47.** $x > 15$ **48.** $x \le 15$ **49.** $-8 < n < 2$ **50.** $4 < n \le 9$

51. $|n| < 5$ **52.** $|x| > 7$ **53.** $|x| \ge 7$ **54.** $|n| \le 5$

55.   **56.**   **57.**

58.   **59.**   **60.**

61 ~ 62 graphs are left to the students

CHAPTER 3 – 2   Solving Inequalities in One Variable   page 68

1. $x < 11$ **2.** $x > 5$

3. $x \ge -9$ **4.** $x \le 3$

(5 ~ 75 graphs are left to the students. You may graph only problems 5, 10, 15, ...)

5. $a > 11$ **6.** $n \ge -6$ **7.** $y < -10$ **8.** $y \le 4$ **9.** $p \le -3$ **10.** $k > 4$ **11.** $x \ge -5$ **12.** $v < 3$

13. $x < 4$ **14.** $x > -4$ **15.** $x \le -4$ **16.** $x > -4$ **17.** $x \ge 4$ **18.** $x \ge 16$ **19.** $x \le -16$

20. $n \ge 3$ **21.** $n < 10$ **22.** $p < -3$ **23.** $a > -3$ **24.** $a > 3$ **25.** $a < -\frac{1}{3}$ **26.** $p \ge -\frac{1}{3}$

27. $y \le -1$ **28.** $x < -10$ **29.** $x > 20$ **30.** $x \le -12$ **31.** $x \le -6$ **32.** $x < -20$

33. $x \ge -9$ **34.** $p > -20$ **35.** $p > 18$ **36.** $a < -40$ **37.** $k \le 21$ **38.** $k \ge 2\frac{1}{2}$ **39.** $x > 2$

40. $x > -4$ **41.** $x \le -\frac{1}{3}$ **42.** $x \ge -\frac{2}{3}$ **43.** $x > 4$ **44.** $x > -1$ **45.** $x < 1$ **46.** $x < 4$

47. $x \ge -4$ **48.** $x \le 4$ **49.** $x > -18$ **50.** $x < -32$ **51.** $x \le 75$

52. All real numbers   **53.** All real numbers   **54.** No solution (The solution set is ϕ.)

55. No solution (The solution set is ϕ.)   **56.** $x < -1\frac{1}{5}$ **57.** $x > -2\frac{1}{2}$ **58.** $n > 4$ **59.** $n \le -21$

60. $p \ge 0$ **61.** $x < 2$ **62.** $x > -\frac{1}{2}$ **63.** $x \le \frac{4}{5}$ **64.** $x \le 3$ **65.** $x > -2$ **66.** $x < -48$ **67.** $y > 6$

68. $x \ge -6$ **69.** $x \ge 7\frac{1}{2}$ **70.** $x \le \frac{3}{5}$ **71.** $x < -10$ **72.** $x < -4\frac{2}{3}$ **73.** $n < -5$ **74.** $p \ge 3\frac{1}{3}$ **75.** $y > 14$

76. All real numbers of c **77.** All positive real numbers of c, or all real numbers of $c > 0$

78. All negative real numbers of c, or all real numbers of $c < 0$.

79. False. (Hint: $-1 > -2$, but $(-1)^2 < (-2)^2$)   **80.** True. (Hint: $-1 > -2$ and $(-1)^3 > (-2)^3$)

81. True. (Hint: They are positive numbers.)

82. All positive real numbers of y, or all real numbers of $y > 0$. (Hint: $3^0 = 2^0$ and $3^{-2} < 2^{-2}$)

In a medical school, a professor asked a student about his major.

Professor : Why do you think a dentist can make more money than a heart surgeon ?

Student: A patient has 36 teeth, but only one heart.

82   A-Plus Notes for Algebra

CHAPTER 3 – 3   Solving Combined Inequalities   page 71 ~ 72

1. B   2. D   3. G   4. C   5. H   6. F (all real numbers)   7. A   8. E   9. E

10. F (all real numbers)   11. B   12. ϕ (empty, no graph)

13. $-2 < x < 5$ 14. $x \geq 8$ 15. $1 < x \leq 4$ 16. $-3 < x \leq 7$ 17. $-5 \leq x \leq -1$ 18. $x \leq 9$

19. $-6 \leq n \leq 4$ 20. $-12 < a \leq -4$

21 ~ 30 graphs are left to the students

21. $x < -5$ 22. $x < 1$ 23. $x \leq 2$ 24. $x \geq 6$ 25. $x < 6$ 26. $x < -6$ 27. $x > -2\frac{1}{2}$

28. $x \geq 2\frac{2}{5}$ 29. $x \leq -18$ 30. $a < 1$

31. ◄─┼─┼─┼─┼─┼─◆─┼─┼─┼─┼─◆─┼─►
      -6   -4   -2   0   2   4   6

32. ◄─┼─┼─┼─┼─○─┼─◆─┼─┼─┼─►
      -6   -4   -2   0   2   4   6

33 ~ 39 graphs are left to the students.   40. No graph (ϕ, empty)

41 ~ 76 graphs are left to he students. The graphs are very useful to write the right answers.

41. $7 < x < 14$ 42. $-7 \leq x < 4$ 43. $-4 < a \leq 3$ 44. $3 < a \leq 6$ 45. $-3 < n \leq 5$

46. $-3 \leq n < 2$ 47. $-8 < x < 4$ 48. No solution (ϕ, empty)   49. $-2 < x < 9$

50. $-4 \leq y \leq 1$ 51. $-1 \leq y < 2$ 52. $3 < b < 5$ 53. $2 < p < 4$ 54. $-3 \leq x < 4$

55. $-2 < x \leq 2$ 56. $-2 < x < 3$ 57. $-3 < x < 2$ 58. No solution (ϕ, empty)

59. $x > -2$ 60. $x < -5$ or $x > 2$ 61. All real numbers of x 62. $x > -5$

63. $x > -3$ 64. $x \leq -3$ or $x \geq 5$ 65. No solution (ϕ, empty)   66. $3 \leq x \leq 5$

67. $-2 \leq n \leq 4$ 68. $n > -2$ 69. $-2 \leq p < 2$ 70. $p > 3$ 71. All real numbers of p

72. $p \geq -2$ 73. $-2 < x \leq 1.5$ 74. $x \leq -3$ or $x \geq 1\frac{2}{3}$ 75. $x < -4.5$ or $x > 3.4$

76. $-18 \leq x \leq 12$

CHAPTER 3 – 4   Solving Word Problems Involving Inequalities   page 74

1. $x < 13$ 2. $x < 5$ 3. $x \geq 46$ 4. $n \leq 43$ 5. $n \leq 10$ 6. $n \geq 12$ 7. $m > 420$

8. $m < 700$ 9. $w > 20$ 10. $w > 3.5$ 11. $x \geq 85$ 12. $x \geq 90$

13. At most 300 miles (300 miles or less)   14. At most 7 days (7 days or less)

15. 180 miles or less   16. 91 (or more)   17. 95 (or more)   18. 47 and 48   19. 49 and 50

20. 48 and 49   21. 42, 43, 44   22. {1, 3, 5} or {3, 5, 7} or {5, 7, 9}

23. At least $19\,cm$ by $7\,cm$ 24. At least $15\,cm$ by $9\,cm$ 25. At most $15\,cm$ by $9\,cm$

26. At least $13\,in.$ by $11\,in.$ 27. At most 20 quarters and 4 dimes in the box

28. At most 13.5 grams (or less) of salt added

CHAPTER 3 – 5   Operations with Absolute Values   page 76

1. 15   2. 15   3. 7.2   4. $\frac{4}{5}$ 5. 1   6. 1   7. 8   8. 0   9. 3   10. 3   11. 9   12. 8   13. 0   14. 18   15. 8

16. True   17. True   18. False   19. False   20. True   21. False   22. True   23. False

24. True   25. False   26. True   27. False   28. True   29. True   30. True

31. D = all real numbers, R = all non-negative real numbers (or all real numbers of $y \geq 0$)

32. D = all real numbers, R = all real numbers of $y \geq 8$ 33. D = all real numbers, R = all non-negative real numbers

34. D = all real numbers, R = all real numbers of $y \geq -8$

35. D = all real numbers, R = all non-positive real numbers (or all real numbers of $y \leq 0$)

36. D = all real numbers, R = all non-negative real numbers   37. D = all real numbers, R = all real numbers of $y \leq 8$

38. D = all real numbers, R = all real numbers of $y \leq -8$ 39. D = all real numbers, R = all non-negative real numbers

40. D = all real numbers, R = all non-positive real numbers   41. D = all real numbers, R = all non-positive real numbers

42. D = all real numbers, R = all non-negative real numbers   43. D = all real numbers, R = all real numbers of $y \geq -1$

44. D = all real numbers, R = all real numbers of $y \geq 1$ 45. D = all real numbers, R = all real numbers of $y \leq 1$

46. $|x| = 3$

$x = 3$ or $x = -3$

47. $y = -|x|$

For $x > 0$, $y = -x$

For $x < 0$, $y = -(-x)$

$y = x$

For $x = 0$, $y = 0$

48. $y = |x - 3|$

For $x - 3 > 0$, $y = x - 3$, $x - y = 3$

For $x - 3 < 0$, $y = -(x - 3)$, $x + y = 3$

For $x - 3 = 0$ (or $x = 3$), $y = 0$

49 ~ 55 graphs are left to the students.

56. True   57. False (Hint: It is false when $x < 0$)   58. $x \geq 0$

59. It is true when a and b are both positive, both negative, or both 0.

60. False (Hint: $|0| = 0$. 0 is a real number. But, 0 is neither positive nor negative.)

CHAPTER 3 – 6   Absolute-Value Equations and Inequalities   page 78

1. $x = 10$ or -10 **2.** $x < -10$ or $x > 10$ **3.** $-10 < x < 10$ **4.** $x \le -10$ or $x \ge 10$

5. $-10 \le x \le 10$ **6.** $x = 14$ or -6 **7.** $x = 6$ or -14 **8.** $x \le -6$ or $x \ge 14$ **9.** $-14 \le x \le 6$

10. $x = 4$ or $-1\frac{1}{5}$ **11.** $x = 1\frac{1}{5}$ or -4 **12.** $-2 < x < \frac{1}{3}$ **13.** $x < -\frac{1}{3}$ or $x > 2$

14. $y < -4$ or $y > -2$ **15.** $-6 \le y \le 0$ **16.** $x = 6$ or -6 **17.** $x = 3$ or -7 **18.** $x < -2$ or $x > 6$

19. $1 < a < 1\frac{2}{3}$ **20.** $a = -2$ or -4 **21.** $n = \frac{1}{2}$ or $-\frac{1}{6}$ **22.** $p \le -1$ or $p \ge 2\frac{1}{3}$ **23.** $x \le 0$ or $x \ge 8$

24. $-8 \le y \le 48$ **25.** $x < -2$ or $x > 1$ **26.** $-1 < x < 2$ **27.** $1 \le x \le 2$ **28.** $x < -2$ or $x > 2$

29. $-3 < x < 3$ **30.** $-6 < x < 10$ **31.** $x < -\frac{1}{2}$ or $x > 3$ **32.** $x \le -1$ or $x \ge 5$ **33.** $x \le -5$ or $x \ge 9$

34. $-8 \le x \le 16$ **35.** $x =$ all real numbers except 0   **36.** No solution   **37.** $x =$ all real numbers

38. $x = 0$ only   **39.** $x =$ all real numbers except 4   **40.** $x =$ all real numbers except -2

41. $n = 4$ only   **42.** No solution   **43.** $-5 \le x \le -1$ or $1 \le x \le 5$ **44.** $-4 \le x < -3$ or $1 < x \le 2$

45. $-2 < x < 2$ or $2 < x < 6$ **46.** $|p - 46| \le 5$ **47.** $41\% \le p \le 51\%$ **48.** $|t - 98.6| \le 1$

49. $a - b \le x \le a + b$ **50.** $x \le -a - b$ or $x \ge -a + b$

CHAPTER 3 EXERCISES   page 79 ~ 80

1. False   **2.** True   **3.** True   **4.** False   **5.** True(Hint: Since $-1 = -1$)   **6.** True   **7.** True

8. False   **9.** False   **10.** True   **11.** $9 > 1$ **12.** $-9 < -1$ **13.** $-3 < 2 < 7$ **14.** $-8 < -4 < 1$

15. $x < -1$ **16.** $n \ge 15$ **17.** $0 < x < 12$ **18.** $n < -3$ or $n > 3$ **19.** $|x| \ge 25$ **20.** $|x - a| \le b$

21. $x \ge 21$ **22.** $x \le 31$ **23.** $x \le -21$ **24.** $x \ge -31$ **25.** $x > 5$ **26.** $x < -5$ **27.** $x < -3$

28. $x > 1$ **29.** $x \ge 30$ **30.** $x \le -30$ **31.** $a > 50$ **32.** $a > -100$ **33.** $x \le 5$ **34.** $x \ge 3$

35. All real numbers   **36.** No solution   **37.** $x < 31\frac{1}{4}$ **38.** $x < -56\frac{1}{4}$ **39.** $x \ge -4$ **40.** $x \ge 3$

41. $n < -\frac{7}{8}$ **42.** $-1 < a < 6$ **43.** $4 \le a < 15$ **44.** $3 < x < 8$ **45.** $-4 < x < 3$ **46.** $-5 \le n < -1$

47. $-2 < x < 3$ **48.** $x < -5$ or $x > 3$ **49.** All real numbers   **50.** No solution

51. $|x| = 2$

   $x = 2$ or $x = -2$

52. $|y| = 2$

   $y = 2$ or $y = -2$

53 ~ 60 graphs are left to the students.

61. No solution   **62.** $x = 15$ or -15 **63.** $x = 5$ or -19 **64.** $-19 < x < 5$

65. $x < -19$ or $x > 5$ **66.** $x = 12$ or -4 **67.** $-4 < x < 12$ **68.** $x < -4$ or $x > 12$

69. $x = -1$ or -3 **70.** $x \le \frac{5}{7}$ or $x \ge 1$ **71.** $\frac{5}{7} \le x \le 1$ **72.** $-1\frac{1}{5} \le x \le 2$ **73.** $k \le \frac{3}{5}$ or $k \ge 1$

74. $a < -6$ or $a > 5$ **75.** $\frac{1}{3} < x < 3\frac{1}{3}$ **76.** $x < -3$ or $x > 3$ **77.** $-2 \le x \le 2$

78. $-1 < x < 5\frac{1}{2}$ **79.** $x \le -16$ or $x \ge 24$ **80.** $-5 < x < 0$ or $2 < x < 5$

81. All negative real numbers of c, or all real numbers of $c < 0$.

82. False. (Hint: $-2 < -1$, but $(-2)^2 > (-1)^2$)

83. D = all real numbers,   R = all non-negative real numbers (or all real numbers of $y \ge 0$)

84. D = all real numbers,   R = all real numbers of $y \ge 5$

85. D = all real numbers,   R = all non-positive real numbers (or all real numbers of $y \le 0$)

86. $x = a + b$ or $x = a - b$ **87.** $a - b < x < a + b$ **88.** 89   **89.** 150 miles or less

90. 72, 74 and 76   **91.** 91 feet by 84 feet   **92.** 4.42 grams

93. $|p - 5| \le 0.2$ (Hint: $4.8\% \le p \le 5.2\%$)

Additional Examples

1. Graph $y = |x| + 2$.

    Solution:

    If $x > 0$, then $y = x + 2$

    If $x < 0$, then $y = -x + 2$

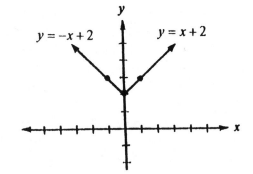

2. Graph $y = |x + 2|$.

    Solution:

    If $(x + 2) > 0$, then $y = x + 2$

        We have: For $x > -2$, $y = x + 2$

    If $(x + 2) < 0$, then $y = -(x + 2)$

        We have: For $x < -2$, $y = -x - 2$

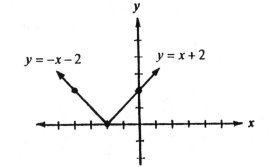

3. Graph $y = |x| - 2x$.

    Solution:

    If $x > 0$, then $y = x - 2x$, $y = -x$

    If $x < 0$, then $y = -x - 2x$, $y = -3x$

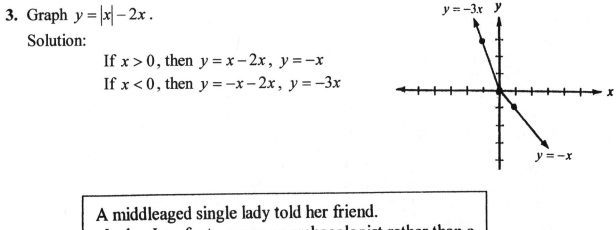

A middleaged single lady told her friend.
  Lady: I prefer to marry an archaeologist rather than a
          doctor or a lawyer.
Friend: Why do you want to marry an archaeologist ?
  Lady: If my husband is an archaeologist, the more I
          get older, the more he is interested in me.

Examples

4. Solve $|2x+1| = 4x-5$.

Solution: 1) Let $2x+1 \geq 0$
$$2x+1 = 4x-5$$
$$-2x = -6$$
$$x = 3$$
It satisfies the equation.

2) Let $2x+1 < 0$
$$2x+1 = -(4x-5)$$
$$2x+1 = -4x+5$$
$$6x = 4, \quad x = \tfrac{2}{3}$$
It does not satisfy the equation.

Ans: $x = 3$.

5. Solve $1 \leq |x-2| \leq 4$.

Solution: 1) $|x-2| \geq 1$
$$x-2 \leq -1 \quad \text{or} \quad x-2 \geq 1$$
$$x \leq 1 \quad \text{or} \quad x \geq 3$$

2) $|x-2| \leq 4$
$$-4 \leq x-2 \leq 4$$
$$-2 \leq x \leq 6$$

Ans: $-2 \leq x \leq 1$ or $3 \leq x \leq 6$.

6. Graph $|x| + |y| = 4$.

Solution: $|x| + |y| = 4$, $|y| = 4 - |x|$

    1) If $y \geq 0$, $y = 4 - |x|$

       If $x \geq 0$, $y = 4 - x$

       If $x < 0$, $y = 4 + x$

    2) If $y < 0$, $y = -4 + |x|$

       If $x \geq 0$, $y = -4 + x$

       If $x < 0$, $y = -4 - x$

7. Graph $y = |x-2| + x$.

Solution:

    1) If $x-2 \geq 0$ (or $x \geq 2$),
$$y = x-2+x$$
$$y = 2x-2$$

    2) If $x-2 < 0$ (or $x < 2$),
$$y = -(x-2)+x$$
$$y = 2$$

11. Find the area bound by $|x| + |y| = 4$

Solution: (See example **6**)

    The length of one side of the square is:
$$r = \sqrt{4^2 + 4^2} = \sqrt{32} = 4\sqrt{2}$$
The area of the square is:

Area $= (4\sqrt{2})^2 = 32$. Ans.

Graphing an absolute-value function by a graphing calculator

Example: Graph $y = |2x - 6|$.

Solution:
Press: **Y1=** → **MATH** → **NUM** → **ENTER**
Enter the function:
   Y1= abs $(2x - 6)$
Press: **GRAPH**
The screen shows:

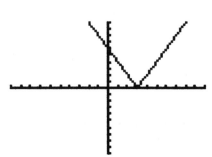

Solving an absolute-value equation by a graphing calculator

Example: Solve $|4x + 2| = 6$.

Solution:
Press: **Y1=** → **MATH** → **NUM** → **ENTER**
Enter the two equations:
   Y1= abs (4x + 2)
   Y2 = 6
Press: **GRAPH**
Find the intersection:
Press: **2nd** → **CALC** → **5: intersect** → **ENTER**
Press: **TRACE**
A flashing trace cursor appears. Move the cursor to the
approximate intersection point. Intersection is displayed
at the bottom of the screen.
The solutions are 1 and −2.

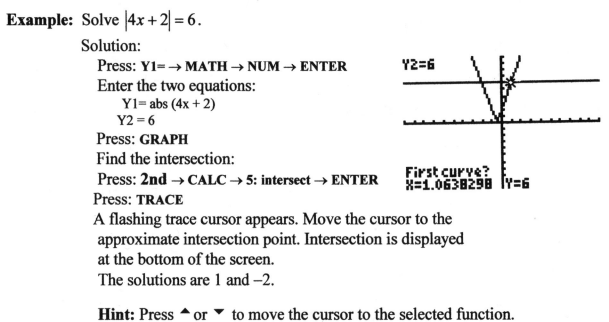

Hint: Press ▲ or ▼ to move the cursor to the selected function.
   Press ◄ or ► to move the cursor to the intersection point.

Graphing a linear inequality
by a graphing calculator

Hint: 1. To graph a linear inequality, we use the **Shade** feature in the
DRAW menu.
Press: **2 nd** → **DRAW** → **7↓ Shade** → **ENTER**
The screen shows the shade command: **Shade (**
   2. In the shade command, we must enter the lower boundary,
the upper boundary, the left boundary, and the right boundary.
Shade (Ymin , Ymax , Xmin , Xmax **)**
   3. If the inequality is " $y \geq$ ", we shade the portion of the plane above
the line graph. The line graph itself is the lower boundary (Ymin).
   4. If the inequality is " $y \leq$ ", we shade the portion of the plane below
the line graph. The line graph itself is the upper boundary (Ymax).

Example: Graph $y \geq -\frac{4}{5}x + 4$.

    Solution:

      Press: **2nd** → **DRAW** → **7↓Shade (** → **ENTER**
      The screen shows: **Shade (**
      Enter: **Shade ((((-)** → **4** → **÷** → **5** → **)** → $\boxed{\text{X,T},\theta,n}$
       + 4 → **,** → **10** → **,** → **(-)** → **10** → **,** → **10)**
      The screen shows: (Ymin is not shaded.)
        Shade ((−4 / 5)x + 4 , 10, −10, 10)

      Press: **ENTER**
      The screen shows:

Solving an inequality in one variable
by a graphing calculator

Example: Solve $8x - 3 > 2x + 9$.
    Solution:
      Press: **Y₁ =** → **8** → $\boxed{\text{X,T},\theta,n}$ → **−** → **3**
        → **2nd** → **TEST** → **3: >** → **ENTER**
        → **2** → $\boxed{\text{X,T},\theta,n}$ → **+** → **9**
      The screen shows:
        $y_1 = 8x - 3 > 2x + 9$

Press: **GRAPH**
The screen shows:
The solutions are given by $x > 2$.

Press: **TRACE**
A flashing trace cursor appears
to scan the x – values along the
graph.

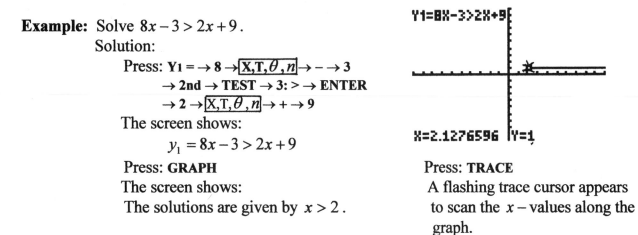

Space for taking Notes

CHAPTER FOUR

Radicals and Complex Numbers

4-1 Square Roots and Radicals

Because $5^2 = 25$, we say that the square root of 25 is 5. We write $\sqrt{25} = 5$.

Because $5^3 = 125$, we say that the cubed root of 125 is 5. We write $\sqrt[3]{125} = 5$.

Because $2^4 = 16$, we say that the 4th root of 16 is 2. We write $\sqrt[4]{16} = 2$.

The symbol $\sqrt[n]{b}$ is called a **radical**. We read $\sqrt[n]{b}$ as " **the nth root of b** ".

b is called the **radicand**, n is called the **index**.

Index "2" of the square root is usually omitted from the radical. $\sqrt{b}$ means $\sqrt[2]{b}$.

To find the square root of a given number, we find a number that, when squared, equals the given number. However, $\sqrt[n]{b}$ means only the **principal root** of b.

$$\sqrt{4} = \sqrt{2 \cdot 2} = \sqrt{2^2} = 2 . \qquad \sqrt{9} = \sqrt{3 \cdot 3} = \sqrt{3^2} = 3 .$$

$$\sqrt{4} = \sqrt{(-2)^2} = -2 \text{ is incorrect.} \qquad \sqrt{9} = \sqrt{(-3)^2} = -3 \text{ is incorrect.}$$

$$\sqrt{(-2)^2} = \sqrt{4} = 2 . \qquad \sqrt{(-3)^2} = \sqrt{9} = 3 .$$

$$\sqrt{0.01} = \sqrt{(0.1)(0.1)} = \sqrt{(0.1)^2} = 0.1 . \qquad \sqrt{1.44} = \sqrt{(1.2)(1.2)} = \sqrt{(1.2)^2} = 1.2$$

To find the nth root of a given number, we follow the same idea as the above examples, and so on.

$$\sqrt[3]{8} = \sqrt[3]{2 \cdot 2 \cdot 2} = \sqrt[3]{2^3} = 2 . \qquad \sqrt[4]{16} = \sqrt[4]{2 \cdot 2 \cdot 2 \cdot 2} = \sqrt[4]{2^4} = 2 .$$

$$\sqrt[3]{-8} = \sqrt[3]{-2 \cdot -2 \cdot -2} = \sqrt[3]{(-2)^3} = -2 . \qquad \sqrt[4]{(-2)^4} = \sqrt[4]{16} = 2, \text{ } \textbf{not} -2.$$

The square root of a given number that is not a perfect square is an **irrational number** (a nonrepeating infinite decimal). We use a calculator to find its decimal approximation, depending on what decimal place we round to.

$$\sqrt{3} \approx 1.732 . \qquad \sqrt{5} \approx 2.236 . \qquad \sqrt{7} \approx 2.646 . \quad \sqrt{8} \approx 2.828 .$$

$$\sqrt{0.15} \approx 0.387 . \qquad \sqrt{17} \approx 4.123 . \quad \sqrt{140} \approx 11.832 .$$

To find the square root of a fraction, we write the square root of the numerator over the square root of the denominator.

$$\sqrt{\frac{4}{9}} = \frac{\sqrt{4}}{\sqrt{9}} = \frac{2}{3} . \quad \text{Or:} \quad \sqrt{\frac{4}{9}} = \sqrt{\frac{2}{3} \cdot \frac{2}{3}} = \frac{2}{3} .$$

$$\sqrt{\frac{12}{35}} = \frac{\sqrt{12}}{\sqrt{35}} \approx \frac{3.464}{5.916} \approx 0.586 . \quad \text{Or:} \quad \sqrt{\frac{12}{35}} \approx \sqrt{0.343} \approx 0.586 .$$

EXERCISES

Evaluate each square root. Check you answers by squaring.
(The symbol " $\pm$ " means "plus or minus", or "positive or negative".)

   Hint: If you cannot find the square root of the given number, you may factor the given
   number by using a perfect-square factor, starting with 4, 9, 16, 25, 36, 49, 64, 81, ….
$$\sqrt{1936} = \sqrt{4 \cdot 484} = \sqrt{4 \cdot 4 \cdot 121} = 4\sqrt{121} = 4\sqrt{11 \cdot 11} = 4 \cdot 11 = 44$$
   Or: $\sqrt{1936} = \sqrt{16 \cdot 121} = \sqrt{16} \cdot \sqrt{121} = 4 \cdot 11 = 44$

1. $\sqrt{16}$ **2.** $\sqrt{25}$ **3.** $-\sqrt{36}$ **4.** $\sqrt{49}$

5. $\sqrt{64}$ **6.** $-\sqrt{81}$ **7.** $\pm\sqrt{144}$ **8.** $\sqrt{10{,}000}$

9. $-\sqrt{196}$ **10.** $\pm\sqrt{400}$ **11.** $-\sqrt{900}$ **12.** $\sqrt{625}$

13. $\sqrt{784}$ **14.** $\sqrt{1{,}296}$ **15.** $\pm\sqrt{2{,}304}$ **16.** $\sqrt{11{,}025}$

17. $\sqrt{0.09}$ **18.** $-\sqrt{0.16}$ **19.** $-\sqrt{0.25}$ **20.** $\pm\sqrt{0.81}$

21. $\sqrt{1.44}$ **22.** $\sqrt{0.0016}$ **23.** $\sqrt{0.0025}$ **24.** $\sqrt{0.0121}$

25. $\sqrt{0.0144}$ **26.** $\pm\sqrt{0.0169}$ **27.** $\sqrt{0.0004}$ **28.** $\sqrt{0.000036}$

29. $\sqrt{\dfrac{9}{4}}$ **30.** $-\sqrt{\dfrac{25}{49}}$ **31.** $\pm\sqrt{\dfrac{1}{36}}$ **32.** $\sqrt{\dfrac{81}{400}}$

33. $\sqrt{\dfrac{100}{121}}$ **34.** $\pm\sqrt{\dfrac{16}{64}}$ **35.** $-\sqrt{\dfrac{24}{54}}$ **36.** $\sqrt{\dfrac{5}{125}}$

Evaluate each radical. If the root is an irrational number, round the decimal to the nearest thousandth.

37. $\sqrt[3]{27}$ **38.** $\sqrt[3]{64}$ **39.** $-\sqrt[4]{16}$ **40.** $\sqrt[4]{81}$

41. $\sqrt[5]{32}$ **42.** $\pm\sqrt{10}$ **43.** $\sqrt{12}$ **44.** $-\sqrt{40}$

45. $\sqrt{75}$ **46.** $\sqrt{27}$ **47.** $-\sqrt{32}$ **48.** $\sqrt{125}$

49. $\pm\sqrt[3]{0.001}$ **50.** $\sqrt[3]{0.008}$ **51.** $\sqrt{0.12}$ **52.** $\pm\sqrt{325}$

53. $2\sqrt[6]{64}$ **54.** $-\sqrt[4]{256}$ **55.** $5\sqrt{70}$ **56.** $3\sqrt{108}$

57. $\sqrt{0.9}$ **58.** $\sqrt{3.5}$ **59.** $\pm 8\sqrt{48}$ **60.** $12\sqrt{4.5}$

61. $\sqrt{\dfrac{15}{25}}$ **62.** $\pm\sqrt{\dfrac{11}{6}}$ **63.** $-\sqrt{\dfrac{54}{8}}$ **64.** $\sqrt{\dfrac{140}{18}}$

4-2 Simplifying Radicals

<u>**Rules of Radicals**</u>

For all nonnegative real numbers a and b :

   1) $\sqrt{a^2} = a$ **2)** $\sqrt{ab} = \sqrt{a} \cdot \sqrt{b}$ **3)** $\sqrt{\dfrac{b}{a}} = \dfrac{\sqrt{b}}{\sqrt{a}}$, where $a \neq 0$

   4) $\sqrt[n]{a} = a^{\frac{1}{n}}$ **5)** $\sqrt[n]{a^m} = a^{\frac{m}{n}}$

For all real numbers a , b , and x :

   6) $\sqrt[n]{a^n} = |a|$ **if** n is even.   **7)** $\sqrt{x^2} = |x|$

     $\sqrt[n]{a^n} = a$ **if** n is odd.   **8)** $\sqrt{x^3} = x\sqrt{x}$

The formula $\sqrt{x^2} = |x|$ shows that when we are finding **square root** of a variable expression, we must use absolute value sign to **ensure** the answer is positive.

$\sqrt{(-3)^2} = \sqrt{9} = 3$, not -3 . $\sqrt[3]{-27} = \sqrt[3]{(-3)^3} = -3$. (Cubed root could be negative.)

$\sqrt{4x^2} = \sqrt{4} \cdot \sqrt{x^2} = 2|x|$. $\sqrt{4x^6} = \sqrt{4} \cdot \sqrt{(x^3)^2} = 2|x^3|$. $\sqrt[3]{8x^3} = \sqrt[3]{8} \cdot \sqrt[3]{x^3} = 2x$.

$\sqrt{4x^4} = \sqrt{4} \cdot \sqrt{(x^2)^2} = 2|x^2| = 2x^2$. (x^2 is always nonnegative. Omit the symbol $|\ |$.)

$\sqrt{4x^8} = \sqrt{4} \cdot \sqrt{(x^4)^2} = 2|x^4| = 2x^4$. (x^4 is always nonnegative. Omit the symbol $|\ |$.)

In the formula $\sqrt{x^3} = x\sqrt{x}$, we must assume that x is nonnegative. The radical $\sqrt{x^3}$ has no meaning (not a real number) if $x < 0$. Therefore, we omit the symbol $|\ |$.

$$\sqrt{x^3} = \sqrt{x^2 \cdot x} = \sqrt{x^2} \cdot \sqrt{x} = |x| \cdot \sqrt{x} = x\sqrt{x} \quad \text{Similarly: } \sqrt[4]{x^5} = x\sqrt[4]{x}$$

To simplify a square root in simplest radical form, we factor the radicand by using a prefect-square factor and leave any factor that is not a perfect-square in the radical form. To simplify a radical, we must **rationalize the denominator** by eliminating the radical from the denominator. No radical is allowed in denominator (see example **14**).

Examples. (Simplify each radical in simplest radical form)

   1. $\sqrt{75} = \sqrt{25 \cdot 3} = 5\sqrt{3}$ **2.** $\sqrt{32} = \sqrt{16 \cdot 2} = 4\sqrt{2}$ **3.** $\sqrt[3]{16} = \sqrt[3]{2^3 \cdot 2} = 2\sqrt[3]{2}$

   4. $\sqrt{25x^2} = 5|x|$ **5.** $\sqrt{81y^2} = 9|y|$ **6.** $\sqrt{24a^2} = \sqrt{4 \cdot 6 \cdot a^2} = 2\sqrt{6}|a|$

   7. $\sqrt{16a^3} = 4\sqrt{a^2 \cdot a} = 4a\sqrt{a}$ **8.** $\sqrt{a^2 - 6a + 9} = \sqrt{(a-3)^2} = |a-3|$

   9. $\sqrt[4]{x^4 y^8} = \sqrt[4]{x^4} \cdot \sqrt[4]{(y^2)^4} = |x|y^2$ **10.** $\sqrt[3]{-8x^6} = \sqrt[3]{(-2x^2)^3} = -2x^2$

   11. $\sqrt[5]{-x^{10}} = \sqrt[5]{(-x^2)^5} = -x^2$ **12.** $\sqrt{18a^4 b^5} = \sqrt{9 \cdot 2 \cdot (a^2)^2 \cdot (b^2)^2 \cdot b} = 3a^2 b^2 \sqrt{2b}$

   13. $\sqrt{\dfrac{18x^2 y^3}{z^2}} = \dfrac{\sqrt{9 \cdot 2 \cdot x^2 \cdot y^2 \cdot y}}{\sqrt{z^2}} = \dfrac{3|x|y\sqrt{2y}}{|z|}$, $z \neq 0$ (Hint: y is nonnegative.)

   14. $\sqrt{\dfrac{2}{5}} = \dfrac{\sqrt{2}}{\sqrt{5}} \cdot \dfrac{\sqrt{5}}{\sqrt{5}} = \dfrac{\sqrt{10}}{\sqrt{25}} = \dfrac{\sqrt{10}}{5}$ **15.** $\sqrt[4]{9} = \sqrt[4]{3^2} = 3^{\frac{2}{4}} = 3^{\frac{1}{2}} = \sqrt{3}$

EXERCISES

Simplify each radical in simplest radical form

1. $\sqrt{12}$ **2.** $\sqrt{18}$ **3.** $-\sqrt{24}$ **4.** $\pm\sqrt{27}$

5. $\sqrt{32}$ **6.** $\sqrt{28}$ **7.** $\sqrt{80}$ **8.** $\sqrt{48}$

9. $\sqrt{150}$ **10.** $\pm\sqrt{108}$ **11.** $\sqrt{120}$ **12.** $-\sqrt{162}$

13. $\sqrt{396}$ **14.** $\sqrt{1800}$ **15.** $\sqrt{3200}$ **16.** $\sqrt{1875}$

17. $5\sqrt{72}$ **18.** $-4\sqrt{44}$ **19.** $\pm 7\sqrt{125}$ **20.** $10\sqrt{363}$

21. $3\sqrt{264}$ **22.** $9\sqrt{648}$ **23.** $8\sqrt{1200}$ **24.** $12\sqrt{1875}$

25. $-\sqrt[3]{54}$ **26.** $\sqrt[3]{-54}$ **27.** $\pm\sqrt[3]{-16}$ **28.** $\sqrt[3]{81}$

29. $\sqrt[3]{-81}$ **30.** $\sqrt[4]{81}$ **31.** $\sqrt[4]{(-3)^4}$ **32.** $\sqrt[5]{(-3)^5}$

33. $\sqrt{\dfrac{16}{4}}$ **34.** $\pm\sqrt{\dfrac{8}{49}}$ **35.** $\dfrac{3}{\sqrt{6}}$ **36.** $-6\sqrt{\dfrac{5}{6}}$

37. $\sqrt{\dfrac{18}{24}}$ **38.** $\sqrt{\dfrac{121}{8}}$ **39.** $3\sqrt{\dfrac{24}{9}}$ **40.** $\sqrt{3\dfrac{2}{5}}$

Simplify each radical in simplest radical form. Assume that all variables are nonnegative real numbers and that all denominators are nonzero.

41. $\sqrt{4a^2}$ **42.** $\sqrt{100x^4}$ **43.** $\pm\sqrt{28a^4}$ **44.** $\sqrt{x^2-8x+16}$

45. $\sqrt{a^2+2ab+b^2}$ **46.** $-\sqrt{32b^2}$ **47.** $\sqrt{18a^3}$ **48.** $\sqrt{24x^5}$

49. $\sqrt{24a^2b^4}$ **50.** $\sqrt{\dfrac{x^4y^5}{4z^2}}$ **51.** $\pm\sqrt{\dfrac{8a^3b^2}{c^4}}$ **52.** $\sqrt{80m^7n^5}$

53. $\sqrt[3]{a^6b^5}$ **54.** $\sqrt[4]{a^{12}b^8}$ **55.** $\sqrt[3]{-27x^3}$ **56.** $\sqrt[5]{-32x^5}$

Simplify each radical in simplest radical form.

57. $\sqrt{4a^2}$ **58.** $\sqrt{100x^4}$ **59.** $\pm\sqrt{28a^4}$ **60.** $\sqrt{x^2-8x+16}$

61. $\sqrt{a^2+2ab+b^2}$ **62.** $-\sqrt{32b^2}$ **63.** $\sqrt{18a^3}$ **64.** $\sqrt{24x^5}$

65. $\sqrt{24a^2b^4}$ **66.** $\sqrt{\dfrac{x^4y^5}{4z^2}}$ **67.** $\pm\sqrt{\dfrac{8a^3b^2}{c^4}}$ **68.** $\sqrt{80m^7n^5}$

69. $\sqrt[3]{a^6b^5}$ **70.** $\sqrt[4]{a^{12}b^8}$ **71.** $\sqrt[3]{-27x^3}$ **72.** $\sqrt[5]{-32x^5}$

73. $\sqrt[4]{4}$ **74.** $\sqrt[6]{216}$ **75.** $\sqrt[4]{x^6}$ **76.** $\sqrt[4]{u^6v^8}$

4-3 Simplifying Radical Expressions

To simplify radical expressions, we use the following steps:
1. Simplify each radical in simplest form.
2. Combine (add or subtract) with like radicands.
3. Use the **Rules of Radicals** to multiply or divide two radicals:

$$\sqrt[n]{a}\cdot\sqrt[n]{b}=\sqrt[n]{ab},\qquad \frac{\sqrt[n]{b}}{\sqrt[n]{a}}=\sqrt[n]{\frac{b}{a}}$$

4. Multiply binomials by using the distributive property or formulas:

$$(a+b)^2=a^2+2ab+b^2,\quad (a-b)^2=a^2-2ab+b^2,\quad (a+b)(a-b)=a^2-b^2$$

5. Rationalize the denominator. (No radicals are in the denominator.)

Examples (Assume that all variables are nonnegative real numbers.)

1. $4\sqrt{2}-\sqrt{2}=3\sqrt{2}$ 2. $6\sqrt{5}-4\sqrt{5}=2\sqrt{5}$ 3. $3\sqrt{7}-2\sqrt{11}+5\sqrt{7}=8\sqrt{7}-2\sqrt{11}$

4. $5\sqrt{3}+\sqrt{12}=5\sqrt{3}+2\sqrt{3}=7\sqrt{3}$ 5. $\sqrt{24}-\sqrt{54}=2\sqrt{6}-3\sqrt{6}=-\sqrt{6}$

6. $3\sqrt{18}+2\sqrt{8}-6\sqrt{50}=3\cdot3\sqrt{2}+2\cdot2\sqrt{2}-6\cdot5\sqrt{2}=9\sqrt{2}+4\sqrt{2}-30\sqrt{2}=-17\sqrt{2}$

7. $5\sqrt{7}+6\sqrt{3}-\sqrt{7}+2\sqrt{3}=(5\sqrt{7}-\sqrt{7})+(6\sqrt{3}+2\sqrt{3})=4\sqrt{7}+8\sqrt{3}$

8. $\sqrt{2}\cdot\sqrt{8}=\sqrt{2\cdot8}=\sqrt{16}=4$ 9. $\sqrt[3]{2}\cdot\sqrt[3]{16}=\sqrt[3]{2\cdot16}=\sqrt[3]{32}=2\sqrt[3]{4}$

10. $2\sqrt{5}\cdot6\sqrt{12}=2\cdot6\cdot\sqrt{5\cdot12}=12\sqrt{60}=12\cdot2\sqrt{15}=24\sqrt{15}$
11. $\sqrt{3}(\sqrt{4}+\sqrt{6})=\sqrt{3}\cdot\sqrt{4}+\sqrt{3}\cdot\sqrt{6}=\sqrt{12}+\sqrt{18}=2\sqrt{3}+3\sqrt{2}$
12. $(4+\sqrt{2})(5-\sqrt{2})=20-4\sqrt{2}+5\sqrt{2}-2=18+\sqrt{2}$
13. $(\sqrt{2}+\sqrt{3})^2=(\sqrt{2})^2+2\cdot\sqrt{2}\cdot\sqrt{3}+(\sqrt{3})^2=2+2\sqrt{6}+3=5+2\sqrt{6}$
14. $(\sqrt{2}-\sqrt{3})^2=(\sqrt{2})^2-2\cdot\sqrt{2}\cdot\sqrt{3}+(\sqrt{3})^2=2-2\sqrt{6}+3=5-2\sqrt{6}$
15. $(\sqrt{2}+\sqrt{3})(\sqrt{2}-\sqrt{3})=(\sqrt{2})^2-(\sqrt{3})^2=2-3=-1$
16. $\sqrt{\frac{5}{3}}\cdot\sqrt{\frac{2}{5}}=\sqrt{\frac{5}{3}\cdot\frac{2}{5}}=\frac{\sqrt{2}}{\sqrt{3}}\cdot\frac{\sqrt{3}}{\sqrt{3}}=\frac{\sqrt{6}}{3}$ 17. $\frac{\sqrt{15}}{\sqrt{6}}=\sqrt{\frac{15}{6}}=\sqrt{\frac{5}{2}}=\frac{\sqrt{5}}{\sqrt{2}}\cdot\frac{\sqrt{2}}{\sqrt{2}}=\frac{\sqrt{10}}{2}$
18. $\sqrt[3]{\frac{5}{3}}\cdot\sqrt[3]{\frac{2}{5}}=\sqrt[3]{\frac{5}{3}\cdot\frac{2}{5}}=\frac{\sqrt[3]{2}}{\sqrt[3]{3}}\cdot\frac{\sqrt[3]{9}}{\sqrt[3]{9}}=\frac{\sqrt[3]{18}}{\sqrt[3]{27}}=\frac{\sqrt[3]{18}}{3}$
19. $4\sqrt{x}+5\sqrt{x}=9\sqrt{x}$ 20. $4x\sqrt{3x}+7x\sqrt{3x}=(4x+7x)\sqrt{3x}=11x\sqrt{3x}$
21. $\sqrt{63x}-\sqrt{7x}=3\sqrt{7x}-\sqrt{7x}=2\sqrt{7x}$ 22. $\sqrt{2x}\cdot\sqrt{9x}=\sqrt{18x^2}=3\sqrt{2}\,x$
23. $\sqrt{ab^2}\cdot\sqrt{4a}=\sqrt{4a^2b^2}=2ab$ 24. $\sqrt{x}(2-\sqrt{x})=2\sqrt{x}-\sqrt{x^2}=2\sqrt{x}-x$
25. $\sqrt{\frac{b}{a}}\cdot\sqrt{\frac{8b}{a}}=\sqrt{\frac{b}{a}\cdot\frac{8b}{a}}=\sqrt{\frac{8b^2}{a^2}}=\frac{\sqrt{8b^2}}{\sqrt{a^2}}=\frac{2\sqrt{2}\,b}{a},\ a\neq0$

EXERCISES

Simplify each radical expression in simplest radical form.

1. $5\sqrt{3} + 2\sqrt{3}$

2. $5\sqrt{3} - 2\sqrt{3}$

3. $2\sqrt{3} - 5\sqrt{3}$

4. $2\sqrt{6} - \sqrt{6} + 7\sqrt{6}$

5. $9\sqrt{5} + 12\sqrt{5} - 4\sqrt{5}$

6. $-\sqrt{18} - \sqrt{8}$

7. $\sqrt{20} + \sqrt{45}$

8. $\sqrt[3]{4} + \sqrt[3]{32}$

9. $5\sqrt{32} - 7\sqrt{50}$

10. $\sqrt{150} + 2\sqrt{96}$

11. $3\sqrt{28} - 2\sqrt{7} + \sqrt{63}$

12. $\sqrt{11} + \sqrt{44} - \sqrt{13}$

13. $\sqrt{5} \cdot \sqrt{10}$

14. $\sqrt[3]{9} \cdot \sqrt[3]{6}$

15. $2\sqrt{5} \cdot 4\sqrt{8}$

16. $6\sqrt{2} \cdot 3\sqrt{18}$

17. $\sqrt{2} \cdot \sqrt{3} \cdot \sqrt{12}$

18. $\sqrt{6} \cdot \sqrt{6} \cdot \sqrt{24}$

19. $(4\sqrt{5})^2$

20. $(4\sqrt[3]{5})^3$

21. $\sqrt{5}(\sqrt{5} - \sqrt{3})$

22. $3\sqrt{2}(2\sqrt{3} + \sqrt{2})$

23. $(3 + \sqrt{2})(5 - \sqrt{2})$

24. $(6 - \sqrt{5})(7 - 2\sqrt{5})$

25. $(\sqrt{5} + \sqrt{4})^2$

26. $(\sqrt{5} - \sqrt{4})^2$

27. $(\sqrt{5} + \sqrt{4})(\sqrt{5} - \sqrt{4})$

28. $\sqrt{\dfrac{3}{2}} - \sqrt{\dfrac{2}{3}}$

29. $\sqrt{\dfrac{7}{11}} + \sqrt{\dfrac{11}{7}}$

30. $\sqrt{\dfrac{4}{5}} - \sqrt{\dfrac{2}{10}}$

31. $\sqrt{\dfrac{2}{5}} + \sqrt{10}$

32. $3\sqrt{6} - \sqrt{\dfrac{2}{3}}$

33. $\sqrt{\dfrac{8}{3}} \cdot \sqrt{\dfrac{5}{4}}$

34. $\sqrt{\dfrac{5}{4}} \cdot \sqrt{\dfrac{4}{5}}$

35. $\dfrac{\sqrt{15}}{\sqrt{10}}$

36. $\dfrac{\sqrt[3]{15}}{\sqrt[3]{10}}$

Simplify each radical expression in simplest radical form. Assume that all variables are nonnegative real numbers and that all denominators are nonzero.

37. $\sqrt{x} + \sqrt{x}$

38. $\sqrt{3x} + \sqrt{3x}$

39. $3\sqrt{x} + 4\sqrt{x}$

40. $5\sqrt{2x} - 2\sqrt{2x}$

41. $7\sqrt{5x} - 11\sqrt{5x}$

42. $\sqrt{3x} - \sqrt{75x}$

43. $3\sqrt{20x} + 4\sqrt{45x}$

44. $7x\sqrt{x} - 5x\sqrt{x}$

45. $\sqrt{2x^2} + 3\sqrt{8x^2}$

46. $2\sqrt{18x^3} - \sqrt{8x^3}$

47. $\sqrt{\dfrac{x^2}{4} + \dfrac{x^2}{16}}$

48. $\sqrt{\dfrac{a}{x}} - \sqrt{\dfrac{x}{a}}$

49. $\sqrt{3m} \cdot \sqrt{8m}$

50. $\sqrt{xy^3} \cdot \sqrt{xy^5}$

51. $(10\sqrt{ab^2})(-2\sqrt{a^3})$

52. $\sqrt{2a}(\sqrt{8a} - \sqrt{18a^3})$

53. $\sqrt{6x}(\sqrt{4x} + 2\sqrt{2x^2})$

54. $(3\sqrt{4x})^2$

55. $(4\sqrt{3x})^3$

56. $\sqrt[3]{56x} + \sqrt[3]{7x}$

57. $x\sqrt[4]{x^{11}} - \sqrt[4]{x^{15}}$

58. $\dfrac{\sqrt{x^2y}}{\sqrt{2xy}}$

59. $\dfrac{\sqrt{2ab}}{\sqrt{6b}}$

60. $\dfrac{\sqrt[3]{2ab}}{\sqrt[3]{6b}}$

4-4 Solving Equations involving Radicals

To find the solution of the equation $x^2 = 25$, we know that $x = 5$ and -5 are the solutions because $5^2 = 25$ and $(-5)^2 = 25$. The numbers, 5 and -5, are the solutions of $x^2 = 25$. We can write the solutions as $x = \pm 5$, which is read as "plus or minus 5", or "positive or negative 5".

Therefore, we solve the equation $x^2 = 25$ by the following steps:

         Example 1: Solve $x^2 = 25$.

                Solution: $x = \pm\sqrt{25} = \pm 5$.   Ans.

Consider next example to find the solution of the equation $x^2 = -9$. Since x^2 is always nonnegative, there is no real-number value of x to be the square root of -9.

         Example 2: Solve $x^2 = -9$.

                Solution: $x = \pm\sqrt{-9}$.   Ans: No real-number solution.

Rules: If $x^2 = a$, where $a \geq 0$ (positive or 0), it has the solutions: $x = \pm\sqrt{a}$.

      If $x^2 = a$, where $a < 0$ (negative), it has no real-number solution.

To solve an equation having a variable in its radicand, we isolate the radical on one side (usually on the left side) of the equation and then remove the radical through raising its power equal to the index of the radical. To solve a radical equation, we must check each solution in the original equation and eliminate the solutions which are **not permissible**.

Example 3: Solve $\sqrt{x} - 3 = 0$.

     Solution:

$$\sqrt{x} = 3 \quad \bigg| \quad \text{Check:}$$
$$(\sqrt{x})^2 = 3^2 \quad \bigg| \quad \sqrt{x} - 3$$
$$\therefore x = 9 \quad \bigg| \quad = \sqrt{9} - 3$$
$$\bigg| \quad = 3 - 3 = 0\ \checkmark$$

        Ans: $x = 9$.

Example 4: Solve $\sqrt{x} + 3 = 0$

     Solution:

$$\sqrt{x} = -3 \quad \bigg| \quad \text{Check:}$$
$$(\sqrt{x})^2 = (-3)^2 \quad \bigg| \quad \sqrt{x} + 3$$
$$\therefore x = 9 \quad \bigg| \quad = \sqrt{9} + 3$$
$$\bigg| \quad = 3 + 3 \neq 0$$
$$\bigg| \quad \text{(not permissible)}$$

     Ans: No real-number solution.

Example 5: Solve $\sqrt{x^2 + 12} - 2x = 0$.

     Solution:

$$\sqrt{x^2 + 12} - 2x = 0 \quad \bigg| \quad \text{Check: } x = 2$$
$$\sqrt{x^2 + 12} = 2x \quad \bigg| \quad \sqrt{2^2 + 12} - 2(2) = 4 - 4 = 0\ \checkmark$$
$$\text{Square both sides: } x^2 + 12 = 4x^2 \quad \bigg| \quad x = -2$$
$$3x^2 = 12 \quad \bigg| \quad \sqrt{(-2)^2 + 12} - 2(-2) = 4 + 4 \neq 0$$
$$x^2 = 4 \quad \bigg| \quad \text{(not permissible)}$$
$$\therefore x = \pm\sqrt{4} = \pm 2. \quad \bigg| \quad \text{Ans: } x = 2.$$

EXERCISES

Solve each equation.

1. $x^2 = 49$

2. $x^2 = 81$

3. $x^2 = 121$

4. $x^2 = 8$

5. $a^2 = 24$

6. $x^2 = 48$

7. $x^2 = -49$

8. $x^2 = -8$

9. $x^2 = -121$

10. $x^2 - 64 = 0$

11. $x^2 - 16 = 0$

12. $x^2 + 16 = 0$

13. $x^2 - 144 = 0$

14. $m^2 - 25 = 0$

15. $x^2 + 4 = 0$

16. $a^2 - 100 = 0$

17. $x^2 - 32 = 0$

18. $y^2 - 72 = 0$

19. $x^2 - 28 = 0$

20. $x^2 - 150 = 0$

21. $2x^2 - 50 = 0$

22. $3x^2 - 48 = 0$

23. $4x^2 - 9 = 0$

24. $9x^2 - 4 = 0$

25. $(x-2)^2 = 16$

26. $(x+4)^2 = 25$

27. $x^2 + 4x + 4 = 9$

28. $x^2 - 4x + 4 = 36$

29. $x^3 = 8$

30. $x^3 = -8$

31. $64x^3 - 27 = 0$

32. $x^4 = 16$

33. $81x^4 - 16 = 0$

Solve each radical equation.

34. $\sqrt{x} = 6$

35. $\sqrt{x} = 8$

36. $\sqrt{x} = -7$

37. $\sqrt{2x} = 4$

38. $\sqrt{3x} = 6$

39. $3\sqrt{x} = 12$

40. $5\sqrt{4x} = 20$

41. $\sqrt{a} + 7 = 9$

42. $\sqrt{x} - 9 = 0$

43. $\sqrt{x+4} - 9 = 0$

44. $\sqrt{x+1} = 3$

45. $\sqrt{2x+3} = 5$

46. $\sqrt{10-x} = 1$

47. $\sqrt{4x-3} = 5$

48. $\sqrt{6-p} = p$

49. $\sqrt{x^2 + 72} = 3x$

50. $\sqrt{x^2 + 24} = 2x$

51. $\sqrt{x+11} = x - 1$

52. $\sqrt{x^2 + 5x - 2} = x$

53. $\sqrt{x^2 - 8x} = 3$

54. $\sqrt{x^2 + 6x} = 4$

55. $\sqrt[3]{a} - 4 = 0$

56. $\sqrt[3]{a-4} = 4$

57. $\sqrt[4]{2a-4} = 2$

58. The square root of two times a number is 6. Find the number.

59. When a number is subtracted from the square root of 2 subtracted from three times the number, the result is 0. Find the number.

60. Find the length of one side of a square field whose area is $50 \, m^2$.

61. In a rectangle, the length is twice as long as the width. The area is 242 cm^2. Find its length and width.

62. The distance d in feet that an object falls in t seconds is given by the formula $d = 16t^2$. How long does it take an object falling from rest to travel 1,024 feet ?

4-5 The Pythagorean Theorem

Right angle: An angle with measure $90°$. It is indicated by a small square.

Right triangle: A triangle having one right (or $90°$) angle.

The hypotenuse: The side opposite the right angle in a triangle.

     The other two sides are the **legs**.

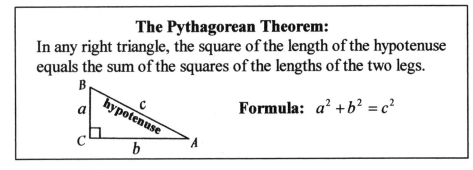

We can use the **Pythagorean Theorem** to find the length of one side of a right triangle when the lengths of the other two sides are known. To apply the Pythagorean Theorem, we need the knowledge of squares and square roots (see Section $4-1$).

To determine whether or not the triangle with the given lengths of its three sides is a right triangle, we apply the Pythagorean Theorem. If the sum of the squares of the lengths of the two shorter sides is equal to the square of the length of the longest side, then it is a right triangle.

Examples

  1. Find the length of the unknown side.

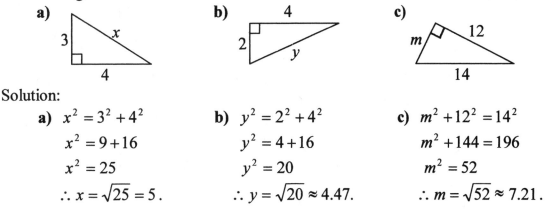

    Solution:

a)	b)	c)
$x^2 = 3^2 + 4^2$	$y^2 = 2^2 + 4^2$	$m^2 + 12^2 = 14^2$
$x^2 = 9 + 16$	$y^2 = 4 + 16$	$m^2 + 144 = 196$
$x^2 = 25$	$y^2 = 20$	$m^2 = 52$
$\therefore x = \sqrt{25} = 5$.	$\therefore y = \sqrt{20} \approx 4.47$.	$\therefore m = \sqrt{52} \approx 7.21$.

  2. Determine whether or not the triangle with the given lengths of its three sides is a right triangle.

      a) 6, 8, 10   **b)** 4, 5, $5\sqrt{2}$ **c)** 6, $6\sqrt{2}$, 6

    Solution:

a)	b)	c)
$6^2 + 8^2 = 100$	$4^2 + 5^2 = 41$	$6^2 + 6^2 = 72$
$10^2 = 100$	$(5\sqrt{2})^2 = 50$	$(6\sqrt{2})^2 = 72$
$6^2 + 8^2 = 10^2$ **Yes**	$4^2 + 5^2 \neq (5\sqrt{2})^2$ **No**	$6^2 + 6^2 = (6\sqrt{2})^2$ **Yes**

EXERCISES

Evaluate. Round to the nearest hundredth.

1. $\sqrt{49}$ 2. $\sqrt{64}$ 3. $\sqrt{121}$ 4. $\sqrt{18}$ 5. $\sqrt{24}$

6. $\sqrt{144}$ 7. $\sqrt{98}$ 8. $\sqrt{75}$ 9. $(2\sqrt{3})^2$ 10. $(3\sqrt{2})^2$

11. $(4\sqrt{5})^2$ 12. $(5\sqrt{2})^2$ 13. $\sqrt{8^2 + 6^2}$ 14. $\sqrt{10^2 - 6^2}$ 15. $\sqrt{(6\sqrt{2})^2 - 6^2}$

Find the length of the unknown side. Round to the nearest hundredth.

16.   17.   18.

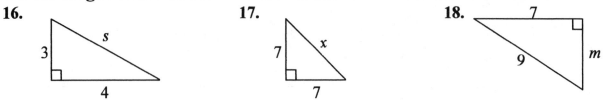

Find the length of the unknown side in the right triangle ABC. Round to the nearest hundredth.

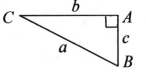

19. $a = 10$, $b = 5$ 20. $a = 12$, $b = 6$

21. $c = 9$, $a = 12$ 22. $b = 15$, $c = 8$

23. $a = 3$, $b = 1.5$ 24. $a = 4\sqrt{5}$, $b = 3\sqrt{3}$

Determine whether or not the triangle with the given lengths of its three sides is a right triangle.

25. 1, 1, 2   26. 1, 1, $\sqrt{2}$ 27. 1, 2, $\sqrt{3}$ 28. 2, 4, $\sqrt{6}$ 29. 5, 5, $5\sqrt{2}$

30. 15, 20, 25   31. 12, 16, 20   32. 20, 7, 25   33. $\sqrt{3}$, $\sqrt{4}$, $\sqrt{7}$ 34. $4\sqrt{3}$, 8, 4

35. A baseball field is shaped like a square diamond. Each side is 100 feet long. Find the distance from home plate to second base.

36. The length and the width of a rectangle field are 40 by 60 meters. Find the length of its diagonal.

37. The length of each leg of an isosceles right triangle is $8\,cm$. Find the length of the hypotenuse.

38. The length of each leg of an isosceles right triangle is s units. Find the length of the hypotenuse in terms of s. Write the answer in simplest radical form.

39. The sides of an equilateral triangle are each 6 inches long. Find the area.

40. The sides of an equilateral triangle are each s units long.
Find the altitude (h) in terms of s.
Write the answer in simplest radical form.

4-6 Distance and Midpoint Formulas

The distance between two points on a horizontal line or on a vertical line is the absolute value of the difference between them. The distance between two points, A and B, is indicated by $\overline{AB}$, which is the segment between A and B.

$\overline{AB} = |3 - (-4)| = |3 + 4| = |7| = 7$

Or: $\overline{AB} = |-4 - 3| = |-7| = 7$

To find the distance between two points not on a horizontal line or a vertical line, we apply the **Pythagorean Theorem**:

Distance Formula: $\overline{p_1 p_2} = \sqrt{(x_2 - x_1)^2 + (y_2 - y_1)^2}$

Midpoint Formula: $M = \left(\dfrac{x_1 + x_2}{2}, \dfrac{y_1 + y_2}{2} \right)$

Examples

1. Find the distance and midpoint between $A(-2, 1)$ and $B(5, 4)$.

Solution:

$$\overline{AB} = \sqrt{[5 - (-2)]^2 + (4 - 1)^2} = \sqrt{49 + 9} = \sqrt{58} \approx 7.62.$$

$$M = \left(\frac{-2 + 5}{2}, \frac{1 + 4}{2} \right) = (1.5,\ 2.5).$$

2. Find the distance and midpoint between $(4, 3\sqrt{3})$ and $(2, -\sqrt{3})$.

Solution:

$$d = \sqrt{(2 - 4)^2 + (-\sqrt{3} - 3\sqrt{3})^2} = \sqrt{(-2)^2 + (-4\sqrt{3})^2} = \sqrt{4 + 48} = \sqrt{52} \approx 7.21.$$

$$M = \left(\frac{4 + 2}{2}, \frac{3\sqrt{3} + (-\sqrt{3})}{2} \right) = (3, \sqrt{3}) = (3,\ 1.73).$$

3. If $M(1, 0)$ is the midpoint of the segment $\overline{AB}$ and $B(-1, -2)$ is the coordinates of point B. Find the coordinates of point A.

Solution: $A(x, y)$, $M(1, 0)$, $B(-1, -2)$

$$\frac{x + (-1)}{2} = 1, \quad x - 1 = 2, \quad \therefore x = 3$$

$$\frac{y + (-2)}{2} = 0, \quad y - 2 = 0, \quad \therefore y = 2$$

Ans: The coordinates of point A is $A(3, 2)$.

EXERCISES

Find the distance between the given two points to the nearest hundredth.

1. A(8, 0), B(2, 0)   **2.** C(−8, 0), D(2, 0)   **3.** M(−8, 0), N(−2, 0)

4. A(0, 6), C(0, −4)   **5.** E(0, −6), F(0, 4)   **6.** G(0, −6), H(0, −4)

7. P(8, 0), Q(0, 2)   **8.** S(−8, 0), T(0, −2)   **9.** P(0, −6), R(−3, 0)

10. A(3, 2), B(7, 8)   **11.** A(2, −3), C(8, 7)   **12.** P(−4, 1), Q(−1, 4)

13. (10, 5), (−7, 0)   **14.** (8, −2), (4, −6)   **15.** (−2, 1), (−5, −4)

16. (−7, −9), (2, −9)   **17.** (−7, 10), (−7, 6)   **18.** (4, −12), (−8, 1)

19. (−1, −1), (−9, −7)   **20.** (−4, −2), (−8, −5)   **21.** $(2, \frac{2}{3})$, (4, 1)

22. $(1, \frac{2}{3})$, $(4, -\frac{1}{2})$ **23.** $(2, \sqrt{3})$, $(6, -2\sqrt{3})$ **24.** $(2, -\sqrt{3})$, $(6, \sqrt{3})$

Find the midpoint between the given two points.

25. A(8, 0), B(2, 0)   **26.** C(−8, 0), D(2, 0)   **27.** E(0, −6), F(0, 4)

28. A(3, 2), B(7, 8)   **29.** P(−4, 1), Q(−1, 4)   **30.** A(2, −3), C(8, 7)

31. (10, 5), (−7, 0)   **32.** (−7, 10), (−7, 6)   **33.** (4, −12), (−8, 1)

34. (−4, −2), (−8, −5)   **35.** $(2, \frac{2}{3})$, (4, 1)   **36.** $(2, \sqrt{3})$, $(6, -2\sqrt{3})$

Given the coordinates of the midpoint (M) and one endpoint (P) of $\overline{PQ}$, find the coordinates of the other endpoint (Q).

37. $M(5, 0)$, $P(2, 0)$ **38.** $M(-3, 0)$, $P(2, 0)$ **39.** $M(5, 2)$, $P(2, -3)$

40. $M(0, -1)$, $P(0, -6)$ **41.** $M(5, 5)$, $P(3, 2)$ **42.** $M(-7, 8)$, $P(-7, 6)$

43. $M(\frac{3}{2}, \frac{5}{2})$, $P(10, 5)$ **44.** $M(3, \frac{5}{6})$, $P(2, \frac{2}{3})$ **45.** $M(4, -\frac{\sqrt{3}}{2})$, $P(2, \sqrt{3})$

Given the three vertices of a triangle, determine whether or not it is a right triangle.

46. A(4, 2), B(2, 2), C(4, 0)   **47.** P(3, 4), Q(3, 1), R(−1, 4)

48. A(−3, 2), B(0, 4), C(0, 2)   **49.** P(2, 3), Q(3, 1), R(4, 5)

50. A(1, 5), B(−1, 3), C(5, 2)   **51.** P(1, 1), Q(−2, −2), R(4, −1)

52. A(0, 5), B(−3, 2), C(2, 3)   **53.** P(0. 5), Q(−2, 3), R(2, 2)

4-7 Imaginary Units and Complex Numbers

To find the solutions of the equation $x^2 - 1 = 0$, we have the solutions $x = \pm\sqrt{1} = \pm 1$. However, there is no real number solution that satisfies the equation $x^2 + 1 = 0$. The solutions of the equation $x^2 + 1 = 0$ is $x = \pm\sqrt{-1}$. The radical $\sqrt{-1}$ is not a real number. There is no real number x whose square is -1. Therefore, we introduce a new mathematical symbol using the **imaginary unit**, denoted by i, which is defined as:

Definition of i: $i = \sqrt{-1}$ and $i^2 = -1$, where i is called "**the square root of** -1".

There is no real number square root of any negative number, such as:
$$\sqrt{-1}, \quad \sqrt{-2}, \quad \sqrt{-4}, \quad \sqrt{-9}, \quad \sqrt{-10}, \ldots\ldots \rightarrow \sqrt[n]{b} \text{ if } n \text{ is even and } b < 0.$$

Examples: 1. $\sqrt{-2} = \sqrt{2(-1)} = \sqrt{2} \cdot \sqrt{-1} = \sqrt{2}\,i$.   2. $\sqrt{-4} = \sqrt{4(-1)} = \sqrt{4} \cdot \sqrt{-1} = 2\,i$.
   3. $\sqrt{-9} = \sqrt{9(-1)} = \sqrt{9} \cdot \sqrt{-1} = 3\,i$.   4. $\sqrt[3]{-8} = \sqrt[3]{-2 \cdot -2 \cdot -2} = -2$.

$i^2 = -1$, $i^3 = i^2 \cdot i = -i$, $i^4 = i^2 \cdot i^2 = (-1)(-1) = 1$, $i^7 = i^4 \cdot i^3 = 1 \cdot -i = -i$.

i **and** $-i$ **are reciprocals:** $\dfrac{1}{i} = \dfrac{1}{i} \cdot \dfrac{i}{i} = \dfrac{i}{i^2} = \dfrac{i}{-1} = -i$, $\dfrac{1}{-i} = \dfrac{1}{-i} \cdot \dfrac{i}{i} = \dfrac{i}{-i^2} = \dfrac{i}{-(-1)} = i$.

Before we simplify radical expressions, we must eliminate negative radicands by i .
$$\sqrt{-2} \cdot \sqrt{-18} = \sqrt{2}\,i \cdot \sqrt{18}\,i = \sqrt{36}\,i^2 = 6(-1) = -6 \quad \text{is correct.}$$
$$\sqrt{-2} \cdot \sqrt{-18} = \sqrt{(-2)(-18)} = \sqrt{36} = 6 \quad \text{is incorrect.}$$
$$\sqrt{a} \cdot \sqrt{b} = \sqrt{ab} \quad \text{does not hold if } a \text{ and } b \text{ are negative.}$$

Complex Numbers: Numbers in the form $a + b\,i$ or $a - b\,i$, such as:
$$1 + 2\,i, \quad 1 - 2\,i, \quad 1 + \sqrt{2}\,i, \quad 1 - \sqrt{2}\,i, \quad 5 - 7\sqrt{3}\,i, \quad 5\sqrt{2} + 7\sqrt{3}\,i, \cdots\cdots.$$

Conjugates: Numbers in the form $a \pm b\,i$, such as: $1 \pm 2\,i,\ 1 \pm \sqrt{2}\,i,\ 5\sqrt{2} \pm 7\sqrt{3}\,i$ $\cdots\cdots$.
   The conjugate of $z = 2 - 5\,i$ is denoted by $\bar{z} = 2 + 5\,i$.

Rules: 1) To add, subtract or multiply two complex numbers, we treat them the same ways as ordinary algebraic expressions.
   2) To divide two complex numbers, we multiply both the numerator and the denominator by the conjugate of the denominator.
   3) The product of two conjugate complex numbers is always a nonnegative real number (no imaginary part). $z\bar{z} = a^2 + b^2$
$$(a + b\,i)(a - b\,i) = a^2 - (b\,i)^2 = a^2 - b^2\,i^2 = a^2 - b^2(-1) = a^2 + b^2 .$$

Examples

   5. $i^{30} = (i^2)^{15} = (-1)^{15} = -1$.   6. $(2 + 3\,i) - (4 - 5\,i) = 2 + 3\,i - 4 + 5\,i = -2 + 8\,i$.

   7. $(2 + 3\,i)(2 - 3\,i) = 2^2 - (3\,i)^2 = 4 - 9\,i^2 = 4 - 9(-1) = 4 + 9 = 13$.

   8. $\dfrac{2 + 3i}{2 - 3i} = \dfrac{2 + 3i}{2 - 3i} \cdot \dfrac{2 + 3i}{2 + 3i} = \dfrac{4 + 12i + 9i^2}{4 - 9i^2} = \dfrac{4 + 12i + 9(-1)}{4 - 9(-1)} = \dfrac{-5 + 12i}{13} = -\dfrac{5}{13} + \dfrac{12}{13}\,i$.

EXERCISES

Simplify each radical in simplest radical form.

1. $\sqrt{16}$ **2.** $\sqrt{8}$ **3.** $\sqrt{-16}$ **4.** $\sqrt{-8}$ **5.** $\sqrt{-18}$

6. $\pm\sqrt{-36}$ **7.** $\pm\sqrt{-10}$ **8.** $-\sqrt{-12}$ **9.** $-\sqrt{-24}$ **10.** $\sqrt{-121}$

11. $\sqrt{-75}$ **12.** $\sqrt{-1.21}$ **13.** $\sqrt{-1.44}$ **14.** $\sqrt{-\frac{1}{25}}$ **15.** $\sqrt{-\frac{25}{81}}$

16. i^8 **17.** i^9 **18.** i^{99} **19.** i^{100} **20.** i^{150}

Simplify each radical expression in the standard form $a+bi$.

21. $5\sqrt{-4}$ **22.** $4\sqrt{-5}$ **23.** $(i^2)(i^3)$ **24.** $(2i^4)(-5i^6)$

25. $2+\sqrt{-4}$ **26.** $2-\sqrt{-5}$ **27.** $4-\sqrt{-7}$ **28.** $5+\sqrt{-16}$

29. $\sqrt{-4}+\sqrt{-9}$ **30.** $\sqrt{-9}-\sqrt{-16}$ **31.** $\sqrt{-8}+\sqrt{-18}$ **32.** $\sqrt{-8}-\sqrt{-18}$

33. $\sqrt{-12}-\sqrt{-75}$ **34.** $5\sqrt{-32}+3\sqrt{-72}$ **35.** $2\sqrt{-20}-3\sqrt{-45}$

Complete each operation. Express all answers in the form $a+bi$.

36. $(3+6i)+(4+2i)$ **37.** $(7+8i)+(5-i)$ **38.** $(6-2i)-(5+4i)$

39. $(2-6i)-(4-5i)$ **40.** $\sqrt{-4}\cdot\sqrt{-16}$ **41.** $\sqrt{-9}\cdot\sqrt{-25}$

42. $\sqrt{-5}\cdot\sqrt{-8}$ **43.** $\sqrt{-7}\cdot\sqrt{-4}$ **44.** $2i(5-4i)$

45. $5i(3i-2)$ **46.** $(3+4i)^2$ **47.** $(3-4i)^2$

48. $(3+4i)(3-4i)$ **49.** $(5+3i)(4+2i)$ **50.** $(2-5i)(4+3i)$

51. $\dfrac{4+7i}{i}$ **52.** $\dfrac{7-4i}{i}$ **53.** $\dfrac{2-3i}{2-i}$

54. $\dfrac{2-3i}{2+3i}$ **55.** $\dfrac{4+5i}{3-2i}$ **56.** $\dfrac{2+\sqrt{3}i}{2-\sqrt{3}i}$

57. $5+i^3$ **58.** i^8-8 **59.** $2i^6-5$

60. $(2+i)^3$ **61.** $3i^4(2-i^2)$ **62.** $i^2-i^3+i^4-i^5$

63. If $z=2-5i$ and $w=3+4i$, find **a)** $\bar{z}$ **b)** $\bar{w}$ **c)** $\bar{\bar{z}}$ **d)** $\overline{z+w}$ **e)** $\bar{z}+\bar{w}$.

64. If $z=a-bi$, find **a)** $z+\bar{z}$ **b)** $z-\bar{z}$.

65. Use complex number to factor x^2+16.

CHAPTER 4   EXERCISES

Evaluate each radical. If the root is an irrational number, round the decimal to the nearest hundredth.

1. $\sqrt{9}$
2. $\sqrt{1}$
3. $\pm\sqrt{121}$
4. $\sqrt{-9}$

5. $\pm\sqrt{-121}$
6. $\sqrt{1600}$
7. $\sqrt{3.61}$
8. $\sqrt{1.21}$

9. $\pm\sqrt{0.0009}$
10. $\sqrt{0.0625}$
11. $\sqrt{15}$
12. $-\sqrt{20}$

13. $\sqrt{\dfrac{16}{9}}$
14. $\sqrt{\dfrac{1}{25}}$
15. $\sqrt{\dfrac{25}{81}}$
16. $\sqrt{\dfrac{20}{45}}$

17. $5\sqrt{25}$
18. $\pm 7\sqrt{12}$
19. $-12\sqrt{4.8}$
20. $8\sqrt{7.2}$

21. $\sqrt{\dfrac{3}{5}}$
22. $\sqrt{\dfrac{5}{3}}$
23. $\sqrt{\dfrac{8}{9}}$
24. $\sqrt{\dfrac{9}{8}}$

25. $\sqrt[3]{125}$
26. $\sqrt[4]{625}$
27. $\sqrt[3]{-125}$
28. $\sqrt[4]{-625}$

Simplify each radical in simplest radical form. Assume that all variables are nonnegative real numbers.

29. $\sqrt{20}$
30. $\sqrt{200}$
31. $\sqrt{75}$
32. $7\sqrt{72}$

33. $\sqrt{\dfrac{3}{5}}$
34. $\sqrt{\dfrac{5}{3}}$
35. $\sqrt{\dfrac{8}{9}}$
36. $\sqrt{\dfrac{9}{8}}$

37. $\sqrt[3]{250}$
38. $\sqrt[4]{405}$
39. $\sqrt[5]{64}$
40. $\sqrt[4]{64}$

41. $\sqrt{9x^2}$
42. $\sqrt{16x^4}$
43. $\sqrt{64x^3}$
44. $\sqrt{18x^3}$

45. $\sqrt{a^2-10a+25}$
46. $\sqrt{4a^2+4a+1}$
47. $\sqrt[3]{24x^5}$
48. $\sqrt[3]{-16n^3}$

49. $\sqrt[4]{32m^4n^5}$
50. $\sqrt[5]{-32m^4n^5}$
51. $\sqrt{\dfrac{x^2y^6}{9z^4}}$
52. $\sqrt{\dfrac{18a^4b}{4c^6}}$

Simplify each radical in simplest radical form.

53. $\sqrt{9x^2}$
54. $\sqrt{16x^4}$
55. $\sqrt{64x^3}$
56. $\sqrt{18x^3}$

57. $\sqrt{a^2-10a+25}$
58. $\sqrt{4a^2+4a+1}$
59. $\sqrt[3]{24x^5}$
60. $\sqrt[3]{-16n^3}$

61. $\sqrt[4]{32m^4n^5}$
62. $\sqrt[5]{-32m^4n^5}$
63. $\sqrt{\dfrac{x^2y^6}{9z^4}}$
64. $\sqrt{\dfrac{18a^4b}{4c^6}}$

-----Continued-----

Simplify each radical expression in simplest radical form. Assume that all variables are nonnegative real numbers.

65. $\sqrt{5} + 4\sqrt{5}$

66. $4\sqrt{6} - 7\sqrt{6} + 5\sqrt{6}$

67. $\sqrt{12} + \sqrt{48} - \sqrt{75}$

68. $\sqrt[3]{16} - \sqrt[3]{54}$

69. $\sqrt{5} \cdot \sqrt{6} \cdot \sqrt{3}$

70. $5\sqrt{10} \cdot 2\sqrt{15}$

71. $(3\sqrt{7})^2$

72. $(3\sqrt[3]{7})^3$

73. $3\sqrt{2}(\sqrt{8} - 8\sqrt{6})$

74. $(2 - \sqrt{5})(3 + \sqrt{5})$

75. $(2\sqrt{3} - \sqrt{2})^2$

76. $(2\sqrt{3} + \sqrt{2})(2\sqrt{3} - \sqrt{2})$

77. $\sqrt{\dfrac{3}{5}} + \sqrt{\dfrac{5}{3}}$

78. $\sqrt{\dfrac{5}{8}} \cdot \sqrt{\dfrac{3}{5}}$

79. $\dfrac{\sqrt[3]{15}}{\sqrt[3]{12}}$

80. $4\sqrt{5x} - 9\sqrt{5x}$

81. $\sqrt{80x} + \sqrt{45x}$

82. $\sqrt{6ab} \cdot \sqrt{12a^5 b}$

83. $\dfrac{\sqrt{2ab^2}}{\sqrt{8ab}}$

84. $\dfrac{\sqrt{5xy}}{\sqrt{2x}}$

85. $\dfrac{\sqrt[3]{5xy}}{\sqrt[3]{2x}}$

Solve each equation. Write the answer in simplest radical form.

86. $x^2 = 5$

87. $\sqrt{x} = 5$

88. $x^2 = -5$

89. $\sqrt{x} = -5$

90. $x^3 = 27$

91. $x^3 = -27$

92. $\sqrt{3x} - 1 = 5$

93. $9x^2 - 4 = 0$

94. $27x^3 + 8 = 0$

95. $\sqrt{5x - 4} = 4$

96. $(x + 4)^2 = 9$

97. $\sqrt{x + 2} = x - 4$

98. $\sqrt{4x^2 + 7x - 3} = 2x$

99. $\sqrt[3]{x} = 3$

100. $\sqrt[4]{3x} = 3$

101. The lengths of two sides of a right triangle are $12\,cm$ and $16\,cm$. Find the length of the hypotenuse.

102. The length of the hypotenuse of a right triangle is 18 feet. The length of one side is 12 feet. Find the length of the unknown side in simplest radical form.

103. A baseball field is shaped like a square diamond. Each side is 90 feet long. Find the length from home plate to second base.

104. Find the distance and midpoint between points $A(4, -3)$ and $B(1, 7)$.

105. Find the distance and midpoint between points $P(6, 4\sqrt{2})$ and $Q(9, 2\sqrt{2})$. Write the answer in simplest radical form.

106. If $M(3, -1)$ is the midpoint of the segment $\overline{AB}$ and $A(8, 12)$ is the coordinates of A. Find the coordinates of point B.

107. Determine whether or not the points $A(5, 5)$, $B(-1, -1)$, and $C(2, -4)$ are the vertices of a right triangle. Explain.

Complete each operation. Rationalize all answers if necessary.

108. $(5 - 7\,i) + (7 - 5\,i)$

109. $(6 - 12\,i) - (12 + 6\,i)$

110. $\sqrt{-81} \cdot \sqrt{-36}$

111. $\sqrt{-27} \cdot \sqrt{-48}$

112. $(5 - 4\,i)(4 - 5\,i)$

113. $i^8 - i^7 + i^3 - i^2$

114. $\dfrac{1}{\sqrt{x} - 2}$

115. $\dfrac{\sqrt{3} - 1}{\sqrt{3} + 1}$

116. $\dfrac{\sqrt{3} - \sqrt{2}}{\sqrt{3} + \sqrt{2}}$

117. $\dfrac{\sqrt{3} - i}{\sqrt{3} + i}$

118. $\dfrac{\sqrt{2} + i}{\sqrt{2} - i}$

Answers

CHAPTER 4 – 1 Square Roots and Radical   page 90

1. 4 **2.** 5 **3.** −6 **4.** 7 **5.** 8 **6.** −9 **7.** ±12 **8.** 100 **9.** −14 **10.** ±20 **11.** −30
12. 25 **13.** 28 **14.** 36 **15.** ±48 **16.** 105 **17.** 0.3 **18.** −0.4 **19.** −0.5 **20.** ±0.9
21. 1.2 **22.** 0.04 **23.** 0.05 **24.** 0.11 **25.** 0.12 **26.** ±0.13 **27.** 0.02 **28.** 0.006
29. $1\frac{1}{2}$ **30.** $-\frac{5}{7}$ **31.** $\pm\frac{1}{6}$ **32.** $\frac{9}{20}$ **33.** $\frac{10}{11}$ **34.** $\pm\frac{1}{2}$ **35.** $-\frac{2}{3}$ **36.** $\frac{1}{5}$
37. 3 **38.** 4 **39.** −2 **40.** 3 **41.** 2 **42.** ±3.162 **43.** 3.464 **44.** −6.325 **45.** 8.660
46. 5.196 **47.** −5.657 **48.** 11.180 **49.** ±0.1 **50.** 0.2 **51.** 0.346 **52.** ±18.028
53. 4 **54.** −4 **55.** 41.833 **56.** 31.177 **57.** 0.949 **58.** 1.871 **59.** ±55.426
60. 25.456 **61.** 0.775 **62.** ±1.354 **63.** −2.598 **64.** 2.789

CHAPTER 4 – 2 Simplifying Radicals   page 92

1. $2\sqrt{3}$ **2.** $3\sqrt{2}$ **3.** $-2\sqrt{6}$ **4.** $\pm3\sqrt{3}$ **5.** $4\sqrt{2}$ **6.** $2\sqrt{7}$ **7.** $4\sqrt{5}$ **8.** $4\sqrt{3}$ **9.** $5\sqrt{6}$
10. $\pm6\sqrt{3}$ **11.** $2\sqrt{30}$ **12.** $-9\sqrt{2}$ **13.** $6\sqrt{11}$ **14.** $30\sqrt{2}$ **15.** $40\sqrt{2}$ **16.** $25\sqrt{3}$
17. $30\sqrt{2}$ **18.** $-8\sqrt{11}$ **19.** $\pm35\sqrt{5}$ **20.** $110\sqrt{3}$ **21.** $6\sqrt{66}$ **22.** $162\sqrt{2}$ **23.** $160\sqrt{3}$
24. $300\sqrt{3}$ **25.** $-3\sqrt[3]{2}$ **26.** $-3\sqrt[3]{2}$ **27.** $\pm2\sqrt[3]{2}$ **28.** $3\sqrt[3]{3}$ **29.** $-3\sqrt[3]{3}$ **30.** 3 **31.** 3
32. −3 **33.** 2 **34.** $\pm\frac{2\sqrt{2}}{7}$ **35.** $\frac{\sqrt{6}}{2}$ **36.** $-\sqrt{30}$ **37.** $\frac{\sqrt{3}}{2}$ **38.** $\frac{11\sqrt{2}}{4}$ **39.** $2\sqrt{6}$ **40.** $\frac{\sqrt{85}}{5}$
41. $2a$ **42.** $10x^2$ **43.** $\pm2\sqrt{7}a^2$ **44.** $x-4$ **45.** $a+b$ **46.** $-4\sqrt{2}\,b$ **47.** $3a\sqrt{2a}$
48. $2x^2\sqrt{6x}$ **49.** $2\sqrt{6}\,ab^2$ **50.** $\frac{x^2y^2\sqrt{y}}{2z}$ **51.** $\pm\frac{2ab\sqrt{2a}}{c^2}$ **52.** $4m^3n^2\sqrt{5mn}$
53. $a^2b\sqrt[3]{b^2}$ **54.** a^3b^2 **55.** $-3x$ **56.** $-2x$
57. $2|a|$ **58.** $10x^2$ (Hint: x^2 is always nonnegative.)
59. $\pm2\sqrt{7}a^2$(Hint: a^2 is always nonnegative.) **60.** $|x-4|$ **61.** $|a+b|$ **62.** $-4\sqrt{2}|b|$
63. $3a\sqrt{2a}$ (Hint: a must be nonnegative.)
64. $2x^2\sqrt{6x}$ (Hint: x must be nonnegative.) **65.** $2\sqrt{6}\,|a|\,b^2$
66. $\frac{x^2y^2\sqrt{y}}{2|z|}$, $z\neq0$ (Hint: y must be nonnegative.)
67. $\pm\frac{2a|b|\sqrt{2a}}{c^2}$, $c\neq0$ (Hint: a must be nonnegative.)
68. $4m^3n^2\sqrt{5mn}$ (Hint: Both m and n must be nonnegative.)
69. $a^2b\sqrt[3]{b^2}$ (Hint: b could be negative.)
70. $\sqrt[4]{a^{12}b^8}=\sqrt[4]{a^{12}}\cdot\sqrt[4]{b^8}=\sqrt[4]{(a^3)^4}\cdot\sqrt[4]{(b^2)^4}=|a^3|\,b^2$ **71.** $-3x$ **72.** $-2x$
73. $\sqrt{2}$ **74.** $\sqrt{6}$ **75.** $|x|\sqrt{|x|}$ **76.** $|u|v^2\sqrt{|u|}$

CHAPTER 4 – 3 Simplifying Radical Expressions   page 94

1. $7\sqrt{3}$ **2.** $3\sqrt{3}$ **3.** $-3\sqrt{3}$ **4.** $8\sqrt{6}$ **5.** $17\sqrt{5}$ **6.** $-5\sqrt{2}$ **7.** $5\sqrt{5}$ **8.** $3\sqrt[3]{4}$
9. $-15\sqrt{2}$ **10.** $13\sqrt{6}$ **11.** $7\sqrt{7}$ **12.** $3\sqrt{11}-\sqrt{13}$ **13.** $5\sqrt{2}$ **14.** $3\sqrt[3]{2}$ **15.** $16\sqrt{10}$
16. 108 **17.** $6\sqrt{2}$ **18.** $12\sqrt{6}$ **19.** 80 **20.** 320 **21.** $5-\sqrt{15}$ **22.** $6\sqrt{6}+6$
23. $13+2\sqrt{2}$ **24.** $52-19\sqrt{5}$ **25.** $9+4\sqrt{5}$ **26.** $9-4\sqrt{5}$ **27.** 1
28. $\frac{\sqrt{6}}{6}$ **29.** $\frac{18\sqrt{77}}{77}$ **30.** $\frac{\sqrt{5}}{5}$ **31.** $\frac{6\sqrt{10}}{5}$ **32.** $\frac{8\sqrt{6}}{3}$ **33.** $\frac{\sqrt{30}}{3}$ **34.** 1 **35.** $\frac{\sqrt{6}}{2}$ **36.** $\frac{\sqrt[3]{12}}{2}$
37. $2\sqrt{x}$ **38.** $2\sqrt{3x}$ **39.** $7\sqrt{x}$ **40.** $3\sqrt{2x}$ **41.** $-4\sqrt{5x}$ **42.** $-4\sqrt{3x}$ **43.** $18\sqrt{5x}$
44. $2x\sqrt{x}$ **45.** $7x\sqrt{2}$ **46.** $4x\sqrt{2x}$ **47.** $\frac{\sqrt{5x}}{4}$ **48.** $\frac{(a-x)\sqrt{ax}}{ax}$ **49.** $2\sqrt{6}\,m$ **50.** xy^4
51. $-20a^2b$ **52.** $4a-6a^2$ **53.** $2\sqrt{6}\,x+4x\sqrt{3x}$ **54.** $36x$ **55.** $192x\sqrt{3x}$ **56.** $3\sqrt[3]{7x}$
57. 0 **58.** $\frac{\sqrt{2x}}{2}$ **59.** $\frac{\sqrt{3a}}{3}$ **60.** $\frac{\sqrt[3]{9a}}{3}$

CHAPTER 4 – 4   Solving Equations involving Radicals   page 96

1. $x = \pm 7$ **2.** $x = \pm 9$ **3.** $x = \pm 11$ **4.** $x = \pm 2\sqrt{2}$ **5.** $a = \pm 2\sqrt{6}$ **6.** $x = 4\sqrt{3}$

7. No real-number solution   **8.** No real-number solution   **9.** No real-number solution

10. $x = \pm 8$ **11.** $x = \pm 4$ **12.** No real-number solution   **13.** $x = \pm 12$ **14.** $m = \pm 5$

15. No real-number solution   **16.** $a = \pm 10$ **17.** $x = \pm 4\sqrt{2}$ **18.** $y = \pm 6\sqrt{2}$ **19.** $x = \pm 2\sqrt{7}$

20. $x = \pm 5\sqrt{6}$ **21.** $x = \pm 5$ **22.** $x = \pm 4$ **23.** $x = \pm 1\frac{1}{2}$ **24.** $x = \pm \frac{2}{3}$ **25.** $x = 6$ or -2

26. $x = 1$ or -9 **27.** $x = 1$ or -5 **28.** $x = 8$ or -4

29. $x = 2$ **30.** $x = -2$ **31.** $x = \frac{3}{4}$ **32.** $x = \pm 2$ **33.** $x = \pm \frac{2}{3}$

34. $x = 36$ **35.** $x = 64$ **36.** No real-number solution   **37.** $x = 8$ **38.** $x = 12$ **39.** $x = 16$

40. $x = 4$ **41.** $a = 4$ **42.** $x = 81$ **43.** $x = 77$ **44.** $x = 8$ **45.** $x = 11$ **46.** $x = 9$

47. $x = 7$ **48.** $p = 2$ **49.** $x = 3$ **50.** $x = 2\sqrt{2}$ **51.** $x = 5$ **52.** $x = \frac{2}{5}$ **53.** $x = 9$ or -1

54. $x = -8$ or 2 **55.** $a = 64$ **56.** $a = 68$ **57.** $a = 10$

58. 18   **59.** 2 or 1 (Hint: $\sqrt{3n-2} - n = 0$)   **60.** 7.07 m **61.** Length = $22\,cm$, width = $11\,cm$

62. 8 seconds

CHAPTER 4 – 5   The Pythagorean Theorem   page 98

1. 7   **2.** 8   **3.** 11   **4.** 4.24   **5.** 4.90   **6.** 12   **7.** 9.90   **8.** 8.66   **9.** 12   **10.** 18   **11.** 80

12. 50   **13.** 10   **14.** 8   **15.** 6   **16.** $s = 5$ **17.** $x = 9.90$ **18.** $m = 5.66$ **19.** $c = 8.66$

20. $c = 10.39$ **21.** $b = 7.94$ **22.** $a = 17$ **23.** $c = 2.60$ **24.** $c = 7.28$ **25.** No   **26.** Yes

27. Yes   **28.** No   **29.** Yes   **30.** Yes   **31.** Yes   **32.** No   **33.** Yes   **34.** Yes   **35.** 141.42 feet

36. 72.11 meters   **37.** 11.31 cm **38.** Hypotenuse = $\sqrt{2}\ s$ **39.** Area = 15.60 $in.^2$ **40.** $h = \frac{\sqrt{3}}{2}s$

CHAPTER 4 – 6   Distance and Midpoint Formulas   page 100

1. $\overline{AB} = 6$ **2.** $\overline{CD} = 10$ **3.** $\overline{MN} = 6$ **4.** $\overline{AC} = 10$ **5.** $\overline{EF} = 10$ **6.** $\overline{GH} = 2$ **7.** $\overline{PQ} \approx 8.25$

8. $\overline{ST} \approx 8.25$ **9.** $\overline{PR} \approx 6.71$ **10.** $\overline{AB} \approx 7.21$ **11.** $\overline{AC} \approx 11.66$ **12.** $\overline{PQ} \approx 4.24$ **13.** $d \approx 17.72$

14. $d \approx 5.66$ **15.** $d \approx 5.83$ **16.** $d = 9$ **17.** $d = 4$ **18.** $d \approx 17.69$ **19.** $d = 10$ **20.** $d = 5$

21. $d \approx 2.03$ **22.** $d \approx 3.22$ **23.** $d \approx 6.56$ **24.** $d \approx 5.29$ **25.** $M(5,0)$ **26.** $M(-3, 0)$

27. $M(0,-1)$ **28.** $M(5, 5)$ **29.** $M(-2.5, 2.5)$ **30.** $M(5, 2)$ **31.** $M(1.5, 2.5)$ **32.** $M(-7, 8)$

33. $M(-2, -5.5)$ **34.** $M(-6, -3.5)$ **35.** $M(3, \frac{5}{6})$ **36.** $M(4, -\frac{\sqrt{3}}{2})$ **37.** $Q(8, 0)$ **38.** $Q(-8,0)$

39. $Q(8,7)$ **40.** $Q(0, 4)$ **41.** $Q(7, 8)$ **42.** $Q(-7, 10)$ **43.** $Q(-7, 0)$ **44.** $Q(4, 1)$ **45.** $Q(6, -2\sqrt{3})$

46 ~ 53 Hint: Find the distance of each side, then apply the **Pythagorean Theorem**.

46. Yes   **47.** Yes   **48.** Yes   **49.** No   **50.** No   **51.** No   **52.** Yes   **53.** No

CHAPTER 4 – 7   Imaginary Unit and Complex Numbers   page 102

1. 4   **2.** 4   **3.** $4i$ **4.** $2\sqrt{2}\,i$ **5.** $3\sqrt{2}\,i$ **6.** $\pm 6i$ **7.** $\pm \sqrt{10}\,i$ **8.** $-2\sqrt{3}\,i$

9. $-2\sqrt{6}\,i$ **10.** $11i$ **11.** $5\sqrt{3}\,i$ **12.** $1.1i$ **13.** $1.2i$ **14.** $\frac{1}{5}i$ **15.** $\frac{5}{9}i$ **16.** 1

17. i **18.** $-i$ **19.** 1   **20.** -1 **21.** $10i$ **22.** $4\sqrt{5}\,i$ **23.** i **24.** 10   **25.** $2 + 2i$

26. $2 - \sqrt{5}\,i$ **27.** $4 - \sqrt{7}\,i$ **28.** $5 + 4i$ **29.** $5i$ **30.** $-i$ **31.** $5\sqrt{2}\,i$ **32.** $-\sqrt{2}\,i$

33. $-3\sqrt{3}\,i$ **34.** $38\sqrt{2}\,i$ **35.** $-5\sqrt{5}\,i$ **36.** $7 + 8i$ **37.** $12 + 7i$ **38.** $1 - 6i$

39. $-2 - i$ **40.** -8 **41.** -15 **42.** $-2\sqrt{10}$ **43.** $-2\sqrt{7}$ **44.** $8 + 10i$ **45.** $-15 - 10i$

46. $-7 + 24i$ **47.** $-7 - 24i$ **48.** 25   **49.** $14 + 22i$ **50.** $23 - 14i$ **51.** $7 - 4i$

52. $-4 - 7i$ **53.** $\frac{7}{5} - \frac{4}{5}i$ **54.** $-\frac{5}{13} - \frac{12}{13}i$ **55.** $\frac{22}{13} + \frac{7}{13}i$ **56.** $\frac{1}{7} + \frac{4\sqrt{3}}{7}i$ **57.** $5 - i$

58. -7 **59.** -7 **60.** $2 + 11i$ **61.** 9   **62.** 0

63. a) $\bar{z} = 2 + 5i$ **b)** $\bar{w} = 3 - 4i$ **c)** $\bar{\bar{z}} = 2 - 5i$ **d)** $\overline{z + w} = 5 + i$ **e)** $\bar{z} + \bar{w} = 5 + i$

   (Hint: It proves that $\overline{z + w} = \bar{z} + \bar{w}$.)

64. a) $2a$ **b)** $-2b\,i$ **65.** $x^2 + 16 = (x + 4i)(x - 4i)$

CHAPTER 4   EXERCISE   Radicals   page 103 ~ 104

1. 3   2. 1   3. ± 11 4. No real-number solution   5. No real-number solution   6. 40   7. 1.9

8. 1.1   9. ± 0.03 10. 0.25   11. 3.87   12. -4.47 13. $1\frac{1}{3}$ 14. $\frac{1}{5}$ 15. $\frac{5}{9}$ 16. $\frac{2}{3}$ 17. 25

18. ± 24.25 19. -26.29 20. 21.47   21. 0.77   22. 1.29   23. 0.94   24. 1.06   25. 5   26. 5

27. -5 28. No real-number solution   29. $2\sqrt{5}$ 30. $10\sqrt{2}$ 31. $5\sqrt{3}$ 32. $42\sqrt{2}$ 33. $\frac{\sqrt{15}}{5}$

34. $\frac{\sqrt{15}}{3}$ 35. $\frac{2\sqrt{2}}{3}$ 36. $\frac{3\sqrt{2}}{4}$ 37. $5\sqrt[3]{2}$ 38. $3\sqrt[4]{5}$ 39. $2\sqrt[5]{2}$ 40. $2\sqrt[4]{4}$ 41. $3x$ 42. $4x^2$

43. $8x\sqrt{x}$ 44. $3x\sqrt{2x}$ 45. $a-5$ 46. $2a+1$ 47. $2x\sqrt[3]{3x^2}$ 48. $-2\sqrt[3]{2}\,n$ 49. $2mn\sqrt[4]{2n}$

50. $-2n\sqrt[5]{m^4}$ 51. $\frac{xy^3}{3z^2}, z\neq 0$ 52. $\frac{3a^2\sqrt{2b}}{2c^3}, \neq 0$ 53. $3|x|$ 54. $4x^2$ 55. $8x\sqrt{x}$

56. $3x\sqrt{2x}$ 57. $|a-5|$ 58. $|2a+1|$ 59. $2x\sqrt[3]{3x^2}$ 60. $-2\sqrt[3]{2}\,n$ 61. $2|m|n\sqrt[4]{2n}$

62. $-2n\sqrt[5]{m^4}$ 63. $\frac{|xy^3|}{3z^2}, z\neq 0$ 64. $\frac{3a^2\sqrt{2b}}{2|c^3|}, c\neq 0$ 65. $5\sqrt{5}$ 66. $2\sqrt{6}$ 67. $\sqrt{3}$ 68. $-\sqrt[3]{2}$

69. $3\sqrt{10}$ 70. $50\sqrt{6}$ 71. 63   72. 189   73. $12-48\sqrt{3}$ 74. $1-\sqrt{5}$ 75. $14-4\sqrt{6}$ 76. 10

77. $\frac{8\sqrt{5}}{15}$ 78. $\frac{\sqrt{6}}{4}$ 79. $\frac{\sqrt[3]{10}}{2}$ 80. $-5\sqrt{5x}$ 81. $7\sqrt{5x}$ 82. $6\sqrt{2}\,a^3b$ 83. $\frac{\sqrt{b}}{2}, b\neq 0$

84. $\frac{\sqrt{10y}}{2}, x\neq 0$ 85. $\frac{\sqrt[3]{20y}}{2}$ 86. $x=\pm 5$ 87. $x=25$ 88. No real-number solution

89. No real-number solution   90. $x=3$ 91. $x=-3$ 92. $x=12$ 93. $x=\pm\frac{2}{3}$ 94. $x=-\frac{2}{3}$

95. $x=4$ 96. $x=-1$ or -7 97. $x=7$ 98. $x=\frac{3}{7}$ 99. $x=27$ 100. $x=3$ 101. $20\,cm$

102. $6\sqrt{5}$ feet   103. 127.28 feet   104. $\overline{AB}\approx 10.44$, M(2.5, 2)

105. $d=\sqrt{17}$, M(7.5, $3\sqrt{2}$)   106. B(-2, -14)   107. Yes   108. $12-12\,i$

109. $-6-18\,i$ 110. -54 111. -36 112. $-41\,i$ 113. 2   114. $\frac{\sqrt{x+2}}{x-4}$ 115. $2-\sqrt{3}$

116. $5-2\sqrt{6}$ 117. $\frac{1}{2}-\frac{\sqrt{3}}{2}i$ 118. $\frac{1}{3}+\frac{2\sqrt{2}}{3}i$

A kid said to his mother.

    Kid: Did you wear contact lenses this morning ?

Mother: No. Why do you ask ?

    Kid: I heard you yell to father this morning:

        "After 20 years of marriage, I have finally

        seen what kind of man you are !".

On graduation day, a high school graduate said to his math teacher.

Student: Thank you very much for giving me a "D" in your class.

        For your kindness, do you want me to do anything for you ?

Teacher: Yes. Please do me a favor.

        Just don't tell anyone that I was your math teacher.

Space for Taking Notes

Proofs of the Pythagorean Theorem

The Pythagorean Theorem is probable the most famous theorem in the history of mathematics and has been used for thousands of years. It can help us solve many real-life problems in sports, sciences, and architecture.

The Pythagorean Theorem was first proved by Pythagoras, a famous Greek philosopher and mathematician who lived about 550 B.C.

There are numerous ways to prove the theorem. Five major proofs are introduced here. Read the following examples and find a new proof on your own. Send your proof to us. Your proof will be included in this book in the next printing.

Example 1: A proof of the Pythagorean Theorem: $a^2 + b^2 = c^2$

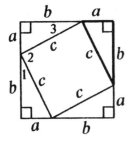

Draw a right triangle with legs a and b, and the hypotenuse c.
Draw a square with side c on the hypotenuse.
Construct a diagram shown at the left.
Since $<1 + <2 = <3 + 90°$ and $<1 = <3$, we have $<2 = 90°$.
The diagram shows a square with side $(a+b)$ units and a smaller square with side c units.
The diagram shows four equal right triangles. The area of each right triangle is $\frac{1}{2}ab$.

The area of the square with side $(a+b)$ units equals the sum of the area of the square with side c units plus the areas of the four right triangles.

$$(a+b)^2 = c^2 + 4(\tfrac{1}{2}ab)$$
$$a^2 + 2ab + b^2 = c^2 + 2ab$$
$$\therefore a^2 + b^2 = c^2 . \text{ The proof is complete.}$$

Example 2: A proof of the Pythagorean Theorem: $a^2 + b^2 = c^2$

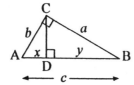

$\triangle ABC$ is a right triangle with legs a and b, and the hypotenuse c. CD is the altitude.

If two triangles are similar, the corresponding sides are proportional.

$\triangle ABC$ is similar to $\triangle BCD$, we have: $\dfrac{c}{a} = \dfrac{a}{y}$ $\therefore a^2 = cy$

$\triangle ABC$ is similar to $\triangle ACD$, we have: $\dfrac{c}{b} = \dfrac{b}{x}$ $\therefore b^2 = cx$

Therefore: $a^2 + b^2 = cy + cx$
$$a^2 + b^2 = c(y + x)$$
$$a^2 + b^2 = c \cdot c \qquad \therefore a^2 + b^2 = c^2 . \text{ The proof is complete.}$$

-----Continued-----

Example 3: A proof of the Pythagorean Theorem: $a^2 + b^2 = c^2$

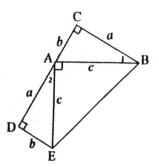

Draw a right triangle ABC with legs a and b, and the hypotenuse c.

Extend AC to point D so that AD = BC = a.

Draw segment DE so that DE $\perp$ CD and DE = AC = b.

Draw segment AE and BE.

Construct a diagram shown at the left.

Based on the **SAS Postulate**, we have $\triangle ABC \cong \triangle ADE$.

Therefore, we have AE = AB = c and $<1 = <2$.

Since $<2 + <BAE = <1 + 90°$, we have $<BAE = 90°$.

The area of the trapezoid BCDE equals the sum of the areas of of the three right triangles. $\frac{1}{2}(a+b)(a+b) = \frac{1}{2}ab + \frac{1}{2}ab + \frac{1}{2}c^2$

$$\frac{1}{2}(a^2 + 2ab + b^2) = ab + \frac{1}{2}c^2$$

$a^2 + \cancel{2ab} + b^2 = \cancel{2ab} + c^2$ $\therefore a^2 + b^2 = c^2$. The proof is complete.

Example 4: A proof of the Pythagorean Theorem: $a^2 + b^2 = c^2$

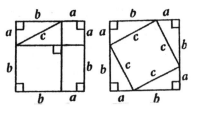

Draw a right triangle with legs a and b, and the hypotenuse c.

Construct two diagrams shown at the left.

Each diagram shows a square with $(a+b)$ units on one side.

The area of the diagram with two squares and two rectangles equals the area of the diagram with one square and four equal right triangles. $a^2 + b^2 + 2ab = c^2 + 4(\frac{1}{2}ab)$

$a^2 + b^2 + \cancel{2ab} = c^2 + \cancel{2ab}$ $\therefore a^2 + b^2 = c^2$. The proof is complete.

Example 5: A proof of the Pythagorean Theorem: $a^2 + b^2 = c^2$

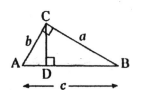

$\triangle ABC$ is a right triangle with legs a and b, and the hypotenuse c.
CD is the altitude.

If two triangles are similar, the ratio of their areas is the square of the ratio of any two corresponding sides. Therefore:

$\triangle ABC$ is similar to $\triangle BCD$, we have: $\dfrac{Area(\triangle BCD)}{Area(\triangle ABC)} = \left(\dfrac{a}{c}\right)^2$

$\triangle ABC$ is similar to $\triangle ACD$, we have: $\dfrac{Area(\triangle ACD)}{Area(\triangle ABC)} = \left(\dfrac{b}{c}\right)^2$

$$\frac{Area(\triangle BCD)}{Area(\triangle ABC)} + \frac{Area(\triangle ACD)}{Area(\triangle ABC)} = \frac{a^2}{c^2} + \frac{b^2}{c^2} \ , \quad \frac{Area(\triangle BCD) + Area(\triangle ACD)}{Area(\triangle ABC)} = \frac{a^2 + b^2}{c^2} \ ,$$

$$1 = \frac{a^2 + b^2}{c^2} \qquad \therefore a^2 + b^2 = c^2. \text{ The proof is complete.}$$

Space for Taking Notes

Space for Taking Notes

CHAPTER FIVE

Polynomial Equations and Functions

5-1 Laws of Exponents

Polynomial: An expression formed by two or more terms.

To simplify a polynomial, we need the formulas of exponents. We have learned the basic formulas of exponents (powers) in the previous chapters.

Now, we will learn more formulas about exponents.

Laws of Exponents (Rules of Powers)

If a and b are real numbers, m and n are positive integers, $a \neq 0$, $b \neq 0$, then:

1) $a^0 = 1$ 2) $a^m \cdot a^n = a^{m+n}$ 3) $\dfrac{a^m}{a^n} = a^{m-n}$ 4) $\dfrac{a^m}{a^n} = \dfrac{1}{a^{n-m}}$

5) $(a^m)^n = a^{mn}$ 6) $(ab)^m = a^m b^m$ 7) $a^{-m} = \dfrac{1}{a^m}$ 8) $\dfrac{1}{a^{-m}} = a^m$

9) $\left(\dfrac{b}{a}\right)^m = \dfrac{b^m}{a^m}$ 10) $\left(\dfrac{b}{a}\right)^{-m} = \left(\dfrac{a}{b}\right)^m$ 11) $a^{\frac{1}{m}} = \sqrt[m]{a}$ 12) $a^{\frac{n}{m}} = (\sqrt[m]{a})^n$

Examples

1. $x^7 \cdot x^3 = x^{7+3} = x^{10}$.

2. $\dfrac{x^7}{x^3} = x^{7-3} = x^4$; $\dfrac{x^7}{x^3} = \dfrac{1}{x^{3-7}} = \dfrac{1}{x^{-4}} = x^4$.

3. $\dfrac{x^3}{x^7} = \dfrac{1}{x^{7-3}} = \dfrac{1}{x^4}$; $\dfrac{x^3}{x^7} = x^{3-7} = x^{-4} = \dfrac{1}{x^4}$.

4. $\dfrac{x^3}{x^3} = 1$; $\dfrac{x^3}{x^3} = x^{3-3} = x^0 = 1$.

5. $a^{\frac{1}{2}} = \sqrt{a}$; $a^{\frac{1}{3}} = \sqrt[3]{a}$; $a^{\frac{1}{5}} = \sqrt[5]{a}$.

6. $1^0 = 1$, $2^0 = 1$, $10^0 = 1$, $x^0 = 1$ if $x \neq 0$.

7. $(a^3)^{-2} = a^{-6} = \dfrac{1}{a^6}$.

8. $\dfrac{ac^5}{a^3 c^2} = a^{1-3} c^{5-2} = a^{-2} c^3 = \dfrac{c^3}{a^2}$.

$\dfrac{ac^5}{a^3 c^2} = \dfrac{c^{5-2}}{a^{3-1}} = \dfrac{c^3}{a^2}$.

9. $\dfrac{35 x^3 y z^6}{55 x^5 y^4 z^2} = \dfrac{7 z^{6-2}}{11 x^{5-3} y^{4-1}} = \dfrac{7 z^4}{11 x^2 y^3}$.

10. $\dfrac{x^2}{y^2}\left(\dfrac{y^3}{x^2}\right)^4 = \dfrac{x^2}{y^2} \cdot \dfrac{(y^3)^4}{(x^2)^4} = \dfrac{x^2}{y^2} \cdot \dfrac{y^{12}}{x^8} = \dfrac{y^{10}}{x^6}$.

11. $25^{\frac{1}{2}} = \sqrt{25} = 5$. ; $32^{\frac{3}{5}} = (\sqrt[5]{32})^3 = 2^3 = 8$.

12. $\left(\dfrac{2}{3}\right)^{-2} = \left(\dfrac{3}{2}\right)^2 = \dfrac{9}{4} = 2\dfrac{1}{4}$.

13. $9^{2-\pi} \cdot 3^{2+2\pi} = (3^2)^{2-\pi} \cdot 3^{2+2\pi} = 3^{4-2\pi} \cdot 3^{2+2\pi}$

$= 3^{4-2\pi+2+2\pi} = 3^6 = 726$.

EXERCISES

Simplify each expression. Leave the product in exponent form.

1. $2^3 \cdot 2^4$ **2.** $(2^3)^4$ **3.** $a^3 \cdot a^4$ **4.** $(a^3)^4$ **5.** $m^{10} \cdot m^5$

6. $(m^{10})^5$ **7.** $a^n \cdot a^2$ **8.** $(a^n)^2$ **9.** $10^3 \cdot 10^5$ **10.** $(10^3)^5$

11. $x^6 \cdot x^7$ **12.** $-(a^3)^2$ **13.** $(-a^3)^2$ **14.** $a^8 \cdot a^{-3}$ **15.** $a^{-8} \cdot a^3$

16. $3^{10} \cdot 3^{-10}$ **17.** $x^a \cdot x^b$ **18.** $a^x \cdot a^y$ **19.** $(a^x)^y$ **20.** $a^{10} \cdot a^{-10}$ if $a \neq 0$

21. $(3a^2)^{-3}$ **22.** $a^{-3} \cdot a^{-3} \cdot a^{-6}$ **23.** $16^{\frac{1}{4}}$ **24.** $32^{\frac{2}{5}}$ **25.** $5^2 x^0$ if $x \neq 0$

26. $(2x^2)(3x^5)$ **27.** $(\frac{3}{4}a^5)(\frac{8}{3}a^2)$ **28.** $(-2a^2)^4$ **29.** $(3xy^2)(2y^3)^4$ **30.** $(\frac{2}{3}x^2)^2(\frac{3}{2}x^3)^4$

Simplify each expression. Assume that no denominator equals 0.

31. $\dfrac{2^8}{2^3}$ **32.** $\dfrac{2^3}{2^8}$ **33.** $\dfrac{a^8}{a^3}$ **34.** $\dfrac{a^3}{a^8}$ **35.** $\dfrac{10^8}{10^3}$

36. $\dfrac{2^{20}}{2^{17}}$ **37.** $\dfrac{5^{15}}{5^{12}}$ **38.** $\dfrac{2^{17}}{2^{20}}$ **39.** $\dfrac{5^{12}}{5^{15}}$ **40.** $\dfrac{x^5}{x^9}$

41. $\dfrac{x^9}{x^5}$ **42.** $\dfrac{x^2}{x^8}$ **43.** $\dfrac{a^6}{a^6}$ **44.** $\dfrac{x^5}{x^5}$ **45.** $\dfrac{x^{12}}{x^n}$

46. $\dfrac{x^n}{x^{12}}$ **47.** $\left(\dfrac{a}{b}\right)^4$ **48.** $\dfrac{x^2 \cdot x^n}{x^{2n}}$ **49.** $\left(\dfrac{a^2}{b^3}\right)^4$ **50.** $\left(\dfrac{x^n}{y^n}\right)^5$

51. $\dfrac{x^4 y^4}{x^4 y}$ **52.** $\dfrac{3a^7 b^2}{21a^4 b^5}$ **53.** $\dfrac{mn^4}{m^5 n^2}$ **54.** $\dfrac{12k^2}{-k^7}$ **55.** $\dfrac{-12x^3 y^3}{4xy}$

56. $\dfrac{(2a^2)^3}{2a}$ **57.** $\dfrac{(3x)^4}{9x^2}$ **58.** $\dfrac{2x^3}{(4x^2)^3}$ **59.** $\dfrac{4m}{(2m)^3}$ **60.** $\dfrac{5p}{15p}$

61. $\dfrac{-2m^5}{10m}$ **62.** $\dfrac{(-2a^2)^3}{(-2a)^2}$ **63.** $\dfrac{-(2x)^2}{(-2x^2)^2}$ **64.** $\dfrac{-24r^7}{-32r^9}$ **65.** $\dfrac{16x^5}{-2x^4}$

66. $\dfrac{2x^2 y^4}{(-xy)^2}$ **67.** $\dfrac{(4ab)^2}{4ab^2}$ **68.** $\dfrac{(2x^2)^4}{(2x^4)^2}$ **69.** $\dfrac{(-a)^{12}}{(-a)^9}$ **70.** $\dfrac{(-m^2)^5}{(-m^4)^3}$

71. $\left(\dfrac{2}{3}\right)^4$ **72.** $\left(\dfrac{2}{3}\right)^{-4}$ **73.** $\left(\dfrac{2a^2}{b^3}\right)^2$ **74.** $\left(\dfrac{xy}{x^3 y^2}\right)^3$ **75.** $\left(\dfrac{xy}{x^3 y^2}\right)^{-3}$

76. $\left(\dfrac{3ab^2}{2a^2 b}\right)^3$ **77.** $\dfrac{a^5 \cdot a^{-5}}{a^3}$ **78.** $\dfrac{a^5}{a^{-3}}$ **79.** $\dfrac{3x^{-4}}{6x^{-6}}$ **80.** $\dfrac{x^5 y^{-3}}{x^{-2} y^2}$

81. $(-x)^{-10}$ **82.** $\dfrac{2^{-2} s^{-1} t}{s^3 t^{-2}}$ **83.** $\left(-\dfrac{x^2}{y}\right)^{-3}$ **84.** $\dfrac{9^0 a^{-3} x}{2^{-1} a^{-1} x^{-2}}$ **85.** $\dfrac{x^{-2}}{2y^{-3}}\left(\dfrac{-2}{x^2 y^{-2}}\right)^{-3}$

86. $\sqrt{8} \cdot \sqrt[3]{8}$ **87.** $27^{-\frac{2}{3}} \cdot 9$ **88.** $9^{3\sqrt{2}} \div 3^{2\sqrt{2}}$ **89.** $9^{\pi+\sqrt{3}} \div 3^{\pi-\sqrt{3}}$ **90.** $10^{2+\pi} \div 5^{2+\pi}$

5-2 Solving Quadratic Equations

Steps for solving a quadratic equation:

1. Solve the equation in the form of perfect-square.

   If $x^2 = k$, then $x = \pm\sqrt{k}$.

2. Solve by factoring if it can be factored to the pattern.

   If $ax^2 + bx + c = 0$ can be factored, then $(px + r)(qx + s) = 0$.

3. Solve by using the quadratic formula if factoring is not possible.

   If $ax^2 + bx + c = 0$, then $x = \dfrac{-b \pm \sqrt{b^2 - 4ac}}{2a}$ **(Quadratic Formula)**

4. Some equations are not quadratic equations, but they can be written in quadratic form by factoring, or by making a substitution. (see Example 3, 5)

Examples

1. Solve $2x^2 - 5x - 3 = 0$ by factoring.
   Solution:

   $$(2x + 1)(x - 3) = 0$$

$2x + 1 = 0$	$x - 3 = 0$
$2x = -1$	$x = 3$
$x = -\frac{1}{2}$	

   Ans: $x = -\frac{1}{2}$ or 3.

2. Solve $2x^2 - 5x - 3 = 0$ by formula.
   Solution: $a = 2$, $b = -5$, $c = -3$

   $$x = \frac{-b \pm \sqrt{b^2 - 4ac}}{2a}$$

   $$= \frac{-(-5) \pm \sqrt{(-5)^2 - 4 \cdot 2 \cdot (-3)}}{2(2)}$$

   $$= \frac{5 \pm \sqrt{25 + 24}}{4} = \frac{5 \pm 7}{4}$$

   $$= \frac{5 + 7}{4} \text{ or } \frac{5 - 7}{4} = 3 \text{ or } -\frac{1}{2}. \text{ Ans.}$$

3. Solve $x^4 - 2x^2 - 8 = 0$.
   Solution: (By factoring)

   $$(x^2 - 4)(x^2 + 2) = 0$$

$x^2 - 4 = 0$	$x^2 + 2 = 0$
$x^2 = 4$	$x^2 = -2$
$x = \pm 2$	$x = \pm\sqrt{-2} = \pm\sqrt{2}\,i$

   Ans: $x = \pm 2$ or $\pm\sqrt{2}\,i$.

4. Solve $x^2 + 4x + 8 = 0$
   Solution: $a = 1$, $b = 4$, $c = 8$

   $$x = \frac{-b \pm \sqrt{b^2 - 4ac}}{2a}$$

   $$= \frac{-4 \pm \sqrt{4^2 - 4 \cdot 1 \cdot 8}}{2(1)} = \frac{-4 \pm \sqrt{16 - 32}}{2}$$

   $$= \frac{-4 \pm \sqrt{-16}}{2} = \frac{-4 \pm 4i}{2}$$

   $$= -2 \pm 2i. \text{ Ans.}$$

5. Solve $2x - 2\sqrt{2x} - 8 = 0$.
   Solution: (By making a substitution)
   Let $u = \sqrt{2x}$, $u^2 = 2x$

   $$u^2 - 2u - 8 = 0$$
   $$(u - 4)(u + 2) = 0$$

$u - 4 = 0$	$u + 2 = 0$
$u = 4$	$u = -2$
$\sqrt{2x} = 4$	$\sqrt{2x} = -2$
$x = 8$	no solution

   Ans: $x = 8$.

EXERCISES

Solve each quadratic equation.

1. $x^2 - 4 = 0$

2. $x^2 + 4 = 0$

3. $(x - 2)^2 = 4$

4. $(x + 2)^2 = 4$

5. $(x + 2)^2 = 6$

6. $(y - 4)^2 - 12 = 0$

7. $(t + 5)^2 - 18 = 0$

8. $(x + 2)^2 = -6$

9. $(y - 4)^2 + 12 = 0$

10. $3x^2 - x - 4 = 0$

11. $6x^2 - 3x - 4 = 0$

12. $12x^2 + 11x - 15 = 0$

13. $3x^2 - 7x + 4 = 0$

14. $9x^2 - 9x - 4 = 0$

15. $x^2 - 2x - 4 = 0$

16. $y^2 - 7y + 10 = 0$

17. $x^2 - 2x + 4 = 0$

18. $y^2 - 7y - 8 = 0$

19. $3x^2 + x - 4 = 0$

20. $9x^2 - 8x - 4 = 0$

21. $t^2 + 4t + 8 = 0$

22. $t^2 + 4t - 9 = 0$

23. $3x^2 + 2x + 1 = 0$

24. $3x^2 - 2x - 2 = 0$

25. $x^4 + 3x^2 - 10 = 0$

26. $a^4 - a^2 - 20 = 0$

27. $3x^3 - 5x^2 - 2x = 0$

28. $3x^3 - 2x^2 - 2x = 0$

29. $x^8 - 5x^4 + 4 = 0$

30. $x^8 - 5x^4 - 24 = 0$

31. $2v(v + 2) = 3(v + 4)$

32. $2v(v - 2) = 3(v - 4)$

33. $x^2 - 5 = 4(x - 1)$

34. $\sqrt{2}\,x^2 - 2x + 2\sqrt{2} = 0$

35. $\sqrt{3}\,x^2 - 3x - 2\sqrt{3} = 0$

36. $p^2 - \sqrt{2}\,p + 5 = 0$

37. $\dfrac{x^2}{2} + \dfrac{x}{4} = 3$

38. $\dfrac{x^2}{2} - 1 = \dfrac{x}{3}$

39. $\dfrac{x^2}{2} + \dfrac{3x}{4} = 1$

40. $\dfrac{1}{t+3} + \dfrac{2}{t-3} = 4$

41. $\dfrac{1}{t+3} - \dfrac{2}{t-3} = 4$

42. $\dfrac{1}{x-1} + \dfrac{2}{x+1} = 2$

43. $\dfrac{1}{x+3} - \dfrac{2}{x-3} = \dfrac{x^2}{x^2-9}$

44. $\dfrac{2}{x-1} - \dfrac{x}{x-2} = \dfrac{1}{x^2-3x+2}$

45. $\dfrac{1}{x} + \dfrac{2}{x^2} = \dfrac{3}{x^3}$

46. $2t - \sqrt{t} - 6 = 0$

47. $\sqrt{x} + 2x - 3 = 0$

48. $2x + 5\sqrt{x} - 3 = 0$

5-3 Graphing Quadratic Functions – Parabolas

Quadratic Equation: An equation of the form $ax^2 + bx + c = 0$ ($a \neq 0$).

Quadratic Function: A quadratic equation of the form $y = ax^2 + bx + c$ ($a \neq 0$).

The graph of a quadratic function $y = ax^2 + bx + c$ ($a \neq 0$) on the coordinate plane is a **parabola**. The central line of a parabola is called **the axis of symmetry**, or simply the **axis**. The point where the parabola crosses its axis is the **vertex**.

General Form of a parabola: $y - k = a(x - h)^2$. The vertex is (h, k).

Intercept Form of a parabola: $y = a(x - p)(x - q)$. The x – intercepts are p and q.

    The x – coordinate of the vertex is the midpoint between the two x – intercepts.

    If $a > 0$ (positive), the parabola opens upward. Vertex is a minimum point.

    If $a < 0$ (negative), the parabola opens downward. Vertex is a maximum point.

$y = x^2$ is the simplest quadratic function. Its vertex is located on the origin (0, 0).

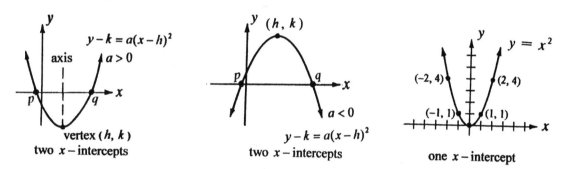

A parabola can have two, one, or no x – intercepts.

To find the x – intercepts of a parabola, we find the **zeros** (roots) of the related quadratic equation (let $y = 0$):

$$\text{Solve } ax^2 + bx + c = 0 \text{ by factoring, or by the formula: } x = \frac{-b \pm \sqrt{b^2 - 4ac}}{2a}.$$

If $b^2 - 4ac > 0$, the equation has two roots. The parabola has two x – intercepts.

If $b^2 - 4ac = 0$, the equation has one root. The parabola has one x – intercept.

If $b^2 - 4ac < 0$, the equation has no real-number root. The parabola has no x – intercept.

The value of $b^2 - 4ac$ indicates the differences among the three cases. Therefore, we call the value of $b^2 - 4ac$ "**the discriminant**" of the quadratic equation.

Steps for graphing a quadratic function: $y = ax^2 + bx + c$ ($a \neq 0$)

    1. Let $y = 0$. Find the zeros (roots). The roots are the x – intercepts.

    2. Find the vertex (h, k) by matching the general form $y - k = a(x - h)^2$.

    3. If $a > 0$, the graph opens upward. If $a < 0$, the graph opens downward.

The graph of a parabola is symmetrical. One half of the graph on one side of the axis is the **mirror image** of the other half. The x – coordinate of the vertex is $-\frac{b}{2a}$.

Examples

1. Graph $y = 2x^2$.

Solution:

One x – intercept $= 0$

Vertex $(0, 0)$

$a = 2 > 0$

Open upward

2. Graph $y = -2x^2$.

Solution:

One x – intercept $= 0$

Vertex $(0, 0)$

$a = -2 < 0$

Open downward

3. Graph $f(x) = x^2 - 2$.

Solution:

$x^2 - 2 = 0$, $x^2 = 2$

$x = \pm\sqrt{2} \approx \pm1.4$

$\therefore$ Two x – intercepts $= 1.4$ and -1.4

$y + 2 = x^2$, $y - (-2) = (x - 0)^2$

$\therefore$ Vertex $(0, -2)$

4. Graph $f(x) = x^2 - 4x$.

Solution:

$x^2 - 4x = 0$, $x(x - 4) = 0$

$x = 0$ or 4

$\therefore$ Two x – intercepts $= 0$ and 4

$y = (x^2 - 4x + 4) - 4$

$y + 4 = x^2 - 4x + 4$

$y - (-4) = (x - 2)^2$ $\therefore$ Vertex $(2, -4)$

5. Graph $f(x) = x^2 - 2x + 1$.

Solution:

$x^2 - 2x + 1 = 0$, $(x - 1)^2 = 0$

$x = 1$

$\therefore$ One x – intercept $= 1$

$y = x^2 - 2x + 1$

$y - 0 = (x - 1)^2$

$\therefore$ Vertex $(1, 0)$

6. Graph $f(x) = (x - 3)^2 + 1$

Solution:

$(x - 3)^2 + 1 = 0$, $(x - 3)^2 = -1$

No real-number root

$\therefore$ No x – intercept

$y = (x - 3)^2 + 1$

$y - 1 = (x - 3)^2$

$\therefore$ Vertex $(3, 1)$

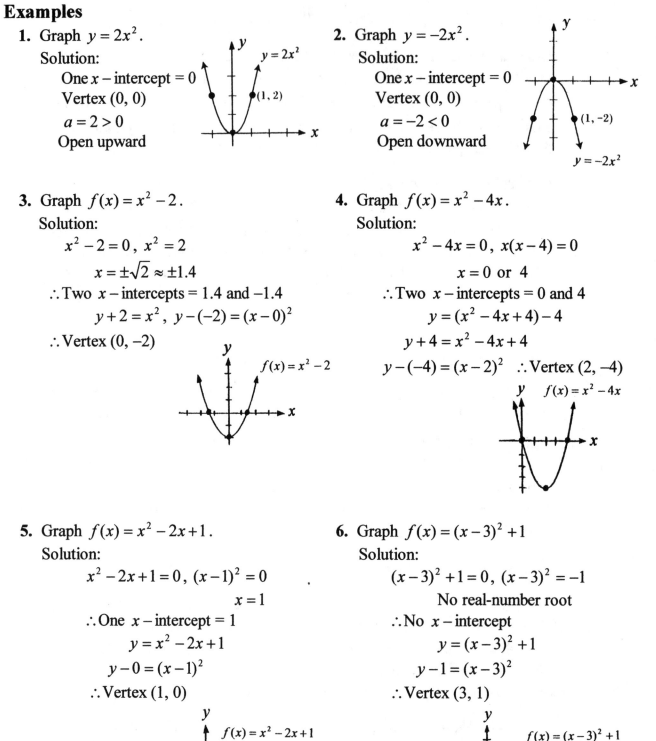

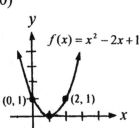

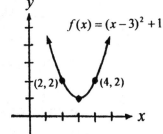

Graphing any Quadratic Function

To graph any quadratic function $f(x) = ax^2 + bx + c \; (a \neq 0)$, we need to rewrite it to match the general form of parabola by the method of **completing the square.**

General Form of Parabola: $f(x) - k = a(x-h)^2$

1) Vertex is (h, k) . **2)** If $a > 0$, open upward. **3)** If $a < 0$, open downward.

In science, we use the quadratic function to find the **extreme values** (maximum or minimum) of its application. There is a maximum value if it is open downward, a minimum value if it is open upward.

Examples

1. Graph $y = -x^2 - 2x + 1$.

Solution:

$$y = -x^2 - 2x + 1$$

$$y - 1 = -(x^2 + 2x)$$

Completing the square

$$y - 1 = -(x^2 + 2x + 1 - 1)$$

$$y - 1 = -(x^2 + 2x + 1) + 1$$

$$y - 1 = -(x+1)^2 + 1$$

$$\therefore \; y - 2 = -(x+1)^2 \text{ is the general form.}$$

Vertex $(-1 , 2)$, $a = -1 < 0$ open downward.

2. Graph $y = 2x^2 - 8x + 4$.

Solution:

$$y = 2x^2 - 8x + 4$$

$$y - 4 = 2x^2 - 8x$$

$$y - 4 = 2(x^2 - 4x)$$

$$y - 4 = 2(x^2 - 4x + 4 - 4)$$

$$y - 4 = 2(x^2 - 4x + 4) - 8$$

$$y + 4 = 2(x^2 - 4x + 4)$$

$$\therefore \; y + 4 = 2(x-2)^2 \text{ is the general form.}$$

Vertex $(2 , -4)$, $a = 2 > 0$ open upward.

3. Graph $f(x) = 4x^2 + 12x + 10$.

Solution:

$$f(x) = 4x^2 + 12x + 10$$

$$f(x) - 10 = 4x^2 + 12x$$

$$f(x) - 10 = 4(x^2 + 3x)$$

$$f(x) - 10 = 4(x^2 + 3x + \tfrac{9}{4} - \tfrac{9}{4})$$

$$f(x) - 10 = 4(x^2 + 3x + \tfrac{9}{4}) - 9$$

$$f(x) - 1 = 4(x^2 + 3x + \tfrac{9}{4})$$

$$\therefore \; f(x) - 1 = 4(x + \tfrac{3}{2})^2 \text{ is the general form.}$$

Vertex $(-\tfrac{3}{2}, 1)$, $a = 4 > 0$ open upward.

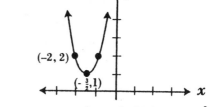

4. Find the extreme value of $f(x) = -2x^2 + 16x - 2$ and determine whether it is a maximum or a minimum.

Solution: $f(x) = -2x^2 + 16x - 2$

$$f(x) + 2 = -2(x^2 - 8x)$$

$$f(x) + 2 = -2(x^2 - 8x + 16 - 16)$$

$$f(x) + 2 = -2(x^2 - 8x + 16) + 32$$

$$f(x) - 30 = -2(x^2 - 8x + 16)$$

$$f(x) - 30 = -2(x - 4)^2 \text{ is the general form.}$$

Vertex $(4 , 30)$, $a = -2 < 0$ open downward.

$\therefore$ There is a maximum value of $f(x)$ when $x = 4$.

Ans: $f(x)_{\text{max.}} = 30$ at $x = 4$.

Examples

5. Find the extreme value of $f(x) = 4x^2 - 20x - 10$ and determine whether it is a maximum or a minimum.

Solution:

$$f(x) = 4x^2 - 20x - 10$$
$$f(x) + 10 = 4x^2 - 20x$$

Completing the square

$$f(x) + 10 = 4(x^2 - 5x)$$
$$f(x) + 10 = 4(x^2 - 5x + \tfrac{25}{4} - \tfrac{25}{4})$$
$$f(x) + 10 = 4(x^2 - 5x + \tfrac{25}{4}) - 25$$
$$f(x) + 35 = 4(x^2 - 5x + \tfrac{25}{4})$$
$$f(x) + 35 = 4(x - \tfrac{5}{2})^2$$

Vertex $(\tfrac{5}{2}, -35)$, $a = 4 > 0$ open upward.

$\therefore$ There is a minimum value of $f(x)$ when $x = \tfrac{5}{2}$.

Ans: $f(x)_{\min} = -35$ at $x = \tfrac{5}{2}$.

6. Find the dimension of the rectangle of greatest area whose perimeter is 24 cm.

Solution: Let $y \rightarrow$ the area.

$x \rightarrow$ the width.

$12 - x \rightarrow$ the length.

$12 - x$

$$y = (12 - x) \cdot x$$
$$y = 12x - x^2$$
$$y = -(x^2 - 12x)$$
$$y = -(x^2 - 12x + 36 - 36)$$
$$y = -(x^2 - 12x + 36) + 36$$
$$y - 36 = -(x^2 - 12x + 36)$$
$$y - 36 = -(x - 6)^2$$

Vertex $(6, 36)$, $a = -1$ open downward.
There is a maximum of y when $x = 6$.

Ans: Maximum area is $36\,cm^2$.
Length is $6\,cm$. Width is $6\,cm$.
(It is a square.)

7. Find the maximum profit if the equation of profit is $f(x) = -38 + 32x - 2x^2$, where $f(x)$ is the profit and x is the number of products sold.

Solution:

$$f(x) = -38 + 32x - 2x^2$$
$$f(x) + 38 = -2x^2 + 32x$$
$$f(x) + 38 = -2(x^2 - 16x)$$

Completing the square

$$f(x) + 38 = -2(x^2 - 16x + 64 - 64)$$
$$f(x) + 38 = -2(x^2 - 16x + 64) + 128$$
$$f(x) - 90 = -2(x^2 - 16x + 64)$$
$$f(x) - 90 = -2(x - 8)^2$$

Vertex $(8, 90)$, $a = -2 < 0$ open downward.

$\therefore$ There is a maximum profit of $f(x)$ when $x = 8$.

Ans: The maximum profit is $90 when the number of products sold are 8.

8. A rocket is fired upward with an initial speed of 160 feet per second. The height of the rocket is given by the formula $h(t) = vt - 16t^2$, where v is the initial speed and t is the time in seconds. What will be the maximum height of the rocket reaches ?

Solution:

$$h(t) = 160t - 16t^2$$
$$h(t) = -16(t^2 - 10t)$$
$$h(t) = -16(t^2 - 10t + 25 - 25)$$
$$h(t) = -16(t^2 - 10t + 25) + 400$$
$$h(t) - 400 = -16(t^2 - 10t + 25)$$
$$h(t) - 400 = -16(t - 5)^2$$

Vertex $(5, 400)$, $a = -16$ open downward.
There is a maximum of $h(t)$ when $t = 5$.

Ans: It reaches the maximum high 400 feet when $t = 5$ seconds.

A formula for finding the vertex of a parabola

We have learned how to find the vertex (maximum or minimum value) of a parabola by using the method of **completing the square**.

Here, a simple formula can be used to find the vertex of a parabola.

Vertex Formula: In a parabola $y = ax^2 + bx + c$, the x − coordinate of vertex occurs at:

$$x = -\frac{b}{2a}$$

Proof:

$$y = ax^2 + bx + c$$

$$y = a(x^2 + \frac{b}{a}x) + c$$

Completing the square:

$$y = a(x^2 + \frac{b}{a}x + \frac{b^2}{4a^2} - \frac{b^2}{4a^2}) + c$$

$$y = a(x^2 + \frac{b}{a}x + \frac{b^2}{4a^2}) - \frac{b^2}{4a} + c$$

$$y - (c - \frac{b^2}{4a}) = a(x^2 + \frac{b}{a}x + \frac{b^2}{4a^2})$$

$$y - (c - \frac{b^2}{4a}) = a(x + \frac{b}{2a})^2$$

We can see that the vertex occurs at $x = -\frac{b}{2a}$.

We have the following rules:

1. If $a > 0$, y has a minimum at $x = -\frac{b}{2a}$.

2. If $a < 0$, y has a maximum at $x = -\frac{b}{2a}$.

We can also prove it by finding the midpoint between the two x − intercepts.

Examples

1. Find the extreme value of $f(x) = -2x^2 + 16x - 2$ and determine whether it is a maximum or a minimum.

Solution: $a = -2$, $b = 16$

The vertex (x − coordinate) of the parabola occurs at $x = -\frac{b}{2a} = -\frac{16}{2(-2)} = 4$.

$a = -2 < 0$, $f(x)$ has a maximum value at $x = 4$.

$f(x)_{max.} = -2x^2 + 16x - 2 = -2(4)^2 + 16(4) - 2 = 30$. Ans.

(Hint: the vertex of the parabola is ($4, 30$).)

Examples

2. Find the extreme value of $f(x) = 4x^2 - 20x - 10$ and determine whether it is a maximum or a minimum.

   Solution: $a = 4$, $b = -20$

   The vertex (x – coordinate) of the parabola occurs at $x = -\frac{b}{2a} = -\frac{-20}{2(4)} = \frac{5}{2}$.

   $a = 4 > 0$, $f(x)$ has a minimum value at $x = \frac{5}{2}$.

   $f(x)_{min.} = 4x^2 - 20x - 10 = 4(\frac{5}{2})^2 - 20(\frac{5}{2}) - 10 = -35$.   Ans.

3. A rocket is fired upward with an initial speed of 160 feet per second. The height of the rocket is given by the formula $h(t) = vt - 16t^2$, where v is the initial speed and t is the time in seconds. What will be the maximum height of the rocket reaches ?

   Solution: $h(t) = 160t - 16t^2$

   $h(t) = -16t^2 + 160t$, $a = -16$, $b = 160$. It is a parabola that opens downward.

   The maximum value of h (height) occurs at the time $t = -\frac{b}{2a} = -\frac{160}{2(-16)} = 5$ seconds.

   The maximum height is: $h(5) = 160(5) - 16(5)^2 = 400$ feet.   Ans.

4. A ball is thrown upward from the top of a tower 96 feet high with an initial upward speed of 80 feet per second. The height of the ball is given by the formula $h = 96 + vt - 16t^2$, where v is the initial speed and t is the time in seconds. What is the maximum height of the ball reaches ?

   Solution: $h = 96 + vt - 16t^2$

   $h = 96 + 80t - 16t^2$, $h = -16t^2 + 80t + 96$, $a = -16$, $b = 80$.

   It is a parabola that opens downward.

   The maximum height of the ball reaches at the time $t = -\frac{b}{2a} = -\frac{80}{2(-16)} = 2.5$ seconds.

   The maximum height is $h_{max.} = -16(2.5)^2 + 80(2.5) + 96 = 196$ feet.   Ans.

5. A projectile is fired from the top of a tower 96 feet high at an initial speed 80 feet per second and at an angle of 45 degrees to the horizontal. The height h in feet of the projectile above the level ground is given by the formula:

$$h(x) = \frac{-32x^2}{v_0^2} + x + h_0 , \qquad v_0 : \text{initial speed.} \qquad h_0 : \text{initial height.}$$

   where x is the horizontal distance from the base of the tower. What is the maximum height reached by the projectile ? How far from the tower will it reach the ground ?

   Solution: $h(x) = \frac{-32x^2}{(80)^2} + x + 96 = -0.005x^2 + x + 96$.

   The maximum height is obtained at $x = -\frac{b}{2a} = -\frac{1}{2(-0.005)} = 100$.

   $h(x)_{max.} = -0.005x^2 + x + 96 = -0.005(100)^2 + 100 + 96 = 146$ feet.

   Let $h(x) = 0$, it reaches the ground at $x = \dfrac{-1 \pm \sqrt{1^2 - 4(-0.005)(96)}}{2(-0.005)} = 271$ feet.

EXERCISES

Determine whether the graph (parabola) of each quadratic function opens upward or downward.

1. $y = -5x^2$

2. $y = 5x^2$

3. $y = x^2 - 2$

4. $f(x) = x^2 + 2$

5. $f(x) = -2(x-3)(x+1)$

6. $f(x) = x^2 - 2x + 1$

7. $f(x) = 3(x+1)(x-2)$

8. $f(x) = -2(x-1)^2$

9. $f(x) = -12 + 3x^2$

Determine how many x – intercepts each quadratic function has by using the discriminant $b^2 - 4ac$.

10. $y = 2x^2 - 4x + 2$

11. $y = x^2 - 4x - 3$

12. $y = x^2 + 2x + 1$

13. $y = x^2 - 3x + 6$

14. $y = x^2 - 6x - 3$

15. $y = 2x^2$

16. $f(x) = -2x^2$

17. $f(x) = 2x^2 + 4$

18. $f(x) = 2x^2 - 4$

19. $f(x) = -2x^2 + 4x + 1$

20. $f(x) = -3x^2 + 6x - 5$

21. $f(x) = 3 - 6x + 3x^2$

Find the x – intercepts (zeros) of each quadratic function. Round answers to the nearest tenth when necessary.

22. $y = x^2 - x - 6$

23. $y = x^2 + x - 6$

24. $y = x^2 - 5x - 6$

25. $y = (x+6)(x-1)$

26. $y = (x-7)(x+4)$

27. $y = x^2 + 12x + 35$

28. $f(x) = 2x^2 - 4x + 2$

29. $f(x) = -2x^2 + 4x + 1$

30. $f(x) = 3(x-2)(x+2)$

31. $f(x) = -2x^2 - 5$

32. $f(x) = x^2 - 4x - 3$

33. $f(x) = 3x^2 - 5x - 1$

Find the vertex (h, k) of each parabola by using the form $f(x) - k = a(x-h)^2$.

34. $y - 5 = 2(x-7)^2$

35. $y - 1 = (x-4)^2$

36. $y + 4 = 3(x-2)^2$

37. $y + 1 = (x+4)^2$

38. $y + 5 = 2(x-7)^2$

39. $y - 4 = 3(x+2)^2$

40. $f(x) = 3(x-4)^2 - 2$

41. $f(x) = (x-5)^2 - 1$

42. $f(x) = 3(x+4)^2 - 2$

43. $f(x) = (x+5)^2 - 1$

44. $f(x) = (x+7)^2 + 9$

45. $f(x) = 5(x-6)^2 + 12$

46. $f(x) - 4 = -6(x+12)^2$

47. $f(x) = -(x-1)^2 - 7$

48. $f(x) = \frac{1}{2}(x-10)^2 - 9$

49. $f(x) = 3x^2 + 12$

50. $f(x) + 7 = -5x^2$

Find the vertex (h, k) of each parabola.

51. $y = x^2 + 4x + 4$

52. $y = x^2 - 4x + 4$

53. $y = x^2 + 4x + 9$

54. $f(x) = x^2 - 4x - 1$

55. $f(x) = x^2 - 6x + 7$

56. $f(x) = x^2 + 6x + 11$

57. $f(x) = x^2 - 2x - 4$

58. $f(x) = x^2 - 10x + 13$

59. $f(x) = x^2 + 14x + 58$

60. $f(x) = x^2 + 10x + 24$

61. $f(x) = 3x^2 - 12x + 8$

62. $f(x) = 2x^2 + 16x + 34$

-----Continued-----

Find the vertex and zeros (x – intercepts). Then graph each parabola.

63. $f(x) = \frac{1}{2}x^2$ **64.** $f(x) = -\frac{1}{2}x^2$ **65.** $f(x) = x^2 + 1$

66. $f(x) = -x^2 + 2$ **67.** $f(x) = (x-1)^2$ **68.** $f(x) = (x+1)^2$

69. $f(x) = 2(x-2)^2$ **70.** $f(x) = (x-3)^2 + 1$ **71.** $f(x) = 2(x-3)^2 + 1$

72. $f(x) = (x+3)(x+1)$ **73.** $f(x) = (x+3)(x-1)$ **74.** $f(x) = x^2 + 2x + 2$

75. Find k so that the equation $9x^2 + 30x + k = 0$ has one real-number root.

76. Find k so that the equation $5x^2 - 3x - k = 0$ has two real-number roots.

77. Find k so that the equation $4kx^2 + 2x - 5 = 0$ has no real-number root.

Compare each graph to the graph of $y = x^2$, describe the shift of the graph.

      78. $y = x^2 + 1$ **79.** $y = (x+1)^2$ **80.** $y = (x-2)^2 - 3$

81. Determine how many x – intercepts the quadratic function has. (Hint: $D = b^2 - 4ac$)

     a. $y = 4x^2 + 4x + 1$ **b.** $y = 4x^2 - 5x + 2$ **c.** $y = 2x^2 - 7x - 3$

82. Find the extreme value (maximum or minimum) of $y = 2x^2 - 8x - 1$.

83. Find the extreme value (maximum or minimum) of $y = -2x^2 + 8x + 10$.

84. Find the dimension of the rectangle of greatest area whose perimeter is $160\,cm$.

85. Find the maximum profit if the equation of profit is $p(x) = 2 + 8x - x^2$, where $p(x)$ is the profit and x is the number of products sold.

86. A projectile is fired vertically upward with an initial speed of 128 feet per second. The height of the projectile is given by $h(t) = vt - 16t^2$, where v is the initial speed and t is the time in seconds. What will be the maximum height of the rocket reaches ?

87. A rocket is fired vertically upward from the top of a tower 224 feet high with an initial upward speed of 208 feet per second. The height of the rocket is given by the formula $h(t) = 224 + vt - 16t^2$, where v is the initial speed and t is the time in seconds. What is the maximum height of the rocket reaches ?

88. A rocket is fired from the top of a tower 100 feet high with an initial speed of 208 feet per second and at an angle of 45 degree to the horizontal. The height h in feet of the rocket above the level ground is given by $h(x) = -0.00074x^2 + x + 100$, where x is the horizontal distance from the base of the tower. a) Find the maximum height of the rocket. b) How far (the x value) will the rocket reach the ground ?

89. An open box is to be made from a cardboard. The width of the cardboard is 5 inches less than the length by cutting 4-inch squares from each corner and folding up the sides. If the volume of the box is 336 cubic inches, find the dimensions of the original cardboard

90. A rectangular garden is to be enclosed by a 80 meters of fencing. One side of which will be against the side of a house. Find the dimensions of the garden to obtain a maximum area.

91. Two adjacent rectangular garden is to be enclosed by a 400 meters of fencing. Find the dimensions of the garden to obtain a maximum area.

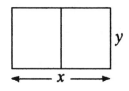

5-4 Solving Polynomial Equations by Factoring

Many polynomial equations can be solved by **factoring** and the **zero-product theory**.
Zero-Product Theory: For all real numbers a and b :
$$a \cdot b = 0 \text{ if and only if } a = 0 \text{ or } b = 0.$$
The zero-product theory tells us when an equation is written as a product of factors, we can find its solutions by setting each of the factors equal to 0 and solve for its roots.
Double Roots: If an equation has two identical factors, the equation is said to have
a double root (see example 4).
The following formulas of special patterns are useful in the process to factor polynomials:

Formulas:

1. $a^2 + 2ab + b^2 = (a+b)^2$	4. $a^3 + b^3 = (a+b)(a^2 - ab + b^2)$
2. $a^2 - 2ab + b^2 = (a-b)^2$	5. $a^3 - b^3 = (a-b)(a^2 + ab + b^2)$
3. $a^2 - b^2 = (a+b)(a-b)$	

An $x-$intercept of a function $y = f(x)$ is a point $(x, 0)$ at which the graph crosses the $x-axis$ ($y = 0$). To find the $x-$intercepts of a function, we find the **zeros** of the function $y = f(x)$ by letting $f(x) = 0$ (see example 1).

Examples

1. Find the zeros of $f(x) = x^2 + x - 2$.

   Solution: Let $f(x) = 0$, $x^2 + x - 2 = 0$
   $$(x+2)(x-1) = 0$$
   $x + 2 = 0 \mid x - 1 = 0$
   $x = -2 \mid x = 1$
   Ans: $x = -2$ and 1.

2. Solve $(2x - 4)(x + 5) = 0$.

   Solution: $(2x-4)(x+5) = 0$
   $2x - 4 = 0 \mid x + 5 = 0$
   $2x = 4 \mid x = -5$
   $x = 2 \mid$
   Ans: $x = 2$ and -5.

3. Solve $2x^2 + 3x = 9$.

   Solution: $2x^2 + 3x - 9 = 0$
   $$(2x-3)(x+3) = 0$$
   $2x - 3 = 0 \mid x + 3 = 0$
   $2x = 3 \mid x = -3$
   $x = 1\frac{1}{2} \mid$ Ans: $x = 1\frac{1}{2}$ and -3.

4. Solve $x^2 - 14x + 49 = 0$.

   Solution: $x^2 - 14x + 49 = 0$
   $$(x-7)(x-7) = 0$$
   $x - 7 = 0 \mid x - 7 = 0$
   $x = 7 \mid x = 7$
   Ans: $x = 7$ (**a double root**).

5. Solve $2a^3 - 6a^2 - 20a = 0$.

   Solution: $2a^3 - 6a^2 - 20a = 0$
   $$2a(a^2 - 3a - 10) = 0$$
   $$2a(a-5)(a+2) = 0$$
   $2a = 0 \mid a - 5 = 0 \mid a + 2 = 0$
   $a = 0 \mid a = 5 \mid a = -2$
   Ans: $a = 0$, 5, and -2.

6. Solve $x^4 - 8x = 0$.

   Solution: $x^4 - 8x = 0$, $x(x^3 - 8) = 0$
   $$x(x-2)(x^2 + 2x + 4) = 0$$
   $x = 0$, $x = 2$, $x^2 + 2x + 4 = 0$
   $$x = \frac{-b \pm \sqrt{b^2 - 4ac}}{2a} = \frac{-2 \pm \sqrt{2^2 - 4 \cdot 4}}{2(1)} = -1 \pm \sqrt{3}\, i$$
   Ans: $x = 0, 2$, and $-1 \pm \sqrt{3}\, i$.

EXERCISES

Find the zeros of each function $y = f(x)$. (Hint: Let $y = 0$)

1. $y = (x+2)(x-1)$

2. $y = (x-10)(x+8)$

3. $y = (x-4)(x-4)$

4. $y = (x-12)(x-12)$

5. $y = (2x+4)(3x-3)$

6. $y = (2x-4)(3x+3)$

7. $y = (3x+3)(3x-3)$

8. $y = (4x-2)(3x-6)$

9. $y = 2x(x-15)$

10. $y = x(2x-10)(3x+8)$

Solve by factoring. Factor out the greatest monomial factor first.

11. $2x^2 + 3x = 0$

12. $2x^2 - 3x = 0$

13. $-4x^2 + 16x = 0$

14. $-12x^2 + 8x = 0$

15. $x^2 - 7x = 0$

16. $35x^2 - 5x = 0$

17. $5x^2 - x^3 = 0$

18. $10a^2 - 20a^3 = 0$

19. $x^2 = 5x$

20. $x^3 = 5x^2$

21. $y^2 = 12y$

22. $4x^2 - 2x^3 = 0$

23. $p^2 = p$

24. $n^3 - n = 0$

25. $10x^2 - 100x = 0$

Solve.

26. $(a-5)(a+7) = 0$

27. $(p+5)(p-5) = 0$

28. $(x-9)(x-9) = 0$

29. $(n+2)(n+12) = 0$

30. $(2x+4)(x-6) = 0$

31. $(3x+6)(4x-2) = 0$

32. $x(x-2)(x+4) = 0$

33. $3y(2y+1)(y-4) = 0$

34. $2x(3x-6)(6x+3) = 0$

35. $x^2 + 3x + 2 = 0$

36. $x^2 - 3x + 2 = 0$

37. $n^2 - 5n - 6 = 0$

38. $x^2 + 16x + 64 = 0$

39. $x^2 - 4x + 4 = 0$

40. $a^2 - 10a + 16 = 0$

41. $2x^2 + 5x + 3 = 0$

42. $2x^2 + 5x - 12 = 0$

43. $4x^2 - 8x + 3 = 0$

44. $6x^2 - 13x - 5 = 0$

45. $4x^2 - 17x - 15 = 0$

46. $6x^2 - 13x + 2 = 0$

47. $a^2 - 9 = 0$

48. $4a^2 - 9 = 0$

49. $2n^3 - 2n = 0$

50. $x^2 + 16 = 8x$

51. $8a^2 - 1 = 7a$

52. $6y^2 + y = 2$

53. $x^3 - x^2 - 6x = 0$

54. $4x^3 + 10x^2 - 6x = 0$

55. $y^4 - 8y^3 + 16y^2 = 0$

56. $p^5 - 13p^3 + 36p = 0$

57. $(x+2)(x-3) = 14$

58. $(x+6)(x-2) = -7$

59. $x(x-3) = 2(x+7)$

60. $3x(x-1) = 2(x+1)$

Solve polynomial equations.

61. $x^3 - 4x = 0$

62. $x^3 + 4x = 0$

63. $x^4 - 3x^2 + 2 = 0$

64. $x^4 + 8x^2 + 15 = 0$

65. $x^4 - 2x^2 - 8 = 0$

66. $x^5 - 16x = 0$

67. $x^4 + x^2 = 0$

68. $x^4 + 2x^3 - x^2 = 0$

69. $x^4 + 2x^3 + x^2 = 0$

70. $x^5 + x^3 - 12x = 0$

71. $x^4 + 3x^3 + 2x^2 = 0$

72. $x^4 - 3x^3 + 2x^2 = 0$

73. $(x-3)^2 + 14x + 7 = 0$

74. $x^4 + 8x = 0$

75. $x^4 - 27x = 0$

76. $x^5 - 8x^2 = 0$

77. $x^4 - 1 = 0$

78. $x^6 - 1 = 0$

5-5 Synthetic Division and Factor Theorem

Dividing a polynomial in x by a binomial of the form $x - c$ is very much like dividing real numbers. When we divide a polynomial by $x - c$, we must arrange the terms of the polynomial in order of decreasing degree of the variable. Keep space (using 0) on the missing terms in degrees of x.

The common process of dividing a polynomial by $x - c$ is shown in Example 1. In this section we use the easier method called **Synthetic Division** to get the same answer.

Synthetic Division (Horner's Algorithm):

Dividing a polynomial by $x - c$, we proceed only the coefficients of the terms of the polynomial.

1) If the divisor is $x + c$, rewrite it as $x - (-c)$.

2) If the divisor is $ax - b$, rewrite it as $a(x - \dfrac{b}{a})$, use the divisor $x - \dfrac{b}{a}$

to proceed the synthetic division. Then, multiply the result by $\frac{1}{a}$.

Examples

1. Divide $\dfrac{2x^3 - 8x^2 - 4x - 6}{x - 3}$.

Solution: (by common division)

$$
\begin{array}{r}
2x^2 - 2x\ -10 \\
x-3{\overline{\smash{\big)}\,2x^3 - 8x^2 - 4x - 6}} \\
\underline{-)\ 2x^3 - 6x^2} \\
-2x^2 - 4x - 6 \\
\underline{-)\ -2x^2 + 6x} \\
-10x - 6 \\
\underline{-)\ -10x + 30} \\
-36
\end{array}
$$

(Remainder)

Ans: $\dfrac{2x^3 - 8x^2 - 4x - 6}{x - 3}$

$= 2x^2 - 2x - 10 - \dfrac{36}{x-3}$.

2. Divide $\dfrac{2x^3 - 8x^2 - 4x - 6}{x - 3}$.

Solution: (by synthetic division)

$$
\begin{array}{r}
\underline{3|}\quad 2 - 8 -\ 4 -\ 6 \\
+)\ \downarrow\ \ 6 -\ 6 - 30 \\
\overline{2 - 2 - 10,\ -36}
\end{array}
$$

└ Remainder

Ans: $\dfrac{2x^3 - 8x^2 - 4x - 6}{x - 3} = 2x^2 - 2x - 10 - \dfrac{36}{x-3}$.

3. Divide $\dfrac{1 + x + 12x^3}{1 + 2x}$.

Solution:

$$
\begin{array}{r}
6x^2 - 3x + 2 \\
2x+1{\overline{\smash{\big)}\,12x^3 + 0\ +\ x + 1}} \\
\underline{-)\ 12x^3 + 6x^2} \\
-6x^2 +\ x \\
\underline{-)\ -6x^2 - 3x} \\
4x + 1 \\
\underline{-)\ 4x + 2} \\
-1\ \leftarrow \text{Remainder}
\end{array}
$$

Ans: $\dfrac{1 + x + 12x^3}{1 + 2x} = 6x^2 - 3x + 2 - \dfrac{1}{2x+1}$.

4. Divide $\dfrac{x^3 + 4x^2 + 6x + 2}{x + 3}$ by synthetic division.

Solution:

$$
\underline{-3\rvert}\quad
\begin{array}{r}
1 + 4 + 6 + 2 \\
+)\;\; -3 - 3 - 9 \\
\hline
1 + 1 + 3, -7 \\
\end{array}
$$

$\underset{\quad\text{Remainder}}{\uparrow}$

Ans: $\dfrac{x^3 + 4x^2 + 6x + 2}{x + 3}$

$= x^2 + x + 3 - \dfrac{7}{x + 3}$.

5. Divide $\dfrac{2x^4 - x^3 + 3x - 5}{x + 3}$ by synthetic division.

Solution:

$$
\underline{-3\rvert}\quad
\begin{array}{r}
2 - 1 + 0 + 3 - 5 \\
+)\;\; -6 + 21 - 63 + 180 \\
\hline
2 - 7 + 21 - 60, + 175 \\
\end{array}
$$

$\underset{\quad\text{Remainder}}{\uparrow}$

Ans: $\dfrac{2x^4 - x^3 + 3x - 5}{x + 3}$

$= 2x^3 - 7x^2 + 21x - 60 + \dfrac{175}{x + 3}$.

Roger was absent yesterday.
Teacher: Roger, why were you absent ?
  Roger: My tooth was aching.
         I went to see my dentist.
Teacher: Is your tooth still aching ?
  Roger: I don't know. My dentist has it.

6. Divide $\dfrac{x^4 + a^4}{x + a}$ by synthetic division.

Solution:

$$
\underline{-a\rvert}\quad
\begin{array}{r}
1 + 0 + 0 + 0 + a^4 \\
+)\;\; -a + a^2 - a^3 + a^4 \\
\hline
1 - a + a^2 - a^3, +2a^4 \\
\end{array}
$$

$\underset{\quad\text{Remainder}}{\uparrow}$

Ans: $\dfrac{x^4 + a^4}{x + a}$

$= x^3 - ax^2 + a^2 x - a^3 + \dfrac{2a^4}{x + a}$.

7. Divide $\dfrac{8x^3 + 2x^2 - 51}{4x - 7}$ by synthetic division.

Solution:

$$\frac{8x^3 + 2x^2 - 51}{4x - 7} = \frac{1}{4} \cdot \frac{8x^3 + 2x^2 - 51}{x - \frac{7}{4}}$$

$$
\underline{\tfrac{7}{4}\rvert}\quad
\begin{array}{r}
8 + 2 + 0 - 51 \\
+)\;\; + 14 + 28 + 49 \\
\hline
8 + 16 + 28, -2 \\
\end{array}
$$

$\underset{\quad\text{Remainder}}{\uparrow}$

$\therefore \dfrac{8x^3 + 2x^2 - 51}{4x - 7}$

$= \dfrac{1}{4}\left(8x^2 + 16x + 28 - \dfrac{2}{x - \frac{7}{4}}\right)$

$= \dfrac{1}{4}\left(8x^2 + 16x + 28 - \dfrac{2}{\frac{4x-7}{4}}\right)$

$= \dfrac{1}{4}\left(8x^2 + 16x + 28 - \dfrac{8}{4x - 7}\right)$

$= 2x^2 + 4x + 7 - \dfrac{2}{4x - 7}$. Ans.

Remainder and Factor Theorems

We already know : $\dfrac{7}{2} = 3 + \dfrac{1}{2}$ where "1" is the remainder.

$$\therefore 7 = 2\,(3 + \tfrac{1}{2}) = 2 \cdot 3 + 1$$

The above example shows the following fact:

Dividend = divisor × quotient + remainder.

In general, when a polynomial $p(x)$ is divided by $x - c$,

$P(x)$ is the dividend, $x - c$ is the divisor, $Q(x)$ is the quotient, R is the remainder,

we have: $P(x) = Q(x)(x - c) + R$

Let $x = c$

$$P(c) = Q(c) \cdot (c - c) + R$$

$\therefore$ $P(c) = R$ --------------- Called **Remainder Theorem.**

Remainder Theorem: The remainder on dividing $P(x)$ by $x - c$ is $P(c)$.

Factor Theorem: If and only if $P(c) = 0$ (i.e. no remainder), then $x - c$ is a factor of $p(x)$ and c is a root of the equation $P(x) = 0$.

This is an easier way to find the value of polynomial $P(x)$ at $x = c$. We apply the **remainder theorem** and use synthetic division to find the remainder on dividing $P(x)$ by $x - c$. Then , the remainder is $P(c)$.

We apply the **factor theorem** to identify and find possible rational roots of a polynomial equation $P(x) = 0$. If $P(c) = 0$, then c is a root of $P(x) = 0$.

Examples

1. Find the value of $P(x) = 2x^3 - 8x^2 - 4x - 6$ when $x = 3$.

   Solution:

   Method 1): Direct substitution.

   $$P(x) = 2x^3 - 8x^2 - 4x - 6$$
   $$P(3) = 2(3)^3 - 8(3)^2 - 4(3) - 6$$
   $$= 54 - 72 - 12 - 6$$
   $$= -36.$$

   Ans: $P(3) = -36$.

   Method 2): Remainder theorem & synthetic division.

   $$\underline{3|\quad 2 - 8 - 4\ - 6}$$
   $$\underline{+)\qquad 6 - 6\ - 30}$$
   $$\quad\ 2 - 2 - 10, -36$$
   $$\qquad\qquad\text{(Remainder)}$$

   Ans: $P(3) = -36$.

2. Divide $\dfrac{2x^3 - 8x^2 - 4x - 6}{x - 3}$, find remainder.

Solution:

Method 1: Common division.

$$
\begin{array}{r}
2x^2 - 2x \ -10 \\
x-3\overline{)2x^3 - 8x^2 - 4x - 6} \\
\underline{-)\ 2x^3 - 6x^2} \\
-2x^2 - 4x - 6 \\
\underline{-)\ -2x^2 + 6x} \\
-10x - 6 \\
\underline{-)\ -10x + 30} \\
-36
\end{array}
$$

(Remainder)

Method 2: Synthetic division:

$$
\begin{array}{r|rrrr}
3 & 2 & -8 & -4 & -6 \\
+) & & 6 & -6 & -30 \\
\hline
& 2 & -2 & -10, & -36
\end{array}
$$

(Remainder)

Method 3: Remainder theorem.

$P(x) = 2x^3 - 8x^2 - 4x - 6$

$P(3) = 2(3)^3 - 8(3)^2 - 4(3) - 6$

$\qquad = 54 - 72 - 12 - 6$

$\qquad = -36$ (Remainder)

3. Is $x - 3$ a factor of

$\qquad P(x) = 2x^3 - 8x^2 - 4x - 6$?

Solution: $P(x) = 2x^3 - 8x^2 - 4x - 6$

$\qquad P(3) = 2(3)^3 - 8(3)^2 - 4(3) - 6$

$\qquad = 54 - 72 - 12 - 6$

$\qquad = -36 \neq 0.$

Ans: $x - 3$ is not a factor of $P(x)$.

(3 is not a root of $P(x) = 0.$)

4. Is $x + 1$ a factor of

$$P(x) = x^3 + 6x^2 + 11x + 6 ?$$

Solution:

$P(x) = x^3 + 6x^2 + 11x + 6$

$P(-1) = (-1)^3 + 6(-1)^2 + 11(-1) + 6$

$\qquad = -1 + 6 - 11 + 6$

$\qquad = 0.$

Ans: $x + 1$ is a factor of $P(x)$.

$\qquad$ (-1 is a root of $P(x) = 0$)

5. Is $x - \sqrt{3}$ a factor of

$$P(x) = x^3 - 3x^2 - 3x + 9 ?$$

Solution:

$P(x) = x^3 - 3x^2 - 3x + 9$

$P(\sqrt{3}) = (\sqrt{3})^3 - 3(\sqrt{3})^2 - 3(\sqrt{3}) + 9$

$\qquad = 3\sqrt{3} - 9 - 3\sqrt{3} + 9$

$\qquad = 0.$

Ans: $x - \sqrt{3}$ is a factor of $P(x)$.

$\qquad$ ($\sqrt{3}$ is a root of $P(x) = 0$)

6. Solve $x^3 - 4x^2 + 8x - 5 = 0$ if one root is 1.

Solution: By synthetic division:

$$\frac{x^3 - 4x^2 + 8x - 5}{x - 1} = x^2 - 3x + 5$$

$x^3 - 4x^2 + 8x - 5 = 0$

$(x - 1)(x^2 - 3x + 5) = 0$

$x - 1 = 0 \ \bigg|\ x^2 - 3x + 5 = 0$

$x = 1 \ \bigg|\ x = \dfrac{3 \pm \sqrt{9 - 20}}{2} = \dfrac{3 \pm \sqrt{11}\, i}{2}$

Ans: $x = 1$ and $\dfrac{3 \pm \sqrt{11}\, i}{2}$.

EXERCISES

Divide: a) by common division. b) by synthetic division if possible.

1. $\dfrac{x^2 - 5x + 6}{x + 2}$

2. $\dfrac{x^2 + 5x - 6}{x - 2}$

3. $\dfrac{x^2 + 3x - 4}{x + 2}$

4. $\dfrac{x^2 - 3x + 4}{x - 2}$

5. $\dfrac{4a^2 + 4a - 12}{2a - 1}$

6. $\dfrac{6a^2 - 5a + 10}{3a + 2}$

7. $\dfrac{x^3 - 5x^2 + 4x - 2}{x - 2}$

8. $\dfrac{x^3 - 5x^2 + 4x - 2}{x + 2}$

9. $\dfrac{2x^3 - 2x + 1}{2x + 4}$

10. $\dfrac{2x^3 - x^2 - 5}{2x - 5}$

11. $\dfrac{-4x^3 + 2x^2 - x + 1}{x + 2}$

12. $\dfrac{2x^4 - 3x^3 + x^2 - x + 5}{x - 5}$

13. $\dfrac{2x^4 + 3x^3 - x^2 + x - 5}{x + 5}$

14. $\dfrac{x^4 + x^2 + 2}{x - 2}$

15. $\dfrac{x^4 - x^2 + 2}{x + 2}$

16. $\dfrac{x^5 - 4x^3 + x}{x + 3}$

17. $\dfrac{x^5 + 4x^3 - x}{x - 3}$

18. $\dfrac{2n^4 - n^3 - 2n + 5}{n^2 + 2n + 1}$

19. $\dfrac{2x^4 - 3x^2 + 7x - 8}{x^2 + x - 3}$

20. $\dfrac{x^3 + 4ax^2 + 4a^2x + a^3}{2a + x}$

21. $\dfrac{4a^4 - 9 - 3a - 2a^3}{2a^2 - a - 3}$

22. $\dfrac{a^4 - 4}{a - 4}$

23. $\dfrac{x^4 + a^4}{x^2 + a^2}$

24. $\dfrac{x^5 + 5}{x + 1}$

Use synthetic division to determine whether $x - c$ is a factor of the given polynomial.

25. $f(x) = x^2 + 4x + 4; \quad x + 2$

26. $f(x) = x^2 + 4x + 4; \quad x - 2$

27. $f(x) = x^2 - 8x + 16; \quad x + 4$

28. $f(x) = x^2 - 8x + 16; \quad x - 4$

29. $f(a) = a^2 - 9; \quad a - 3$

30. $f(a) = a^2 - 9; \quad a + 3$

31. $f(x) = x^3 - 8; \quad x - 2$

32. $f(x) = x^3 + 8; \quad x - 2$

33. $f(x) = 4x^2 - 9; \quad x - \frac{3}{2}$

34. $f(x) = 4x^2 - 9; \quad x + \frac{3}{2}$

35. $f(x) = x^3 - x^2 - 6x; \quad x - 3$

36. $f(x) = x^3 - x^2 - 6x; \quad x + 3$

37. $f(x) = 4x^3 + 10x^2 - 14x; \quad x - 1$

38. $f(x) = 4x^3 + 10x^2 - 14x; \quad x + 1$

39. $f(p) = p^5 - 13p^3 + 36p; \quad p + 3$

40. $f(t) = t^5 + t^3 + t + 1; \quad t + 1$

41. $p(x) = x^3 - 2x^2 - 3x + 6; \quad x - \sqrt{3}$

42. $p(x) = 2x^3 + x^2 - 10x - 5; \quad x + \sqrt{5}$

43. $p(x) = x^4 - x^3 - 5x^2 - x - 6; \quad x - i$

44. $p(a) = a^3 + a^2 + 4a - 2; \quad a + 2i$

-----Continued-----

If one root of the equation is given. Solve the equation for all roots.

45. $x^3 + 6x^2 + 11x + 6 = 0$; $x = -2$

46. $2x^3 + 9x^2 + 7x - 6 = 0$; $x = -3$

47. $2x^3 + x^2 - 10x - 5 = 0$; $x = \sqrt{5}$

48. $x^3 - 3x^2 + 3x - 9 = 0$; $x = -\sqrt{3}\,i$

49. $x^3 - 4x^2 + 21x - 34 = 0$; $x = 2$

50. $x^3 + x + 10 = 0$; $x = -2$

51. $2x^3 - x^2 - 7x + 6 = 0$; $x = 1$

52. $2a^3 + a^2 - 8a + 3 = 0$; $a = \frac{3}{2}$

Find the polynomial equation with the given roots.

53. $-1, -2, -3$

54. $-2, -3, \frac{1}{2}$

55. $-\frac{1}{2}, \pm\sqrt{5}$

56. $3, \pm\sqrt{3}\,i$

57. $2, 1 \pm 4\,i$

58. $-2, 1 \pm 2i$

59. $1, -2, \frac{3}{2}$

60. $\frac{3}{2}, -1 \pm \sqrt{2}$

Use the Factor Theorem to determine whether $x - c$ is a factor of the given polynomial.

61. $f(x) = x^2 + 4x + 4$; $x + 2$

62. $f(x) = x^2 + 4x + 4$; $x - 2$

63. $f(x) = x^2 - 8x + 16$; $x + 4$

64. $f(x) = x^2 - 8x + 16$; $x - 4$

65. $f(a) = a^2 - 9$; $a - 3$

66. $f(a) = a^2 - 9$; $a + 3$

67. $f(x) = x^3 - 8$; $x - 2$

68. $f(x) = x^3 + 8$; $x - 2$

69. $f(x) = 4x^2 - 9$; $x - \frac{3}{2}$

70. $f(x) = 4x^2 - 9$; $x + \frac{3}{2}$

71. $f(x) = x^3 - x^2 - 6x$; $x - 3$

72. $f(x) = x^3 - x^2 - 6x$; $x + 3$

73. $f(x) = 4x^3 + 10x^2 - 14x$; $x - 1$

74. $f(x) = 4x^3 + 10x - 14x$; $x + 1$

75. $f(p) = p^5 - 13p^3 + 36p$; $p + 3$

76. $f(t) = t^5 + t^3 + t + 1$; $t + 1$

77. $p(x) = x^3 - 2x^2 - 3x + 6$; $x - \sqrt{3}$

78. $p(x) = 2x^3 + x^2 - 10x - 5$; $x + \sqrt{5}$

79. $p(x) = x^4 - x^3 - 5x^2 - x - 6$; $x - i$

80. $p(a) = a^3 + a^2 + 4a - 2$; $a + 2\,i$

81. Find k such that $f(x) = 2x^3 - x^2 - kx + 6$ has the factor $x - 1$.

82. Find k such that $f(x) = 2x^3 + kx^2 - 7x + 6$ has the factor $x + 2$.

83. Find k such that $f(x) = 2x^3 - x^2 - 7x + k$ has the factor $x - \frac{3}{2}$.

84. Find k such that $f(x) = x^3 + kx^2 - 3x - 9$ has the factor $x - \sqrt{3}$.

85. If $f(x) = 3x^{20} - 5x^{15} + x^2 - 2$ is divided by $x + 1$, find the remainder.

86. If $f(x) = 3x^{20} - 5x^{15} + x^2 - 2$ is divided by $x - 1$, find the remainder.

87. Use synthetic division to solve the equation $2x^4 - 5x^3 + 5x - 2 = 0$ given that two roots are $-1, 2$.

88. Is $\frac{1}{2}$ a root (zero) of the equation $2x^3 + 9x^2 + 7x - 6 = 0$?

89. Is $-\sqrt{5}$ a root (zero) of the equation $2x^3 + x^2 - 10x - 5 = 0$?

90. When a polynomial $f(x)$ is divided by $x - 2$, the quotient is $3x^2 + x + 3$ and the remainder is 4. Find $f(x)$.

5-6 Finding Zeros of Polynomial Functions with Degree n

Polynomial Function: It is an algebraic function in the following form: $y = p(x)$

$$p(x) = a_n x^n + a_{n-1} x^{n-1} + \cdots\cdots + a_1 x + a_0.$$

(Where a_n, a_{n-1}, $\cdots\cdots$, a_1, a_0 are real numbers and n is a nonnegative integer.)

Finding the zeros of a polynomial function is to find any value of x that makes $p(x) = 0$. Therefore, finding the zeros of a polynomial function is to solve the following polynomial equation.   Solve $a_n x^n + a_{n-1} x^{n-1} + \cdots\cdots + a_1 x + a_0 = 0$

The **real zeros** of a polynomial function $y = p(x)$ (the real roots of a polynomial equation $p(x) = 0$) are also the x–intercepts of the graph of $y = p(x)$ on the coordinate plane.

Rules for finding the zeros of a polynomial function with degree n ($n \geq 3$):
  1) **Fundamental Theorem of Polynomial :**
       Every polynomial function of positive degree n has exactly n zeros.
  2) **Conjugate Pairs Theorem:** a and b are real numbers, $b \neq 0$.
       If $a + bi$ is a zero of polynomial function, then $a - bi$ is also a zero.
  3) **Newton's Rule (Rational Zeros Theorem):** It is used to find the rational zeros of a polynomial function $p(x) = a_n x^n + a_{n-1} x^{n-1} + \cdots\cdots + a_1 x + a_0$.
       a) If $a_n = 1$ and r is an integer zero, then r must be a factor of a_0.
       b) If $a_n \neq 1$ and $\frac{b}{a}$ is a zero (a, b are integers, $x = \frac{b}{a}$ and $ax - b = 0$),
          then b must be a factor of a_0, a must be a factor of a_n.

To solve a polynomial equation in degree n, we find certain roots by Newton's Rule and factor it as a product of its factors in degree 1 and quadratic forms.

If a polynomial $p(x)$ has the factor $(x - c)^n$, we say that c is a zero of **multiplicity** n of $p(x)$, or a root of multiplicity n of $p(x) = 0$. It has an x–intercept at $x = c$.

1. Solve $x(x+3)^2(x-1) = 0$.
   Solution:
   $$x = 0 \mid (x+3)^2 = 0 \mid x - 1 = 0$$
   $$ x = -3 \mid x = 1$$
   Ans: $x = 0, 1$,
           and -3 (multiplicity 2,
                  or a double root)

2. If two roots of a quadratic equations

   are -3 and 1, find such an equation.
   Solution:
   The equation is:
   $$(x+3)(x-1) = 0$$
   $\therefore x^2 + 2x - 3 = 0$ is the equation. Ans.

3. Solve $x^3 + 6x^2 + 11x + 6 = 0$.
   Solution: by Newton's Rule: (factors of 6)
   Possible roots: $\pm 1, \pm 2, \pm 3, \pm 6$
   By factor theorem:
   $$P(-1) = (-1)^3 + 6(-1)^2 + 11(-1) + 6 = 0$$
   $$\therefore x + 1 \text{ is a factor.}$$
   By synthetic division:
   $$\frac{x^3 + 6x^2 + 11x + 6}{x + 1} = x^2 + 5x + 6$$
   $$x^3 + 6x^2 + 11x + 6 = 0$$
   $$(x+1)(x^2 + 5x + 6) = 0$$
   $$(x+1)(x+2)(x+3) = 0$$
   Ans: $x = -1, -2$, and -3

Descartes' Rule of Signs

Descartes' Rule of Signs provides information about the number and location of the real roots of a polynomial equation (or the real zeros of a polynomial function).

If we write a polynomial equation $p(x) = 0$ in decreasing power of x and ignore missing terms, we can determine the number of the real roots and the nature of the roots from **Descartes' rule of signs.**

Descartes's Rule of Signs can reduce the search for the real zeros of a polynomial function.

Descartes' rules of signs:

1. The number of positive real roots of $p(x) = 0$ is equal to the number of variations (changes) in sign between terms in $p(x) = 0$, or else is less than this number by a positive even integer.

2. The number of negative real roots of $p(x) = 0$ is equal to the number of variations (changes) in sign between terms in $p(-x) = 0$, or else is less than this number by a positive even integer.

Examples

1. Determine the nature of the roots of $2x^4 + 2x^3 + x^2 - 3x - 6 = 0$.
   Solution:

   $p(x) = 2x^4 + 2x^3 + x^2 - 3x - 6$ has 1 change in sign. It has exactly 1 positive real root.

   $p(-x) = 2x^4 - 2x^3 + x^2 + 3x - 6$ has 3 changes in sign. It has either 3 or 1 negative real roots.

   Ans. The equation must have 4 roots.
   There are two possibilities for the nature of the roots:
   1). 1 positive real root, 3 negative real roots, 0 imaginary root.
   2). 1 positive real root, 1 negative real root, 2 imaginary roots.

2. Determine the possible number of roots of $x^4 - 4x^3 + 13x^2 - 32x + 40 = 0$.
   Solution:

   $p(x) = x^4 - 4x^3 + 13x^2 - 32x + 40$ has 4 changes in sign. It has 4, 2 or 0 positive real roots.

   $p(-x) = x^4 + x^3 + 13x^2 + 32x + 40$ has no changes in sign. It has no negative real roots.

   Ans: The equation must have 4 roots.
   There are three possibilities for the nature of the roots:
   1). 4 positive real roots.
   2). 2 positive real roots, 2 imaginary roots.
   3). 4 imaginary roots.

Examples

3. How many positive real roots does $2x^4 + x^3 - 3x^2 - 8x = 0$ have ?
    Solution:

    $$p(x) = 2x^4 + x^3 - 3x^2 - 8x \text{ has 1 change in sign.}$$

    Ans: It has only 1 positive real root.

4. Determine the possible number of real roots of $x^3 + 2x^2 - 1 = 0$.
    Solution:

    $$p(x) = x^3 + 2x^2 - 1 \text{ has 1 change in sign. It has exactly 1 positive real root.}$$

    $$p(-x) = -x^3 + 2x^2 - 1 \text{ has 2 changes in sign. It has either 2 negative real roots or no negative real root.}$$

    Ans: It has 1 positive real root, either 2 negative real roots or 0 negative real root.

5. What is the maximum number of points can the graph of $y = 4x^5 + 3x^3 - 4x + 2$ cross the x-axis ?
    Solution:

    Find the real roots of $4x^5 + 3x^3 - 4x + 2 = 0$.

    $$p(x) = 4x^5 + 3x^3 - 4x + 2 \text{ has 2 changes in sign. It has either 2 positive real roots or 0 positive real root.}$$

    $$p(-x) = -4x^5 - 3x^3 + 4x + 2 \text{ has 1 change in sign. It has exactly 1 negative real root.}$$

    Ans: It has a maximum of 3 real roots.
          The graph crosses the x-axis at a maximum of 3 different points.

A patient complained because everyone said he was stupid. He went to see several doctors and asked medicines for his stupidity. One doctor gave him ten days of medicines.
Ten days later, he went to see his doctor again.

Patient: You fooled me. The medicines you gave me did not cure my stupidity. They must be vitamins only.

Doctor: You are right. They are all vitamins.
        You are very smart now!

Theorem on Bounds (Upper and Lower Bounds on $p(x) = 0$)

The information of the upper and lower bounds for the roots of $p(x) = 0$ is useful for the search for the real roots because it restricts the possible roots that lie between the upper and the lower bounds.

Rules of the upper and lower bounds for the roots on $p(x) = 0$:

$p(x)$ is divided by $x - r$ from synthetic division.

1. If r is positive and the numbers in the last row (including the remainder) are all positive or zero, then we can conclude that r is an upper bound of all roots. there is no root greater than r.

2. If r is negative and the numbers in the last row (including the remainder) are alternatively positive and negative (0 is counted as either positive or negative), then we can conclude that r is a lower bound of all roots. there is no root less than r.

3. If $x = r$ is an upper bound, then $x = r + n$ (n is any positive real number) is also an upper bound. To find the smallest positive integer which is an upper bound, we try each of the successive integers of r ($r = 1, 2, 3, 4, ...$).
   If $x = r$ is a lower bound, then $x = r - n$ (n is any positive real number) is also a lower bound. To find the largest negative integer which is a lower bound, we try each of the successive integers of r ($r = -1, -2, -3, -4, ...$).

   Hint: The largest (smallest) root may or may not be tested as the upper (lower) bound by the synthetic division.

Example

1. Use synthetic division to determine if each value of x is an upper bound, lower bound, or neither of the roots of $x^4 - x^3 - 5x^2 - x - 6 = 0$.

   a) $x = -2$ b) $x = 1$ c) $x = 3$ d) $x = 4$

   Solution: (Hint: Its roots are 3, -2, $\pm i$)

   a)
   $$\begin{array}{r} -2\underline{\big|\ 1-1-5-1-6} \\ +)\ \ -2+6-2+6 \\ \hline 1-3+1-3,\ 0 \end{array}$$
   -2 is a lower bound.

   b)
   $$\begin{array}{r} 1\underline{\big|\ 1-1-5-1-6} \\ +)\ \ +1+0-5-6 \\ \hline 1+0-5-6,-12 \end{array}$$
   Neither.

   c)
   $$\begin{array}{r} 3\underline{\big|\ 1-1-5\ -1-6} \\ +)\ \ +3+6+3+6 \\ \hline 1+2+1+2,\ 0 \end{array}$$
   3 is an upper bound.

   d)
   $$\begin{array}{r} 4\underline{\big|\ 1-1\ -5\ -1\ -6} \\ +)\ \ +4+12+28+108 \\ \hline 1+3+\ 7+27,\ 102 \end{array}$$
   4 is an upper bound.

   $x = 3$ is the smallest upper bound. There is no root greater than 3.

   Hint: After some trial-and-error tests, we can find the smallest and largest integers that are upper and lower bounds, respectively, for all real roots of the equation. It shows that all real roots are in the closed interval [−2, 3].

Even and Odd Functions

An **even function** is a function $f(x)$ if $f(-x) = f(x)$ for all real numbers of x in the domain of $f(x)$. The graph of an even function is symmetric about the $y-axis$.

An **odd function** is a function $f(x)$ if $f(-x) = -f(x)$ for all real numbers of x in the domain of $f(x)$. The graph of an odd function is symmetric about the origin (it rotates $180°$ about the origin).

The concept of **even and odd functions** helps us to identify the graph of $y = f(x)$.

To determine whether an equation (a relation) is even or odd, we use the same steps as determining a function by examining $f(-x)$. If the graph of an equation (a relation) is symmetric about the $x-axis$, the relation is not a function. A function is a relation in which each value of x is paired with exactly one value of y.

Examples

1. Is $f(x) = 2x^2$ an even or an odd function ?
Solution:
$$f(x) = 2x^2$$
$$f(-x) = 2(-x)^2 = 2x^2$$
We have $f(x) = f(-x)$.
Therefore, it is an even function.

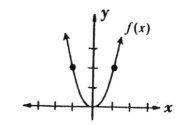

2. Is $f(x) = 2x^3$ an even or an odd function ?
Solution:
$$f(x) = 2x^3$$
$$f(-x) = 2(-x)^3 = -2x^3$$
We have $f(x) = -f(-x)$.
Therefore, it is an odd function.

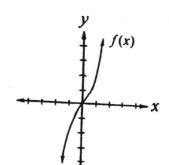

3. Is $f(x) = |x|$ an even or an odd function ?
Solution:
$$f(x) = |x|$$
$$f(-x) = |-x| = |x|$$
We have $f(x) = f(-x)$.
Therefore, it is an even function.

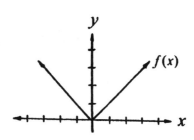

Examples

4. Is $y = 5$ an even or an odd function ?

Solution:

$$f(x) = 5 \text{ and } f(-x) = 5$$

We have $f(x) = f(-x)$.

Therefore, it is an even function.

5. Is $f(x) = x^3 - 4$ an even or an odd function ?

Solution:

$$f(x) = x^3 - 4$$
$$f(-x) = (-x)^3 - 4 = -x^3 - 4$$

We have $f(x) \neq f(-x)$, $f(x) \neq -f(-x)$.

Therefore, It is neither an even nor an odd function.

6. Is $x^2 + y^2 = 4$ an even or an odd relation ?

Solution:

$$y^2 = 4 - x^2 , \quad y = \pm\sqrt{4 - x^2} \quad \therefore f(x) = \pm\sqrt{4 - x^2}$$
$$f(-x) = \pm\sqrt{4 - (-x)^2} = \pm\sqrt{4 - x^2}$$

We have $f(x) = f(-x)$, $f(x) = -f(-x)$

Therefore, it is either an even or an odd relation.

(Hint: It is not a function.)

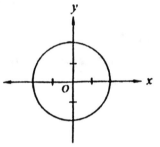

7. Is $f = \{(2, 1), (3, 2), (0, 0), (-1, -3), (-3, -2)\}$ an even or an odd function ?

Solution:

It is not symmetric about either the y-axis or the origin.

Therefore, it is neither an even nor an odd function.

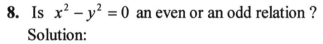

8. Is $x^2 - y^2 = 0$ an even or an odd relation ?

Solution:

$$y^2 = x^2 , \quad y = \pm\sqrt{x^2} \quad \therefore f(x) = \pm\sqrt{x^2}$$
$$f(-x) = \pm\sqrt{(-x)^2} = \pm\sqrt{x^2}$$

We have $f(x) = f(-x)$, $f(x) = -f(-x)$

Therefore, it is either an even or an odd relation.

(Hint: It is not a function.)

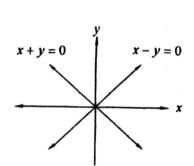

Rules of identifying the general shape of the graph of $y = f(x)$:

1. If the largest exponent in $y = f(x)$ is an even number, the two ends of the graph in the coordinate plane are in the same direction (either upward or downward).

2. If the largest exponent in $y = f(x)$ is an odd number, the two ends of the graph in the coordinate plane are at the opposite direction.

3. If all exponents in $y = f(x)$ are even numbers (it is an even function), the graph is symmetric about the $y-axis$.

$$y = x^4 - x^2 - 1$$

x	-2	-1	$-\frac{1}{2}$	0	$\frac{1}{2}$	1	2
y	11	-1	$-\frac{19}{16}$	-1	$-\frac{19}{16}$	-1	11

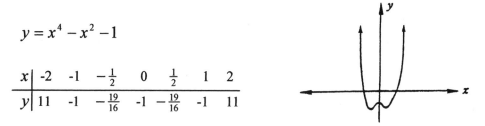

4. If all exponents in $y = f(x)$ are odd numbers (it is an odd function) and there is no constant term, the graph is symmetric about the origin.

$$y = x^3 - x$$

x	-2	-1	0	1	2
y	-6	0	0	0	6

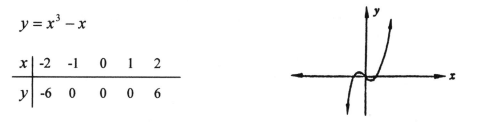

Examples

1. Identify the graph of $y = \frac{1}{2}x^4 - x^2$.

   Solution:

x	-2	-1	0	1	2
y	4	$-\frac{1}{2}$	0	$-\frac{1}{2}$	4

   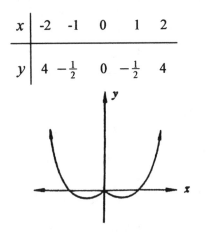

   The largest exponent is an even number. Therefore,
   The two ends of the graph are in the same direction.
      a. $f(x)$ increases without bound as x increases
         without bound.
      b. $f(x)$ increases without bound as x decreases
         without bound.
   It is an even function. Therefore, it is symmetric
   about the $y - axis$.

2. Identify the graph of $y = -x^3 + 2x + 1$.

   Solution:

x	-2	-1	0	1	2
y	5	0	1	2	-3

   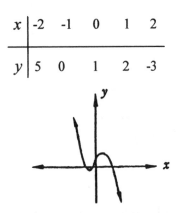

   The largest exponent is an odd number. Therefore,
   The two ends of the graph are in the opposite direction.
      a. $f(x)$ decreases without bound as x increases
         without bound.
      b. $f(x)$ increases without bound as x decreases
         without bound.

3. $p(x) = ax^5 - bx^4 - cx^3 + x^2 + 2$. If $p(x)$ decreases without bound as x increases without
   bound, identify the graph of $p(x)$ as x decreases without bound.

   Solution: The largest exponent is an odd number.
           The two ends of its graph in the coordinate plane are at the opposite direction.
           If $p(x)$ decreases without bound as x increases without bound, then, as x
           decreases without bound, $p(x)$ increases without bound.
           (Hint: The direction of the graph is similar to Example 2.)

Finding Complex Zeros of Polynomial Functions with Degree *n*

We have learned how to find the real zeros of a polynomial function. Here, we will learn how to find the **complex zeros** in the form $a + b\,i$ of a polynomial function with degree n.

In the complex number $a + b\,i$, it is a real number if $b = 0$.

Therefore, Finding the complex zeros of a polynomial function requires finding all zeros of the form $a + b\,i$ including the real zeros ($b = 0$).

We have also learned to solve a polynomial equation in degree n by factoring it as a product of its factors in degree 1 and quadratic forms.

In the real number system, a quadratic equation is said to be **no solution** (no zeros) if the equation has no real number solution.

In the complex number system, every quadratic equation has a solution (zeros), either real or complex.

In the complex number system, every polynomial function $p(x)$ of degree $n \geq 1$ can be factored into n linear factors of the form:

$$p(x) = a_n(x - r_1)(x - r_2)\cdots\cdots(x - r_n): \quad r_1, \ r_2, \cdots\cdots, \ r_n \text{ are the zeros, either real or complex.}$$

Fundamental Theorem of Algebra: In the complex number system, a polynomial function $p(x)$ of degree $n \geq 1$ has at least one zero.

In the complex number system, every polynomial function of degree $n \geq 1$ has exactly n zeros, either real or complex.

Conjugate Pairs Theorem: In the complex number system, if $a + b\,i\,(b \neq 0)$ is a complex zero of a polynomial function $p(x)$ whose coefficients are real numbers, then the conjugate $a - b\,i$ is also a complex zero of $p(x)$.

Examples

1. Finding all zeros of the function $f(x) = x^4 - 5x^3 + 3x^2 + 19x - 30$.

    Solution:

    Using the Newton's Rule (Rational Zeros Theorem), we can find two rational zeros of $f(x)$, 3 and –2.

    Using synthetic division, we have:
    $$f(x) = x^4 - 5x^3 + 3x^2 + 19x - 30$$
    $$= (x - 3)(x + 2)(x^2 - 4x + 5)$$
    To find zeros(roots), let $f(x) = 0$.
    $$(x - 3)(x + 2)(x^2 - 4x + 5) = 0$$
    We have the following four zeros(roots):
    $$x = 3, \ -2$$
    $$x = \frac{-(-4) \pm \sqrt{(-4)^2 - 4 \cdot 1 \cdot 5}}{2(1)} = \frac{4 \pm \sqrt{-4}}{2} = \frac{4 \pm 2i}{2} = 2 \pm i$$

    Ans: $x = 3, -2$, and $2 \pm i$.

Examples

2. Solve $x^4 - x^3 - 5x^2 - x - 6 = 0$.

Solution: Find two roots by Newton's Rule.

Possible roots:(factors of –6)

$\pm 1, \pm 2, \pm 3, \pm 6$

By factor theorem:

$P(3) = 3^4 - 3^3 - 5(3)^2 - 3 - 6 = 0$

$P(-2) = (-2)^4 - (-2)^3 - 5(-2)^2$
$\qquad -(-2) - 6 = 0$

$\therefore \ x - 3 \ $ and $\ x + 2 \ $ are factors.

$$\frac{x^4 - x^3 - 5x^2 - x - 6}{(x-3)(x+2)} = x^2 + 1$$

$x^4 - x^3 - 5x^2 - x - 6 = 0$

$(x-3)(x+2)(x^2+1) = 0$

$$\begin{array}{c|c|c} x - 3 = 0 & x + 2 = 0 & x^2 + 1 = 0 \\ x = 3 & x = -2 & x^2 = -1 \\ & & x = \pm i \end{array}$$

Ans: $x = 3, -2, \pm i$.

3. Solve $2x^3 + 9x^2 + 7x - 6 = 0$.

Solution: Find one root by Newton's Rule.

Factors of –6: $\pm 1, \pm 2, \pm 3, \pm 6 \rightarrow b$

Factors of 2: $\pm 1, \pm 2 \qquad \rightarrow a$

Possible roots ($\frac{b}{a}$):

$\pm 1, \pm 2, \pm 3, \pm 6, \pm \frac{1}{2}, \pm \frac{3}{2}$

By factor theorem:

$p(-2) = 2(-2)^3 + 9(-2)^2 + 7(-2) - 6 = 0$

$\therefore \ x + 2$ is a factor.

$$\frac{2x^3 + 9x^2 + 7x - 6}{x+2} = 2x^2 + 5x - 3$$

$2x^3 + 9x^2 + 7x - 6 = 0$

$(x+2)(2x^2 + 5x - 3) = 0$

$(x+2)(2x-1)(x+3) = 0$

$$\begin{array}{c|c|c} x + 2 = 0 & 2x - 1 = 0 & x + 3 = 0 \\ x = -2 & x = \frac{1}{2} & x = -3 \end{array}$$

Ans: $x = -2, \frac{1}{2}, -3$.

4. Solve $2x^3 + x^2 - 10x - 5 = 0$.

Solution: Find one root by Newton's Rule.

Factors of –5: $\pm 1, \pm 5 \rightarrow b$

Factors of 2: $\pm 1, \pm 2 \rightarrow a$

Possible roots ($\frac{b}{a}$):

$\pm 1, \pm 5, \pm \frac{1}{2}, \pm \frac{5}{2}$

By factor theorem:

$P(-\frac{1}{2}) = 2(-\frac{1}{2})^3 + (-\frac{1}{2})^2 - 10(-\frac{1}{2})$
$\qquad\qquad -5 = 0$

$\therefore \ x + \frac{1}{2}$ is a factor.

$$\frac{2x^3 + x^2 - 10x - 5}{x + \frac{1}{2}} = 2x^2 - 10$$

$2x^3 + x^2 - 10x - 5 = 0$

$(x + \frac{1}{2})(2x^2 - 10) = 0$

$$\begin{array}{c|c} x + \frac{1}{2} = 0 & 2x^2 - 10 = 0 \\ x = -\frac{1}{2} & x = \pm\sqrt{5} \end{array}$$

Ans: $x = -\frac{1}{2}, \pm\sqrt{5}$

5. Solve $x^3 - 3x^2 + 3x - 9 = 0$ if one root is $\sqrt{3}\,i$.

Solution:

If $\sqrt{3}\,i$ is a root, then $-\sqrt{3}\,i$ is also a root.

$\therefore \ x - \sqrt{3}\,i \ $ and $\ x + \sqrt{3}\,i \ $ are factors of the equation.

$(x - \sqrt{3}\,i)(x + \sqrt{3}\,i) = x^2 - 3i^2 = x^2 + 3$

$$\frac{x^3 - 3x^2 + 3x - 9}{x^2 + 3} = x - 3$$

$x^3 - 3x^2 + 3x - 9 = 0$

$(x^2 + 3)(x - 3) = 0$

$$\begin{array}{c|c} x^2 + 3 = 0 & x - 3 = 0 \\ x^2 = -3 & x = 3 \\ x^2 = \pm\sqrt{3}\,i & \end{array}$$

Ans: $x = \pm\sqrt{3}\,i$, 3.

6. Solve $x^4 - 4x^3 + 13x^2 - 32x + 40 = 0$
if one root is $2 - i$.
Solution:
If $2 - i$ is a root, then $2 + i$ is also a root.
$\therefore$ $x - (2 - i)$ and $x - (2 + i)$ are factors of the equation.

$$(x - 2 + i)(x - 2 - i) = (x - 2)^2 - i^2$$
$$= x^2 - 4x + 4 + 1 = x^2 - 4x + 5$$

$$\frac{x^4 - 4x^3 + 13x^2 - 32x + 40}{x^2 - 4x + 5} = x^2 + 8$$

$$x^4 - 4x^3 + 13x^2 - 32x + 40 = 0$$
$$(x^2 - 4x + 5)(x^2 + 8) = 0$$

$x^2 - 4x + 5 = 0$	$x^2 + 8 = 0$
$x = \dfrac{4 \pm \sqrt{16 - 20}}{2}$	$x^2 = -8$
$= 2 \pm i$	$x = \pm 2\sqrt{2}\, i$

Ans: $x = 2 \pm i$, $\pm 2\sqrt{2}\, i$.

7. If two roots of a cubic equation with real coefficients are 2 and $1 + 4i$, what is the third root? Find such an equation.
Solution:
The third root is $1 - 4i$.
The equation is:
$$(x - 2)[x - (1 + 4i)][x - (1 - 4i)] = 0$$
$$(x - 2)(x - 1 - 4i)(x - 1 + 4i) = 0$$
$$(x - 2)[(x - 1)^2 - (4i)^2] = 0$$
$$(x - 2)(x^2 - 2x + 1 + 16) = 0$$
$$(x - 2)(x^2 - 2x + 17) = 0$$
$$x^3 - 2x^2 + 17x - 2x^2 + 4x - 34 = 0$$

$\therefore$ $x^3 - 4x^2 + 21x - 34 = 0$ is the equation.
Ans.

8. A quadratic equation with real coefficients has one root $1 - 2i$. What is the other root? Find such an equation.
Solution:
The other root is $1 + 2i$.
The equation is:
$$[x - (1 - 2i)][x - (1 + 2i)] = 0$$
$$(x - 1 + 2i)(x - 1 - 2i) = 0$$
$$(x - 1)^2 - (2i)^2 = 0$$
$$x^2 - 2x + 1 + 4 = 0$$

$\therefore$ $x^2 - 2x + 5 = 0$ is the equation.
Ans.

9. Solve $2x^4 + 2x^3 + x^2 - 3x - 6 = 0$
if one root is $\dfrac{-1 + \sqrt{7}i}{2}$.
Solution:
If $\frac{-1+\sqrt{7}i}{2}$ is a root, $\frac{-1-\sqrt{7}i}{2}$ is also a root.
$$(x - \tfrac{-1+\sqrt{7}i}{2})(x - \tfrac{-1-\sqrt{7}i}{2})$$
$$= (x + \tfrac{1}{2} - \tfrac{\sqrt{7}i}{2})(x + \tfrac{1}{2} + \tfrac{\sqrt{7}i}{2})$$
$$= (x + \tfrac{1}{2})^2 - (\tfrac{\sqrt{7}i}{2})^2 = x^2 + x + \tfrac{1}{4} + \tfrac{7}{4}$$
$$= x^2 + x + 2$$
$$\frac{2x^4 + 2x^3 + x^2 - 3x - 6}{x^2 + x + 2} = 2x^2 - 3$$

$$2x^4 + 2x^3 + x^2 - 3x - 6 = 0$$
$$(x^2 + x + 2)(2x^2 - 3) = 0$$

$x^2 + x + 2 = 0$	$2x^2 - 3 = 0$
$x = \dfrac{-1 \pm \sqrt{1 - 8}}{2}$	$x = \pm\sqrt{\dfrac{3}{2}}$
$= \dfrac{-1 \pm \sqrt{7}i}{2}$	$= \pm\dfrac{\sqrt{6}}{2}$

Ans: $x = \dfrac{-1 \pm \sqrt{7}i}{2}$ and $\pm\dfrac{\sqrt{6}}{2}$.

Finding a polynomial function with indicated zeros

Examples

10. Find the polynomial function of least degree with real coefficient, a leading coefficient of 1, and the given zeros are -3 and $1+2\,i$.

Solution:.

By Conjugate Pairs Theorem: $1-2\,i$ is also a root.

We have the function:
$$f(x) = (x+3)[x-(1+2\,i)][\,x-(1-2\,i)]$$
$$= (x+3)[x-1-2\,i][\,x-1+2\,i]$$
$$= (x+3)[(x-1)^2 - 4\,i^2]$$
$$= (x+3)[x^2 - 2x+1+4]$$
$$= (x+3)(x^2 - 2x+5)$$

Therefore
$$f(x) = x^3 - 7x^2 + 16x - 10. \text{ Ans.}$$

11. Find the polynomial function in factored form that has zeros 2, 4(multiplicity 3), and -1 (multiplicity 2); and satisfied $f(0) = 8$.

Solution:

We have the function:
$$f(x) = a(x-2)(x-4)^3(x+1)^2$$

Since $f(0) = 8$, we can find a:
$$8 = a(0-2)(0-4)^3(0+2)^2$$
$$8 = 128a$$
$$\therefore a = \tfrac{1}{16}$$

Therefore
$$f(x) = \tfrac{1}{16}(x-2)(x-4)^3(x+1)^2. \text{ Ans.}$$

A high school boy is playing the violin.
His girlfriend sits next to him.
Boy: Do you like the song I played ?
Girl: Yes. It makes me miss my father.
Boy: Was your father a musician ?
Girl: No. He was a sawmill worker.

Estimating Real Zeros of Polynomial Functions with degree n

By this time we have learned many methods to find the zeros of polynomial functions. The real zeros of a polynomial function are the $x-$intercepts of the graph of $y = p(x)$ on the coordinate plane. They are the roots of the polynomial equation $p(x) = 0$.

However, it is not easy to find the zeros of a polynomial function of degree three or greater if it is not factorable. Its zeros are irrational numbers. We can find its irrational zeros by an approximation technique called **iterative process**. In the process, we obtain better and better approximations by using the Intermediate Value Theorem.

Intermediate Value Theorem: $p(x)$ is a polynomial function. If $a < b$ and if $p(a)$ and $p(b)$ are of opposite sign, then there is at least one zero of $p(x)$ between a and b.

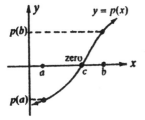

The Intermediate Value Theorem states that if there are two numbers a and b in the function $p(x)$, such that $p(a)$ is negative and $p(b)$ is positive, then the function $p(x)$ must have at least one number c between a and b such that $p(c) = 0$. The number c is a zero of $p(x)$.

Example

1. Estimate the real zeros of $f(x) = x^3 + x^2 - 1$ to the nearest hundredth.

   Solution:

   Based on the Intermediate Value Theorem, since $f(0) = -1$ and $f(1) = 1$, we conclude that there is at least one zero between 0 and 1.
   Based on Descartes' Rule of Signs, there is exactly one positive real zero.
   Based on the Newton's Rule (Rational Zeros Theorem), there is no positive or negative rational zero since $f(1) \neq 0$ and $f(-1) \neq 0$.
   Therefore, there is only one zero between 0 and 1, and it is irrational.
   We divide the interval between 0 and 1 into 10 equal subintervals and evaluate each endpoint of the subintervals until we can apply the Intermediate Value Theorem again to reach the desired accuracy.

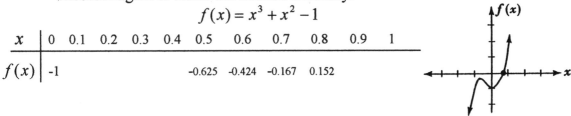

$$f(x) = x^3 + x^2 - 1$$

x	0	0.1	0.2	0.3	0.4	0.5	0.6	0.7	0.8	0.9	1
$f(x)$	-1						-0.625	-0.424	-0.167	0.152	

   There is an irrational zero between 0.7 and 0.8.
   Apply the Intermediate Value Theorem again between 0.7 and 0.8.
   Since $f(0.75) = -0.115$ and $f(0.76) = 0.17$, there is a zero between 0.75 and 0.76.

   Ans: $x \approx 0.75$ (approximately) and two imaginary roots (conjugates).

Space for taking notes

EXERCISES

Find all the zeros of the function. State the multiplicity of each zero.

1. $f(x) = x(x-1)(x+2)$

2. $f(x) = -x(x+5)(x-4)$

3. $f(x) = x^2(2x+1)(3x-6)^3$

4. $f(x) = (x+4)^2(x^2-9)^4$

5. $p(x) = (x^2+x-12)^3(9x^2-1)^2$

6. $p(x) = (2x+3)(4x^2-9)^5$

7. $p(x) = (x^2+3x+2)^2(x+2)^3$

8. $p(x) = x^4 - 5x^2 - 36$

9. $f(x) = x^2(x^2-1)(x^2+1)^3$

10. $f(x) = 3x^3(x-4)^2(x^2-16)$

Find the polynomial equation $f(x) = 0$ that has the indicated zeros.

11. $x = 1, -4$

12. $x = -1, 4$

13. $x = -2, 5$

14. $x = 2, -5$

15. $x = 0, 2, -3$

16. $x = 0, 3, -2$

17. $x = -4, -5, 2$

18. $x = 2i, -2i, 1$

19. $x = 4i, -4i, -3$(multiplicity 2)

20. $x = 1+i, 1-i, 2$(multiplicity 3)

Find the polynomial function $f(x)$ that has the indicated zeros and satisfies the given condition. (Hint: Find a from $f(x) = a(x-r_1)(x-r_2)(x-r_3)\cdots$).

21. $x = 1, -4$; $f(2) = 18$

22. $x = -1, 4$; $f(-3) = 28$

23. $x = -2, 5$; $f(1) = -24$

24. $x = 2, -5$; $f(-2) = -6$

25. $x = 0, 2, -3$; $f(3) = -6$

26. $x = 0, 3, -2$; $f(4) = 24$

27. $x = -4, -5, 2$; $f(-1) = 72$

28. $x = 2i, -2i, 1$; $f(2) = 48$

29. $x = 4i, -4i, -3$(multiplicity 2); $f(-2) = 40$

30. $x = 1+i, 1-i, 2$(multiplicity 3); $f(3) = 30$

Use Descartes' Rule of Signs to determine the number of possible positive, negative, and complex solutions (with imaginary roots) of the equation.

31. $x^3 - x^2 - 6x - 8 = 0$.

32. $x^3 + 6x^2 + 11x + 6 = 0$.

33. $x^4 - x^3 - 5x^2 - x - 6 = 0$.

34. $x^5 - 2x^3 - x + 6 = 0$.

35. $t^7 - 6 = 0$.

36. $y^8 - 9 = 0$.

37. What is the maximum number of points can the graph of $y = x^5 - x^4 + 3x^3 - 2x^2 + x$ cross the $x-axis$?

Use synthetic division and the rules on upper and lower bounds to determine if each value of x is an upper bound, lower bound, or neither of the roots of the equation.

38. $x^2 + 2x - 3 = 0$.   a) $x = -3$ b) $x = -1$ c) $x = 1$

39. $x^3 + 6x^2 + 11x + 6 = 0$.   a) $x = -7$ b) $x = -6$ c) $x = 1$

40. $2x^3 + 9x^2 + 7x - 6 = 0$.   a) $x = -5$ b) $x = -3$ c) $x = 0$

41. $x^4 - 2x^3 - 23x^2 - 12x + 36 = 0$.   a) $x = -7$ b) $x = -3$ c) $x = 6$

-----Continued-----

Find the smallest and largest integers that are upper and lower bound on the real roots (zeros).

42. $x^3 + 7x^2 + 2x - 40 = 0$ **43.** $x^4 - 2x^3 - 23x^2 - 12x + 36 = 0$

44. $p(x) = x^4 - 4x^2 - 5$ **45.** $f(x) = 2x^3 + 5x^2 - x - 4$

Identify whether each function or relation is even, odd, or neither.

46. $f(x) = 4x^2$ **47.** $f(x) = 5x^3$ **48.** $f(x) = \frac{5}{x}$ **49.** $y - x = 0$ **50.** $f(x) = 4x^2 - 4$

51. $f(x) = 5x^3 - 4$ **52.** $x^2 - y^2 = 4$ **53.** $f = \{(1, 2), (3, 6), (0, 0), (-1, -2), (-3, -6)\}$

Identify the general shape of the graph of $y = f(x)$.

54. $y = x^3$ **55.** $y = -x^3 + 3x^2 - 3x + 1$ **56.** $y = -x^4$ **57.** $y = x^3 + x^2 - 1$

58. $y = x^4 + x^2 - 1$ **59.** $y = \frac{1}{4}x^4$ **60.** $y = x^5 + x^3 - x$

Solve each polynomial equation in the complex number system.

61. $x^2 - 9 = 0$ **62.** $x^2 + 9 = 0$ **63.** $x^2 - 16 = 0$ **64.** $x^2 + 16 = 0$

65. $x^2 + 64 = 0$ **66.** $x^2 - 64 = 0$ **67.** $2x^2 - 16 = 0$ **68.** $x^2 - 4x = 0$

69. $3x^2 + 5x - 2 = 0$ **70.** $x^2 + 4x + 8 = 0$ **71.** $x^2 - 6x + 10 = 0$ **72.** $x^2 + x - 1 = 0$

73. $x^2 + x + 1 = 0$ **74.** $x^2 + x + 5 = 0$ **75.** $x^2 + 6x + 11 = 0$ **76.** $x^2 - 2x + 5 = 0$

77. $3x^2 + 2x - 2 = 0$ **78.** $3x^2 + 2x + 2 = 0$ **79.** $5x^2 - 2x + 1 = 0$ **80.** $x^3 - 8 = 0$

81. $x^2 + 8 = 0$ **82.** $x^4 - 16 = 0$ **83.** $x^4 - 1 = 0$ **84.** $x^3 - 7x^2 + 16x - 10 = 0$

85. $x^3 + x^2 - x + 15 = 0$ **86.** $3x^3 + 5x^2 - 2 = 0$ **87.** $5x^3 - 12x^2 + 5x - 2 = 0$

88. $x^4 + 3x^2 - 4 = 0$ **89.** $x^4 + 4x^3 + 2x^2 - x + 6 = 0$ **90.** $2x^4 + 11x^3 - 6x^2 + 64x + 32 = 0$

91. Solve $x^3 - 4x^2 + 2x - 8 = 0$ if one root is $\sqrt{2}\,i$.

92. Solve $x^3 - 3x^2 + 7x - 5 = 0$ if one root is $1 - 2\,i$.

93. Solve $x^4 + 2x^2 - 8 = 0$ if one root is $2\,i$.

94. Solve $x^4 - x^3 + 4x^2 - 3x + 3 = 0$ if one root is $\frac{1 + \sqrt{3}i}{2}$.

95. Is 3 a zero of $f(x) = x^5 - 3x^4 + 5x^3 + 2x^2 - x + 7$?

96. Is $\frac{2}{5}$ a zero of $f(x) = 2x^4 + 5x^3 - 2x^2 - 4x + 1$?

97. A quadratic equation with real coefficients has one root $4 + i$. Find such an equation.

98. A cubic equation with real coefficients has two roots 3 and $1 - 2\,i$. Find such an equation.

99. Prove that the function $f(x) = x^3 + x^2 + x - 5$ has a zero in the interval [1, 2].

100. Prove that the function $f(x) = x^4 + x^2 - 1$ has a zero in the interval [-1, 0], and a zero in the interval [0, -1].

101. Estimate the zeros of $f(x) = x^3 + x^2 + x - 5$ correct to two decimal places.

102. Estimate the zeros of $f(x) = x^4 + x^2 - 1$ correct to two decimal places.

5-7 Graphing Polynomial Functions

We have already learned how to graph the most commonly used polynomial functions in algebra. Understanding the basic shapes of the following graphs will help us to analyze and graph more complicated polynomial functions.

1. $f(x) = 0$.   It is a zero function. The graph is the x-axis.

2. $f(x) = c$.   It is a constant function.

The graph is a horizontal line with y-intercept c.

3. $f(x) = x$.   It is an identity function.

The graph is a straight line passing through the origin.

4. $f(x) = mx + b$.   It is a linear function.

The graph is a straight line with slope m and y-intercept b.

5. $f(x) = x^2$.   It is a quadratic function. It is a parabola.

The graph is symmetric with respect to the y-axis.

6. $f(x) = x^3$.   It is a cubic function.

The graph is symmetric with respect to the origin.

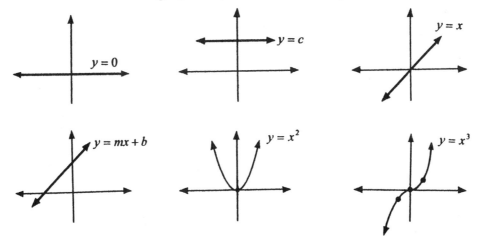

Graphing Power Functions of the form $f(x) = x^n$

As n increases, the graph tends to be flat near the origin and closer to the x-axis in the interval $[-1, 1]$. The graph increases rapidly when $x < -1$ or $x > 1$.

A. $f(x) = x^n$ if n is even   **B.** $f(x) = x^n$ if n is odd

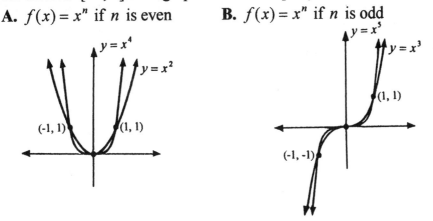

Transformations of Graphs

Knowing the graphs of common functions $y = f(x)$ can help us to graph a wide variety of polynomial functions by transformations (shifts, reflects, stretches, compressions).
Shift and reflection change only the position of the graph. The basic shape is unchanged.
Stretch and compression change (distort) the basic shape of the original graph.
The transformations of the original graph of $y = f(x)$ have the following types:

   1. $y = f(x) + c$. It moves (vertical shift) the original graph c units upward.
   2. $y = f(x) - c$. It moves (vertical shift) the original graph c units downward.
   3. $y = f(x + c)$. It moves (horizontal shift) the original graph c units to the left.
   4. $y = f(x - c)$. It moves (horizontal shift) the original graph c units to the right.
   5. $y = -f(x)$.   It reflects the original graph over the $x - axis$.
   6. $y = f(-x)$.   It reflects the original graph over the $y - axis$.
   7. $y = c\,f(x)$.   It stretches vertically if $c > 1$ and compressed vertically if $0 < c < 1$.
   8. $y = f(cx)$.   It compressed horizontally if $c > 1$ and stretches horizontally if $0 < c < 1$.

Examples (Show how the graphs are transformed with the above types.)

1. Graph: $y = x + 3$ and $y = x - 3$.
   Solution:
      $y = x + 3$. It moves the graph of $y = x$,
         3 units upward.
      $y = x - 3$. It moves the graph of $y = x$,
         3 units downward.

2. Graph: $y = x^2 + 1$ and $y = (x + 1)^2$.
   Solution:
      $y = x^2 + 1$. It moves the graph of $y = x^2$,
         1 unit upward.
      $y = (x + 1)^2$. It moved the graph of $y = x^2$,
         1 unit to the left.

3. Graph: $y = (x - 2)^2 - 3$.
   Solution:
      It moves the graph of $y = x^2$, 2 units
      to the right and 3 units downward.

4. Graph: $y = -x^3$ and $y = (-x)^3$
   Solution:
      $y = -x^3$. It reflects the graph of $y = x^3$
         over the $x - $axis.
      $y = (-x)^3$. It reflects the graph of $y = x^3$
         over the $y - $axis.

5. Graph: $y = 2x^2$ and $y = \frac{1}{2}x^2$.

Solution:

$y = 2x^2$. It stretches the graph of $y = x^2$ vertically
from (2, 4) to (2, 8) and give a narrower parabola.

$y = \frac{1}{2}x^2$. It compresses the graph of $y = x^2$ vertically
from (2, 4) to (2, 2) and gives a wider parabola.

6. Graph: $y = (2x)^2$ and $y = (\frac{1}{2}x)^2$.

Solution:

$y = (2x)^2$. It compresses the graph of $y = x^2$ horizontally from (2, 4) to (1, 4) and give a narrower parabola.

$y = (\frac{1}{2}x)^2$. It stretches the graph of $y = x^2$ horizontally from (2, 4) to (4, 4) and give a wider parabola.

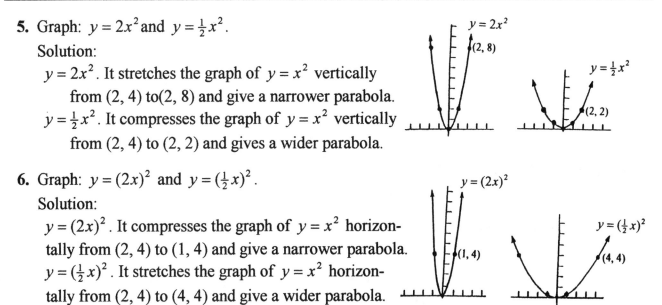

Graphing polynomial functions of higher degree

1. The graph of every polynomial function $f(x) = a_n x^n + a_{n-1} x^{n-1} + \cdots\cdots + a_1 x + a_0$ is a smooth and continuous curve. Its turning points are rounded. It has no holes, gaps, or sharp turns.

2. The points at which a graph changes direction are called **turning points**. The turning points are the **local maxima or minima** (**relative maxima or minima**) of the function. The maximum number of turning points of the graph is $n-1$.

3. The maximum number of real zeros ($x-$intercepts) is n.

4. If r is a repeated zero of **even** multiplicity (the sign of the function does not change from one side to the other side of r), the graph **touches** the $x-$intercept at r.

5. If r is a repeated zero of **odd** multiplicity (the sign of the function changes from one side to the other side of r), the graph **crosses** the $x-$intercept at r.

6. **End Behavior**: If n is even and the **leading coefficient** is positive ($a_n > 0$), the graph rises to the left and right. If n is even and the leading coefficient is negative ($a_n < 0$), the graph falls to the left and right.

7. **End Behavior**: If n is odd and the **leading coefficient** is positive ($a_n > 0$), the graph falls to the left and rises to the right. If n is odd and the leading coefficient is negative ($a_n < 0$), the graph rises to the left and falls to the right.

$$f(x) = ax^4 + bx^3 + cx^2 + dx + e \ (a \neq 0)$$

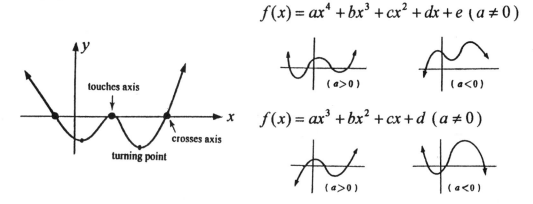

$$f(x) = ax^3 + bx^2 + cx + d \ (a \neq 0)$$

Steps for graphing a polynomial function $y = f(x)$

1. Solve the equation $f(x) = 0$ for real zeros (x – intercepts).
2. Decide whether the graph crosses or touches the x – axis at each x – intercept.
3. Decide the maximum number of turning points.
4. Apply the Leading Coefficient Test to decide whether the graph rises or falls. (Optional)
5. Test each interval between two x – intercepts to find each interval on which the graph is either above the x – axis or below the x – axis.
6. Plot a few additional points and connect the points with a smooth and continuous curve.

Examples

7. Graph: $f(x) = x^3 - 8x^2 + 20x - 16$.

Solution:

Factor the function and solve for real zeros:

$$f(x) = (x-2)^2(x-4) = 0$$

The x – intercepts are 2 (multiplicity 2) and 4.
The graph touches the x – axis at 2 and crosses the x – axis at 4.
It has at most two turning points.
Test each interval: The two x – intercepts divide the graph into 3 intervals.

$$x < 2, \ f(x): \ + \cdot - = - \ \text{below } x - \text{axis}$$
$$2 < x < 4, \ f(x): \ + \cdot - = - \ \text{below } x - \text{axis}$$
$$x > 4, \ f(x): \ + \cdot + = + \ \text{above } x - \text{axis}$$

Compute a few additional points:

x	1	1.5	2.5	3	3.5	4.5
$f(x)$	-3	-0.63	-0.38	-1	-1.13	3.13

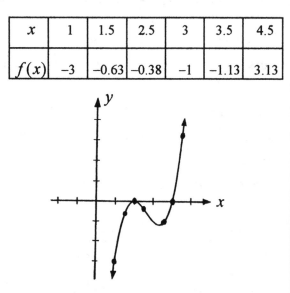

Hint: In calculus, we can find the location of the turning points, a local maximum at $x = 2$ and a local minimum at $x = 3.3$.

8. Graph: $f(x) = x^3(x-4)^2$.

Solution:

$$f(x) = x^3(x-4)^2 = 0$$

The x – intercepts are 0 (multiplicity 3) and 4 (multiplicity 2).
The graph crosses the origin and touches the x – axis at 4
It has at most four turning points.
Test each interval: (3 intervals)

$$x < 0, \ f(x): - \cdot + = - \ \text{below } x - \text{axis}$$
$$0 < x < 4, \ f(x): + \cdot + = + \ \text{above } x - \text{axis}$$
$$x > 4, \ f(x): + \cdot + = + \ \text{above } x - \text{axis}$$

Compute a few additional points:

x	-1	1	2	3	5
$f(x)$	-25	9	32	27	125

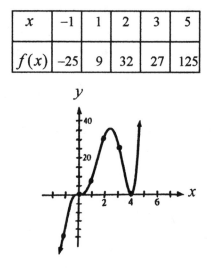

Applications of polynomial functions
Example
9. An open box is to be made from a cardboard that is 20 inches by 15 inches by cutting equal squares from each corner and folding up the sides. Find the x – value for which the volume of the box is maximum.

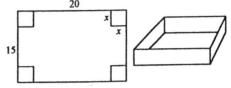

Solution:

The volume function is:

$$V(x) = (20 - 2x)(15 - 2x)x$$
$$= (300 - 40x - 30x + 4x^2)x$$
$$= (4x^2 - 70x + 300)x$$
$$\therefore V(x) = 4x^3 - 70x^2 + 300x$$

The x – intercepts are 0, 7.5, and 10. The graph crosses the x – axis at 0, 7.5, and 10.
Consider only the interval $0 < x < 7.5$ and compute a few points:

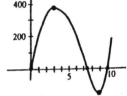

x	2	2.7	**2.8**	2.9	3
$V(x)$	352	378	**380**	379	378

$$V(2.8) \approx 380 \text{ is a maximum.}$$

The maximum volume is about 380 cubic inches when $x \approx 2.8$ inches. Ans.
(Hint: We can easily find the answer by the method in calculus. $V(2.83)_{\text{max.}} = 379.06$)

Finding a polynomial model of best-fitting for a set of data
In real-life situation, we want to organize and analyze numerical data and make prediction. Enter the data in a graphing calculator having **regression feature** and make a scatter diagram, we can obtain a polynomial function of best fit to these data.

Example
10. The table below shows the raw materials sold in tons from year 1 to 7. Enter the data in a graphing calculator and make a scatter diagram, we have the following cubic function of best fit to these data. Use this function to predict the sale for 10th year.

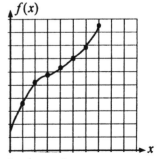

Years(x)	1	2	3	4	5	6	7
Sales (tons)	18.6	25.1	29.2	32.1	35	39.1	45.6

$$f(x) = 0.2x^3 - 2.4x^2 + 12.3x + 8.5$$

Solution:

$$f(10) = 0.2(10)^3 - 2.4(10)^2 + 12.3(10) + 8.5 = 91.5 \text{ tons. Ans.}$$

Graphing a polynomial function containing an absolute value
Example

11. Graph $f(x) = \left| x^2 - 4 \right|$.

Solution:

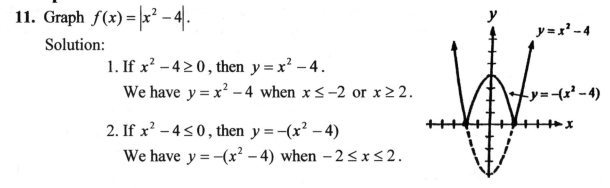

1. If $x^2 - 4 \geq 0$, then $y = x^2 - 4$.

   We have $y = x^2 - 4$ when $x \leq -2$ or $x \geq 2$.

2. If $x^2 - 4 \leq 0$, then $y = -(x^2 - 4)$

   We have $y = -(x^2 - 4)$ when $-2 \leq x \leq 2$.

Hint: Graphing $y = \left| f(x) \right|$, we graph $y = f(x)$ and reflect the part that is below the x-axis through the x-axis.

A teacher is testing a seven-year old boy.
Teacher: What is the answer of 5×6 ?
Boy: 15
Teacher: That is not right.
               The answer should be 30.
Boy: At least my answer is half right.

EXERCISES

Use the graph of $f(x) = x^2$ to write each function that is graphed after the indicated transformations.

1. Shift up 5 units
2. Shift down 5 units
3. Shift right 5 units
4. Shift left 5 units
5. Reflect about x – axis
6. Reflect about y – axis
7. Stretch vertically by a factor of 5
8. Stretch horizontally by a factor of 5

9. Shift left 5 units
   Reflect about x – axis

10. Shift up 3 units
    Reflect about y – axis

11. Three transformations: Shift up 2 units, Reflect about the y – axis, and Shift right 1 unit

12. The graph of a function is made after the following three transformations applied to the graph of $f(x) = x^3$. Write the function.

   1. Reflect about the x – axis   2. Shift up 4 units   3. Shift right 6 units

13. The graph of a function is made after the following three transformations applied to the graph of $y = \sqrt{x}$. Write the function.

   1. Shift left 6 units   2. Reflect about the y – axis   3. Shift down 4 units

Match each of the function with its graph.

14. $y = x^2 + 1$
15. $y = x^2 - 1$
16. $f(x) = (x+1)^2$
17. $f(x) = -(x+2)^2 + 5$
18. $y = -x^3 + 1$
19. $y = (x+2)^6 - 4$
20. $f(x) = (x+2)^2(x-2)$
21. $f(x) = x^2(x^2 - 4)$

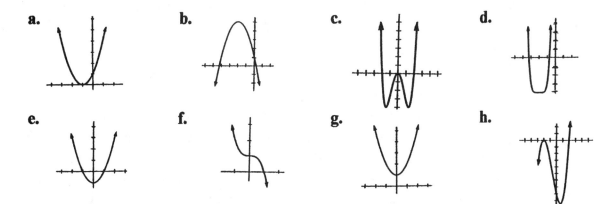

Graph each of the following polynomial functions.

22. $y = x^3 - 2x$
23. $f(x) = -x^3 + 3x^2$
24. $f(x) = x^4 - 4x^2$
25. $y = (x+2)^2(x-2)$
26. $f(x) = (x^2 - 4)(x^2 - 2)$
27. $f(x) = -2x^3 - x^2 + x$
28. $y = -x^4 + 3x^3 - x^2 - 3x + 2$
29. $y = x^5 - 5x^3 + 4x$
30. $f(x) = x^3 + 6x^2 + 9x$
31. $f(x) = |9 - x^2|$
32. $f(x) = |x^3 - 1|$
33. $f(x) = |x^3 - 2x|$

-----Continued-----

34. The formula for finding the falling distance (d) of an object dropped toward the ground in t seconds is $d = 16t^2$.

    a. Find the formula for finding the time (t) that it takes an object dropped from a height of d feet to reach the ground.

    b. If a ball drops off a height of 144 feet above the ground, how long will it take the ball to reach the ground.

35. The volume of a rectangular box is 1800 cubic inches. The length is 5 inches greater than the width. The height is 3 inches less than the length. Find the dimensions of the box.

36. An open box is to be made from a square cardboard of 42 meters on a side by cutting equal squares from each corner and folding up the sides. Find the x – value for which the volume of the box is maximum and find the maximum volume.

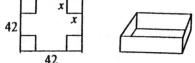

37. An open box has a volume of 20 cubic feet with the length x of a side of the square base.

    a. Write the amount A of material needed to make the box as a function of x.

    b. How much material is needed for a base 2 feet by 2 feet ?

    c. Find x – value for which the material needed is minimum.

38. The population of a kind of alligator is 144. Suppose that the number (n) of the alligator after t years is given by the function $n(t) = -t^4 + 32t^2 + 144$. When does the population of the alligator become extinct ?

39. The total cost C (in dollars) to produce calculators for a company is given by the function

$$C(x) = 0.25x^2 - 20x + 600 \quad \text{where } x \text{ is the number of units produced.}$$

How many units should be produced to obtain a minimum cost ?

40. The total profit P (in dollars) to sell calculators for a company is given by the function

$$P(x) = -0.25x^2 + 20x - 100 \quad \text{where } x \text{ is the number of units sold.}$$

How many units should be sold to obtain a maximum profit ?

41. An open box is to be made from a cardboard that is 30 inches by 20 inches by cutting equal squares of length x inches of each side from each corner and folding up the sides. Find the x – value for which the volume of the box is maximum.

42. The total profit P (in dollars) that a company makes depends on the amount x (in hundreds of dollars) the company spends on advertising according to the following function.

$$P(x) = -1.5x^3 + 50x^2 - 5,000$$

Find the advertising amounts that obtain a maximum profit.

43. The table below shows the total cost (C) of printing A-Plus math book (in thousands of dollars) and the number (x) of copies printed (in thousands). We have a model of the following cubic function of best fit to these data.

x	0	1	2	3	4	5	6	7	8	9	10
Cost	5.20	8.51	11.28	13.57	15.44	16.95	18.16	19.13	19.92	20.59	21.20

$$C(x) = 0.01x^3 - 0.3x^2 + 3.6x + 5.2$$

    a. Use a graphing calculator to verify the function given is the cubic function of best fit.

    b. Predict the cost of printing 20,000 copies.

5-8 Polynomial Inequalities and Absolute Values

In chapter 3, we have learned how to solve inequalities in one variable of degree 1, and how to solve absolute-value equations and inequalities.

To solve a polynomial inequality which can be factored into factors of degree less than two, we find the critical points and test each interval between two sets of critical points. If the original inequality is satisfied by testing the interval, the interval is the answer.

A sign graph is useful in the process.

Examples

1. Solve $x^2 + 2x - 15 > 0$.

Solution:
$$x^2 + 2x - 15 > 0$$
$$(x+5)(x-3) > 0$$
$$\text{Let } y = (x+5)(x-3) = 0$$

We have the **critical points**: $x = -5$ and 3

Test each interval between two sets of critical points if the inequality is satisfied. (Choose a convenient x-value in each interval.)

$$y = (x+5)(x-3) > 0 \quad \text{(It is positive.)}$$

$$x < -5, \quad - \cdot - = + \quad \text{yes}$$
$$-5 < x < 3, \quad + \cdot - = - \quad \text{no}$$
$$x > 3, \quad + \cdot + = + \quad \text{yes}$$

Ans: $x < -5$ or $x > 3$. (disjunction inequality)

2. Solve $x^3 + x^2 - 2x \le 0$.

Solution:
$$x^3 + x^2 - 2x \le 0$$
$$x(x^2 + x - 2) \le 0$$
$$x(x+2)(x-1) \le 0$$
$$\text{Let } y = x(x+2)(x-1) = 0.$$

We have the **critical points**: $x = -2, 0, 1$

Test each interval between two sets of critical points if the original inequality is satisfied: (Choose a convenient x-value in the interval.)

$$y = x(x+2)(x-1) \le 0. \quad \text{(It is negative.)}$$

$$x < -2, \quad - \cdot - \cdot - = - \quad \text{yes}$$
$$-2 < x < 0, \quad - \cdot + \cdot - = + \quad \text{no}$$
$$0 < x < 1, \quad + \cdot + \cdot - = - \quad \text{yes}$$
$$x > 1, \quad + \cdot + \cdot + = + \quad \text{no}$$

The critical points ($x = -2, 0, 1$) also satisfy the original inequality.

Ans: $x \le -2$ or $0 \le x \le 1$. (disjunction inequality)

Solving polynomial inequalities with Absolute Values

In Section 3-6, we have learned how to solve simple inequalities involving absolute values. To solve a polynomial inequality with absolute value, we may use the following facts:

1. $|x|^2 = x^2$.   **2.** If $x^2 > a$, then $|x| > \sqrt{a}$.   **3.** If $x^2 < a$, then $|x| < \sqrt{a}$.

Examples

3. Solve $|x + 2| < 5$.

Solution:

Method 1: $|x + 2| < 5$

$$-5 < x + 2 < 5$$

$$\therefore -7 < x < 3. \text{ Ans.}$$

Method 2: $|x + 2| < 5$

$$(x + 2)^2 < 5^2$$

$$x^2 + 4x + 4 < 25$$

$$x^2 + 4x - 21 < 0$$

$$(x + 7)(x - 3) < 0$$

We have the critical points:

$$x = -7 \text{ and } 3$$

Test: $y = (x + 7)(x - 3) < 0$

$$x < -7, \quad - \cdot - = + \quad \text{no}$$

$$-7 < x < 3, \quad + \cdot - = - \quad \text{yes}$$

$$x > 3, \quad + \cdot + = + \quad \text{no}$$

Ans: $-7 < x < 3$.

4. Solve $|x^2 - 2x| < x$.

Solution:

$$(x^2 - 2x)^2 < x^2$$

$$x^4 - 4x^3 + 4x^2 - x^2 < 0$$

$$x^4 - 4x^3 + 3x^2 < 0$$

$$x^2(x^2 - 4x + 3) < 0$$

$$x^2 - 4x + 3 < 0$$

$$(x - 3)(x - 1) < 0$$

We have the critical points:

$$x = 3 \text{ and } 1$$

Test: $y = (x - 3)(x - 1) < 0$

$$x < 1, \quad - \cdot - = + \quad \text{no}$$

$$1 < x < 3, \quad - \cdot + = - \quad \text{yes}$$

$$x > 3, \quad + \cdot + = + \quad \text{no}$$

Ans: $1 < x < 3$.

5. Solve $|x^2 + 4x + 4| > 5$.

Solution:

$$\left|(x + 2)^2\right| > 5$$

$$(x + 2)^2 > 5$$

$$|x + 2| > \sqrt{5}$$

$$x + 2 < -\sqrt{5} \text{ or } x + 2 > \sqrt{5}$$

Ans: $x < -1 - \sqrt{5}$ or $x > -2 + \sqrt{5}$.

6. Solve $|x^2 + 6x + 9| < 25$

Solution:

$$\left|(x + 3)^2\right| < 25$$

$$(x + 3)^2 < 25$$

$$|x + 3| < 5$$

$$-5 < x + 3 < 5$$

Ans: $-8 < x < 2$.

Solving inequalities if factoring is not possible, we apply the method of completing the square.

Examples

7. Solve $x^2 + 6x + 9 < 25$.

Solution:

$$(x+3)^2 < 25$$

$$|x+3| < 5$$

$$-5 < x+3 < 5$$

Ans: $-8 < x < 2$.

(Hint: Same answer as example 6.)

8. Solve $x^2 + 2x + 1 \geq \frac{9}{4}$.

Solution:

$$(x+1)^2 \geq \frac{9}{4}$$

$$|x+1| \geq \frac{3}{2}$$

$$x+1 \leq -\frac{3}{2} \quad \text{or} \quad x+1 \geq \frac{3}{2}$$

$$x \leq -\frac{3}{2} - 1 \qquad x \geq \frac{3}{2} - 1$$

Ans: $x \leq -\frac{5}{2}$ or $x \geq \frac{1}{2}$ Ans.

9. Solve $x^2 - 3x + 1 \leq 0$.

Solution:

Completing the square:

$$(x-\tfrac{3}{2})^2 - \tfrac{9}{4} + 1 \leq 0$$

$$(x-\tfrac{3}{2})^2 - \tfrac{5}{4} \leq 0$$

$$(x-\tfrac{3}{2})^2 \leq \tfrac{5}{4}$$

$$\left|x-\tfrac{3}{2}\right| \leq \tfrac{\sqrt{5}}{2}$$

$$-\tfrac{\sqrt{5}}{2} \leq x - \tfrac{3}{2} \leq \tfrac{\sqrt{5}}{2}$$

Ans: $\frac{3-\sqrt{5}}{2} \leq x \leq \frac{3+\sqrt{5}}{2}$.

10. Solve $2x^2 - x - 5 \geq 0$.

Solution:

Completing the square:

$$2(x^2 - \tfrac{1}{2}x) - 5 \geq 0$$

$$2[(x-\tfrac{1}{4})^2 - \tfrac{1}{16}] - 5 \geq 0$$

$$2(x-\tfrac{1}{4})^2 - \tfrac{1}{8} - 5 \geq 0$$

$$2(x-\tfrac{1}{4})^2 \geq \tfrac{41}{8} \ , \ (x-\tfrac{1}{4})^2 \geq \tfrac{41}{16}$$

$$\left|x-\tfrac{1}{4}\right| \geq \tfrac{\sqrt{41}}{4}$$

$$x - \tfrac{1}{4} \leq -\tfrac{\sqrt{41}}{4} \text{ or } x - \tfrac{1}{4} \geq \tfrac{\sqrt{41}}{4}$$

Ans: $x \leq \frac{1-\sqrt{41}}{4}$ or $x \geq \frac{1+\sqrt{41}}{4}$.

11. Solve $3x^2 - 6x \leq 4$.

Solution:

Completing the square:

$$3(x^2 - 2x) \leq 4$$

$$3[(x-1)^2 - 1] \leq 4$$

$$3(x-1)^2 - 3 \leq 4$$

$$3(x-1)^2 \leq 7$$

$$(x-1)^2 \leq \tfrac{7}{3}$$

$$|x-1| \leq \sqrt{\tfrac{7}{3}}$$

$$-\sqrt{\tfrac{7}{3}} \leq x - 1 \leq \sqrt{\tfrac{7}{3}}$$

Ans: $1 - \sqrt{\tfrac{7}{3}} \leq x \leq 1 + \sqrt{\tfrac{7}{3}}$.

12. Solve $\left|x^2 - 2x - 5\right| > 2$.

Solution:

$$x^2 - 2x - 5 < -2 \quad \text{or} \quad x^2 - 2x - 5 > 2$$

Completing the squares:

$$(x-1)^2 - 1 - 5 < -2 \qquad (x-1)^2 - 1 - 5 > 2$$

$$(x-1)^2 < 4 \qquad\qquad (x-1)^2 > 8$$

$$|x-1| < 2 \qquad\qquad |x-1| > \sqrt{8}$$

$$-2 < x - 1 < 2 \quad x-1 < -\sqrt{8} \text{ or } x-1 > \sqrt{8}$$

$$-1 < x < 3 \quad\quad x < 1 - \sqrt{8} \text{ or } x > 1 + \sqrt{8}$$

Ans: $x < 1 - \sqrt{8}$, or $-1 < x < 3$,

or $x > 1 + \sqrt{8}$.

EXERCISES

Solve each inequality.

1. $x^2 - 2x < 8$ **2.** $x^2 - 2x > 8$ **3.** $x^3 - 2x^2 - 8x \le 0$

4. $x^3 - 2x^2 - 8x \ge 0$ **5.** $x(x+2)(x-1) \le 0$ **6.** $(x-5)(x-1)(x+1)(x+2) > 0$

7. $|x-3| < 1$ **8.** $|x-3| > 1$ **9.** $|x-3| \ge 5$

10. $|x-3| \le 5$ **11.** $|x-5| > 0.5$ **12.** $|2x+3| > 8$

13. $|8-3x| < 2$ **14.** $|2-5x| > 3$ **15.** $|3x-2| \le 7$

16. $|1-4a| < 5$ **17.** $|1-4a| > 5$ **18.** $|4-a| > 0$

19. $|x^2 - 3x| < 2x$ **20.** $|x^2 - 2x| > x$ **21.** $|x^2 - 3x| > 2x$

22. $|x^2 - 2x - 3| \ge 2$ **23.** $|(x+1)(x-1)| \ge 2$ **24.** $2|x-1| < x^2$

25. $x^2 + 4x + 4 > 16$.   **26.** $x^2 + 4x - 16 < 0$.   **27.** $x^2 - x - 1 \le 0$.

28. $-x^2 - x + 6 \le 0$.   **29.** $2x^2 - x - 2 \ge 0$.   **30.** $3x^2 - 6x - 8 \le 0$.

31. $-3x^2 + x + 1 \ge 0$.   **32.** $4x^2 + 16x + 7 \ge 0$.   **33.** $x^2 - 4x + 4 \ge 0$.

34. $2x^2 - x + 1 > 0$.   **35.** $x^2 - 4x + 4 > 0$.   **36.** $4x^2 + 4x + 1 < 0$.

37. $x^2 - 2x + 3 < 0$.   **38.** $4x^2 + 4x + 1 \le 0$.   **39.** $(x+2)^2 < 25$

40. $(x+2)^2 > 25$ **41.** $x^3 - 4x \ge 0$ **42.** $x^3 - 4x < 0$

43. $x^5 - 3x^4 \le 0$ **44.** $x^2 + 4x - 5 \ge 0$ **45.** $x^2 - 6x - 7 < 0$

46. $x^3 + 6x^2 + 11x + 6 < 0$ **47.** $x^3 - 7x - 6 > 0$ **48.** $2x^3 + 9x^2 + 7x - 6 \le 0$

49. A projectile is fired vertically upward from the ground with an initial speed of 224 feet per second. The high of the projectile is given by $h(t) = vt - 16t^2$, where v is the initial speed and t is the time in seconds.

     a) During what time period will its height exceed 768 feet ?

     b) During what time periods will its height not exceed 768 feet ?

50. The normal human body temperature is $37\,^\circ C$. A temperature x that differs from normal by more than $0.84\,^\circ C$ is considered to be sick.

     a) Write an inequality involving absolute value to describe the unhealthy temperature x.

     b) Solve the inequality for x.

5-9 Inverse Functions

One to One function: A function f is a one-to-one function if and only if for each range value there corresponds exactly one domain value. An increasing (or decreasing) function is a one-to-one function. A horizontal line intersects the graph of a one-to-one function in at most one point. Note that, both $y = x^3$ and $y = x^2$ are functions. However, $y = x^3$ is a one-to-one function, $y = x^2$ is not a one-to-one function.

Inverse of a function: The inverse of a function f is obtained by interchanging the variables x and y in f. The inverse of f has the set of ordered pairs (y, x).

Inverse Functions: If $y = f(x)$ is a one-to-one function, a function $y = g(x)$ is the inverse function of f **if and only if** $f(g(x)) = g(f(x)) = x$.

$y = x^2$ has an inverse $x = y^2$ (or $y = \pm\sqrt{x}$) which is not a function.

For the function $y = f(x)$ to have an inverse function, f must be one-to-one.

The inverse function of the function f is denoted by f^{-1}, which is read " f inverse ".

If the functions $f(x)$ and $g(x)$ are inverses (functions) of each other, then:

1. the ordered pairs (x, y) of one equation are obtained by interchanging the variables x and y in the other equation.
2. their graphs are **mirror image** of each other with respect to the line $y = x$.
3. If $f(x)$ and $g(x)$ are inverses, then they have the property : $f(g(x)) = g(f(x)) = x$
   Example: $f(x) = x + 2$ and $g(x) = x - 2$ are inverse functions of each other.
   $$f(g(x)) = f(x - 2) = x - 2 + 2 = x \quad ; \quad g(f(x)) = g(x + 2) = x + 2 - 2 = x$$

A function f must have an inverse (a relation) which can be a function or not a function. In certain textbooks, they conventional state that "this function has no inverse" . It means "this function has no inverse function".

Examples

1. Let $f(x) = x + 2$, find $f^{-1}(x)$.

Solution: $y = x + 2$

Interchange x and y: $x = y + 2$

We have: $y = x - 2$

$\therefore f^{-1}(x) = x - 2$. Ans.

2. Let $f(x) = \frac{2}{3}x + 1$, find $f^{-1}(x)$.

Solution: $y = \frac{2}{3}x + 1$

Interchange x and y: $x = \frac{2}{3}y + 1$

We have: $y = \frac{3}{2}x - \frac{3}{2}$

$\therefore f^{-1}(x) = \frac{3}{2}x - \frac{3}{2}$. Ans.

Examples

3. Does the function $f(x) = x^2 - 1$ have an inverse function ?

Solution:

$$f(1) = 0, \ f(-1) = 0$$

We have: $f(1) = f(-1)$.

Ans: $f(x)$ does not have an inverse function.

(Hint: $f(x)$ has an inverse $y = \pm\sqrt{x+1}$. It is a relation, not a function.)

4. Let $f(x) = x^2$, find the inverse. If $f(x)$ has no inverse function, so state.

Solution:

$$y = x^2$$

Interchange x and y

We have: $x = y^2$ (It is a relation, but not a function.)

$$y = \pm\sqrt{x}$$

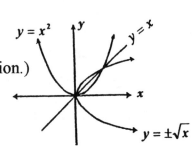

Ans: $f(x)$ has an inverse $y = \pm\sqrt{x}$.
It is a relation, not a function.

5. Let $f(x) = x^2$ and $x \geq 0$, find $f^{-1}(x)$.

Solution:

$$y = x^2 \text{ and } x \geq 0$$

Interchange x and y

We have: $x = y^2$ and $y \geq 0$ (It is a function.)

$$y = \sqrt{x} \text{ where } x \geq 0$$

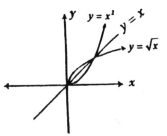

Ans: $f^{-1}(x) = \sqrt{x}$ where $x \geq 0$.

6. Let $f(x) = x^2$ and $x \leq 0$, find $f^{-1}(x)$.

Solution:

$$y = x^2 \text{ and } x \leq 0$$

Interchange x and y

We have: $x = y^2$ and $y \leq 0$ (It is a function.)

$$y = -\sqrt{x} \text{ where } x \geq 0$$

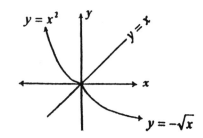

Ans: $f^{-1}(x) = -\sqrt{x}$ where $x \geq 0$.

Examples

7. If $f(x) = \sqrt{x+1}$, find the inverse function by limiting the domain of $f^{-1}(x)$.

Solution:

$$y = \sqrt{x+1}$$
$$y^2 = x+1 \text{ and } x \geq -1$$

Interchange x and y:

$$x^2 = y+1 \text{ and } y \geq -1 \text{ (it is a function)}$$
$$y = x^2 -1 \text{ and } x \geq 0$$

Therefore $f^{-1}(x) = x^2 - 1$ where $x \geq 0$.

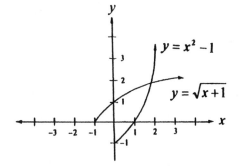

8. If $f = \{(3,1),(4,2),(5,2)\}$, find the inverse.

Solution:

The inverse is $g = \{(1,3),(2,4),(2,5)\}$.

(Hint: It is an inverse, a relation but not a function.)

(The function f has no inverse function.)

9. If $f = \{(3,1),(4,2),(5,2)\}$, find the inverse function by limiting the domain of f.

Solution:

The domain of the original function must be limited to "3 and 4" or "3 and 5"

Therefore $f^{-1} = \{(1,3),(2,4)\}$ by limiting the domain of f to 3 and 4.

Or $f^{-1} = \{(1,3),(2,5)\}$ by limiting the domain of f to 3 and 5.

10. Are the function $f(x) = \frac{2}{3}x + 1$ and $g(x) = \frac{3}{2}x - \frac{3}{2}$ inverses of each other ?

Solution:

$$f(g(x)) = \frac{2}{3}(\frac{3}{2}x - \frac{3}{2}) + 1 = x - 1 + 1 = x$$
$$g(f(x)) = \frac{3}{2}(\frac{2}{3}x + 1) - \frac{3}{2} = x + \frac{3}{2} - \frac{3}{2} = x$$

$$f(g(x)) = g(f(x)) = x \quad \therefore f(x) \text{ and } g(x) \text{ are inverses of each other.}$$

EXERCISES

Determine whether the given function is one-to-one.

1. $f(x) = 5x + 9$

2. $f(x) = \frac{2-x}{3}$

3. $g(x) = x^2 - 3$

4. $h(x) = \sqrt{9 - x^2}$

5. $f(x) = (x+2)^2$

6. $g(x) = x^2 - 3$ and $x \geq 0$

7. $p(x) = \frac{1}{x+2}$

8. $f(x) = \sqrt{x}$

9. $f(x) = |x|$

10. $f(x) = 5x^{\frac{2}{5}}$

11. $v(x) = \sqrt[3]{x}$

12. $f(x) = 10$

13. $f(x) = \frac{1}{x^2}$

14. $f(x) = 5x^3 - 2$

15. $f(x) = x^5 - 9$

Find the inverse function. If it has no inverse function, so state.

16. $f(x) = x - 5$

17. $f(x) = x + 5$

18. $f(x) = \frac{x}{5}$

19. $f(x) = 5x$

20. $f(x) = 2x - 4$

21. $f(x) = \frac{x}{2} + 2$

22. $f(x) = 5x + 1$

23. $f(x) = \frac{x-1}{5}$

24. $f(x) = \frac{1}{x}$

25. $f(x) = 4 - 5x$

26. $f(x) = \frac{4-x}{5}$

27. $f(x) = 4 - \frac{1}{2}x$

28. $f(x) = \frac{3-5x}{4}$

29. $f(x) = \frac{3-4x}{5}$

30. $f(x) = \frac{6+2x}{7}$

31. $f(x) = x^2 + 5$

32. $f(x) = x^2 + 5$ and $x \geq 0$

33. $f(x) = 2x^2 + 4$

34. $f(x) = 1 - x^2$ and $x \geq 0$

35. $f(x) = 9 - x^2$ and $x \geq 0$

36. $f(x) = x^3$

37. $f(x) = \sqrt[3]{x}$

38. $f(x) = x^3 + 1$

39. $f(x) = x^{\frac{2}{5}}$

40. $f(x) = x^5 - 4$

41. $f(x) = (x-4)^5$

42. $f(x) = \sqrt[5]{x-4}$

43. $f(x) = \sqrt{x^5 - 4}$

44. $f(x) = |x-1|$ and $x \leq 1$

45. $f(x) = |x+4|$ and $x \geq -4$

46. If $f = \{(1, 4), (2, 5), (4, 6), (5, 7)\}$, find the inverse and determine whether the inverse is a function.

47. If $f = \{(1, 4), (2, 5), (4, 6), (5, 6)\}$, find the inverse and determine whether the inverse is a function.

Verify that the functions f and g are inverse functions.

48. $f(x) = 5x$ and $g(x) = \frac{x}{5}$.

49. $f(x) = 5x + 1$ and $g(x) = \frac{x-1}{5}$.

50. $f(x) = x^2 + 5$; $x \geq 0$ and $g(x) = \sqrt{x-5}$; $x \geq 5$.

51. $f(x) = (x-2)^2$; $x \geq 2$ and $g(x) = \sqrt{x} + 2$; $x \geq 0$.

52. $f(x) = x^3 + 1$ and $g(x) = \sqrt[3]{x-1}$.

53. $f(x) = x^3 - 12$ and $g(x) = \sqrt[3]{x+12}$

54. $f(x) = \frac{5}{9}(x - 32)$ and $g(x) = \frac{9}{5}x + 32$

55. $f(x) = ax + b$ and $g(x) = \frac{x-b}{a}$

56. The cost (in dollars) to buy a certain product is given by the function $y = 250 - 15x$, where x is the number of the product purchased. a) Write the inverse of the function.
b) What does each variable represent in the inverse function ? c) Find the number of product purchased if the total cost is $160.

5-10 Composition of Functions

When we apply one function after another function's values, such as $f(g(x))$, is called " **the composition of** the function f with the function g ". It is denoted as $f \circ g$.

$f(g(x))$ **indicates that the domain of $f \circ g$ is all x-values such that x is in the range of g and g is in the domain of f**.

The composition of f with g is: $(f \circ g)(x) = f(g(x))$

The composition of g with f is: $(g \circ f)(x) = g(f(x))$

The composition of a function $f(x)$ with its inverse $f^{-1}(x)$ is equal to x.

$$(f \circ f^{-1})(x) = f(f^{-1}(x)) = x, \quad (f^{-1} \circ f)(x) = f^{-1}(f(x)) = x$$

Examples

1. If $f(x) = 3x^2 - 6$ and $g(x) = x^2 - 4$, find $(f \circ g)(x)$ and $(g \circ f)(x)$.

   Solution:

   $$(f \circ g)(x) = f(g(x)) = 3(g(x))^2 - 6 = 3(x^2 - 4)^2 - 6 = 3(x^4 - 8x^2 + 16) - 6$$
   $$= 3x^4 - 24x^2 + 42.$$
   $$(g \circ f)(x) = g(f(x)) = (f(x))^2 - 4 = (3x^2 - 6)^2 - 4 = 9x^4 - 36x^2 + 36 - 4$$
   $$= 9x^4 - 36x^2 + 32.$$

2. Given $f(x) = \frac{2}{3}x + 1$ and $f^{-1}(x) = \frac{3}{2}x - \frac{3}{2}$, find $f \circ f^{-1}$ and $f^{-1} \circ f$.

   Solution:

   $$f \circ f^{-1} = f(f^{-1}(x)) = \frac{2}{3}(\frac{3}{2}x - \frac{3}{2}) + 1 = x - 1 + 1 = x.$$
   $$f^{-1} \circ f = f^{-1}(f(x)) = \frac{3}{2}(\frac{2}{3}x + 1) - \frac{3}{2} = x + \frac{3}{2} - \frac{3}{2} = x.$$

3. If $f(g(x)) = 2x - 8$ and $f(x) = 2x$, find $g(x)$ and $g(2)$.

   Solution: $f(g(x)) = 2(g(x)) = 2x - 8 \quad \therefore g(x) = x - 4$ and $g(2) = 2 - 4 = -2$.

4. If $g(f(x)) = 2x - 1$ and $g(x) = x - 4$, find $f(x)$ and $f(2)$.

   Solution: $g(f(x)) = f(x) - 4 = 2x - 1 \quad \therefore f(x) = 2x + 3$ and $f(2) = 2(2) + 3 = 7$.

5. The radius (in meters) of a hot-hydrogen spherical balloon is changing with the time (in seconds) according to the function $r(t) = 1.5t^2$, $t \geq 0$. The volume (in cubic meters) of the balloon is given by the function $V(r) = \frac{4}{3}\pi r^3$. Find the volume V of the balloon as a function of time t.

   Solution:

   We have a composite function relationship in which V is a function of r and r is a function of t.

   $$V = \frac{4}{3}\pi r^3 = \frac{4}{3}\pi (\frac{3}{2}t^2)^3 = \frac{4}{3}\pi (\frac{27}{8}t^6) = \frac{9}{2}\pi t^6$$
   $$\therefore V(t) = \frac{9}{2}\pi t^6. \text{ Ans.}$$

EXERCISES

Operations on the compositions of the given functions.

1. If $f(x) = 2x - 4$ and $g(x) = 3x + 7$, find $(f \circ g)(x)$ and $(g \circ f)(x)$.

2. If $f(x) = 2x + 4$ and $g(x) = 3x - 7$, find $(f \circ g)(x)$ and $(g \circ f)(x)$.

3. If $f(x) = x^2$ and $g(x) = x - 1$, find $(f \circ g)(x)$ and $(g \circ f)(x)$.

4. If $f(x) = 3x$ and $g(x) = x^2 - 4$, find $(f \circ g)(x)$ and $(g \circ f)(x)$.

5. If $f(x) = 2x^2 + x - 1$ and $g(x) = 2x + 1$, find $(f \circ g)(x)$ and $(g \circ f)(x)$.

6. If $f(x) = 2x$ and $g(x) = 3x^3 - 4x$, find $(f \circ g)(x)$ and $(g \circ f)(x)$.

7. If $f(x) = |x|$ and $g(x) = -15$, find $(f \circ g)(x)$ and $(g \circ f)(x)$.

8. If $f(x) = -15$ and $g(x) = -x^2$, find $(f \circ g)(x)$ and $(g \circ f)(x)$.

9. If $f(x) = x^2 - 2x$ and $g(x) = \sqrt{x + 3}$, find $(f \circ g)(x)$ and $(g \circ f)(x)$.

10. If $f(x) = \dfrac{1}{x + 1}$ and $g(x) = x + 1$, find $(f \circ g)(x)$ and $(g \circ f)(x)$.

11. If $f(x) = 2x - 6$ and $f^{-1}(x) = \frac{1}{2}x + 3$, find $f \circ f^{-1}$ and $f^{-1} \circ f$.

12. If $f(x) = x^2$ and $x \geq 0$, find $f \circ f^{-1}$ and $f^{-1} \circ f$.

13. If $f(g(x)) = 4x - 4$ and $f(x) = x - 1$, find $g(x)$.

14. If $f(g(x)) = 4x^2 - 4x$ and $f(x) = x^2 - 1$, find $g(x)$.

15. If $g(f(x)) = 4x - 7$ and $g(x) = 4x - 3$, find $f(x)$ and $f(3)$.

16. If $f(x) = x^2 + 9$ and $g(x) = \sqrt{x}$, find the domain of $f \circ g$.

17. If $f(x) = x^2 + 9$ and $g(x) = \sqrt{x}$, find the domain of $g \circ f$.

18. If $f(x) = x^2 - 9$ and $g(x) = \sqrt{x}$, find the domain of $g \circ f$.

19. Several values of two functions $f(x)$ and $g(x)$ are listed below.
    Find $(f \circ g)(0)$, $(f \circ g)(3)$, $(g \circ f)(0)$, and $(g \circ f)(3)$.

x	−14	−7	−2	0	1	2	3	4
$f(x)$	−13	−6	−1	1	2	3	4	5

x	0	1	2	3	4
$g(x)$	2	1	−2	−7	−14

20. If $f = \{(1, 4), (2, 5), (3, 6), (4, 7), (5, 8)\}$ and $g = \{(4, 3), (5, 4), (6, 5)\}$,
    find **a)** $f \circ g$ **b)** $g \circ f$.

21. The cost (c) of manufacturing x units of a refrigerator is given by the function
    $c(x) = 1200 + 30x$. The number of x units manufacturing in t hours is given by
    the function $x(t) = 20t$. Write the composite function $(c \circ x)(t)$ that represents
    the cost of manufacturing x units after t hours.

22. A forest fire is spreading in the form of a circle. If the radius of this circle increases
    at a rate of 16 feet per minute. Find the total area (A) as a function of time (t) (in minutes).

23. A container has the shape of a circular cylinder. The volume is 16π cubic feet. Write the
    total surface area (A) of this container as a function of the radius.

5-11 Greatest Integer Functions

The greatest-integer function is the most common type of the **step functions.**
The greatest-integer function is in the form:

$$f(x) = \text{int}(x) \quad \text{or} \quad y = [x] = i \;, \quad i \text{ is an integer and } \; i \le x < i+1$$

(i is the greatest integer that is less than or equal to the real number x.)
We have:

$$\text{int}(1)=1 \text{ or } [1]=1, \; \text{int}(2.7)=2 \text{ or } [2.7]=2, \; [3]=3, \; [1.2]=1, \; [2.8]=2,$$
$$[0.4]=0, \; [0.9]=0, \; [\tfrac{7}{3}]=2, \; [-1]=-1, \; [-2]=-2, \; [-3]=-3, \; [-0.5]=-1,$$
$$[-1.5]=-2, \; [-2.1]=-3, \; [-3.2]=-4, \; [-3.5]=-4, \; [-3.9]=-4.$$

To graph $f(x) = \text{int}(x)$, we plot several points. For values of $-3 \le x < -2$, the value
of $f(x)$ is -3. The x-intercepts are in the interval $[0, 1)$. A solid dot is used to indicate
the value of $f(x) = -3$ at $x = -3$. The graph has an infinite number of line segments
(or steps) and suddenly steps from one value to another without continuity.
The domain of $f(x) = \text{int}(x)$ is all real numbers and its range is all integers.

Examples:

1. Graph $y = [x]$.

Solution:

$-3 \le x < -2 \;, \; y = -3$

$-2 \le x < -1 \;, \; y = -2$

$-1 \le x < 0 \;\;\; , \; y = -1$

$0 \le x < 1 \;\;\; , \; y = 0$

$1 \le x < 2 \;\;\; , \; y = 1$

$2 \le x < 3 \;\;\; , \; y = 2$

$3 \le x < 4 \;\;\; , \; y = 3$

2. Graph $y = [2x]$.

Solution:

$-1.5 \le x < -1 \;\;, \; y = -3$

$-1 \le x < -0.5 \;, \; y = -2$

$-0.5 \le x < 0 \;\;\;\; , \; y = -1$

$0 \le x < 0.5 \;\; , \; y = 0$

$0.5 \le x < 1 \;\;\;\; , \; y = 1$

$1 \le x < 1.5 \;\; , \; y = 2$

$1.5 \le x < 2 \;\;\;\; , \; y = 3$

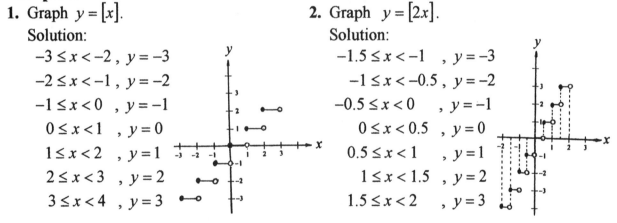

3. The cost of using a long-distance call is $1.25 for the first minutes and $0.35 for each
additional minute or portion of a minute.

   a) Use the greatest integer function to write the cost (c) for a call lasting t minutes.

   b) Find the cost of a call lasting 16 minutes and 15 seconds.

   c) Graph the function.

Solution:

   a) $c(t) = 1.25 - 0.35 [1 - t]$ c)

   b) $c(16\tfrac{15}{60}) = 1.25 - 0.35 [1 - 16.25]$

$$= 1.25 - 0.35 [-15.25]$$
$$= 1.25 - 0.35(-16)$$
$$= 1.25 + 5.60 = \$6.85$$

Hint: $c(t) = 1.25 + 0.35[t - 1]$ is incorrect.

EXERCISES

Evaluate
1. $[9]$ 2. $[9.5]$ 3. $[9.543]$ 4. $[-9]$ 5. $[-9.5]$
6. $[-9.543]$ 7. $[-0.12]$ 8. $[0.12]$ 9. $[1.99]$ 10. $[-1.99]$

Evaluate each greatest integer function.

11. If $f(x) = \text{int}(x)$, find $f(1.8)$.   12. If $f(x) = \text{int}(3x)$, find $f(1.8)$.

13. If $f(x) = \text{int}(3x)$, find $f(-1.8)$.   14. If $f(x) = \left[\frac{x}{3}\right]$, find $f(1.8)$.

15. If $f(x) = \left[\frac{x}{5}\right]$, find $f(-1.8)$.   16. If $f(x) = 2[x-1]$, find $f(7.2)$.

17. If $f(x) = 2[x-1]$, find $f(-7.2)$ 18. If $f(x) = 2[x-3]+5$, find $f(2)$.

19. If $f(x) = 2[x-3]+5$, find $f(-2)$ 20. If $f(x) = [-x]$, find $f(2.5)$.

Graph each greatest integer function.

21. $f(x) = [x]-2$.   22. $f(x) = [x-2]$.   23. $f(x) = [x-1]$.

24. $f(x) = [x]-1$.   25. $f(x) = [x+5]$.   26. $f(x) = [x]+5$.

27. $f(x) = 2[x]$.   28. $f(x) = 2[x+1]$.   29. $f(x) = 2[x-3]+5$.

30. $f(x) = [-x]$.   31. $f(x) = x-[x]$.   32. $f(x) = x+[x]$.

33. Find the range of $y = [x]$.
34. Find the range of $y = \frac{[x]}{x}$ if x is an integer.
35. Find the range of $y = \frac{[x]}{x}$ if x is not an integer and $x > 0$.
36. Find the range of $y = \left[\frac{[x]}{x}\right]$ with domain $x > 0$.

37. The cost of using an international long-distance call is $1.95 for under 1 minute plus $0.65 for 1 minute or more under an additional minute.
    a) Use the greatest integer function to write the cost (c) for a call lasting t minutes.
    b) Find the cost of a call lasting 12 minutes and 24 seconds.
    c) Find the cost of a call lasting 12 minutes.
    d) Graph the function.
38. The cost of using an international long-distance call is $1.95 for the first minute and $0.65 for each additional minute or portion of a minute.
    a) Use the greatest integer function to write the cost (c) for a call lasting t minutes.
    b) Find the cost of a call lasting 12 minutes and 24 seconds.
    c) Find the cost of a call lasting 12 minutes.
    d) Graph the function.

5-12 Piecewise-defined Functions

The piecewise-defined function is a function described by two or more equations over a specified domain (the x-values).

$$f(x) = \begin{cases} x^2 - 2, & x \le 0 \\ x - 1, & x > 0 \end{cases}$$

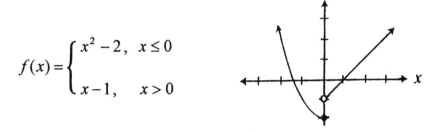

A close circle "•" shows "included". An open circle "o" shows "not included".

Examples

1. Evaluate the function when $x = -2, 0, 1$, and 2.

$$f(x) = \begin{cases} x^2 - 2, & x \le 0 \\ x - 1, & x > 0 \end{cases}$$

Solution:

$f(x) = x^2 - 2$ when $x = -2$: $f(-2) = (-2)^2 - 2 = 2$.

$f(x) = x^2 - 2$ when $x = 0$: $f(0) = 0^2 - 2 = -2$.

$f(x) = x - 1$ when $x = 1$: $f(1) = 1 - 1 = 0$.

$f(x) = x - 1$ when $x = 2$: $f(2) = 2 - 1 = 1$.

2. A publishing company pays income tax at a rate of 30% on the first income $50,000, and all income over $50,000 is taxed at a rate of 40%.

   a) Write a piecewise-defined function $T(x)$ that shows the total tax on an income of x dollars.

   b) What is the total tax on an income $95,000 ?

Solution:

   a) If $0 < x \le 50000$, $T(x) = 0.30x$

   If $x > 50000$, $T(x) = 0.30(50,000) + 0.40(x - 50,000)$

   $= 15,000 + 0.40x - 20,000$

   $= 0.40x - 5,000$

   We have the piecewise-defined function:

$$f(x) = \begin{cases} 0.30x & \text{if } 0 < x \le 50,000 \\ 0.40x - 5000 & \text{if } x > 50,000 \end{cases}$$

   b) $T(95,000) = 0.40(95,000) - 5,000 = \$33,000$.

EXERCISES

Evaluate each function when $x = -2, -1, 0, 1, 2,$ **and** 3.

1. $f(x) = \begin{cases} x+3, & x \leq 0 \\ x-3, & x > 0 \end{cases}$

2. $f(x) = \begin{cases} x^2 - 2, & x < 1 \\ -2x+4, & x \geq 1 \end{cases}$

3. $f(x) = \begin{cases} 4-x^2, & x < 2 \\ x+3, & x \geq 2 \end{cases}$

4. $f(x) = \begin{cases} x^2 + 3, & x \leq 1 \\ 2x^2 - 2, & x > 1 \end{cases}$

5. $f(x) = \begin{cases} x^2 - 3, & x < -1 \\ 4, & -1 \leq x \leq 0 \\ 5x-1, & x > 0 \end{cases}$

6. $f(x) = \begin{cases} 2x^2, & x \leq 1 \\ 3, & 1 < x \leq 3 \\ -2x+10, & x > 3 \end{cases}$

Sketch the graph of $f(x)$.

7. $f(x) = \begin{cases} 4 & \text{if } x \leq -2 \\ -3 & \text{if } x > -2 \end{cases}$

8. $f(x) = \begin{cases} 4 & \text{if } x < -2 \\ -x & \text{if } -2 \leq x \leq 3 \\ -2 & \text{if } x > 3 \end{cases}$

9. $f(x) = \begin{cases} x^2 - 2 & \text{if } x \leq 2 \\ \frac{1}{2}x + 2 & \text{if } x > 2 \end{cases}$

10. $f(x) = \begin{cases} 2 & \text{if } x \leq -1 \\ x^3 & \text{if } -1 < x \leq 1 \\ -2x+4 & \text{if } x > 1 \end{cases}$

11. $f(x) = \begin{cases} -2x^2 & \text{if } -1 \leq x < 1 \\ -0.5 & \text{if } x = 1 \\ x & \text{if } x > 1 \end{cases}$

Find the domain and range of $f(x)$.

12. $f(x) = \begin{cases} 2x & \text{if } x < 0 \\ 3 & \text{if } x \geq 0 \end{cases}$

13. $f(x) = \begin{cases} 2x & \text{if } x < 1 \\ 3 & \text{if } x \geq 1 \end{cases}$

14. $f(x) = \begin{cases} x^2 - 2 & \text{if } x \leq 2 \\ \frac{1}{2}x + 2 & \text{if } x > 2 \end{cases}$

15. $f(x) = \begin{cases} -2x^2 & \text{if } -2 \leq x < 1 \\ -0.5 & \text{if } x = 1 \\ x & \text{if } x > 1 \end{cases}$

16. $f(x) = \begin{cases} 2x^2 & \text{if } x \leq 1 \\ 3 & \text{if } 1 < x \leq 3 \\ -2x+10 & \text{if } x > 3 \end{cases}$

Write the piecewise-defined function whose graph is shown.

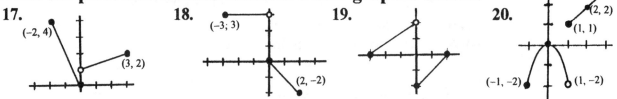

17. (-2, 4) (3, 2)

18. (-3; 3) (2, -2)

19.

20. (2, 2) (1, 1) (-1, -2) (1, -2)

21. Write the greatest integer function $f(x) = [x]$, $0 \leq x < 4$ in the form of the piecewise-defined function.

22. A publishing company pays income tax at a rate of 20% on the first income $50,000, and all income over $50,000 is taxed at a rate of 30%. a) Write a piecewise-defined function $T(x)$ that shows the total tax on an income of x dollars. b) What is the total tax on an income $120,000 ?

23. A publishing company pays income tax at a rate of 20% on the first income $50,000, 30% on the next $30,000, and 40% on all income over $80,000. a) Write a piecewise-defined function $T(x)$ that shows the total tax on an income of x dollars. b) What is the total tax on an income $120,000 ?

5-13 Square Root and Cubic Root Functions

We have learned how to graph the quadratic functions ($y = cx^2$) in Section 5-3 and the cubic functions ($y = cx^3$) in Section 5-7.

In this section we will learn to graph two types of radical functions, **the square root and cubic root functions**.

Understanding the basic shapes of the following two graphs will help us to analyze and graph more complicated square root and cubic root functions. (See page 150)

The graph of $y = \sqrt{x}$	The graph of $y = \sqrt[3]{x}$
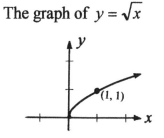	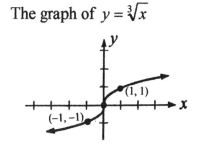
Starts at (0, 0) and passes through the point (1, 1).	Passes through (0, 0) and the points (−1, −1) and (1, 1).
Domain: $x \geq 0$. Range: $y \geq 0$	Domain: x are all real numbers.
It is the upper half of the parabola $x = y^2$ having a horizontal axis.	Range: y are all real numbers.

1. The graph of $y = a\sqrt{x}$ starts at (0, 0) and passes through the point (1, a).

2. The graph of $y = a\sqrt[3]{x}$ passes through (0, 0) and the points (−1, −a) and (1, a).

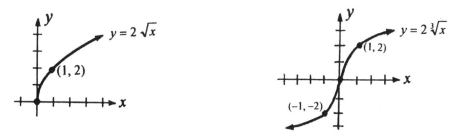

3. The graph of $y = a\sqrt{x - h} + k$ moves (shifts) the graph of $y = a\sqrt{x}$ by h units to the right and k units upward.

4. The graph of $y = a\sqrt[3]{x - h} + k$ moves (shifts) the graph of $y = a\sqrt[3]{x}$ by h units to the right and k units upward.

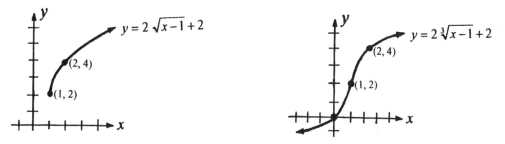

EXERCISES

Match the function with its graph.

1. $f(x) = \sqrt{x} + 2$ 2. $f(x) = \sqrt{x+3}$ 3. $f(x) = \sqrt{x-2}$ 4. $f(x) = \sqrt{x+1}$

5. $y = \sqrt[3]{x} + 2$ 6. $y = \sqrt[3]{x} - 1$ 7. $y = -\sqrt[3]{x-2}$ 8. $y = 3\sqrt[3]{x}$

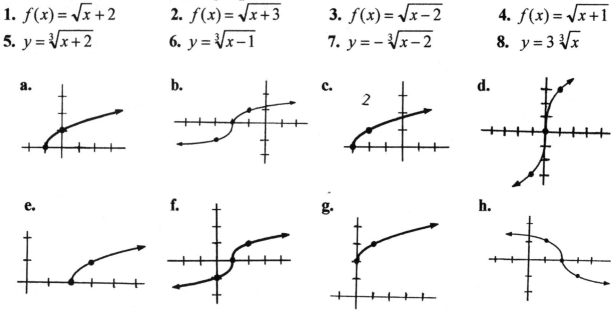

Use the graph of $y = \sqrt{x}$ **to compare and describe how each graph is moved.**

9. $f(x) = \sqrt{x} + 4$ 10. $f(x) = \sqrt{x} - 4$ 11. $f(x) = \sqrt{x+4}$ 12. $f(x) = \sqrt{x-4}$

13. $y = \sqrt{x-5} + 3$ 14. $y = \sqrt{x+5} + 3$ 15. $y = \sqrt{x+5} - 3$ 16. $y = \sqrt{x-5} - 3$

Use the graph of $y = \sqrt[3]{x}$ **to compare and describe how each graph is moved.**

17. $f(x) = \sqrt[3]{x} - 6$ 18. $f(x) = \sqrt[3]{x} + 6$ 19. $f(x) = \sqrt[3]{x+6}$ 20. $f(x) = \sqrt[3]{x-6}$

21. $y = \sqrt[3]{x-1} + 3$ 22. $y = \sqrt[3]{x+1} - 3$ 23. $y = \sqrt[3]{x+1} + 3$ 24. $y = \sqrt[3]{x-1} - 3$

Graph each of the square root and cubic root function.

25. $f(x) = \sqrt{x+4}$ 26. $f(x) = \sqrt{x-2}$ 27. $f(x) = \sqrt{x+2} + 1$ 28. $f(x) = \sqrt{2x}$

29. $y = 3\sqrt{x}$ 30. $y = 3\sqrt{x+1} - 2$ 31. $y = -2\sqrt{x-3}$ 32. $y = 2\sqrt[3]{x} + 2$

33. $y = 2\sqrt[3]{x+1} + 2$ 34. $y = 3\sqrt[3]{x+1}$ 35. $y = 3\sqrt[3]{x-1} + 3$

36. The time (t) in seconds that an object falls the distance (d) in feet is given by the formula $t(d) = \frac{1}{4}\sqrt{d}$. Find the time of an object falls 1,024 feet.

37. The radius (r) in meters of a spherical balloon is given by the formula $r = \sqrt[3]{\frac{3v}{4\pi}}$, where v is the volume of the balloon in cubic meters. Find the radius of the balloon if the volume is 36π cubic meters.

CHAPTER 5 EXERCISES

Simplify each expression. Assume that no denominator equals 0.

1. $2^{12} \cdot 2^{15}$ **2.** $3^8 \cdot 3^{14}$ **3.** $(4^9)^7$ **4.** $5^{-14} \cdot 5^{40}$ **5.** $-(4a^2b^3)^2$

6. $(-3a^3b^2)^2$ **7.** $x^5 \cdot x^{-12} \cdot x^{20}$ **8.** $y^{-3} \cdot y^{10} \cdot y^{12}$ **9.** $x^{2a} \cdot x^{-b} \cdot x$ **10.** $(2x^3)^2(3x^4)^3$

11. $(7x^{-2})^{-4}$ **12.** $64^{\frac{1}{3}}$ **13.** $64^{\frac{2}{3}}$ **14.** $(16x^4)^{\frac{3}{4}}$ **15.** $(\frac{2}{3}x^3)^2(\frac{3}{2}x^2)^4$

16. $\dfrac{3a^5b^6}{18a^2b^9}$ **17.** $\dfrac{x^6y^{-4}}{x^{-2}y^7}$ **18.** $\left(\dfrac{ab^2}{a^{-4}b^5}\right)^{-4}$ **19.** $\sqrt[3]{9} \cdot \sqrt{27}$ **20.** $\dfrac{4^{3\sqrt{2}}}{8^{2\sqrt{2}}}$

Solve each quadratic equation.

21. $x^2 - 52 = 0$ **22.** $(x+3)^2 = 52$ **23.** $(x-3)^2 + 52 = 0$

24. $4(x+3)^2 - 52 = 0$ **25.** $3x^2 - 5x - 2 = 0$ **26.** $2y^2 + 9y - 5 = 0$

27. $x^2 - x - 1 = 0$ **28.** $y^2 + 3y - 5 = 0$ **29.** $a^2 + 3a + 5 = 0$

30. $2x^2 - 3x + 2 = 0$ **31.** $x^3 - 2x^2 - 3x = 0$ **32.** $x^3 + 2x^2 + 5x = 0$

33. $x^4 + 2x^2 - 8 = 0$ **34.** $x^8 - 13x^4 + 36 = 0$ **35.** $2a + 5\sqrt{a} - 12 = 0$

Find the vertex and zeros ($x-$ intercepts). Then graph each parabola.

36. $f(x) = x^2 - 4$ **37.** $f(x) = -x^2 + 4$ **38.** $f(x) = -(x-3)^2 - 5$

39. $f(x) = 4(x+5)^2 - 16$ **40.** $f(x) = x^2 + 6x - 1$ **41.** $f(x) = (x+6)(x-4)$

42. Find the minimum cost if the equation of cost is $c(x) = 2x^2 - 32x + 218$, where $c(x)$ is the cost and x is the number of product manufactured ?

43. A rocket is thrown upward from the top of a tower 90 feet high with an initial upward speed of 89.6 feet per second. The height of the rocket is given by the formula $h = 90 + 89.6t - 16t^2$, where t is the time in seconds. What is the maximum height that the rocket reaches ?

Solve polynomial equations by factoring.

44. $5x^3 + 20x = 0$ **45.** $2n^3 - 8n = 0$ **46.** $y^4 - y^3 - 12y^2 = 0$

47. $p^5 + 3p^3 - 4p = 0$ **48.** $x^5 + 27x^2 = 0$ **49.** $x^5 - 27x^2 = 0$

50. $3x^3 - 12x^2 + 12x = 0$ **51.** $2x^3 + 16x^2 + 32x = 0$ **52.** $x^6 - 64 = 0$

Use synthetic division, factor theorem, or remainder theorem to determine whether $x - c$ is a factor of the given polynomial

53. $f(x) = x^4 + 3x^3 - 8x^2 + x - 10; \ x - 2$ **54.** $f(x) = x^5 + x^4 - x^2 - 4x + 12; \ x + 2$

55. $f(x) = x^5 - 7x^3 - 20x + 9; \ x - 3$ **56.** $f(x) = x^6 - 3x^3 + 4x^2 - 3x + 1; \ x - i$

57. $f(x) = 4x^{30} - 2x^{20} + x^2 + x - 4; \ x - 1$ **58.** $f(x) = 5x^{23} + 3x^{40} - x^{19} + x + 3; \ x + 1$

59. Find the polynomial equation with the given roots -4 and $\pm 2i$.

60. Find the polynomial equation with the given roots 5 and $\pm\sqrt{6}$.

61. Is $\sqrt{3}\,i$ a root (zero) of the equation $x^3 - 3x^3 + 3x - 9 = 0$? Explain.

62. Is $\sqrt{5}$ a root (zero) of the equation $2x^3 + x^2 - 10x - 5 = 0$? Explain. -----Continued-----

Find all zeros of the polynomial function. State the multiplicity of each root.

63. $f(x) = x(x+5)(x-6)$ **64.** $f(x) = x^4 - 2x^2 - 48$ **65.** $f(x) = x^2(x-2)^3(x+4)$

66. $f(x) = 8x^4(x-6)^2(x^2+36)$ **67.** $f(x) = 8x^4(x-6)^2(x^2-36)$

68. Find the polynomial equation that has the roots 2(multiplicity 3) and $\pm 2i$.

69. Find the polynomial function that has the zeros -3(multiplicity 2), $\pm 3i$ and satisfied the condition $f(3) = 216$.

70. Use Descartes' Rule of Signs to determine the nature of the roots of
$$x^5 - 6x^4 + 16x^3 - 32x^2 + 48x + 32 = 0$$

71. Use Descartes' Rule of Signs to determine the maximum number of points can the graph of
$y = 4x^4 + 24x^3 + 72x^2 + 216x + 324$ cross the x-axis.

72. Use synthetic division and theorem on bounds to determine if each value of x is an upper bound, lower bound, or neither of the roots of the equation $x^4 - 2x^3 - 25x^2 + 26x + 120 = 0$.

73. Is $f(x) = 5x^2 - 2$ an even or an odd function ?

74. Is $f(x) = 5x^3 - 2x$ an even of an odd function ?

75. $f(x) = 5x^3 - 2$ an even or an odd function ?

76. Identify the general shape of the graph of $y = -x^4 + 3x^3 - x^2 - 3x + 2$.

77. Identify the general shape of the graph of $y = x^5 - 5x^3 + 4x$.

Solve each polynomial equation in the complex number system.

78. $x^2 - 49 = 0$ **79.** $2x^2 + 98 = 0$ **80.** $6x^2 + x - 12 = 0$ **81.** $x^2 + 3x - 2 = 0$

82. $x^2 + 3x + 2 = 0$ **83.** $5x^2 + 2x - 1 = 0$ **84.** $5x^2 + 2x + 1 = 0$ **85.** $x^4 + 27x = 0$

86. $x^4 - 27x = 0$ **87.** $x^3 + x^2 - 11x + 10 = 0$ **88.** $4x^3 + 9x^2 + x - 2 = 0$

89. $x^5 + 2x^4 + 3x^3 + 6x^2 - 4x - 8 = 0$ **90.** $x^5 + 4x^4 + x^3 - 10x^2 - 4x + 8 = 0$

91. A quadratic equation with real coefficients has the given root $2 - i$. Find such an equation.

92. A cubic equation with real coefficients has the given roots -2 and $-\frac{1}{8} + \frac{\sqrt{17}}{8}$. Find such an equation.

93. Find the polynomial function of least degree with real coefficients, a leading coefficient of 1, and the given zeros are -1, 1 and $2i$.

94. Find the polynomial function in factored form that has -2(multiplicity 4), 1(multiplicity 3) and 3(a double root); and satisfied $f(0) = 16$.

95. Use the Intermediate Value Theorem to show that the function
$f(x) = x^5 + 2x^4 + 3x^3 + 6x^2 - 4x - 8$ has a zero in the interval [0, 2].

96. Estimate the zeros (roots) of $f(x) = x^3 + 3x - 1$ correct to three decimal places.

97. The graph of a function is made after the two transformations applied to the graph of $f(x) = x^2$. Write the function.   1. Shift up 3 units   2. Shift left 4 units

98. The graph of a function is made after the three transformations applied to the graph of $f(x) = 2x^3$. Write the function.
   1. Shift down 1 unit   2. Shift right 4 units   3. Reflect about the y-axis.   **----Continued----**

99. The volume of a rectangular box is 3024 cubic inches. The length is 6 inches greater than the width. The height is 4 inches less than the length. Find the dimension of the box.

100. The population of a kind of snake is 320. Suppose the number (n) of the snake after t years is given by the function $n(t) = -t^4 + 59t^2 + 320$. When does the population of the snake become extinct ?

101. The total cost c (in dollars) to print math books is given by the function $c(x) = 0.004x^2 - 34x + 100000$ where x is the number of copies printed. How many copies should be printed to obtain a minimum cost ?

102. The total profit p (in dollars) to sell math books is given by the function $p(x) = -0.004x^2 + 34x - 30000$ where x is the number of copies sold. How many copies should be sold to obtain a maximum profit ?

103. Use a graphing calculator to find the cubic function of best-fit to the following data:

x	1	2	3	4	5	6	7
$f(x)$	16.6	19.2	16.6	9.4	−1.8	−16.4	−33.8

Graph each of the following polynomial functions.

104. $y = x^3 + 2x$ 105. $y = x^4 + 4x^2$ 106. $y = (x^2 - 1)(x^2 - 4)$ 107. $f(x) = 2x^3 + x^2 - x$

108. $f(x) = -x^5 + 5x^3 - 4x$ 109. $f(x) = x^3 + 4x^2 + 4x$ 110. $y = |x^2 + 2|$

Solve each inequality.

111. $x^2 - 3x - 4 > 0$ 112. $x^2 - 3x \le 4$ 113. $x^3 - 3x^2 - 4x \ge 0$ 114. $x^3 - 5x^2 - x + 5 > 0$

115. $|x - 2| > 6$ 116. $|x - 2| \le 6$ 117. $|x^2 + 2x + 1| > 9$ 118. $|x^2 + 2x + 1| \le \frac{25}{4}$

119. $x^2 + 2x - 4 \ge 2$ 120. $|x^2 + 2x - 4| \ge 2$ 121. $2x^2 - 6x + 1 < 0$ 122. $x^2 - 8x + 21 < 0$

123. A rocket is fired vertically upward from the ground with an initial speed of 256 feet per second. The height of the rocket is given by $h(t) = vt - 16t^2$, where v is the initial speed and t is the time in seconds. During what time period will its height exceed 960 feet ?

Find the inverse function. If it has no inverse function, so state.

124. $f(x) = 2x + 10$ 125. $f(x) = \frac{x-10}{2}$ 126. $f(x) = 2x^2 - 5$

127. $f(x) = 2x^2 - 5, x \ge 0$ 128. $f(x) = \sqrt[5]{x - 2}$ 129. $f(x) = \frac{3x-2}{x+1}$

130. $f(x) = |2x + 2|, x \le -1$ 131. $f(x) = \sqrt{a^2 - x^2}, 0 \le x \le a$

132. If $f = \{(2, 1), (3, 5), (4, 5)\}$, find the inverse and determine whether the inverse is a function.

133. The Celsius temperature (y) is given by the formula $y = \frac{5}{9}(x - 32)$, where x is the Fahrenheit temperature.
   a) By interchanging x and y, find the formula of the Fahrenheit temperature (y).
   b) Prove that these two formula are inverse function of each other.

-----Continued-----

134. If $f(x) = 3x^2 - x + 1$ and $g(x) = x - 1$, find $(f \circ g)(x)$ and $(g \circ f)(x)$.

135. If $f(x) = x^2 - 1$ and $x \geq 0$, find f^{-1}, $f \circ f^{-1}$ and $f^{-1} \circ f$.

136. John bought a new car that was original priced at $21,500. He was given a factory rebate of $2,000, followed by a dealer's discount of 10%. Let x represents the original price, $r(x)$ represents the price after rebate, and $p(x)$ represents the price after the dealer's discount. a) Express $r(x)$ in function notation.   b) Express $p(x)$ in function notation.
c) Express the composite function $p(r(x))$.   d) Find the price John paid.

137. The cost of mailing a first-class letter is $0.37 for the first ounce and $0.23 for each additional ounce or portion of an ounce.
   a) Write the function that describes the relationship between the number of ounces (x) and the cost (c) of postage.
   b) Find the cost of postage weighing 3.4 ounce.

138. A school is planning to rent buses for the students to attend a field trip. Each bus can accommodate a maximum of 50 students. Use the greatest integer function to write the relationship between the number (n) of the students and the number (b) of buses needed.

139. The monthly charge of long-distance calls for x minutes is $5.40 plus $0.15 per minute for the first 100 minutes in the month and $0.12 per minute for all calls over 100 minutes.
   a) Write a piecewise-defined function that shows the total monthly charge $c(x)$ for using x minutes.
   b) What is the monthly charge for using 90 minutes in a month ?
   c) What is the monthly charge for using 150 minutes in a month ?

140. A telephone company have the following rate schedule for monthly customer charge:

Monthly service charge	$19.50
For the 1st 50 minutes	$0.10/minute
Next 50 minutes	$0.06/minute
Over 100 minutes	$0.04/minute

   a) Write a piecewise-defined function that shows the total monthly charge $c(x)$ for using x minutes.
   b) What is the charge for using 80 minutes in a month ?
   c) What is the charge for using 120 minutes in a month ?

Graph each of the following square root and cubic root function.

141. $f(x) = \sqrt{x-4}$ **142.** $f(x) = \sqrt{x-4} + 1$ **143.** $f(x) = 2\sqrt{x+4}$

144. $f(x) = \sqrt[3]{x-4}$ **145.** $f(x) = \sqrt[3]{x-4} + 1$ **146.** $f(x) = 2\sqrt[3]{x+4}$

A mother asks her 6-year old son about his new eyeglasses.
Mother: Can you see very well with your new eyeglasses when you are in the class.
   Son: Yes. I can see very clear all of my teacher's writing on the blackboard.
Mother: Will you get better grades ?
   Son: No. I still have the same brain.

Answers

CHAPTER 5 – 1   Laws of Exponents   page 114

1. 2^7 **2** 2^{12} **3.** a^7 **4.** a^{12} **5.** m^{15} **6.** m^{50} **7.** a^{n+2} **8.** a^{2n} **9.** 10^8 **10.** 10^{15} **11.** x^{13}

12. $-a^6$ **13.** a^6 **14.** a^5 **15.** a^{-5} **16.** 3^0, or 1 **17.** x^{a+b} **18.** a^{x+y} **19.** a^{xy} **20.** a^0, or 1

21. $-27a^{-6}$ **22.** a^{-12} **23.** 2 **24.** 4 **25.** 25 **26.** $6x^7$ **27.** $2a^7$ **28.** $16a^8$ **29.** $48xy^{14}$

30. $\frac{9}{4}x^{16}$ **31.** 32 **32.** $\frac{1}{32}$ **33.** a^5 **34.** a^{-5} **35.** 100,000 **36.** 8 **37.** 125 **38.** $\frac{1}{8}$ **39.** $\frac{1}{125}$

40. x^{-4} **41.** x^4 **42.** x^{-6} **43.** 1 **44.** 1 **45.** x^{12-n} **46.** x^{n-12} **47.** $\frac{a^4}{b^4}$ **48.** x^{2-n} **49.** $\frac{a^8}{b^{12}}$

50. $\frac{x^{5n}}{y^{5n}}$ **51.** y^3 **52.** $\frac{a^3}{7b^3}$ **53.** $\frac{n^2}{m^4}$ **54.** $-\frac{12}{k^5}$ **55.** $-3x^2y^2$ **56.** $4a^5$ **57.** $9x^2$ **58.** $\frac{1}{32x^3}$

59. $\frac{1}{2m^2}$ **60.** $\frac{1}{3}$ **61.** $-\frac{m^4}{5}$ **62.** $-2a^4$ **63.** $-\frac{1}{x^2}$ **64.** $\frac{3}{4r^2}$ **65.** $-8x$ **66.** $2y^2$ **67.** $4a$

68. 4 **69.** $-a^3$ **70.** $\frac{1}{m^2}$ **71.** $\frac{16}{81}$ **72.** $\frac{81}{16}$ **73.** $\frac{4a^4}{b^6}$ **74.** $\frac{1}{x^6 y^3}$ **75.** $x^6 y^3$ **76.** $\frac{27b^3}{8a^3}$

77. $\frac{1}{a^3}$ **78.** a^8 **79.** $\frac{x^2}{2}$ **80.** $\frac{x^7}{y^5}$ **81.** $\frac{1}{x^{10}}$ **82.** $\frac{t^3}{4s^4}$ **83.** $-\frac{y^3}{x^6}$ **84.** $\frac{2x^3}{a^2}$ **85.** $-\frac{x^4}{16y^3}$

86. $8^{5/6}$ **87.** 1 **88.** $3^{4\sqrt{2}}$ **89.** $3^{\pi+3\sqrt{3}}$ **90.** $4\cdot 2^\pi$

CHAPTER 5 – 2   Solving Quadratic Equations   page 116

1. $x=\pm 2$ **2.** $x=\pm 2i$ **3.** $x=0,4$ **4.** $x=0,-4$ **5.** $x=-2\pm\sqrt{6}$ **6.** $y=4\pm 2\sqrt{3}$

7. $x=-5\pm 3\sqrt{2}$ **8.** $x=-2\pm\sqrt{6}i$ **9.** $y=4\pm 2\sqrt{3}$ **10.** $x=1\frac{1}{3},-1$ **11.** $x=\frac{3\pm\sqrt{105}}{12}$

12. $x=\frac{3}{4},-1\frac{2}{3}$ **13.** $x=1\frac{1}{3},1$ **14.** $x=-\frac{1}{3},1\frac{1}{3}$ **15.** $x=1\pm\sqrt{5}$ **16.** $y=2,5$ **17.** $x=1\pm\sqrt{3}\,i$

18. $y=-1,8$ **19.** $x=-1\frac{1}{3},1$ **20.** $x=\frac{4\pm 2\sqrt{13}}{9}$ **21.** $t=-2\pm 2i$ **22.** $t=-2\pm\sqrt{13}$ **23.** $x=\frac{-1\pm\sqrt{2}i}{3}$

24. $x=\frac{1\pm\sqrt{7}}{3}$ **25.** $x=\pm\sqrt{5}\,i,\pm\sqrt{2}$ **26.** $a=\pm\sqrt{5},\pm 2i$ **27.** $x=0,-\frac{1}{3},2$ **28.** $x=0,\frac{1\pm\sqrt{7}}{3}$

29. $x=\pm\sqrt{2}\,i,\pm\sqrt{2},\pm i,\pm 1$ **30.** $x=\pm\sqrt[4]{8},\pm\sqrt[4]{8}\,i,\pm\sqrt[4]{3}\sqrt{i},\pm\sqrt[4]{3}\sqrt{-i}$ **31.** $v=\frac{-1\pm\sqrt{97}}{4}$

32. $v=\frac{7\pm\sqrt{47}}{4}$ **33.** $x=2\pm\sqrt{5}$ **34.** $x=\frac{\sqrt{2}\pm\sqrt{6}i}{2}$ **35.** $x=\frac{\sqrt{3}\pm\sqrt{11}}{2}$ **36.** $p=\frac{\sqrt{2}\pm 3\sqrt{2}i}{2}$ **37.** $x=\frac{-1\pm\sqrt{97}}{4}$

38. $x=\frac{1\pm\sqrt{19}}{3}$ **39.** $x=\frac{-3\pm\sqrt{41}}{4}$ **40.** $x=\frac{3\pm\sqrt{633}}{8}$ **41.** $t=\frac{-1\pm\sqrt{433}}{8}$ **42.** $x=\frac{3\pm\sqrt{17}}{4}$ **43.** $x=\frac{-1\pm\sqrt{35}i}{2}$

44. $x=\frac{3\pm\sqrt{11}i}{2}$ **45.** $x=-3,1$ **46.** $t=4$ **47.** $x=1$ **48.** $x=\frac{1}{4}$

CHAPTER 5 – 3   Graphing Quadratic Functions-Parabolas   page 123 ~ 124

1. Opens downward **2.** Opens upward **3.** Opens upward **4.** Opens upward
5. Opens downward **6.** Opens upward **7.** Opens upward **8.** Opens downward
9. Open upward **10.** One **11.** Two **12.** One **13.** None **14.** Two **15.** One **16.** One
17. None **18.** Two **19.** Two **20.** None **21.** One
22. x – intercepts = 3 and –2 **23.** x – intercepts = –3 and 2 **24.** x – intercepts = 6 and –1
25. x – intercepts = –6 and 1 **26.** x – intercepts = 7 and –4 **27.** x – intercepts = –5 and –7
28. x – intercept = 1 only **29.** x – intercepts = 2.2 and –0.2 **30.** x – intercepts = 2 and –2
31. x – intercepts = none **32.** x – intercepts = 4.6 and –0.6 **33.** x – intercept = 1.9 and –0.2
34. (7, 5) **35.** (4, 1) **36.** (2, –4) **37.** (–4, –1) **38.** (7, –5) **39.** (–2, 4) **40.** (4, –2)
41. (5, –1) **42.** (–4, –2) **43.** (–5, –1) **44.** (–7, 9) **45.** (6, 12) **46.** (–12, 4) **47.** (1, –7)
48. (10, –9) **49.** (0, 12) **50.** (0, –7)

51. $y=(x+2)^2$, vertex (–2, 0)   **52.** $y=(x-2)^2$, vertex (2, 0)
53. $y-5=(x+2)^2$, vertex (–2, 5)   **54.** $y+5=(x-2)^2$, vertex (2, –5)
55. $y+2=(x-3)^2$, vertex (3, –2)   **56.** $y-2=(x+3)^2$, vertex (–3, 2)
57. $y+5=(x-1)^2$, vertex (1, –5)   **58.** $y+12=(x-5)^2$, vertex (5, –12)
59. $y-9=(x+7)^2$, vertex (–7, 9)   **60.** $y+1=(x+5)^2$, vertex (–5, –1)
61. $y+4=3(x-2)^2$, vertex (2, –4)   **62.** $y-2=2(x+4)^2$, vertex (–4, 2)

178   A-Plus Notes for Algebra

CHAPTER 5 – 3   Graphing Quadratic Function – Parabola   page 123 ~ 124 (continued)

63. $y=\frac{1}{2}x^2$ **64.** $y=-\frac{1}{2}x^2$ **65.** $y=x^2+1$ **66.** $y=-x^2+2$

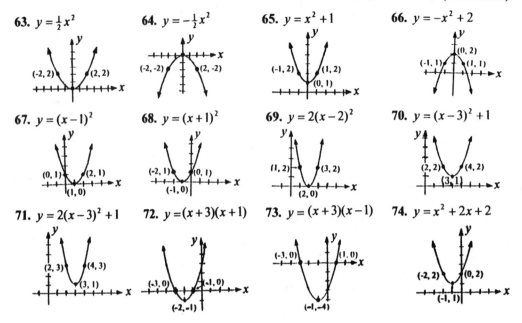

67. $y=(x-1)^2$ **68.** $y=(x+1)^2$ **69.** $y=2(x-2)^2$ **70.** $y=(x-3)^2+1$

71. $y=2(x-3)^2+1$ **72.** $y=(x+3)(x+1)$ **73.** $y=(x+3)(x-1)$ **74.** $y=x^2+2x+2$

75. $k=25$ **76.** $k>-\frac{9}{20}$ **77.** $k<-\frac{1}{20}$ **78.** It moves the graph of $y=x^2$, one unit upward.

79. It moves the graph of $y=x^2$, 1 unit to the left

80. It moves the graph of $y=x^2$, 2 units to the right and 3 units downward.

81. a. 1 x–intercept **b.** None **c.** 2 x–intercepts **82.** $y_{min.}=-9$ at $x=2$ **83.** $y_{max.}=18$ at $x=2$

84. $40cm\times 40cm=1600\ cm^2$ **85.** $P_{max.}=18$ at $x=4$ **86.** 256 feet **87.** 896 feet

88. a. 438 feet **b.** 1445 feet **89.** 20 by 15 inches **90.** 40 by 20 inches **91.** 100 by $66\frac{2}{3}$ meters

CHAPTER 5 – 4   Solving Polynomial Equations by Factoring   page 126

1. $x=-2,1$ **2.** $x=10,-8$ **3.** $x=4$ (a double root) **4.** $x=12$ (a double root) **5.** $x=-2,1$
6. $x=2,-1$ **7.** $x=-1,1$ **8.** $x=\frac{1}{2},2$ **9.** $x=0,15$ **10.** $x=0,5,-2\frac{2}{3}$ **11.** $x=0,-1\frac{1}{2}$
12. $x=0,1\frac{1}{2}$ **13.** $x=0,4$ **14.** $x=0,x=\frac{2}{3}$ **15.** $x=0,7$ **16.** $x=0,\frac{1}{7}$ **17.** $x=0$ (a double root), 5
18. $a=0$ (a double root), $\frac{1}{2}$ **19.** $x=0,5$ **20.** $x=0$ (a double root), 5 **21.** $y=0,12$
22. $x=0$ (a double root),2 **23.** $p=0,1$ **24.** $n=0$ (a double root),1,–1 **25.** $x=0,10$ **26.** $a=5,-7$
27. $p=-5,5$ **28.** $x=9$ (a double root) **29.** $n=-2,-12$ **30.** $x=-2,6$ **31.** $x=-2,\frac{1}{2}$
32. $x=0,2,-4$ **33.** $y=0,-\frac{1}{2},4$ **34.** $x=0,2,-\frac{1}{2}$ **35.** $x=-2,-1$ **36.** $x=2,1$ **37.** $n=6,-1$
38. $x=-8$ (a double root) **39.** $x=2$ (a double root) **40.** $x=8,2$ **41.** $x=-1\frac{1}{2},-1$ **42.** $x=1\frac{1}{2},-4$
43. $x=\frac{1}{2},1\frac{1}{2}$ **44.** $x=-\frac{1}{3},2\frac{1}{2}$ **45.** $x=-\frac{3}{4},5$ **46.** $x=\frac{1}{6},2$ **47.** $a=3,-3$ **48.** $a=-1\frac{1}{2},1\frac{1}{2}$
49. $n=0,1,-1$ **50.** $x=4$ (a double root) **51.** $a=-\frac{1}{8},1$ **52.** $y=\frac{1}{2},-\frac{2}{3}$ **53.** $x=0,3,-2$
54. $x=0,-3,\frac{1}{2}$ **55.** $y=0$ (a double root), 4(a double root) **56.** $p=0,3,-3,2,-2$ **57.** $x=-4,5$
58. $x=1,-5$ **59.** $x=7,-2$ **60.** $x=-\frac{1}{3},2$ **61.** $x=0,\pm2$ **62.** $x=0,x=\pm2i$ **63.** $x=\pm1,\pm\sqrt{2}$
64. $x=\pm\sqrt{3}i,\pm\sqrt{5}i$ **65.** $x=\pm2,\pm\sqrt{2}i$ **66.** $x=0,\pm2,\pm2i$ **67.** $x=0$ (a double root), $\pm i$
68. $x=0$ (a double root), $-1\pm\sqrt{2}$ **69.** $x=0$ (a double root),–1(a double root) **70.** $x=0,\pm2i,\pm\sqrt{3}$
71. $x=0$ (a double root),–1,–2 **72.** $x=0$ (a double root), 1, 2 **73.** $x=-4$ (a double root)
74. $x=0,-2,1\pm\sqrt{3}i$ **75.** $x=0,3,\dfrac{-3\pm\sqrt{3}i}{2}$ **76.** $x=0$ (a double root), 2, $-1\pm\sqrt{3}i$

77. $x=\pm1,\pm i$ **78.** $x=-1,1,\dfrac{1\pm\sqrt{3}i}{2},\dfrac{-1\pm\sqrt{3}i}{2}$

CHAPTER 5 – 5   Synthetic Division and Factor Theorem   page 131 ~ 132

1. $x - 7 + \frac{20}{x+2}$ **2.** $x + 7 + \frac{8}{x-2}$ **3.** $x + 1 - \frac{6}{x+2}$ **4.** $x - 1 + \frac{2}{x-2}$ **5.** $2a + 3 - \frac{9}{2a-1}$ **6.** $2a - 3 + \frac{16}{3a+2}$

7. $x^2 - 3x - 2 - \frac{6}{x-2}$ **8.** $x^2 - 7x + 18 - \frac{38}{x+2}$ **9.** $x^2 - 2x + 3 - \frac{11}{2x+4}$ **10.** $x^2 + 2x + 5 + \frac{20}{2x-5}$

11. $-4x^2 + 10x - 21 + \frac{43}{x+2}$ **12.** $2x^3 + 7x^2 + 36x + 179 + \frac{900}{x-5}$ **13.** $2x^3 - 7x^2 + 34x - 169 + \frac{840}{x+5}$

14. $x^3 + 2x^2 + 5x + 10 + \frac{22}{x-2}$ **15.** $x^3 - 2x^2 + 3x - 6 + \frac{14}{x+2}$ **16.** $x^4 - 3x^3 + 5x^2 - 15x + 46 - \frac{138}{x+3}$

17. $x^4 + 3x^3 + 13x^2 + 39x + 116 + \frac{348}{x-3}$ **18.** $2n^2 - 5n + 8 - \frac{13n+3}{n^2+2n+1}$ **19.** $2x^2 - 2x + 5 - \frac{4x-7}{x^2+x-3}$

20. $x^2 + 2ax + \frac{a^3}{x+2a}$ **21.** $2a^2 + 3$ **22.** $a^3 + 4a^2 + 16a + 64 + \frac{252}{a-4}$ **23.** $x^2 - a^2 + \frac{2a^4}{x^2+a^2}$

24. $x^4 - x^3 + x^2 - x + 1 + \frac{4}{x+1}$ **25.** yes   **26.** no   **27.** no   **28.** yes   **29.** yes   **30.** yes   **31.** yes   **32.** no

33. yes   **34.** yes   **35.** yes   **36.** no   **37.** yes   **38.** no   **39.** yes   **40.** no   **41.** yes   **42.** yes   **43.** yes   **44.** No

45. $x = -3, -2, -1$ **46.** $x = -3, -2, \frac{1}{2}$ **47.** $x = -\frac{1}{2}, \pm\sqrt{5}$ **48.** $x = 3, \pm\sqrt{3}\,i$ **49.** $x = 2, 1 \pm 4i$

50. $x = -2, 1 \pm 2i$ **51.** $x = -2, 1, \frac{3}{2}$ **52.** $x = \frac{3}{2}, 1 \pm 2i$

Compare the answers of 53 ~ 60 to the equations of 45 ~ 52:   **53.** $x^3 + 6x^2 + 11x + 6 = 0$

54. $2x^3 + 9x^2 + 7x - 6 = 0$ **55.** $2x^3 + x^2 - 10x - 5 = 0$ **56.** $x^3 - 3x^2 + 3x - 9 = 0$

57. $x^3 - 4x^2 + 21x - 34$ **58.** $x^3 + x + 10 = 0$ **59.** $2x^3 - x^2 - 7x + 6 = 0$ **60.** $2x^3 + x^2 - 8x + 3 = 0$

Compare the answers of 61 ~ 80 to the answers of 25 ~ 44:   **61.** yes   **62.** no   **63.** no   **64.** yes

65. yes   **66.** yes   **67.** yes   **68.** no   **69.** yes   **70.** yes   **71.** yes   **72.** no   **73.** yes   **74.** no   **75.** yes

76. no   **77.** yes   **78.** yes   **79.** yes   **80.** no   **81.** $k = 7$ **82.** $k = -1$ **83.** $k = 6$ **84.** $k = 3$

85. 1   **86.** 3   **87.** $x = -1, 1, \frac{1}{2}, 2$ **88.** yes   **89.** yes   **90.** $f(x) = 3x^3 - 5x^2 + x - 2$

CHAPTER 5 – 6   Finding Zeros of Polynomial Functions with Degree n   page 147 ~ 148

1. $x = 0, 1, -2$ **2.** $x = -5, 0, 4$ **3.** $x = -\frac{1}{2}, 0(\text{a double root}), 2(\text{multiplicity 3})$

4. $x = -2(\text{a double root}), -3(\text{multiplicity 4}), 3(\text{multiplicity 4})$

5. $x = -4(\text{multiplicity 3}), 3(\text{multiplicity 3}), \frac{1}{3}(\text{a double root}), -\frac{1}{3}(\text{a double root})$

6. $x = -\frac{3}{2}(\text{multiplicity 6}), \frac{3}{2}(\text{multiplicity 5})$ **7.** $x = -2(\text{multiplicity 5}), -1(\text{multiplicity 2})$

8. $x = \pm 2i, \pm 3$ **9.** $x = 0, \pm 1, \pm i (\text{multiplicity 3})$ **10.** $x = 0(\text{multiplicity 3}), -4, 4(\text{multiplicity 3})$

11. $x^2 + 3x - 4 = 0$ **12.** $x^2 - 3x - 4 = 0$ **13.** $x^2 - 3x - 10 = 0$ **14.** $x + 3x - 10 = 0$

15. $x^3 + x^2 - 6x = 0$ **16.** $x^3 - x^2 - 6x = 0$ **17.** $x^3 + 7x^2 + 2x - 40 = 0$ **18.** $x^3 - x^2 + 4x - 4 = 0$

19. $x^4 + 6x^3 + 25x^2 + 96x + 144 = 0$ **20.** $x^5 - 8x^4 + 26x^3 - 48x^2 + 40x - 16 = 0$

21. $f(x) = 3x^2 + 9x - 12$ **22.** $f(x) = 2x^2 - 6x - 8$ **23.** $f(x) = 2x^2 - 6x - 20$

24. $f(x) = \frac{1}{2}x^2 + \frac{3}{2}x - 5$ **25.** $f(x) = -\frac{1}{3}x^3 - \frac{1}{3}x^2 + 2x$ **26.** $f(x) = x^3 - x^2 - 6x$

27. $f(x) = -2x^3 - 14x^2 - 4x + 8$ **28.** $f(x) = 6x^3 - 6x^2 + 24x - 24$

29. $f(x) = 2x^4 + 12x^3 + 50x^2 + 192x + 288$ **30.** $f(x) = 6x^5 - 48x^4 + 156x^3 - 264x^2 + 240x - 96$

31. a. 1 positive real root, 2 negative real roots, 0 imaginary root.

  b. 1 positive real root, 0 negative real root, 2 imaginary roots.

32. a. 3 negative real roots, 0 imaginary root.   b. 1 negative real root, 2 imaginary roots.

33. a. 1 positive real root, 3 negative real roots, 0 imaginary root.

  b. 1 positive real root, 1 negative real root, 2 imaginary roots.

34. a. 2 positive real roots, 1 negative real root, 2 imaginary roots.

  b. 0 positive real root, 1 negative real root, 4 imaginary roots.

35. 1 positive real root, 6 imaginary roots.

36. 1 positive real root, 1 negative real root, 6 imaginary roots.

37. The graph crosses the $x - axis$ at a maximum of 4 different points.

38. a) lower bound b) neither c) upper bound   **39.** a) lower bound b) lower bound c) upper bound

40. a) lower bound b) neither c) neither   **41.** a) lower bound b) neither c) neither

42. 2 is the smallest upper bound of the roots.   **43.** 7 is the smallest upper bound of the roots.

  −8 is the largest lower bound of the roots.   −4 is the largest lower bound of the roots.

44. 3 is the smallest upper bound of the roots.   **45.** 1 is the smallest upper bound of the roots.

  −3 is the largest lower bound of the roots.   −3 is the largest lower bound of the roots.

CHAPTER 5 – 6   page 147 ~ 148 (continued)

46. Even **47.** Odd **48.** Odd **49.** Odd **50.** Even **51.** Neither **52.** Either **53.** Odd

54. $f(x)$ increases without bound as x increases without bound, decreases without bound as x decreases without bound, and symmetric about the origin.

55. $f(x)$ decreases without bound as x increases without bound, and increases without bound as x decreases without bound.

58. $f(x)$ increases without bound as x increases without bound, increases without bound as x decreases without bound, and symmetric about the $y-axis$.

60. $f(x)$ increases without bound as x increases without bound, decreases without bound as x decreases without bound, and symmetric about the origin.

61. $x=\pm 3$ **62.** $x=\pm 3i$ **63.** $x=\pm 4$ **64.** $x=\pm 4i$ **65.** $x=\pm 8i$ **66.** $x=\pm 8$ **67.** $x=\pm 2\sqrt{2}$

68. $x=0$ or 4 **69.** $x=\frac{1}{3}$ or -2 **70.** $x=-2\pm 2i$ **71.** $x=3\pm i$ **72.** $x=-\frac{1}{2}\pm \frac{\sqrt{5}}{2}$ **73.** $x=-\frac{1}{2}\pm \frac{\sqrt{3}}{2}i$

74. $x=-\frac{1}{2}\pm \frac{\sqrt{19}}{2}i$ **75.** $x=-3\pm \sqrt{2}i$ **76.** $x=1\pm 2i$ **77.** $x=-\frac{1}{3}\pm \frac{\sqrt{7}}{3}$ **78.** $x=-\frac{1}{3}\pm \frac{\sqrt{5}}{3}i$

79. $x=\frac{1}{3}\pm \frac{2}{3}i$ **80.** $x=2$ or $-1\pm \sqrt{3}i$ **81.** $x=-2$ or $1\pm \sqrt{3}i$ **82.** $x=\pm 2$ or $\pm 2i$ **83.** $x=\pm 1$ or $\pm i$

84. $x=1$ or $3\pm i$ **85.** $x=-3$ or $1\pm 2i$ **86.** $x=-1$ or $-\frac{1}{3}\pm \frac{\sqrt{7}}{3}i$ **87.** $x=2$ or $\frac{1}{5}\pm \frac{2}{5}i$

88. $x=-1, 1,$ or $\pm 2i$ **89.** $x=-3, -2,$ or $\frac{1}{2}\pm \frac{\sqrt{3}}{2}i$ **90.** $x=-4, \frac{1}{2},$ or -4 (a double root)

91. $x=4, \pm \sqrt{2}i$ **92.** $x=1$ or $1\pm 2i$ **93.** $x=\pm \sqrt{2}$ or $\pm 2i$ **94.** $x=\frac{1\pm \sqrt{3}i}{2}$ or $\pm \sqrt{3}i$ **95.** no **96.** no

97. $x^2-8x+17=0$ **98.** $x^3-5x^2+11x-15=0$ **99.** $f(-1)=-5, f(2)=9$, yes

100. $f(-1)=1, f(0)=-1$, yes ; $f(0)=-1, f(-1)=1$, yes **101.** 1.27 **102.** ± 0.78

CHAPTER 5 – 7   Graphing Polynomial Functions   page 155 ~ 1156

1. $y=x^2+5$ **2.** $y=x^2-5$ **3.** $y=(x-5)^2$ **4.** $y=(x+5)^2$ **5.** $y=-x^2$ **6.** $y=x^2$ **7.** $y=5x^2$

9. $y=-(x+5)^2$ **10.** $y=x^2+3$ **11.** $y=(x-1)^2+2$ **12.** $y=-(x-6)^3+4$ **13.** $y=\sqrt{6-x}-4$

14. g **15.** e **16.** a **17.** b **18.** f **19.** d **20.** h **21.** c

22.   **24.**   **26.**   **28.**   **31.**

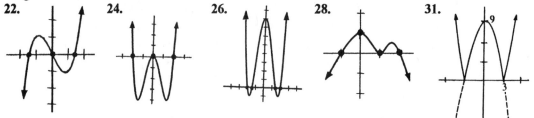

Hint: Graphing $y=|f(x)|$, we graph $y=f(x)$ and reflect the part that is below the $x=axis$ through the $x-axis$.

34. a. $t=\dfrac{\sqrt{d}}{4}$ b. 3 seconds **35.** length = 15 inches, width = 10 inches, height = 12 inches

36. maximum volume = 5488 cubic meters when $x=7$ meters

37. a. $A(x)=x^2+\frac{80}{x}$ b. 44 square feet c. $x\approx 3.45$ feet **38.** after 6 years **39.** 40 units

40. 40 units **41.** about 3.9 inches **42.** about $2,220

43. a. left to the students to verify b. $C(20)=\$37,200$

Roger was absent yesterday.
Teacher: Roger, Why were you absent ?
Roger: My tooth was aching.
        I went to see my dentist.
Teacher: Is your tooth still aching ?
Roger: I don't know. My dentist has it.

CHAPTER 5 – 8   Polynomial Inequalities and Absolute Values   page 160

1. $-2 < x < 4$ **2.** $x < -2$ or $x > 4$ **3.** $x \leq -2$ or $0 \leq x \leq 4$ **4.** $-2 \leq x \leq 0$ or $x \geq 4$

5. $-3 < x < 2$ or $x > 5$ **6.** $-5 \leq x \leq -2$ or $1 \leq x \leq 3$ **7.** $2 < x < 4$ **8.** $x < 2$ or $x > 4$

9. $x \leq -2$ or $x \geq 8$ **10.** $-2 \leq x \leq 8$ **11.** $4.5 < x < 5.5$ **12.** $x < -5\frac{1}{2}$ or $x > 2\frac{1}{2}$ **13.** $2 < x < \frac{10}{3}$.

14. $x < -\frac{1}{5}$ or $x > 1$.   **15.** $-\frac{5}{3} \leq x \leq 3$. **16.** $-1 < a < \frac{3}{2}$ **17.** $a < -1$ or $a > \frac{3}{2}$

18. a are all real numbers except 4.   **19.** $1 < x < 5$ **20.** $x < 1$ or $x > 3$ **21.** $x < 1$ or $x > 5$

22. $x \leq 2 - \sqrt{2}$ or $x \geq 2 + \sqrt{2}$ **23.** $x \leq 1 - \sqrt{6}$, $1 - \sqrt{2} \leq x \leq 1 + \sqrt{2}$ or $x \geq 1 + \sqrt{6}$.

24. $x < -1 - \sqrt{3}$ or $x > -1 + \sqrt{3}$.   **25.** $x < -6$ or $x > 2$.   **26.** $-2 - 2\sqrt{5} < x < -2 + 2\sqrt{5}$.

27. $\frac{1 - \sqrt{5}}{2} \leq x \leq \frac{1 + \sqrt{5}}{2}$.   **28.** $x \leq -3$ or $x \geq 2$.   **29.** $x \leq \frac{1 - \sqrt{17}}{4}$ or $x \geq \frac{1 + \sqrt{17}}{4}$.   **30.** $1 - \frac{\sqrt{33}}{3} \leq x \leq 1 + \frac{\sqrt{33}}{3}$.

31. $\frac{1 - \sqrt{13}}{6} \leq x \leq \frac{1 + \sqrt{13}}{6}$.   **32.** $x \leq -\frac{7}{2}$ or $x \geq -\frac{1}{2}$.   **33.** x are all real numbers.   **34.** x are all real numbers.

35. x are all real numbers except 2.   **36.** No solution.   **37.** No solution.   **38.** $x = -\frac{1}{2}$ only.

39. $-7 < x < 3$ **40.** $x < -7$ or $x > 3$ **41.** $-2 \leq x \leq 0$ or $x \geq 2$ **42.** $x < -2$ or $0 < x < 2$ **43.** $x \leq 3$

44. $x \leq -5$ or $x \geq 1$ **45.** $-1 < x < 7$ **46.** $x < -3$ or $-2 < x < -1$ **47.** $-2 < x < -1$ or $x > 3$

48. $x \leq -3$ or $-2 \leq x \leq \frac{1}{2}$ **49.** a) $6 < t < 8$ b) $t \leq 6$ seconds or $t \geq 8$

50. a) $|x - 37| > 0.84$ b) $x < 36.16°C$ or $x > 37.84°C$

CHAPTER 5 – 9   Inverse Functions   page 164

1. yes   **2.** yes   **3.** no   **4.** no   **5.** no   **6.** yes   **7.** yes   **8.** yes   **9.** no   **10.** no   **11.** yes   **12.** no

13. no   **14.** yes   **15.** yes   **16.** $f^{-1}(x) = x + 5$ **17.** $f^{-1}(x) = x - 5$ **18.** $f^{-1}(x) = 5x$ **19.** $f^{-1}(x) = \frac{x}{5}$

20. $f^{-1}(x) = \frac{x}{2} + 2$ **21.** $f^{-1}(x) = 2x - 4$ **22.** $f^{-1}(x) = \frac{x-1}{5}$ **23.** $f^{-1}(x) = 5x + 1$ **24.** $f^{-1}(x) = \frac{1}{x}$

25. $f^{-1}(x) = \frac{4-x}{5}$ **26.** $f^{-1}(x) = 4 - 5x$ **27.** $f^{-1}(x) = 8 - 2x$ **28.** $f^{-1}(x) = \frac{3-4x}{5}$ **29.** $f^{-1}(x) = \frac{3-5x}{4}$

30. $f^{-1}(x) = \frac{7x-6}{2}$ **31.** It has no inverse function.   **32.** $f^{-1}(x) = \sqrt{x-5}$ and $x \geq 5$

33. It has no inverse function.   **34.** $f^{-1}(x) = \sqrt{1-x}$ and $x \leq 1$ **35.** $f^{-1}(x) = \sqrt{9-x}$ and $x \leq 9$

36. $f^{-1}(x) = \sqrt[3]{x}$ **37.** $f^{-1}(x) = x^3$ **38.** $f^{-1}(x) = \sqrt[3]{x-1}$ **39.** It has no inverse function

40. $f^{-1}(x) = \sqrt[5]{x+4}$ **41.** $f^{-1}(x) = \sqrt[5]{x} + 4$ **42.** $f^{-1}(x) = x^5 + 4$ **43.** $f^{-1}(x) = \sqrt[5]{x^2 + 4}$

44. $f^{-1}(x) = 1 - x$ and $x \geq 0$ **45.** $f^{-1}(x) = x - 4$ and $x \geq 0$

46. The inverse is $g = \{(4, 1), (5, 2), (6, 4), (7, 5)\}$. The inverse is a function.

47. The inverse is $g = \{(4, 1), (5, 2), (6, 4), (6, 5)\}$. It is a relation but not a function.

48 ~ 55 are left to the students. Hint: verified by $f(g(x)) = g(f(x)) = x$

56. a) $y = \frac{250 - x}{15}$ b) y = number of product purchased   c) 6

x = total cost

CHAPTER 5 – 10   Composition of Functions   page 166

1. $(f \circ g)(x) = 6x + 10$, $(g \circ f)(x) = 6x - 5$ **2.** $(f \circ g)(x) = 6x - 10$, $(g \circ f)(x) = 6x + 5$

3. $(f \circ g)(x) = (x-1)^2$, $(g \circ f)(x) = x^2 - 1$ **4.** $(f \circ g)(x) = 3x^2 - 12$, $(g \circ f)(x) = 9x^2 - 4$

5. $(f \circ g)(x) = 8x^2 + 10x + 2$, $(g \circ f)(x) = 4x^2 + 2x - 1$

6. $(f \circ g)(x) = 6x^3 - 8x$, $(g \circ f)(x) = 24x^3 - 8x$ **7.** $(f \circ g)(x) = 15$, $(g \circ f)(x) = -15$

8. $(f \circ g)(x) = -15$, $(g \circ f)(x) = -225$ **9.** $(f \circ g)(x) = x + 3 - 2\sqrt{x+3}$, $(g \circ f)(x) = \sqrt{x^2 - 2x + 3}$

10. $(f \circ g)(x) = \frac{1}{x+2}$, $(g \circ f)(x) = \frac{x+2}{x+1}$ **11.** $f \circ f^{-1} = x$, $f^{-1} \circ f = x$ **12.** $f \circ f^{-1} = x$, $f^{-1} \circ f = x$

13. $g(x) = 4x - 3$ **14.** $g(x) = 2x - 1$ **15.** $f(x) = x - 1$, $f(3) = 2$

16. All real numbers of $x \geq 0$ **17.** All real numbers   **18.** All real numbers of $x \leq -3$ or $x \geq 3$

19. $(f \circ g)(0) = 3$, $(f \circ g)(3) = -6$, $(g \circ f)(0) = -2$, $(g \circ f)(3) = -14$

20. a. $f \circ g = \{(4, 6), (5, 7), (6, 8)\}$ b. $g \circ f$ does not exist.   **21.** $(c \circ x)(t) = 1200 + 600t$

22. $A(t) = \pi t^2$ **23.** $A(r) = \frac{2\pi(r^3 + 16)}{r}$

CHAPTER 5 – 11   Greatest Integer Functions   page 168
1. 9   2. 9   3. 9   4. –9   5. –10   6. –10   7. –1   8. 0   9. 1   10. –2   11. 1   12. 5   13. –6
14. 0   15. –1   16. 12.   17. –18   18. 3   19. –5   20. –3
21. $f(x) = [x] - 2$ 23. $f(x) = [x-1]$ 25. $f(x) = [x+5]$ 27. $f(x) = 2[x]$

29. $f(x) = 2[x-3]+5$ 30. $f(x) = [-x]$ 31. $f(x) = x - [x]$ 32. $f(x) = x + [x]$

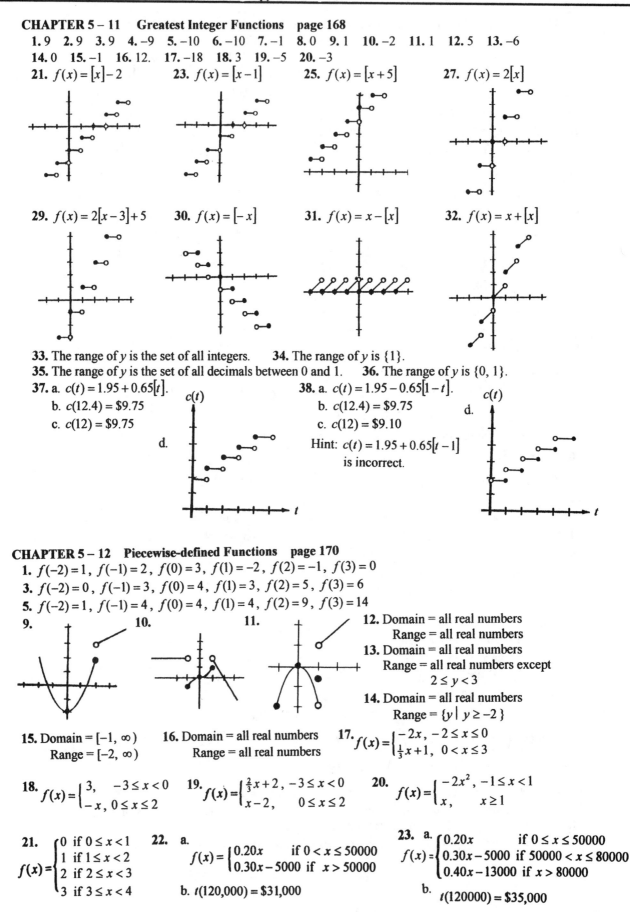

33. The range of y is the set of all integers.   **34.** The range of y is {1}.
35. The range of y is the set of all decimals between 0 and 1.   **36.** The range of y is {0, 1}.
37. a. $c(t) = 1.95 + 0.65[t]$.   **38.** a. $c(t) = 1.95 - 0.65[1-t]$.
    b. $c(12.4) = \$9.75$ b. $c(12.4) = \$9.75$
    c. $c(12) = \$9.75$ c. $c(12) = \$9.10$
    d.   Hint: $c(t) = 1.95 + 0.65[t-1]$
             is incorrect.
    d.

CHAPTER 5 – 12   Piecewise-defined Functions   page 170
1. $f(-2) = 1$, $f(-1) = 2$, $f(0) = 3$, $f(1) = -2$, $f(2) = -1$, $f(3) = 0$
3. $f(-2) = 0$, $f(-1) = 3$, $f(0) = 4$, $f(1) = 3$, $f(2) = 5$, $f(3) = 6$
5. $f(-2) = 1$, $f(-1) = 4$, $f(0) = 4$, $f(1) = 4$, $f(2) = 9$, $f(3) = 14$
9.   **10.**   **11.**

12. Domain = all real numbers
     Range = all real numbers
13. Domain = all real numbers
     Range = all real numbers except
          $2 \le y < 3$
14. Domain = all real numbers
     Range = $\{y \mid y \ge -2\}$

15. Domain = $[-1, \infty)$ **16.** Domain = all real numbers   **17.** $f(x) = \begin{cases} -2x, & -2 \le x \le 0 \\ \frac{1}{3}x + 1, & 0 < x \le 3 \end{cases}$
     Range = $[-2, \infty)$ Range = all real numbers

18. $f(x) = \begin{cases} 3, & -3 \le x < 0 \\ -x, & 0 \le x \le 2 \end{cases}$ **19.** $f(x) = \begin{cases} \frac{2}{3}x + 2, & -3 \le x < 0 \\ x - 2, & 0 \le x \le 2 \end{cases}$ **20.** $f(x) = \begin{cases} -2x^2, & -1 \le x < 1 \\ x, & x \ge 1 \end{cases}$

21. $f(x) = \begin{cases} 0 \text{ if } 0 \le x < 1 \\ 1 \text{ if } 1 \le x < 2 \\ 2 \text{ if } 2 \le x < 3 \\ 3 \text{ if } 3 \le x < 4 \end{cases}$ **22.** a. $f(x) = \begin{cases} 0.20x & \text{if } 0 < x \le 50000 \\ 0.30x - 5000 & \text{if } x > 50000 \end{cases}$
     b. $t(120,000) = \$31,000$

23. a. $f(x) = \begin{cases} 0.20x & \text{if } 0 \le x \le 50000 \\ 0.30x - 5000 & \text{if } 50000 < x \le 80000 \\ 0.40x - 13000 & \text{if } x > 80000 \end{cases}$
     b. $t(120000) = \$35,000$

CHAPTER 5 – 13 Square Roots and Cubic Root Functions page 172

1. g   2. c   3. e   4. a   5. b   6. f   7. h   8. d

9. by 4 units upward   10. by 4 units downward   11. by 4 units to the left   12. by 4 units to the right

13. by 4 units to the right and 3 units upward   14. by 5 units to the left and 3 units upward

15. by 5 units to the left and 3 units downward   16. By 5 units to the right and 3 units downward.

17. by 6 units downward   18. by 6 units upward   19. by 6 units to the left   20. by 6 units to the right

21. by 1 unit to the right and 3 units upward   22. by 1 unit to the left and 3 units downward

23. by 1 unit to the left and 3 units upward   24. By 1 unit to the right and 3 units downward.

25. $y = \sqrt{x+4}$ 27. $y = \sqrt{x+2}+1$ 29. $y = 3\sqrt{x}$ 32. $y = 2\sqrt[3]{x}+2$ 35. $y = 3\sqrt[3]{x-1}+3$

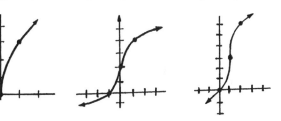

36. $t = 8$ seconds   37. $r = 3$ meters

Chapter 5 Exercises   page 173 ~ 176

1. 2^{27} 2. 3^{22} 3. 4^{63} 4. 5^{26} 5. $-16a^4b^6$ 6. $9a^6b^4$ 7. x^{13} 8. y^{19} 9. x^{2a-b+1} 10. $108x^{18}$

11. $\frac{1}{2401}x^8$ 12. 4   13. 16   14. $8x^3$ 15. $\frac{9}{4}x^{14}$ 16. $\frac{1}{6}a^3b^{-3}$ 17. x^8y^{-11} 18. $a^{-20}b^{12}$ 19. $9\sqrt[6]{3}$

20. 1   21. $x = \pm2\sqrt{13}$ 22. $x = -3\pm2\sqrt{13}$ 23. $x = 3\pm2\sqrt{13}\,i$ 24. $x = -3\pm\sqrt{13}$ 25. $x = -\frac{1}{3}, 2$

26. $y = \frac{1}{2}, -5$ 27. $x = \frac{1}{2}\pm\frac{\sqrt{5}}{2}$ 28. $x = -\frac{3}{2}\pm\frac{\sqrt{29}}{2}$ 29. $x = -\frac{3}{2}\pm\frac{\sqrt{11}}{2}i$ 30. $x = \frac{3}{4}\pm\frac{\sqrt{7}}{4}i$ 31. $x = -1, 0, 3$

32. $x = -1\pm2i$ 33. $x = \pm2i, \pm\sqrt{2}$ 34. $x = \pm\sqrt{3}, \pm\sqrt{3}\,i, \pm\sqrt{2}, \pm\sqrt{2}\,i$ 35. $a = \frac{9}{4}$

36. vertex $(0, -4)$ 38. vertex $(3, -5)$ 40. vertex $(-3, -10)$ 41. vertex $(-1, -25)$

x-intercepts $= \pm2$ no x-intercepts x-intercepts $= -6.16, 0.16$ x-intercepts $= -6, 4$

open upward   open downward   open upward   open upward

36 ~ 41 graphs are left to the students

42. Minimum cost = \$90 when $x = 8$ units   43. Maximum height = 215.44 feet   44. $x = 0, \pm2i$

45. $n = 0, \pm2$ 46. $y = 0$(double root), $4, -3$ 47. $p = 0, \pm1, \pm2\,i$ 48. $x = 0$(double root), $-3, \frac{3}{2}\pm\frac{3\sqrt{3}}{2}i$

49. $x = 0, 3, -\frac{3}{2}\pm\frac{3\sqrt{5}}{2}$ 50. $x = 0, 2$(double root)   51. $x = 0, -4$(double root)

52. $x = \pm2, 1\pm\sqrt{3}\,i, -1\pm\sqrt{3}\,i$ 53. yes   54. yes   55. no   56. yes   57. yes   58. no

59. $x^3+4x^2+4x+16 = 0$ 60. $x^3+4x^2+4x+16 = 0$ 61. yes   62. yes   63. $x = -5, 0, 6$

64. $x = \pm\sqrt{6}\,i, \pm2\sqrt{2}$ 65. $x = -4, 0$(multiplicity 2), 2(multiplicity 3)

66. $x = 0$(multiplicity 4), 6(multiplicity 2), $\pm6\,i$ 67. $x = 0$(multiplicity 4), 6(multiplicity 3), -6

68. $x^5-6x^4+16x^3-32x^2+48x-32 = 0$ 69. $f(x) = 4x^4+24x^3+72x^2+216x+324$

70. Three possibilities (see problem 68):   71. It has a maximum of 4 negative real roots.

  a. 5 positive real roots   (see problem 69. It touches the x-axis at $x = -3$.)

  b. 3 positive real roots, 2 imaginary roots   72. a. lower bound   b. neither   c. upper bound

  c. 1 positive real roots, 4 imaginary roots   (All roots are in the closed interval $[-5, 7]$.)

73. an even function   74. an odd function   75. neither

76. y decreases without bound as x increases without bound.   78. $x = \pm7$ 79. $\pm7\,i$

  y decreases without bound as x decreases without bound.   80. $x = -\frac{3}{2}, \frac{4}{3}$ 81. $x = -\frac{3}{2}\pm\frac{\sqrt{17}}{2}$

77. y increases without bound as x increases without bound.   82. $x = -2, -1$ 83. $x = -\frac{1}{5}\pm\frac{\sqrt{6}}{5}$

  y decreases without bound as x increases without bound.   84. $x = -\frac{1}{5}\pm\frac{2}{5}i$

85. $x = -3, 0, \frac{3}{2}\pm\frac{3\sqrt{3}}{2}i$ 86. $x = 0, 3, -\frac{3}{2}\pm\frac{3\sqrt{3}}{2}i$ 87. $x = 2, -\frac{3}{2}\pm\frac{\sqrt{29}}{2}$ 88. $x = -2, -\frac{1}{8}\pm\frac{\sqrt{17}}{8}$

89. $x = -2, -1, 1, \pm1, \pm2\,i$ 90. $x = -2$(mutliplicity 3), 1(a double root)   91. $x^2-4x+5 = 0$

92. $4x^3+9x^2+x-2 = 0$ 93. $f(x) = x^4+3x^2-4$ 94. $f(x) = -\frac{1}{9}(x+2)^4(x-1)^3(x-3)^2$

95. $f(0) = -8$, $f(2) = 96$, yes   96. $x \approx 0.322$ 97. $f(x) = (x+4)^2+3$ 98. $f(x) = -2(x+4)^3-1$

99. $18\times12\times14$ cubic inches   100. $t = 8$ years   101. Min. $c(4250) = \$27,750$

CHAPTER 5 **Exercises (continued)** **page 173 ~ 176**

102. Max. $p(4250) = \$42,250$ **103.** $f(x) = 0.1x^3 - 3.2x^2 + 11.5x + 8.2$

104 ~110 graphs are left to the students. Send your graphs to the author for free consultation.

111. $x < -1$ or $x > 4$ **112.** $-1 \leq x \leq 4$ **113.** $-1 \leq x \leq 0$ or $x \geq 4$ **114.** $-1 < x < 1$ or $x > 5$

115. $x < -4$ or $x > 8$ **116.** $-4 \leq x \leq 8$ **117.** $x < -4$ or $x > 2$ **118.** $-\frac{7}{2} \leq x \leq \frac{3}{2}$

119. $x \leq -1 - \sqrt{7}$ or $x \geq -1 + \sqrt{7}$ **120.** $x \leq -1 - \sqrt{7}$, or $-1 - \sqrt{3} \leq x \leq -1 + \sqrt{3}$, or $x \geq -1 + \sqrt{7}$

121. $2x^2 - 6x + 1 < 0$ **122.** no solution **123.** $6 < t < 10$ seconds **124.** $f^{-1}(x) = \frac{x}{2} - 5$

125. $f^{-1}(x) = 2x + 10$ **126.** no inverse function **127.** $f^{-1}(x) = \sqrt{\dfrac{x+5}{2}}$, $x \geq 5$ **128.** $f^{-1}(x) = x^5 + 2$

129. $f^{-1}(x) = \dfrac{-x-2}{x-3}$ **130.** $f^{-1}(x) = -\frac{1}{2}x - 1$, $x \geq 0$ **131.** $f^{-1}(x) = \sqrt{a^2 - x^2}$, $0 \leq x \leq a$

132. $g = \{(1, 2), (5, 3), (5, 4)\}$. It is not a function.

133. a) $y = \frac{9}{5}x + 32$ b) Proof is left to the students. Hint: $f(f^{-1}(x)) = f^{-1}(f(x)) = x$

134. $(f \circ g)(x) = 3x^2 - 7x + 5$, $(g \circ f)(x) = 3x^2 - x$

135. $f^{-1} = \sqrt{x+1}$ where $x \geq -1$, $f \circ f^{-1} = x$, $f^{-1} \circ f = x$

136. a) $r(x) = x - 2000$ b) $p(x) = 0.9x$ c) $p(r(x)) = 0.9(x - 2000)$ d) $\$17,550$

137. a) $c(x) = 0.37 - 0.23\lceil 1 - x \rceil$ b) $\$1.06$ Hint: $c(x) = 0.37 + 0.23\lceil x - 1 \rceil$ is incorrect.

138. $b(n) = \left\lceil \dfrac{n-1}{50} \right\rceil + 1$; or $b(n) = \text{int}\left\lceil \dfrac{n-1}{50} \right\rceil + 1$

139. a) $c(x) = \begin{cases} 0.15x + 5.40 & \text{if } 0 \leq x \leq 100 \\ 0.12x + 8.40 & \text{if } x > 100 \end{cases}$

b) $\$18.90$ c) $\$26.40$

140. $c(x) = \begin{cases} 0.10x + 19.50 & \text{if } 0 \leq x \leq 50 \\ 0.06x + 21.50 & \text{if } 50 < x \leq 100 \\ 0.04x + 23.50 & \text{if } x > 100 \end{cases}$ b) $\$26.30$ c) $\$28.30$

141~146 Graphs are left to the students. Send your graphs to the author for free consultation.

A father has invited all his friends for a party. He has two swimming pools in the back yard, one is filled with water and one is empty.

Son: How can your friends swim in this pool
without water ?

Father: Not all my friends know how to swim.

Graphing a quadratic function by a graphing calculator

Examples:

1. Graph $y = x^2$.

Solution:

Press: **Y=** $\rightarrow$ $\boxed{\text{X,T,}\theta\text{,}n}$ $\rightarrow \wedge \rightarrow 2 \rightarrow$ **GRAPH**

The screen shows:

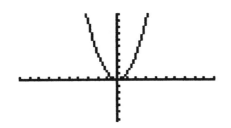

2. Graph $y = \frac{1}{2}(x-2)^2 - 4$

Solution:

Press: **Y=** $\rightarrow$ ($1 \div 2$) $\rightarrow$ ($\rightarrow \boxed{\text{X,T,}\theta\text{,}n} \rightarrow -2 \rightarrow$)

$\rightarrow \wedge \rightarrow 2 \rightarrow -4 \rightarrow$ **GRAPH**

The screen shows:

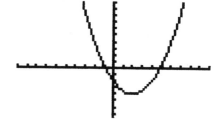

Graphing a quadratic function with horizontal axis by a graphing calculator

Examples:

3. Graph $x = y^2$.

Solution:

Solve for y: $y = \pm\sqrt{x}$

Press: **Y=**

Enter:

Y1= $\rightarrow$ **2nd** $\rightarrow \sqrt{\ }$ ($\rightarrow \boxed{\text{X,T,}\theta\text{,}n} \rightarrow$)

Y2= $\rightarrow$ (-) $\rightarrow$ **2nd** $\rightarrow \sqrt{\ }$ ($\rightarrow \boxed{\text{X,T,}\theta\text{,}n} \rightarrow$)

The screen shows:

$y_1 = \sqrt{x}$

$y_2 = -\sqrt{x}$

Press: **GRAPH**

The screen shows:

Hint: We can simply enter **Y2** as $-$**Y1** :

Press: **Y2=** $\rightarrow$ (–) $\rightarrow$ **VARS** $\rightarrow$ **Y-VARS**

$\rightarrow$ **1:FUNCTION** $\rightarrow$ **ENTER**

4. Graph $x = \frac{1}{2}(y+2)^2 - 4$

Solution:

Solve for y: $y = -2 \pm \sqrt{2(x+4)}$

Press: **Y=**

Enter:

Y1= $\rightarrow$ (-) $\rightarrow$ **2** $\rightarrow$ + $\rightarrow$ **2nd** $\rightarrow \sqrt{\ }$ ($\rightarrow$ 2

$\rightarrow$ ($\rightarrow \boxed{\text{X,T,}\theta\text{,}n} \rightarrow$ + 4 $\rightarrow$) $\rightarrow$)

Y2= $\rightarrow$ (-) $\rightarrow$ **2** $\rightarrow$ – $\rightarrow$ **2nd** $\rightarrow \sqrt{\ }$ ($\rightarrow$ 2

$\rightarrow$ ($\rightarrow \boxed{\text{X,T,}\theta\text{,}n} \rightarrow$ +4 $\rightarrow$) $\rightarrow$)

The screen shows:

$y_1 = -2 + \sqrt{2(x+4)}$

$y_2 = -2 - \sqrt{2(x+4)}$

Press: **GRAPH**

The screen shows:

Graphing a polynomial function by a graphing calculator

Example 1: Graph $y = x^3 + x^2 - 12x$.

Solution:

Press: **Y=** $\rightarrow$ $\boxed{\text{X,T,}\theta\text{,}n}$ $\rightarrow$ ^ $\rightarrow$ **3** + $\rightarrow$ $\boxed{\text{X,T,}\theta\text{,}n}$
$\rightarrow x^2 \rightarrow$ **–12** $\rightarrow$ $\boxed{\text{X,T,}\theta\text{,}n}$ $\rightarrow$ **GRAPH**

If the screen does not show a complete graph, change the viewing window:

Press: **WINDOR**

Change the viewing window to:

$$x_{min} = -5 , \quad x_{max} = 5$$
$$y_{min} = -50 , \quad y_{max} = 50$$

Change the scale factors to:

$$x_{scl} = 1 , \quad y_{scl} = 10$$

Press: **GRAPH**

The screen shows:

Graphing a polynomial function containing absolute value by a graphing calculator

Example 2: Graph $y = \left| x^2 - 4 \right|$.

Solution:

Press: **Y=** $\rightarrow$ **MATH** $\rightarrow$ **NUM** $\rightarrow$ **1:abs(** $\rightarrow$ **ENTER**
$\rightarrow \boxed{\text{X,T,}\theta\text{,}n} \rightarrow x^2 \rightarrow - \rightarrow$ **4** $\rightarrow$) $\rightarrow$ **GRAPH**

The screen shows:

Estimating zeros of a polynomial function by a graphing calculator

Example 3: Estimating the zeros of $f(x) = x^3 + x^2 - 12x$.

Solution: Graph the function (see Example 1).

Press: **2nd** $\rightarrow$ **CALC** $\rightarrow$ **2:zero** $\rightarrow$ **ENTER**

The screen shows the graph.

Move the blinking cursor ◁ or ▷ to the three
x-intercepts, we have the approximate values
$x = -4.042553$, 0, and 2.9787234
(Hint: The exact values are $x = -4$, 0, and 3)

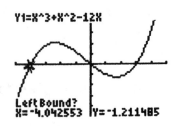

Finding a quadratic function (a regression curve) of best-fit by a graphing calculator

Example: Use a graphing calculator to find the quadratic function of best-fit to the following data.

x	0	1	2	3	4	5	6
$f(x)$	7	2	−1	−2	−1	2	7

Solution:

1. To enter the data:
   Press: **STAT → 1:EDIT → ENTER**
   Move the cursor and enter the data
   on the screen:

L1	L2
0	7
1	2
2	−1
3	−2
4	−1
5	2
6	7

2. Find the cubic function:
   Press: **2nd → QUIT**
   We have the home screen.
   Press: **STAT**
   Move the cursor to:
   CALC and **QuadReg → ENTER**
   Screen shows **QuadReg**
   Select L1 as the x list and L2 as the y list:
   Press: **2nd → L1 →** , **→ 2nd → L2 →** ,
   Screen shows: **QuadReg L1, L2,**
   Press: **VARS** and move cursor to **Y-VARS**
   Screen shows: **1: Function → ENTER**
   Screen shows: **1: Y1 → ENTER**
   Screen shows: **QuadReg L1, L2, Y1 → ENTER**
   Screen shows: $y = ax^2 + bx + c$
   $$a = 1, b = -6, c = 7$$
   $y = x^2 - 6x + 7$ is the answer.

3. Graph the function:
   Press: **Y= → GRAPH**
   Screen shows:

   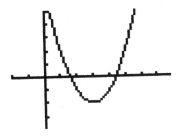

4. Check the answer:
   Press: **VARS**
   Move cursor to **Y-VARS → ENTER**
   Screen shows: **1: Y1 → ENTER**
   Screen shows the function and **Y1**
   Type: **Y1(0) → ENTER**
   Screen shows: 7
   Press: **2nd → ENTRY**
   Screen shows: Y1(0) = 7
         Y1(0)
   Type: Y1(0) = 7
         Y1(1) → **ENTER**
   Screen shows: Y1(0) = 7
         Y1(1) = 2
   Press: **2nd → ENTRY**
   (Continue the entry of other data)

Find a cubic function (a regression curve) of best-fit by a graphing calculator

Example: Use a graphing calculator to find the cubic function of best-fit to the following data.

x	1	2	3	4	5	6	7
$f(x)$	18.6	25.1	29.2	32.1	35	39.1	45.6

Solution:

1. To enter the data:
   Press: **STAT → 1:EDIT → ENTER**
   Move the cursor and enter the data
   on the screen:

L₁	L₂
1	18.6
2	25.1
3	29.2
4	32.1
5	35
6	39.1
7	45.6

2. Find the cubic function:
   Press: **2nd → QUIT**
   We have the home screen.
   Press: **STAT**
   Move the cursor to:
      CALC and **CubicReg → ENTER**
   Screen shows: **CubicReg**
   Select L₁ as the x list and L₂ as the y list:
   Press: **2nd → L1 → , → 2nd → L2 → ,**
   Screen shows: **CubicReg L1, L2,**
   Press: **VARS** and move cursor to **Y-VARS**
   Screen shows: **1: Function → ENTER**
   Screen shows: **1: Y1 → ENTER**
   Screen shows: **CubicReg L1, L2, Y1 → ENTER**
   Screen shows: $y = ax^3 + bx^2 + cx + d$
       $a = 0.2$, $b = -2.4$, $c = 12.3$, $d = 8.5$
   $y = 0.2x^3 - 2.4x^2 + 12.3x + 8.5$ is the answer.

3. Graph the function:
   Press: **Y= → GRAPH**
   Change the viewing window:
   Press: **WINDOW**
     $x_{min} = 0$, $x_{max} = 10$, $x_{scl} = 1$
     $y_{min} = 0$, $y_{max} = 50$, $y_{scl} = 5$
   Press: **GRAPH**
   Screen shows:

   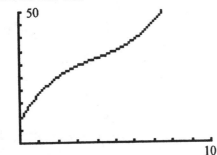

4. Check the answer:
   Press: **VARS**
   Move cursor to **Y-VARS → ENTER**
   Screen shows: **1: Y1 → ENTER**
   Screen shows the function and **Y1**
   Type: **Y1(1) → ENTER**
   Screen shows: 18.6
   Press: **2nd → ENTRY**
   Screen shows: Y1(1) = 18.6
            Y1(1)
   Type: Y1(1) = 18.6
          Y1(2) **→ ENTER**
   Screen shows: Y1(1) = 18.6
            Y1(2) = 25.1
   Press: **2nd → ENTRY**
     (Continue the entry of other data)

Graphing an inverse function
by a graphing calculator

Examples:

1. Graph the inverse of $y = 2x - 4$.

   Solution:

   Press: **Y=** → 2 → $\boxed{\textbf{X,T,θ,} n}$ → − → 4
   → **2nd** → **DRAW** → **8↓DrawInv**
   → **ENTER**

   The screen shows: **DrawInv**

   Press: **VARS** → **Y-VARS** → **1:Function**
   → **ENTER** → **1: Y₁** → **ENTER**

   The screen shows: **DrawInv Y₁**

   Press: **ENTER**

   The screen shows:

   (Hint: $y = \frac{x}{2} + 2$ is the inverse.)

2. Graph the inverse of $y = x^3 + 1$.

   Solution:

   Press: **Y=** → $\boxed{\textbf{X,T,θ,} n}$ → ∧ → 3 → + → 1
   → **2nd** → **DRAW** → **8↓DrawInv**
   → **ENTER**

   The screen shows: **DrawInv**

   Press: **VARS** → **Y-VARS** → **1:Function**
   → **ENTER** → **1:Y₁** → **ENTER**

   The screen shows: **DrawInv Y₁**

   Press: **ENTER**

   The screen shows:

   (Hint: $y = \sqrt[3]{x-1}$ is the inverse.)

Graphing a composite function
by a graphing calculator

Example:

3. Let $f(x) = x^2 - 9$ and $g(x) = \sqrt{x}$. Graph $y = (f \circ g)(x)$.

   Solution:

   Assign $y_1 = \sqrt{x}$, $y_2 = (y_1)^2 - 9$

   Press: **Y1=** → **2nd** → $\sqrt{}$ (→ $\boxed{\textbf{X,T,θ,} x}$ →) → **ENTER**
   → **VARS** → **Y-VARS** → **1:function** → **ENTER**
   → **1: Y1**→ **ENTER**

   The screen shows: **Y1=** $\sqrt{x}$

   Y2= Y1   Press: x^2 → − → **9**

   Change the viewing window to:

   $$x_{\min} = -10, \ x_{\max} = 40, \ x_{scl} = 4$$
   $$y_{\min} = -10, \ y_{\max} = 20, \ y_{scl} = 4$$

   Press: **GRAPH**   The screen shows:

   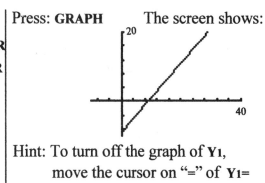

   Hint: To turn off the graph of **Y1**,
   move the cursor on "=" of **Y1=**
   and press **DEL → ENTER**
   Highlight is deleted.

Graphing a greatest integer function by a graphing calculator

Example: Graph $y = \lfloor 2x \rfloor$.

Solution:

Press: **Y= → MATH → NUM → int (→ ENTER**
→ 2 → $\boxed{X,T,\theta,n}$ →) → GRAPH

Change the viewing window to:

$x_{min} = -5$, $x_{max} = 5$, $x_{scl} = 1$
$y_{min} = -5$, $y_{max} = 5$, $y_{scl} = 1$

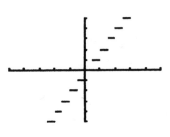

Press: **GRAPH**

Remove the connecting lines between selected intervals by changing **MODE** setting to **Dot**.

Press: **MODE → Dot → ENTER → GRAPH**

The screen shows:

Graphing a piecewise-defined function by a graphing calculator

Example: Graph

$$y = \begin{cases} 2 & \text{if } x \le -1 \\ x^3 & \text{if } -1 < x \le 1 \\ -2x+4 & \text{if } x > 1 \end{cases}$$

Solution:

Enter the following equation:

$$Y_1 = 2(x \le -1) + x^3(-1 < x)(x \le 1) + (-2x+4)(x > 1)$$

Press: **Y1= → 2 → (→ $\boxed{X,T,\theta,n}$ → 2nd → TEST → 6: ≤ → ENTER**
→ (–) → 1 →) → + → $\boxed{X,T,\theta, n}$ → ∧ → 3 → (→ (–) → 1 → 2nd
→ TEST → 5: < → ENTER → $\boxed{X,T,\theta, n}$ →) → (→ $\boxed{X,T,\theta,n}$ → 2nd
→ TEST → 6: ≤ → ENTER → 1 →) → + → (→ (–) → 2 → $\boxed{X,T,\theta,n}$
→ + → 4 →) → (→ $\boxed{X,T,\theta,n}$ → 2nd → TEST → 3: > → ENTER
→ 1 →) → GRAPH

Change the viewing window to:

$x_{min} = -5$, $x_{max} = 5$, $x_{scl} = 1$
$y_{min} = -5$, $y_{max} = 5$, $y_{scl} = 1$

Remove connecting lines between selected intervals by changing **MODE** setting to **Dot**.

Press: **MODE → Dot →ENTER → GRAPH**

The screen shows: (Hint: the calculator does not show the
                                     including and excluding end points.

Graphing a square root function by a graphing calculator

Example: Graph $y = \sqrt{1-x}$.

    Solution:

        Press: **Y=** $\rightarrow$ **2nd** $\rightarrow$ $\sqrt{\ }$ $(\rightarrow 1 \rightarrow - \rightarrow$ **X,T,θ,n** $\rightarrow)$

           $\rightarrow$ **GRAPH**

        The screen shows:

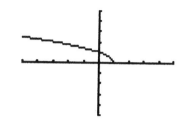

Graphing a cubic root function by a graphing calculator

Example: Graph $y = \sqrt[3]{x-1}$.

    Solution:

$$y = (x-1)^{\frac{1}{3}}$$

        Press: **Y=** $\rightarrow (\rightarrow$ **X,T,θ,n** $\rightarrow - \rightarrow 1 \rightarrow) \rightarrow \wedge \rightarrow ($

           $\rightarrow (\rightarrow 1 \rightarrow \div \rightarrow 3 \rightarrow) \rightarrow$ **GRAPH**

        The screen shows:

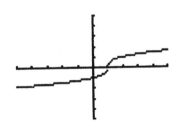

A patient said to his dentist.
Patient: Why are you charging me $300 ?
        Last time, you charged me only $100.
Dentist: Two patients in the waiting room ran
        away when they heard you screaming.

Space for Taking Notes

Space for Taking Notes

Space for Taking Notes

Rational Equations and Functions

6-1 Simplifying Rational Expressions

Rational Expression (Algebraic fraction):

An algebraic expression that is a quotient (ratio) of two polynomials.
It is also referred to as **algebraic fraction**.

$\dfrac{x+1}{2x}$, $\dfrac{2x^2+3x-5}{x^2+4}$, and $\dfrac{xy^2}{x^2+xy}$ are rational expressions.

A rational expression is in simplest form if the numerator and the denominator have no common factors (except 1 and -1).

To simplify a rational expression, we reduce the fraction to lowest terms by using their greatest common factor (GCF) or the laws of exponents (see Section $5 \sim 1$).
To simplify a rational expression, we factor the numerator and the denominator and then cancel any common factor.
To simplify a rational expression, we also **restrict** the variables by **excluding** any values that make the denominator equal to zero. It is **undefined** if the denominator equals 0.

Important: If factors of the numerator and the denominator are opposites of one another, we take the negative of the numerator or the denominator.
(See examples 4 and 5)

Examples (simplify each expression)

1. $\dfrac{16x^2}{24x} = \dfrac{16x^2/8x}{24x/8x} = \dfrac{2x}{3}$, $x \neq 0$ **or:** $\dfrac{16x^2}{24x} = \dfrac{16}{24} \cdot x^{2-1} = \dfrac{2}{3}x$, $x \neq 0$

2. $\dfrac{18x^4y^2}{27xy^6} = \dfrac{18}{27} \cdot \dfrac{x^4}{x} \cdot \dfrac{y^2}{y^6} = \dfrac{2x^3}{3y^4}$, $x \neq 0, y \neq 0$ 3. $\dfrac{3xy}{3x+3y} = \dfrac{\cancel{3xy}}{\cancel{3}(x+y)} = \dfrac{xy}{x+y}$, $x \neq -y$

4. $\dfrac{x-1}{1-x} = \dfrac{\cancel{x-1}}{-\cancel{(x-1)}} = -1$, $x \neq 1$ 5. $\dfrac{3-x}{4x-12} = \dfrac{3-x}{4(x-3)} = \dfrac{-\cancel{(x-3)}}{4\cancel{(x-3)}} = -\dfrac{1}{4}$, $x \neq 3$

6. $\dfrac{x^2-4}{x+2} = \dfrac{\cancel{(x+2)}(x-2)}{\cancel{x+2}} = x-2$, $x \neq -2$

7. $\dfrac{x^2-5x+6}{x^2-x-2} = \dfrac{(x-2)(x-3)}{(x+1)(x-2)} = \dfrac{x-3}{x+1}$, $x \neq -1, x \neq 2$

195

EXERCISES

Determine all values of variables for which the given rational expression is undefined.

1. $\dfrac{2}{x}$

2. $\dfrac{5}{x-1}$

3. $\dfrac{15}{a-5}$

4. $\dfrac{y-3}{y+4}$

5. $\dfrac{4x}{2x-6}$

6. $\dfrac{3}{(n-2)(n+1)}$

7. $\dfrac{n+4}{(n+5)(n-1)}$

8. $\dfrac{t-1}{t^2+2t}$

9. $\dfrac{20}{x^2+5}$

10. $\dfrac{x-8}{4}$

11. $\dfrac{p-3}{p^2-4p-5}$

12. $\dfrac{y+9}{y^3-y^2-2y}$

Simplify each rational expression. Give all restrictions on the variables.

13. $\dfrac{9x}{3x}$

14. $\dfrac{6x^5}{2x^2}$

15. $\dfrac{9x^2}{6x^5}$

16. $\dfrac{15a^2b^4}{25a^6b}$

17. $\dfrac{4x+2y}{2}$

18. $\dfrac{2}{4x+2y}$

19. $\dfrac{6}{9y-12}$

20. $\dfrac{9y-12}{6}$

21. $\dfrac{4(x+y)}{8(x-y)}$

22. $\dfrac{x-5}{5-x}$

23. $\dfrac{2a-6}{3-a}$

24. $\dfrac{a^2-4}{2-a}$

25. $\dfrac{a-2}{4-a^2}$

26. $\dfrac{x-8}{(x+8)(x-8)}$

27. $\dfrac{14+7x}{7x}$

28. $\dfrac{12+4y}{3+y}$

29. $\dfrac{2(x+2y)}{(x+2y)(x-y)}$

30. $\dfrac{3(x+y)(x-y)}{9(x+y)}$

31. $\dfrac{(a-2)^2}{a-2}$

32. $\dfrac{5(a-1)}{15(a-1)^2}$

33. $\dfrac{(a-b)^2}{(a+b)(a-b)}$

34. $\dfrac{x^2-8x+16}{x-4}$

35. $\dfrac{x^2-8x+15}{x-5}$

36. $\dfrac{3-n}{n^2-2n-3}$

37. $\dfrac{2a-12}{a^2-4a-12}$

38. $\dfrac{k^2+k-12}{5k+20}$

39. $\dfrac{10-2p}{p^2-4p-5}$

40. $\dfrac{x^2+5x-14}{x^2+3x-28}$

41. Determine all values of x for which the rational expression is undefined: $\dfrac{2x^2+8x}{x^2-3x-18}$

42. Determine all values of x for which the rational expression equals zero: $\dfrac{2x^2+8x}{x^2-3x-18}$

43. Solve the equation $ax^2-a^2=-bx-b^2$ for x in terms of a and b.

44. The area of a rectangle is given by $x^2+10x+21$. The length is given by $x+7$. Find the width in terms of x.

6-2 Adding and Subtracting Rational Expressions

1) To add or subtract rational expressions (algebraic fractions) having the same denominators, we combine their numerators.

$$\frac{b}{a}+\frac{c}{a}=\frac{b+c}{a} \quad ; \quad \frac{b}{a}-\frac{c}{a}=\frac{b-c}{a}$$

2) To add or subtract rational expressions (algebraic fractions) having different denominators, we rewrite each fraction having the least common denominator (**LCD**), and then combine the resulting fractions. **LCD** is the least common multiple (**LCM**) of their denominators.

$$\frac{c}{a}+\frac{d}{b}=\frac{bc}{ab}+\frac{ad}{ab}=\frac{bc+ad}{ab} \quad ; \quad \frac{c}{a}-\frac{d}{b}=\frac{bc}{ab}-\frac{ad}{ab}=\frac{bc-ad}{ab}$$

Examples

1. $\dfrac{3x}{5}+\dfrac{7x}{5}=\dfrac{3x+7x}{5}=\dfrac{10x}{5}=2x$

2. $\dfrac{x}{5}-\dfrac{3x}{2}=\dfrac{2x}{10}-\dfrac{15x}{10}=\dfrac{2x-15x}{10}=-\dfrac{13x}{10}$

3. $\dfrac{5a+3}{4}-\dfrac{a-1}{4}=\dfrac{5a+3-(a-1)}{4}=\dfrac{5a+3-a+1}{4}=\dfrac{4a+4}{4}=\dfrac{\cancel{4}(a+1)}{\cancel{4}}=a+1$

4. $\dfrac{1}{x}+\dfrac{1}{2x}=\dfrac{2}{2x}+\dfrac{1}{2x}=\dfrac{2+1}{2x}=\dfrac{3}{2x}, \; x\neq 0$

5. $\dfrac{1}{x^2}-\dfrac{2}{x^3}=\dfrac{x}{x^3}-\dfrac{2}{x^3}=\dfrac{x-2}{x^3}, \; x\neq 0$

6. $\dfrac{1}{2x^2}-\dfrac{5}{6x}=\dfrac{3}{6x^2}-\dfrac{5x}{6x^2}=\dfrac{3-5x}{6x^2}, \; x\neq 0$

7. $\dfrac{3}{x-1}+\dfrac{5}{x-1}=\dfrac{3+5}{x-1}=\dfrac{8}{x-1}, \; x\neq 1$

8. $\dfrac{7}{x-2}+\dfrac{2}{2-x}=\dfrac{7}{x-2}-\dfrac{2}{x-2}=\dfrac{5}{x-2}, \; x\neq 2$

9. $\dfrac{3}{x}-\dfrac{2}{x-2}=\dfrac{3(x-2)}{x(x-2)}-\dfrac{2x}{x(x-2)}=\dfrac{3(x-2)-2x}{x(x-2)}=\dfrac{3x-6-2x}{x(x-2)}=\dfrac{x-6}{x^2-2x}, \; x\neq 0,2$

10. $\dfrac{2a}{a^2-4}-\dfrac{5}{a+2}=\dfrac{2a}{(a+2)(a-2)}-\dfrac{5(a-2)}{(a+2)(a-2)}=\dfrac{2a-5(a-2)}{(a+2)(a-2)}=\dfrac{2a-5a+10}{(a+2)(a-2)}$

$$=\dfrac{-3a+10}{a^2-4}, \; a\neq -2,2$$

11. $\dfrac{-2}{x^2-5x+6}+\dfrac{2}{x-3}=\dfrac{-2}{(x-3)(x-2)}+\dfrac{2}{x-3}=\dfrac{-2}{(x-3)(x-2)}+\dfrac{2(x-2)}{(x-3)(x-2)}$

$$=\dfrac{-2+2(x-2)}{(x-3)(x-2)}=\dfrac{-2+2x-4}{(x-3)(x-2)}=\dfrac{2x-6}{(x-3)((x-2)}=\dfrac{2\cancel{(x-3)}}{\cancel{(x-3)}(x-2)}=\dfrac{2}{x-2}, \; x\neq 3,2$$

EXERCISES

Find the LCM of each of the following.

1. 3, 5

2. 4, 9

3. 2, 6

4. 5, 15

5. 6, 10

6. 6, 8

7. 2, 5x

8. 3, 4x

9. 4, 6x

10. 8, 10x

11. 4a, 8ab

12. 8a, 4ab

13. 3x, 6x^2

14. x, y

15. 7y^3, 28y

16. $x+3$, $x-3$

17. 12x^2y, 16xy^2

18. a^2-9, $a-3$

19. 5$(a-2)$, 6$(a-2)^2$

20. $x(x-1)$, x^2-1

Find the LCD of the fractions.

21. $\dfrac{1}{3}, \dfrac{1}{5}$

22. $\dfrac{1}{5}, \dfrac{1}{15}$

23. $\dfrac{1}{2x}, \dfrac{1}{5x}$

24. $\dfrac{1}{8x}, \dfrac{1}{10x}$

25. $\dfrac{1}{4a}, \dfrac{1}{8ab}$

26. $\dfrac{1}{3x}, \dfrac{1}{6x^2}$

27. $\dfrac{1}{x+3}, \dfrac{1}{x-3}$

28. $\dfrac{1}{a^2-9}, \dfrac{1}{a-3}$

29. $x, \dfrac{1}{x}$

30. $\dfrac{1}{a}, 5$

31. $\dfrac{1}{x}, \dfrac{1}{y}$

32. $\dfrac{1}{x}, \dfrac{1}{x^2y}$

33. $\dfrac{1}{x}, \dfrac{1}{x-5}$

34. $\dfrac{1}{3c}, \dfrac{1}{2(c-2)}$

35. $\dfrac{1}{t+1}, \dfrac{1}{(t+1)^2}$

36. $\dfrac{1}{x-2}, \dfrac{1}{x+1}$

Simplify. Give all restrictions on the variables.

37. $\dfrac{x}{3}+\dfrac{x}{5}$

38. $\dfrac{4a}{5}-\dfrac{a}{15}$

39. $\dfrac{1}{2x}-\dfrac{3}{5x}$

40. $\dfrac{1}{8x}+\dfrac{9}{10x}$

41. $\dfrac{3}{4a}-\dfrac{1}{8ab}$

42. $\dfrac{2}{3x}-\dfrac{5}{6x^2}$

43. $\dfrac{4}{x+3}+\dfrac{3}{x-3}$

44. $\dfrac{8a}{a^2-9}-\dfrac{4}{a-3}$

45. $x+\dfrac{2}{x}$

46. $\dfrac{1}{a}-5$

47. $\dfrac{2}{x}+\dfrac{5}{y}$

48. $\dfrac{2}{x}+\dfrac{9}{x^2y}$

49. $\dfrac{2}{x}-\dfrac{1}{x-5}$

50. $\dfrac{2}{3c}-\dfrac{5}{2(c-2)}$

51. $\dfrac{3}{t+1}-\dfrac{3t}{(t+1)^2}$

52. $\dfrac{1}{2(x+1)}+\dfrac{2}{x+1}$

53. $\dfrac{3}{x-2}+\dfrac{7}{x+1}$

54. $\dfrac{2x}{x-4}-\dfrac{1}{3x-12}$

55. $\dfrac{3}{x^2-9}-\dfrac{1}{2x-6}$

56. $\dfrac{x}{x-2}+\dfrac{1}{2-x}$

57. $\dfrac{4n}{2n-1}+\dfrac{2}{1-2n}$

58. $\dfrac{a-2}{a-1}-\dfrac{a+1}{a+2}$

59. $\dfrac{4}{x^2+2x-3}+\dfrac{1}{x+3}$

60. $\dfrac{4}{y-3}-\dfrac{5y-12}{y^2-6y+9}$

61. $\dfrac{1}{b^2+4b+3}+\dfrac{1}{b^2-2b-15}$

62. $\dfrac{5}{3+c}-\dfrac{3}{3-c}+\dfrac{2c+7}{9-c^2}$

63. $\dfrac{x-1}{x^2+5x}-\dfrac{x+1}{x^2-25}+\dfrac{1}{x^2-5x}$

6-3 Multiplying and Dividing Rational Expressions

1) To multiply two rational expressions (algebraic fractions), we multiply their numerators and multiply their denominators. We can multiply first and then simplify, or simplify first and then multiply.

$$\frac{x^2}{3y}\cdot\frac{6}{5x}=\frac{6x^2}{15xy}=\frac{2x}{5y}\ ;\qquad \text{Or:}\ \frac{x^2}{3y}\cdot\frac{6}{5x}=\frac{2x}{5y}$$

2) To divide two rational expressions (algebraic fractions), we multiply the reciprocal of the divisor.

$$\frac{4x}{y}\div\frac{8x^2}{3y}=\frac{4x}{y}\cdot\frac{3y}{8x^2}=\frac{3}{2x}$$

3) To divide long polynomials, we follow the same ways of dividing real numbers. We must arrange the terms in both polynomials in order of decreasing degree of one variable. Keep space (using 0) on missing terms in degree (see example **10**).

Examples (Assume that no variable has a value for which the denominator is zero.)

1. $\dfrac{x^2}{4}\cdot\dfrac{x}{2}=\dfrac{x^2\cdot x}{4\cdot 2}=\dfrac{x^3}{8}$

2. $\dfrac{x^2}{4}\div\dfrac{x}{2}=\dfrac{x^2}{4}\cdot\dfrac{2}{x}=\dfrac{x}{2}$

3. $\dfrac{2x}{3y}\cdot\dfrac{xy}{8}=\dfrac{x^2}{12}$

4. $\dfrac{2x}{3y}\div\dfrac{xy}{8}=\dfrac{2x}{3y}\cdot\dfrac{8}{xy}=\dfrac{16}{3y^2}$

5. $\dfrac{a-3}{a}\cdot\dfrac{a^3}{a^2-9}=\dfrac{a-3}{a}\cdot\dfrac{a^2}{(a+3)(a-3)}=\dfrac{a^2}{a+3}$

6. $\dfrac{a^2-5a+6}{a^2}\div\dfrac{a^2-9}{a}=\dfrac{a^2-5a+6}{a^2}\cdot\dfrac{a}{a^2-9}=\dfrac{(a-2)(a-3)}{a^2}\cdot\dfrac{a}{(a+3)(a-3)}=\dfrac{a-2}{a(a+3)}$

7. $\dfrac{x^2-4x-5}{x}\cdot\dfrac{x^2+x}{x-5}=\dfrac{(x+1)(x-5)}{x}\cdot\dfrac{x(x+1)}{x-5}=(x+1)^2$

8. $\dfrac{x^2-4x-5}{x}\div\dfrac{x^2+x}{x-5}=\dfrac{x^2-4x-5}{x}\cdot\dfrac{x-5}{x^2+x}=\dfrac{(x+1)(x-5)}{x}\cdot\dfrac{x-5}{x(x+1)}=\dfrac{(x-5)^2}{x^2}$

9. Divide: $\dfrac{x^3+4x^2+6x+2}{x+3}$

Solution:

```
              x² +  x +3
     x+3 ) x³ +4x² +6x +2
          -) x³ +3x²
              x² +6x
           -) x² +3x
                  3x +2
               -) 3x +9
                    - 7  ←Remainder
```

Ans: $\dfrac{x^3+4x^2+6x+2}{x+3}=x^2+x+3-\dfrac{7}{x+3}$.

10. Divide: $\dfrac{1+x+12x^3}{1+2x}$

Solution:

```
              6x² -3x +2
     2x+1 ) 12x³ +0  + x +1
          -) 12x³ +6x²
              -6x² +  x
           -) -6x² -3x
                    4x +1
                 -) 4x +2
                      -1 ← Remainder
```

Ans: $\dfrac{1+x+12x^3}{1+2x}=6x^2-3x+2-\dfrac{1}{2x+1}$.

EXERCISES

Simplify. Give all restrictions on the variables.

1. $\dfrac{3}{x^3}\cdot\dfrac{x^2}{6}$

2. $\dfrac{8}{x}\cdot\dfrac{x^3}{2}$

3. $\dfrac{3y}{2}\cdot\dfrac{6}{y^2}$

4. $\dfrac{5y^4}{4}\cdot\dfrac{2}{y}$

5. $\dfrac{6x^2}{y}\cdot\dfrac{y^4}{2x^5}$

6. $\dfrac{12a}{5b^3}\cdot\dfrac{15b}{4a^5}$

7. $\dfrac{7a^4}{b^2}\cdot\dfrac{3b^6}{14a}$

8. $\dfrac{cd^2}{3c}\cdot\dfrac{18}{d^7}$

9. $\left(\dfrac{2x}{y}\right)^3\cdot\dfrac{y}{2x}$

10. $\left(\dfrac{4x}{y}\right)^2\cdot\dfrac{y}{2x}$

11. $\dfrac{x+2}{x-5}\cdot\dfrac{x-5}{x-2}$

12. $\dfrac{2x-4y}{3y}\cdot\dfrac{6y^2}{x-2y}$

13. $\dfrac{a-3}{3-a}$

14. $\dfrac{5a-10b}{2b-a}$

15. $\dfrac{n^2-16}{2n}\cdot\dfrac{n}{n+4}$

16. $\dfrac{n^2-4n-5}{n^2-25}$

17. $\dfrac{3x}{10}\div\dfrac{x^2}{5}$

18. $\dfrac{x^3}{y^4}\div\dfrac{x}{y^2}$

19. $\dfrac{5n^2}{2p}\div\dfrac{n^4}{4p^3}$

20. $\dfrac{ab^3}{5}\div\dfrac{a^3b}{10}$

21. $4x\div\left(\dfrac{2x}{3}\right)^2$

22. $\left(\dfrac{3y}{2}\right)^3\div15y^2$

23. $\dfrac{x+2}{x-5}\div\dfrac{x-2}{x-5}$

24. $\dfrac{2x-4y}{3y}\div\dfrac{x-2y}{6y^2}$

25. $\dfrac{3+3a}{8}\div\dfrac{1+a}{4a}$

26. $\dfrac{x^2-4}{3}\div\dfrac{x-2}{9}$

27. $\dfrac{2}{x-y}\div\dfrac{4}{y-x}$

28. $\dfrac{10-2a}{5}\div\dfrac{15-3a}{10}$

Simplify. Assume that no variable has a value for which the denominator is zero.

29. $\dfrac{x^2+2x-3}{x^2+x-2}\cdot\dfrac{4x+8}{x+4}$

30. $\dfrac{x^2-3x+2}{x+2}\cdot\dfrac{2x+4}{x^2+2x-3}$

31. $\dfrac{x^2-9}{15x}\cdot\dfrac{3x}{x+3}$

32. $\dfrac{x+1}{x^2-5x+6}\cdot\dfrac{x-2}{x^2-1}$

33. $\dfrac{m^2-2m-8}{m+5}\div\dfrac{m^2-4}{m^2-25}$

34. $\dfrac{x^2-y^2}{x^3}\div\dfrac{4x-4y}{x}$

35. $\dfrac{a^2-2a-15}{a^2+6a+8}\div\dfrac{a^2-6a+5}{a^2+3a-4}$

36. $\dfrac{3a+9}{5a+10}\cdot\dfrac{3a+12}{a+3}\div\dfrac{a^2-16}{a^2-4a-12}$

Divide polynomials.

37. $\dfrac{x^2-5x+6}{x+2}$

38. $\dfrac{x^2+5x-6}{x-2}$

39. $\dfrac{4a^2+4a-12}{2a-1}$

40. $\dfrac{6a^2-5a+10}{3a+2}$

41. $\dfrac{2x^3-2x+1}{2x+4}$

42. $\dfrac{2x^3-x^2-5}{2x-5}$

43. $\dfrac{a^4-4}{a-4}$

44. $\dfrac{2n^4-n^3-2n+5}{n^2+2n+1}$

6-4 Solving Rational Equations

If a rational equation consists of one fraction equal to another (it is a proportion), then the **cross products** are equal (see example 1).

To solve a rational equation consisting of two or more fractions on one side, we multiply both sides of the equation by their **least common denominator (LCD)**.

Example 1:

Solve $\dfrac{2}{x} = \dfrac{4}{5}$.

Solution:

$$2 \cdot 5 = 4 \cdot x$$

$$10 = 4x$$

$$\therefore x = \tfrac{10}{4} = 2\tfrac{1}{2}. \text{ Ans.}$$

Example 2:

Solve $\dfrac{x}{3} + \dfrac{x}{4} = 7$.

Solution: LCD = 12

$$12\left(\dfrac{x}{3} + \dfrac{x}{4}\right) = 12(7)$$

$$4x + 3x = 84$$

$$7x = 84$$

$$\therefore x = 12. \text{ Ans.}$$

Example 3:

Solve $\dfrac{3}{x} - \dfrac{1}{2x} = \dfrac{1}{3}$.

Solution: LCD = 6x

$$6x\left(\dfrac{3}{x} - \dfrac{1}{2x}\right) = 6x\left(\dfrac{1}{3}\right)$$

$$18 - 3 = 2x$$

$$15 = 2x$$

$$\therefore x = \tfrac{15}{2} = 7\tfrac{1}{2}. \text{ Ans.}$$

To solve a rational equation, we always test each root in the original equation. A root is **not permissible** if it makes the denominator of the original equation equal to 0.

Example 4: Solve $\dfrac{18}{x^2 - 9} = \dfrac{x}{x-3}$.

Solution: $18(x-3) = x(x^2 - 9)$

$$18(\cancel{x-3}) = x(x+3)(\cancel{x-3})$$

$$18 = x(x+3)$$

$$18 = x^2 + 3x$$

$$0 = x^2 + 3x - 18$$

$$x^2 + 3x - 18 = 0$$

$$(x+6)(x-3) = 0$$

$$x = -6 \mid x = 3 \,(\textbf{not permissible}) \quad \therefore x = -6. \text{ Ans.}$$

If the final statement is a "**true**" statement, the equation has roots for all real numbers.
If the final statement is a "**false**" statement, the equation has no root.

Example 5: Solve $\dfrac{4}{2x-2} = \dfrac{2}{x-1}$.

Solution:
$$4(x-1) = 2(2x-2)$$
$$4x - 4 = 4x - 4$$
$$0 = 0 \,(\textbf{true})$$
Ans: The equation has roots for all real numbers except 1.

Example 6: Solve $\dfrac{4}{2x+1} = \dfrac{2}{x}$.

Solution:
$$4x = 2(2x+1)$$
$$4x = 4x + 2$$
$$0 = 2 \,(\textbf{false})$$
Ans: No solution.

EXERCISES

Solve each rational equation (fractional equation).

1. $\dfrac{x}{3} = \dfrac{4}{5}$

2. $\dfrac{x}{6} = \dfrac{3}{5}$

3. $\dfrac{2}{5} = \dfrac{7}{3x}$

4. $\dfrac{5}{2x} = \dfrac{12}{7}$

5. $\dfrac{2y}{3} = \dfrac{7}{6}$

6. $\dfrac{12}{5} = \dfrac{4y}{9}$

7. $\dfrac{15}{x} = 3$

8. $\dfrac{3x}{7} = -6$

9. $\dfrac{a-3}{8} = \dfrac{3}{4}$

10. $\dfrac{a-7}{7} = \dfrac{7-a}{3}$

11. $\dfrac{n-5}{4} = 12$

12. $12 = \dfrac{2n+4}{3}$

13. $\dfrac{4}{2x-1} = \dfrac{3}{x}$

14. $\dfrac{5n-2}{3} = \dfrac{5n+2}{4}$

15. $\dfrac{2+3x}{2-3x} = -5$

16. $\dfrac{12}{t+3} = \dfrac{24}{t+14}$

17. $\dfrac{2a-2}{a-4} = \dfrac{1}{4}$

18. $\dfrac{5}{3a+2} = \dfrac{5}{3a-2}$

19. $\dfrac{3}{5y+2} = \dfrac{4}{2+5y}$

20. $\dfrac{3x}{x+2} - \dfrac{1}{x} = 0$

21. $\dfrac{8}{4x+5} = \dfrac{2}{x}$

22. $\dfrac{2}{4x-6} = \dfrac{1}{2x-3}$

23. $\dfrac{5}{y^2-16} = \dfrac{2}{y+4}$

24. $\dfrac{4}{y^2-4} = \dfrac{9}{y-2}$

25. $\dfrac{x}{5} + \dfrac{x}{4} = 2$

26. $\dfrac{2x}{7} - \dfrac{x}{3} = 2$

27. $\dfrac{2}{x} + \dfrac{1}{3x} = 6$

28. $\dfrac{5}{2x} + \dfrac{4}{3x} = 7$

29. $\dfrac{5}{x} - \dfrac{3}{4x} = \dfrac{1}{2}$

30. $\dfrac{3}{2x} - \dfrac{2}{3x} = \dfrac{5}{6}$

31. $\dfrac{9}{2y} - \dfrac{2}{3} = \dfrac{4}{y}$

32. $\dfrac{3y}{8} - y = \dfrac{5}{6}$

33. $\dfrac{4}{n+1} - \dfrac{1}{n} = 1$

34. $\dfrac{3}{n-1} + \dfrac{2}{n} = \dfrac{3}{2}$

35. $\dfrac{12}{a(a-2)} - \dfrac{6}{a-2} = 1$

36. $\dfrac{8}{x^2-4} = \dfrac{x}{x-2}$

37. $\dfrac{2c+5}{3} - \dfrac{c-4}{5} = \dfrac{24}{5}$

38. $\dfrac{3c+1}{10} + \dfrac{2c-1}{5} = 2$

39. $\dfrac{m-4}{16} + \dfrac{m+4}{48} = \dfrac{3}{2}$

40. $\dfrac{1}{n-4} = \dfrac{8}{n^2-16}$

41. $\dfrac{4n-8}{n-2} = 4$

42. $\dfrac{1}{a-3} + \dfrac{1}{a^2-9} = \dfrac{8}{a+3}$

43. Find the ratio of x to y : $\dfrac{x+y}{2} = \dfrac{x-y}{3}$.

44. Find the ratio of x to y : $\dfrac{2x+3y}{5} = \dfrac{3x+2y}{4}$.

45. Find the slope of the graph of the given equation: $\dfrac{2x-1}{3} + \dfrac{2y+1}{2} = \dfrac{5}{6}$.

6-5 Solving Word Problems by Rational Equations

Rational equations (fractional equations) are often used in solving word problems.

Examples:

1. One fifth of a number is 7 less than two thirds of the number. Find the number.

Solution:

    Let n = the number

    The equation is:

$$\frac{n}{5} + 7 = \frac{2n}{3}$$

    LCD = 15

$$3n + 105 = 10n$$

$$105 = 7n \quad \therefore n = 15$$

Ans: 15.

2. The sum of two numbers is 48 and their quotient is $\frac{3}{5}$. Find the numbers.

Solution:

    Let $\quad n = 1^{st}$ number

    $48 - n = 2^{nd}$ number

    The equation is: $\dfrac{n}{48 - n} = \dfrac{3}{5}$

$$5n = 3(48 - n)$$

$$5n = 144 - 3n$$

$$8n = 144 \quad \therefore n = 18$$

Ans: 18 and 30.

3. If two pounds of beef cost $5.80, how much do 8 pounds cost ?

Solution:

    Let x = cost for 8 pounds

    The equation is:

$$\frac{5.80}{2} = \frac{x}{8}$$

$$5.80(8) = 2x$$

$$46.40 = 2x \quad \therefore x = 23.20$$

Ans: $23.20.

4. A bus uses 4.5 gallons of gasoline to travel 75 miles. How many gallons would the bus use to travel 95 miles ?

Solution:

    Let x = the number of gallons

    The equation is: $\dfrac{75}{4.5} = \dfrac{95}{x}$

$$75x = 4.5(95)$$

$$75x = 427.5 \quad \therefore x = 5.7$$

Ans: 5.7 gallons.

5. John can finish a job in 5 hours. Steve can finish the same job in 4 hours. How many hours do they need to finish the job if they work together ?

Solution:

    Let x = number of hours needed to do the job together

    John can do $\frac{1}{5}$ of the job per hour.

    Steve can do $\frac{1}{4}$ of the job per hour.

    The equation is: $\dfrac{x}{5} + \dfrac{x}{4} = 1$

$$4x + 5x = 20, \quad 9x = 20$$

$$\therefore x = 2\tfrac{2}{9} \quad \text{Ans: } 2\tfrac{2}{9} \text{ hours.}$$

6. A 16-gallon salt-water solution contains 25% pure salt. How much water should be added to produce the solution to 20% salt ?

Solution:

    Let x = gallons of water added

    Pure salt in the original solution:

$$16 \times 25\% = 16 \times 0.25 = 4 \text{ gal.}$$

    The equation is:

$$\frac{4}{16 + x} = 20\%, \quad \frac{4}{16 + x} = \frac{1}{5}$$

$$20 = 16 + x \quad \therefore x = 4$$

Ans: 4 gallons of water.

EXERCISES

1. One sixth of a number is 7 less than three fourths of the number. Find the number.
2. Three fourths of a number is 7 more than one sixth of the number. Find the number.
3. The sum of two numbers is 44 and their quotient is $\frac{3}{8}$. Find the numbers.
4. The sum of the reciprocals of two consecutive positive even integers is $\frac{5}{12}$. Find the integers.
5. The sum of a number and its reciprocal is $\frac{25}{12}$. Find the number.
6. The sum of two numbers is 12 and the sum of their reciprocals is $\frac{12}{35}$. Find the numbers.
7. If three pounds of beef cost $4.92, how much do 10 pounds cost ?
8. If five cans of dog food cost $3.45, how much do 12 cans cost ?
9. A car uses 6 gallons of gasoline to travel 96 miles. How many gallons would the car use to travel 160 miles ?
10. A car uses 6 gallons of gasoline to travel 96 miles. How many miles can the car travel on a full tank of 17 gallons ?
11. In a random survey taken in a city election, 25 of the 120 voters surveyed voted John for mayor. Based on the result of this survey, estimate how many of 24,000 total voters in the city should vote John for mayor ?
12. A map shows the length and the width of a rectangular field with the drawing 1.2 m by 0.8 m. 1 m represents 250 meters. Find the actual dimensions of the field.
13. Tom can paint the garage in 4 hours. Jack can paint the same garage in 2.4 hours. How long would it take them to paint the garage if they work together ?
14. Tom can paint the garage in 4 hours. Working together, Tom and Jack can paint the garage in 1.5 hours. How long would it take Jack to paint the garage alone ?
15. John can finish a job in 5 hours. Steve can finish the same job in 4 hours. After working together for 1 hour, Steve leaves. How long will it take John to complete the job ?
16. Nancy can ride 15 miles on her bicycle in the same time that it takes her to walk 6 miles. If her riding speed is 5 miles per hour faster than her walking speed, how fast does she walk ?
17. Nancy drives 20 miles to her school each day. If she drives 10 miles per hour faster, it takes her 4 minutes less to get to school. Find her new speed.
18. Flying in still air, an airplane travels 550 km/h. It can travel 1200 km with the wind in the same time that it travels 1000 km against the wind. Find the speed of the wind.
19. A boat travels 25 miles per hour in still water. It takes $3\frac{1}{3}$ hours to travel 40 miles up a river and then to return by the same route. What is the speed of the current in the river ?
20. A 12-gallon salt-water solution contains 25% pure salt. How much water should be added to produce the solution to 15% salt ?
21. A 18-gallon salt-water solution contains 15% pure salt. How much water should be added to produce the solution to 12% salt ?
22. A 18-gallon salt-water solution contains 15% pure salt. How much pure salt should be added to produce the solution to 20% salt ?

6-6 Inverse Variations

In Section 2-8, we learned that a linear equation in the form $y = kx$ is called an equation having a **direct variation**. The graph of an equation having a direct variation is a straight line.

An equation in the form $y = \frac{k}{x}$ (where $x \neq 0$), or $xy = k$ is called an equation having an **inverse variation**. k is a nonzero constant. k is called the **constant of variation**.

For example, in the equation $y = \frac{3}{x}$, if x is increased, then y is decreased. We say that y varies inversely as x, or y is inversely proportional to x. Therefore, $y = \frac{3}{x}$ is an example of an inverse variation. The graph of $y = \frac{3}{x}$, or $xy = 3$ is not a straight line. It is a **hyperbola**.

If (x_1, y_1) and (x_2, y_2) are two ordered pairs of an equation having an inverse variation, we have: $x_1 y_1 = k$ and $x_2 y_2 = k$.

$$\therefore x_1 y_1 = x_2 y_2$$

In science, there are many word problems which involve the concept of inverse variations.

Examples

1. If y varies inversely as x, and if $y = 6$ when $x = 2$, find the constant of variation.

   Solution: Let $xy = k$ $\therefore k = xy = 2 \cdot 6 = 12$. Ans.

2. If y varies inversely as x, and if $y = 6$ when $x = 2$, find y when $x = 3$.

   Solution:

   Let $xy = k$ $\therefore k = xy = 2 \cdot 6 = 12$, $xy = 12$ is the equation.

   When $x = 3$, $y = \frac{12}{x} = \frac{12}{3} = 4$. Ans.

3. If (x_1, y_1) and (x_2, y_2) are ordered pairs of the same inverse variation, find y_1.

   $x_1 = 3$, $y_1 = ?$, $x_2 = 12$, $y_2 = 8$

   Solution:

   Let $x_1 y_1 = x_2 y_2$, $3y_1 = 12 \cdot 8$, $3 y_1 = 96$ $\therefore y_1 = 32$. Ans.

   Or: $k = x_2 y_2 = 12 \cdot 8 = 96$ $\therefore x_1 y_1 = 96$, $y_1 = \frac{96}{x_1} = \frac{96}{3} = 32$. Ans.

4. The number of days needed to build a house varies inversely as the number of workers working on the job. It takes 140 days for 5 workers to finish the job. If the job has to be finished in 100 days, how many workers are needed?

   Solution: Let $d_1 = 140$, $w_1 = 5$, $d_2 = 100$, $w_2 = ?$

   We have: $d_1 w_1 = d_2 w_2$

   $140 \cdot 5 = 100 \cdot w_2$ $\therefore w_2 = \frac{700}{100} = 7$ Ans: 7 workers.

EXERCISES

Find the constant of variation and write an equation of inverse variation.
(Hint: "varies inversely as" and "is inversely proportional to" have the same meaning.

1. If y varies inversely as x, and if $y = 10$ when $x = 2$.
2. If y is inversely proportional to x, and if $y = 10$ when $x = 5$.
3. If n is inversely proportional to m, and if $n = 200$ when $m = 20$.
4. If m varies inversely as n, and if $m = 30$ when $n = 90$.
5. If y is inversely proportional to $(x + 8)$, and if $y = 150$ when $x = 2$.

For each inverse variation described, find each missing value.

6. If y varies inversely as x, and if $y = 10$ when $x = 2$, find y when $x = 12$.
7. If y varies inversely as x, and if $y = 10$ when $x = 2$, find x when $y = 12$.
8. If y is inversely proportional to x, and if $y = 2$ when $x = 10$, find y when $x = 12$.
9. If n varies inversely as m, and if $n = 150$ when $m = 6$, find n when $m = 20$.
10. If n varies inversely as m, and if $n = 150$ when $m = 6$, find m when $n = 20$.

If (x_1, y_1) and (x_2, y_2) are ordered pairs of the same inverse variation, find each missing value.

11. $x_1 = 1$, $y_1 = ?$, $x_2 = 3$, $y_2 = 12$
12. $x_1 = 3$, $y_1 = ?$, $x_2 = 4$, $y_2 = 80$
13. $x_1 = 6$, $y_1 = ?$, $x_2 = 8$, $y_2 = 9.6$
14. $x_1 = 6$, $y_1 = 4$, $x_2 = 9$, $y_2 = ?$
15. $x_1 = ?$, $y_1 = 1.5$, $x_2 = 15$, $y_2 = 9$
16. $x_1 = 1$, $y_1 = 7$, $x_2 = ?$, $y_2 = 17.5$
17. $x_1 = \frac{4}{5}$, $y_1 = ?$, $x_2 = \frac{2}{5}$, $y_2 = \frac{1}{10}$
18. $x_1 = \frac{1}{4}$, $y_1 = \frac{1}{5}$, $x_2 = \frac{5}{3}$, $y_2 = ?$
19. $x_1 = \frac{7}{8}$, $y_1 = ?$, $x_2 = \frac{1}{2}$, $y_2 = \frac{1}{7}$
20. $x_1 = \frac{1}{5}$, $y_1 = 5$, $x_2 = 5$, $y_2 = ?$

21. Does the equation $xy = 5$ define an inverse variation? Explain.
22. Are the ordered pairs (6, 4), (3, 8), (2, 12) in the same inverse variation? Explain.
23. The number of days needed to finish a job varies inversely as the number of workers working on the job. It takes 40 hours for 3 workers to finish the job. If the job has to be finished in 15 hours, how many workers are needed?
24. The time required to travel a given distance is inversely proportional to the speed of a car. If it takes 5.6 hours to travel from city A to city B at an average speed of 60 miles/h, how long will it take to make the same trip at an average speed of 70 miles/h.
25. The length of a rectangle of given area varies inversely as the width. A rectangle field has length $24\,m$ and width $18\,m$. Find the width of another rectangle field of equal area whose length is $12\,m$.
26. At a fixed water pressure, the speed of the water varies inversely as the diameter of the pipe. If the water flows at 30 miles/h through a pipe with a $2\,cm$ diameter, what would be the speed of the water through a pipe with a $1.5\,cm$?
27. If z varies directly as x and inversely as y, and if $z = 9$ when $x = 15$ and $y = 3$, find z when $x = 45$ and $y = 2$.

6-7 Fractional Inequalities and Absolute Values

Since we don't know whether its variable is positive or negative, a fractional inequality in which a variable appears in a denominator can change signs of inequality if we multiply (or divide) each side by its common denominator. Therefore, we rewrite the inequality with **0** on the right side and consider two cases. **Testing the intervals** of the critical points is a good method to identify the answers. (See Section 5-8.)

Examples

1. Solve $\dfrac{3}{x} < 5$.

Solution: **Method 1:** $\frac{3}{x} - 5 < 0$, $\therefore \frac{3-5x}{x} < 0$

$$x > 0 \text{ and } 3-5x < 0 \quad \Big| \quad x < 0 \text{ and } 3-5x > 0$$
$$-5x < -3 \qquad\qquad\quad -5x > -3$$
$$x > \tfrac{3}{5} \qquad\qquad\qquad x < \tfrac{3}{5}$$
$$\therefore x > \tfrac{3}{5} \qquad\qquad\qquad \therefore x < 0$$

Ans. $x < 0$ or $x > \tfrac{3}{5}$.

Method 2: $\frac{3-5x}{x} < 0$. It is negative.

Critical points: $x = 0, \ \tfrac{3}{5}$

Test intervals: $x < 0, \ \frac{(+)}{(-)} = -$ yes

$0 < x < \tfrac{3}{5}, \ \frac{(+)}{(+)} = +$ no

$x > \tfrac{3}{5}, \ \frac{(-)}{(+)} = -$ yes

Ans. $x < 0$ or $x > \tfrac{3}{5}$.

2. Solve $\dfrac{x+1}{x} > 3$.

Solution:

$$\tfrac{x+1}{x} - 3 > 0, \ \tfrac{x+1-3x}{x} > 0$$
$$\therefore \tfrac{1-2x}{x} > 0. \text{ It is positive.}$$

Critical points: $x = 0, \ \tfrac{1}{2}$

Test intervals:

$x < 0, \ \frac{(+)}{(-)} = -$ no

$0 < x < \tfrac{1}{2}, \ \frac{(+)}{(+)} = +$ yes

$x > \tfrac{1}{2}, \ \frac{(-)}{(+)} = -$ no

Ans: $0 < x < \tfrac{1}{2}$.

In Section 3 – 6, we have learned how to solve simple inequalities involving absolute values. To solve a fractional inequality involving absolute value, we use this fact:

$$|x|^2 = x^2$$

3. Solve $\left|\dfrac{x-2}{4x+1}\right| \geq 1$.

Solution:

$$|x-2| \geq |4x+1|$$
$$(x-2)^2 \geq (4x+1)^2$$
$$x^2 - 4x + 4 \geq 16x^2 + 8x + 1$$
$$-15x^2 - 12x + 3 \geq 0$$
$$5x^2 + 4x - 1 \leq 0$$
$$(5x-1)(x+1) \leq 0$$

$5x-1 \geq 0$ and $x+1 \leq 0$ | $5x-1 \leq 0$ and $x+1 \geq 0$

$x \geq \tfrac{1}{5}$, $x \leq -1$ | $x \leq \tfrac{1}{5}$, $x \geq -1$

Not permissible | $\therefore -1 \leq x \leq \tfrac{1}{5}, \ x \neq -\tfrac{1}{4}$. Ans.

(We can also test the intervals to find the answer.)

EXERCISES

Solving the following fractional inequalities.

1. $\dfrac{2}{x} < 4$

2. $\dfrac{2}{x} \geq 4$

3. $\dfrac{1}{x} - 2 < 0$

4. $\dfrac{1}{x} - 2 > 0$

5. $\dfrac{1}{x} - x > 0$

6. $\dfrac{1}{x} - x < 0$

7. $\dfrac{1}{x} - x \geq 0$

8. $\dfrac{x+1}{x} - 3 < 0$

9. $\dfrac{x+4}{x+2} > \dfrac{1}{3}$

10. $\dfrac{2x+3}{x} < 1$

11. $\dfrac{2x-1}{x-2} \leq 3$

12. $\dfrac{2x-3}{x+2} < \dfrac{1}{3}$

13. $\dfrac{5}{x-6} > \dfrac{3}{x+2}$

14. $\dfrac{x-2}{x} < \dfrac{x-4}{x-6}$

15. $\dfrac{3}{y+2} + \dfrac{7}{15} \geq \dfrac{23}{3y+6}$

16. $\left| \dfrac{x}{2} \right| > 7$

17. $\left| \dfrac{x}{2} \right| < 7$

18. $\left| \dfrac{x-4}{2} \right| \geq 7$

19. $\left| \dfrac{x-4}{2} \right| \leq 7$

20. $\left| \dfrac{x-2}{3} \right| < 5$

21. $\left| \dfrac{x-2}{3} \right| > 5$

22. $\left| \dfrac{1-x}{x+2} \right| \leq 1$

23. $\left| \dfrac{2x-5}{x-6} \right| < 3$

24. $\left| \dfrac{2x-5}{x-6} \right| < 4$

25. $\left| \dfrac{x-2}{4x+1} \right| \leq 1$

26. $\left| \dfrac{x-2}{4x+1} \right| > 1$

27. $\left| \dfrac{2x-5}{x-6} \right| \geq 4$

A mother asks her 6-year old son about his new eyeglasses.
Mother: Can you see very well with your new eyeglasses
when you are in the class.
Son: Yes. I can see very clear all of my teacher's writing
on the blackboard.
Mother: Will you get better grades ?
Son: No. I still have the same brain.

6-8 Complex Fractions

Complex Fraction: A fraction whose numerator or denominator, or both, contains:
　　　　　　　　　1) one or more fractions.　2) powers with negative exponents.

To simplify a complex fraction, we simplify its numerator and denominator independently. Then, multiply the numerator by the reciprocal of the denominator.

Examples

1. $\dfrac{\frac{1}{2}}{\frac{1}{3}} = \dfrac{1}{2} \cdot \dfrac{3}{1} = \dfrac{3}{2} = 1\dfrac{1}{2}$.

2. $\dfrac{1+\frac{1}{2}}{1-\frac{1}{3}} = \dfrac{\frac{2+1}{2}}{\frac{3-1}{3}} = \dfrac{\frac{3}{2}}{\frac{2}{3}} = \dfrac{3}{2} \cdot \dfrac{3}{2} = \dfrac{9}{4} = 2\dfrac{1}{4}$.

3. $\dfrac{\frac{1}{x}}{\frac{2}{x}} = \dfrac{1}{\cancel{x}} \cdot \dfrac{\cancel{x}}{2} = \dfrac{1}{2}$, $x \neq 0$

4. $\dfrac{a+a^{-2}}{1+a^{-1}} = \dfrac{a+\frac{1}{a^2}}{1+\frac{1}{a}} = \dfrac{\frac{a^3+1}{a^2}}{\frac{a+1}{a}} = \dfrac{a^3+1}{a^2} \cdot \dfrac{a}{a+1}$

$= \dfrac{a^3+1}{a(a+1)} = \dfrac{(\cancel{a+1})(a^2-a+1)}{a(\cancel{a+1})} = \dfrac{a^2-a+1}{a}$,

$a \neq 0, -1$

5. $1 - \dfrac{1}{1-\frac{1}{x-2}} = 1 - \dfrac{1}{\frac{(x-2)-1}{x-2}} = 1 - \dfrac{1}{\frac{x-3}{x-2}}$

$= 1 - \dfrac{x-2}{x-3} = \dfrac{x-3-(x-2)}{x-3} = \dfrac{-1}{x-3}$,

$x \neq 2, 3$

6. $\dfrac{x+\frac{x-3}{x+1}}{x-\frac{2}{x+1}} = \dfrac{\frac{x(x+1)+(x-3)}{x+1}}{\frac{x(x+1)-2}{x+1}}$

$= \dfrac{\frac{x^2+2x-3}{x+1}}{\frac{x^2+x-2}{x+1}} = \dfrac{x^2+2x-3}{\cancel{x+1}} \cdot \dfrac{\cancel{x+1}}{x^2+x-2}$

$= \dfrac{(x+3)(\cancel{x-1})}{(x+2)(\cancel{x-1})} = \dfrac{x+3}{x+2}$, $x \neq -2, -1, 1$

7. $1 + \dfrac{1}{1+\frac{1}{2}} = 1 + \dfrac{1}{\frac{3}{2}} = 1 + \dfrac{2}{3} = \dfrac{5}{3} = 1\dfrac{2}{3}$.

8. $\dfrac{a-b}{a^{-1}-b^{-1}} = \dfrac{a-b}{\frac{b-a}{ab}} = (a-b) \cdot \dfrac{ab}{b-a} = -ab$.

$a \neq 0,\ b \neq 0,\ a \neq b$

9. $x + \dfrac{1}{x-\frac{1}{x}} = x + \dfrac{1}{\frac{x^2-1}{x}} = x + \dfrac{x}{x^2-1}$

$= \dfrac{x(x^2-1)+x}{x^2-1} = \dfrac{x^3}{x^2-1}$, $x \neq 0, \pm 1$

EXERCISES

Simplify each complex fraction.

1. $\dfrac{\frac{1}{3}+2}{\frac{1}{2}+3}$

2. $\dfrac{\frac{1}{5}-3}{\frac{2}{5}+4}$

3. $\dfrac{\frac{1}{2}+1}{\frac{1}{3}+1}$

4. $\dfrac{1+\frac{1}{3}}{1-\frac{1}{2}}$

5. $\dfrac{\frac{2}{5}+\frac{4}{5}}{\frac{3}{10}-\frac{1}{5}}$

6. $\dfrac{\frac{1}{2}+\frac{1}{3}}{\frac{1}{5}+\frac{1}{6}}$

7. $\dfrac{\frac{5}{x}+1}{\frac{4}{x}-1}$

8. $\dfrac{\frac{6}{x^2}-2}{\frac{3}{x^2}-1}$

9. $\dfrac{\frac{x^2-5}{4}+1}{\frac{x+1}{8}}$

10. $\dfrac{\frac{b}{a}+1}{\frac{a}{b}+1}$

11. $\dfrac{1-\frac{b}{a}}{1-\frac{a}{b}}$

12. $\dfrac{\frac{b}{a}-\frac{a}{b}}{\frac{1}{a}-\frac{1}{b}}$

13. $\dfrac{\frac{x}{2}+3}{1+\frac{6}{x}}$

14. $\dfrac{\frac{x}{2}-5}{1+\frac{x}{4}}$

15. $\dfrac{\frac{a^2}{b^2}-1}{\frac{1}{a}+\frac{1}{b}}$

16. $\dfrac{\frac{y}{x^2}-\frac{x}{y^2}}{\frac{1}{x^2}-\frac{1}{y^2}}$

17. $\dfrac{\frac{x^2}{(x-1)^2}-1}{\frac{2x-1}{(x-1)^2}}$

18. $\dfrac{x-\frac{1}{x}}{x+2+\frac{1}{x}}$

19. $1-\dfrac{1}{1-\frac{1}{x-3}}$

20. $x+\dfrac{1}{x-\frac{1}{x-3}}$

21. $\dfrac{\frac{1}{x+h}-\frac{1}{x}}{h}$

22. $\dfrac{\frac{1}{(x+h)^2}-\frac{1}{x^2}}{h}$

23. $\dfrac{a-a^{-2}}{1-a^{-1}}$

24. $\dfrac{x^{-2}-y^{-2}}{x^{-1}+y^{-1}}$

25. $\dfrac{\frac{1}{x+y}}{x^{-1}+y^{-1}}$

26. $\dfrac{a-a^{-2}}{1-a^{-1}}$

27. $\dfrac{a+b}{a^{-1}+b^{-1}}$

28. $2+\dfrac{2}{2+\frac{1}{2}}$

29. $2+\dfrac{2}{2+\frac{2}{2+\frac{1}{2}}}$

30. $x+\dfrac{1}{x+\dfrac{1}{x+\frac{1}{x+\frac{1}{x}}}}$

6-9 Decomposing Rational Expressions (Partial Fractions)

Decomposing a rational expression into its partial fractions is useful in the study of calculus.

Rational expression: A quotient of polynomials.

We have learned how to combine (add or subtract) rational expressions.

$$\frac{3}{x+2}+\frac{5}{x-2}=\frac{3(x-2)+5(x+2)}{x^2-4}=\frac{3x-6+5x+10}{x^2-4}=\frac{8x+4}{x^2-4}$$

To decompose a rational expression into its partial fractions, we factor the denominator into the product of linear and irreducible quadratic factors.

Rules:

1. If the denominator has the form $(ax+b)^m$, the decomposition of its partial fractions must have:

$$\frac{N(x)}{(ax+b)^m}=\frac{A_1}{ax+b}+\frac{A_2}{(ax+b)^2}+\cdots\cdots+\frac{A_m}{(ax+b)^m}$$

2. If the denominator has the form of the irreducible factor $(ax^2+bx+c)^m$, the decomposition of its partial fractions must have:

$$\frac{N(x)}{(ax^2+bx+c)^m}=\frac{A_1x+C_1}{ax^2+bx+c}+\frac{A_2x+C_2}{(ax^2+bx+c)^2}+\cdots\cdots+\frac{A_mx+C_m}{(ax^2+bx+c)^m}$$

3. If the rational expression is an **improper fraction** (the degree of the polynomial in the numerator is **greater than** or **equal to** the degree of the polynomial in the denominator), we divide the denominator into the numerator first (see Example 5 and Exercises 11, 12, 13, 14).

Examples

1. Decompose into its partial fractions: $\dfrac{8x+4}{x^2-4}$

   Solution:

   $$\frac{8x+4}{x^2-4}=\frac{A}{x+2}+\frac{B}{x-2}$$

   Multiplying both sides by $(x+2)(x-2)$, we have:
   $$8x+4=A(x-2)+B(x+2)$$

Let $x=2$,	Let $x=-2$
$8(2)+4=B(2+2)$	$8(-2)+4=A(-2-2)$
$20=4B \quad \therefore B=5$	$-12=-4A \quad \therefore A=3$

   $$\therefore \quad \frac{8x+4}{x^2-4}=\frac{3}{x+2}+\frac{5}{x-2} \quad \text{Ans.}$$

Examples

2. Decompose $\dfrac{5x^2 + 13x + 14}{(x+3)(x-2)(x+1)}$ into its partial fractions.

Solution:

$$\frac{5x^2 + 13x + 14}{(x+3)(x-2)(x+1)} = \frac{A}{x+3} + \frac{B}{x-2} + \frac{C}{x+1}$$

$$5x^2 + 13x + 14 = A(x-2)(x+1) + B(x+3)(x+1) + C(x+3)(x-2)$$

Let $x = 2$, $5(2)^2 + 13(2) + 14 = B(2+3)(2+1)$

$$60 = 15B \quad \therefore B = 4$$

Let $x = -1$, $5(-1)^2 + 13(-1) + 14 = C(-1+3)(-1-2)$

$$6 = -6C \quad \therefore C = -1$$

Let $x = -3$, $5(-3)^2 + 13(-3) + 14 = A(-3-2)(-3+1)$

$$20 = 10A \quad \therefore A = 2$$

$$\therefore \frac{5x^2 + 13x + 14}{(x+3)(x-2)(x+1)} = \frac{2}{x+3} + \frac{4}{x-2} - \frac{1}{x+1}. \text{ Ans}$$

3. Decompose $\dfrac{5x - 17}{(x-2)^2}$ into its partial fractions.

Solution:

$$\frac{5x-17}{(x-2)^2} = \frac{A}{x-2} + \frac{B}{(x-2)^2}$$

$$5x - 17 = A(x-2) + B$$

Let $x = 2$, $5(2) - 17 = B$

$$\therefore -7 = B$$

Let $x = 0$, $5(0) - 17 = -2A + B$

$$-17 = -2A + (-7)$$

$$-10 = -2A$$

$$\therefore 5 = A$$

$$\therefore \frac{5x-17}{(x-2)^2} = \frac{5}{x-2} - \frac{7}{(x-2)^2}. \text{ Ans.}$$

Examples

4. Decompose into its partial fractions for $\dfrac{4x^3+5x^2-6x+8}{x(x+1)(x^2+4)}$.

Solution:

$$\frac{4x^3+5x^2-6x+8}{x(x+1)(x^2+4)} = \frac{A}{x} + \frac{B}{x+1} + \frac{Cx+D}{x^2+4}$$

$$4x^3+5x^2-6x+8 = A(x+1)(x^2+4) + Bx(x^2+4) + (Cx+D)x(x+1)$$

Let $x = -1$, $\;4(-1)^3 + 5(-1)^2 - 6(-1) + 8 = B(-1)[(-1)^2+4]$

$$15 = -5B \quad \therefore B = -3$$

Let $x = 0$, $\;\;8 = A(0+1)(0+4)$, $\;8 = 4A \quad \therefore A = 2$

We have no other convenient choices for $x-$ value. To find the values of C and D, we substitute $A = 2$ and $B = -3$ into the expression.

$$4x^3+5x^2-6x+8 = 2(x+1)(x^2+4) - 3x(x^2+4) + (Cx+D)x(x+1)$$

$$= 2(x^3+x^2+4x+4) - 3x^3 - 12x + Cx^3 + Cx^2 + Dx^2 + Dx$$

$$= 2x^3 + 2x^2 + 8x + 8 - 3x^3 - 12x + Cx^3 + Cx^2 + Dx^2 + Dx$$

$$= -x^3 + Cx^3 + 2x^2 + Cx^2 + Dx^2 - 4x + Dx + 8$$

$$\therefore 4x^3 + 5x^2 - 6x + 8 = (-1+C)x^3 + (2+C+D)x^2 + (D-4)x + 8$$

Comparing the coefficients of like terms on opposite sides, we have:

$$-1+C = 4 \quad \therefore C = 5$$

$$D - 4 = -6 \quad \therefore D = -2$$

$$\therefore \frac{4x^3+5x^2-6x+8}{x(x+1)(x^2+4)} = \frac{2}{x} - \frac{3}{x+1} + \frac{5x-2}{x^2+4}. \quad \text{Ans.}$$

5. Decompose into its partial fractions for $\dfrac{8x^3+4x^2+4x-12}{x^2-4}$.

Solution:

It is an improper fraction. We divide the denominator into the numerator.

$$\frac{8x^3+4x^2+4x-12}{x^2-4} = 8x+4+\frac{36x+4}{x^2-4}$$

$$\frac{36x+4}{x^2-4} = \frac{A}{x+2} + \frac{B}{x-2}$$

$$36x+4 = A(x-2) + B(x+2)$$

Let $x = 2$, $\;\;36(2)+4 = B(2+2)$, $\;76 = 4B \quad \therefore B = 19$

Let $x = -2$, $36(-2)+4 = A(-2-2)$, $\;-68 = -4A$, $\;\;\therefore A = 17$

$$\therefore \frac{8x^3+4x^2+4x-12}{x^2-4} = 8x+4+\frac{17}{x+2} + \frac{19}{x-2}. \quad \text{Ans.}$$

EXERCISES

Decompose into its partial fractions for each fractional expression.

1. $\dfrac{1}{x^2 + x}$

2. $\dfrac{2}{x^2 - 2x}$

3. $\dfrac{1}{x^2 - 1}$

4. $\dfrac{6}{x^2 + x - 2}$

5. $\dfrac{21}{x^2 + x - 12}$

6. $\dfrac{2}{4x^2 - 9}$

7. $\dfrac{-x - 5}{(x+1)(x-1)}$

8. $\dfrac{9x + 24}{x^2 + 2x - 8}$

9. $\dfrac{x + 12}{x^2 - x - 6}$

10. $\dfrac{x + 1}{x^2 - 4x}$

11. $\dfrac{x^2 + 1}{x(x-1)}$

12. $\dfrac{x^2 + 2}{x^2 - x}$

13. $\dfrac{x^2 - x}{x^2 + x - 2}$

14. $\dfrac{x^2 - 2x}{x^2 + x + 1}$

15. $\dfrac{3x^2 - 7x - 2}{x(x+1)(x-1)}$

16. $\dfrac{-x + 8}{2x^2 + x - 1}$

17. $\dfrac{x + 5}{(x-1)^2}$

18. $\dfrac{2x}{(x-3)^2}$

19. $\dfrac{2x - 5}{(x-1)^2}$

20. $\dfrac{3x^2 - 4x + 2}{(x-1)^3}$

21. $\dfrac{3x^2 - 21x + 6}{(x+2)(x-4)(x-3)}$

22. $\dfrac{5x^2 + 15x + 12}{x(x+2)^2}$

23. $\dfrac{x^3}{(x-1)^3}$

24. $\dfrac{x^4}{(x-1)^3}$

25. $\dfrac{4x}{(x-1)(x^2+1)}$

26. $\dfrac{x^2 + 5}{(x-1)(x^2+x+1)}$

27. $\dfrac{7x^2 - 11x + 26}{(x-3)(x^2+5)}$

28. $\dfrac{x^3 + 2x^2 + x - 2}{x^2 - 1}$

29. $\dfrac{x^3 - 3x^2 - 11x + 16}{x^2 + x - 6}$

30. $\dfrac{3x^2 + 2x + 8}{x^2 + 4x}$

31. $\dfrac{5x^3 + 4x}{(x^2 + 2)^2}$

32. $\dfrac{x^4 + 2x^3 + 5x^2 + 12x + 5}{x^3 + 2x^2 + x}$

6-10 Graphing Rational Functions

A **rational expression (algebraic fraction)** is a quotient (ratio) of two polynomials. A **rational (fractional) function** is a function with the quotient (ratio) of two polynomials.

$y = \frac{1}{x}$ is the most popular and simplest rational function. Its graph is a **hyperbola**.

The graph of a rational function has a **vertical asymptote** at $x = a$ if the values for y increase (or decrease) without bound as x approaches a, either from the right or from the left.

The graph of a rational function has a **horizontal asymptote** at $y = b$ if the values for y approach b as x increases (or decreases) without bound.

The graph of a rational function may have several vertical asymptotes, but at most one horizontal asymptote.

Rules for graphing a rational function:

$$y = f(x) = \frac{a_n x^n + a_{n-1} x^{n-1} + \cdots\cdots + a_0}{b_m x^m + b_{m-1} x^{m-1} + \cdots\cdots + b_0}$$

1. Find x – intercept (let $y = 0$) and y – intercept (let $x = 0$).
2. Find the vertical asymptotes by setting the denominator equal to zero.
3. If $n < m$, then $y = 0$ (it is the $x - axis$) is a horizontal asymptote.
4. If $n > m$, then there is no horizontal asymptote.
5. If $n = m$, then $y = \frac{a_n}{b_m}$ is a horizontal asymptote.
6. If n is more than m only by 1 degree, then there is a slant (oblique) asymptote which can be determined by dividing the denominator into the numerator.
7. Complete the missing portions of the graph by plotting a few additional points.
8. The graphs of all rational functions of the form $y = \frac{ax + a_0}{bx + b_0}$ are hyperbolas.

Examples

1. Graph $y = \frac{1}{x}$.

   Solution:
   1. $f(-x) = -f(x)$. It is symmetric about the origin.
   2. There is no x – intercept and y – intercept.
   3. $x = 0$ (It is the $y - axis$) is a vertical asymptote.
   4. The degree of the numerator is less than the degree of denominator. Therefore $y = 0$ (It is the $x - axis$) is a horizontal asymptote.
   5. The domain and range are all nonzero real numbers.

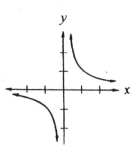

-----Continued-----

Examples

2. Graph $y = \dfrac{3}{x-2}$.

Solution:

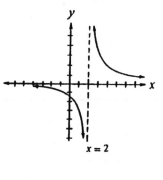

1. Let $x = 0$, $y = -\frac{3}{2}$, $(0, -\frac{3}{2})$ is the y – intercept.
2. $x = 2$ is a vertical asymptote.
3. The degree of the numerator is less than the degree of the denominator. Therefore, $y = 0$ (it is the $x - axis$) is a horizontal asymptote.
4. The domain is all real numbers except 2. The range is all nonzero real numbers.

3. Graph $y = \dfrac{1}{x^2}$.

Solution:

1. $f(-x) = f(x)$. It is symmetric about the $y - axis$.
2. There is no x – intercept and y – intercept.
3. $x = 0$ (it is the $y - axis$) is a vertical asymptote.
4. $y = 0$ (it is the $x - axis$) is a horizontal asymptote.
5. The domain is all nonzero real numbers. The range is all real numbers of $y > 0$.

4. Graph $y = \dfrac{1}{(x-1)^2} - 4$.

Solution:

1. Let $x = 0$, then $y = -3$. $(0, -3)$ is the y – intercept.
   Let $y = 0$, then $x = \frac{1}{2}$ and $1\frac{1}{2}$. $(\frac{1}{2}, 0)$ and $(1\frac{1}{2}, 0)$ are the x – intercepts.
2. The line $x = 1$ is a vertical asymptote.
3. The line $y = -4$ is a horizontal asymptote because y approaches -4 as x increases (or decreases) without bound.
4. The domain is all real numbers except 1. The range is all real numbers of $y > -4$.

Other method: It shifts the graph of $y = \frac{1}{x^2}$

         by 1 unit to the right and 4 units downward.

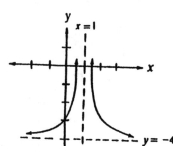

Examples

5. Graph $y = \dfrac{2}{x^2 - 2x + 2}$.

Solution:

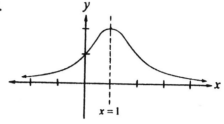

1. Let $x = 0$, then $y = 1$. $(0, 1)$ is the y – intercept.
2. $x^2 - 2x + 2 = x^2 - 2x + 1 + 1 = (x-1)^2 + 1 \neq 0$.
   There is no vertical asymptote.
3. The degree of the numerator is less than the degree of the denominator. Therefore, $y = 0$ is the horizontal asymptote.
4. $y = \dfrac{2}{(x-1)^2 + 1} > 0$. The graph is above the x – axis.
5. $f(0) = f(2)$. The graph is symmetric about the line $x = 1$.

6. Graph $y = \dfrac{x^2 - 7x + 12}{x - 3}$.

Solution:

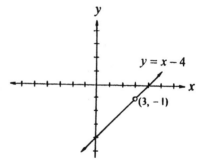

1. $y = \dfrac{x^2 - 7x + 12}{x - 3} = \dfrac{(x-3)(x-4)}{x-3} = x - 4$, $x \neq 3$
2. The value of y at $x = 3$ is undefined. The graph of the function is a straight line that has an open circle (hole) at $(3, -1)$.

7. Graph $y = \dfrac{x - 3}{x^2 + x - 2}$.

Solution:

$$y = \dfrac{x - 3}{x^2 + x - 2} = \dfrac{x - 3}{(x+2)(x-1)} .$$

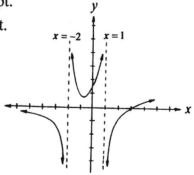

1. Let $x = 0$, then $y = \frac{-3}{-2} = \frac{3}{2}$. $(0, \frac{3}{2})$ is the y – intercept.
   Let $y = 0$, then $x = 3$. $(3, 0)$ is the x – intercept.
2. Two vertical asymptotes: $x = -2$ and $x = 1$.
3. One horizontal asymptote: $y = 0$
4. Critical points: $x = -2, 1, 3$
   $-\infty < x < -2$, y is negative.
   $-2 < x < 1$, y is positive.
   $1 < x < 3$, y is negative.
   $3 < x < \infty$, y is positive.

Examples

8. Graph $y = \dfrac{1}{|x|}$.

Solution:

If $x > 0$, $y = \dfrac{1}{x}$.

If $x < 0$, $y = -\dfrac{1}{x}$.

1. $f(-x) = f(x)$. It is symmetric about the $y-axis$.
2. There is no $x-$intercept and $y-$intercept.
3. $x = 0$ is a vertical asymptote.
4. $y = 0$ is a horizontal asymptote.
5. $y > 0$.

9. Graph $y = \dfrac{x^2}{x+1}$.

Solution:

1. Let $x = 0$, then $y = 0$. It passes the origin. There is no $y-$intercept.
    Let $y = 0$, then $x = 0$. It passes the origin. There is no $x-$intercept.
2. $x = -1$ is the vertical asymptote.
    There is no horizontal asymptote.
3. Dividing the denominator into the numerator, we have:

$$y = \dfrac{x^2}{x+1} = x - 1 + \dfrac{1}{x+1}.$$

$y = x - 1$ is the slant asymptote.

(oblique asymptote)

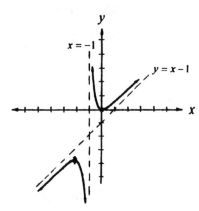

Before a test, a student asks his math teacher.
    Student: Could you please tell me the answers
             to the test questions ?
    Teacher: Do you promise to keep the answers
             secret ?
    Student: Yes, I do.
    Teacher: So do I.

EXERCISES

Graph each of the following rational functions. Use the graph to state the domain and range of each function.

1. $y = \dfrac{1}{x+1}$

2. $y = \dfrac{x+2}{x-1}$

3. $y = \dfrac{1}{2x}$

4. $y = \dfrac{1}{(x-3)^2}$

5. $y = \dfrac{1}{(x+2)^2} - 4$

6. $y = \dfrac{x}{x-2}$

7. $f(x) = \dfrac{x^2-1}{x-1}$

8. $f(x) = \dfrac{x-1}{x^2-1}$

9. $f(x) = \dfrac{x^2+1}{x+1}$

10. $p(x) = \dfrac{1}{|x-3|}$

11. $Q(x) = \dfrac{3}{x-2}$

12. $R(x) = \dfrac{x}{x^2-1}$

Find the horizontal, vertical, and slant (oblique) asymptotes, if any, of each rational function.

13. $y = \dfrac{3}{x}$

14. $y = \dfrac{1}{2x}$

15. $y = \dfrac{3}{x+1}$

16. $y = \dfrac{3}{x-2}$

17. $y = \dfrac{3}{x} + 4$

18. $y = \dfrac{2}{x+3} - 4$

19. $y = \dfrac{x+2}{x+5}$

20. $y = \dfrac{x+2}{x-1}$

21. $y = \dfrac{2x+2}{3x-1}$

22. $y = \dfrac{3x+1}{x-5}$

23. $y = \dfrac{5x}{x+4}$

24. $y = \dfrac{3x+6}{2x-4}$

25. $y = \dfrac{x+1}{2x-6}$

26. $y = \dfrac{5}{x-9} + 10$

27. $y = \dfrac{1}{(x-3)^2}$

28. $y = \dfrac{12x-3}{8x+4}$

29. $y = \dfrac{1}{(x+2)^2} - 4$

30. $y = \dfrac{x}{x-2}$

31. $R(x) = \dfrac{x-1}{x^2-1}$

32. $H(x) = \dfrac{x^2}{x^4-1}$

33. $P(x) = \dfrac{x^3+2x+1}{x^2+1}$

34. $Q(x) = \dfrac{x^2+1}{x-6}$

35. $P(x) = \dfrac{3x^6}{x^3-1}$

36. $K(x) = \dfrac{x^2+1}{x+1}$

37. $Q(x) = \dfrac{3x^4+2}{x^3+2x}$

38. $R(x) = \dfrac{x^3-1}{x-x^2}$

39. $P(x) = \dfrac{6x^2+2x+3}{3x^2-5x-2}$

40. $N(x) = \dfrac{6x^4}{3x^2-5x-2}$

41. A closed box that has a square base of length x and a volume of 4,000 cubic inches is designed. a) Find the function $S(x)$ for the surface area (material needed) of the box.
b) Using a graphing calculator to find the dimensions of the box that minimize the surface area.

Space for Taking Notes

CHAPTER 6   EXERCISES

Simplify each rational expression. Give all restrictions on the variables.

1. $\dfrac{6x}{3x^2}$

2. $\dfrac{12x^6}{4x^4}$

3. $\dfrac{24a^3b^2}{32ab^5}$

4. $\dfrac{8x-12}{4}$

5. $\dfrac{2-4x}{2x-1}$

6. $\dfrac{n^2-9}{n-3}$

7. $\dfrac{16(a+3)^4}{6(a+3)^2}$

8. $\dfrac{x-4}{x^2+x-20}$

9. $\dfrac{a^2-25}{a^2-3a-10}$

10. $\dfrac{t^2-4}{t^2+2t}$

11. $\dfrac{y^2+2y-24}{y^2+3y-28}$

12. $\dfrac{x^2+2x-3}{x^3+5x^2+6x}$

Simplify. Assume that no variable has a value for which the denominator is 0.

13. $\dfrac{x}{2}+\dfrac{x}{7}$

14. $\dfrac{y}{2}-\dfrac{y}{7}$

15. $\dfrac{1}{2x}-\dfrac{1}{7x}$

16. $\dfrac{1}{2y}+\dfrac{1}{7y}$

17. $\dfrac{a}{5}\cdot\dfrac{10}{a^4}$

18. $\dfrac{9}{x^2}\cdot\dfrac{x}{3}$

19. $\dfrac{15n^2}{2}\div\dfrac{3n^5}{4}$

20. $\dfrac{7}{n^3}\div\dfrac{21}{12n^6}$

21. $\dfrac{1}{x^2y}+\dfrac{1}{xy^2}$

22. $\dfrac{2}{ab}+\dfrac{4}{a^2b}$

23. $\dfrac{2}{ab}\cdot\dfrac{a^2b}{4}$

24. $\dfrac{4}{x^2y}\div\dfrac{8}{xy^2}$

25. $\dfrac{1}{a+2}+\dfrac{2}{a+3}$

26. $\dfrac{2}{n-2}-\dfrac{3}{n+3}$

27. $\dfrac{2}{n-2}\cdot\dfrac{2-n}{4}$

28. $\dfrac{12}{3-y}\div\dfrac{9}{y-3}$

29. $\dfrac{3}{a}+\dfrac{4}{a+2}$

30. $\dfrac{3}{a-2}-\dfrac{4}{a+2}$

31. $\dfrac{4}{a-2}\cdot\dfrac{a^2-4}{2}$

32. $\dfrac{4}{a+2}\div\dfrac{2}{a^2-4}$

33. $\dfrac{5}{x^2-25}-\dfrac{4}{x+5}$

34. $\dfrac{6}{x^2-25}+\dfrac{2}{x-5}$

35. $\dfrac{12}{x^2-25}\cdot\dfrac{4x-20}{8}$

36. $\dfrac{6}{2x-8}\div\dfrac{9}{x^2-16}$

37. $\dfrac{x^2+5x-14}{x-2}\cdot\dfrac{x}{x+7}$

38. $\dfrac{x^2-36}{x+6}\div\dfrac{x-6}{x+6}$

39. $\dfrac{a^2-b^2}{a}\div\dfrac{2a-2b}{a^4}$

40. $\dfrac{m^2+2m-3}{m}\cdot\dfrac{m^2}{m-1}$

41. $\dfrac{1}{c+1}+\dfrac{2}{c-1}-\dfrac{4}{c^2-1}$

42. $\dfrac{4}{x^2+2x-3}-\dfrac{1}{x-1}$

43. $\dfrac{4}{x^2-49}\div\dfrac{x+2}{x+7}\cdot\dfrac{x^2-5x-14}{8}$

Divide polynomials.

44. $\dfrac{x^2-4x-5}{x+2}$

45. $\dfrac{x^3-2x^2+4x-6}{x-4}$

46. $\dfrac{4x^3-3x+4}{2x+1}$

47. $\dfrac{x^3-a^3}{x+a}$

-----Continued-----

Solve each rational equation (Fractional equation).

48. $\dfrac{x}{5} = \dfrac{3}{4}$ 49. $\dfrac{5}{x} = \dfrac{2}{9}$ 50. $\dfrac{12}{5y} = \dfrac{2}{5}$ 51. $\dfrac{3y}{4} = \dfrac{9}{2}$

52. $\dfrac{n-4}{3} = \dfrac{2}{7}$ 53. $\dfrac{2n+1}{2} = \dfrac{n}{3}$ 54. $\dfrac{3x+5}{4} = 5$ 55. $\dfrac{4x-2}{3x-6} = 2$

56. $\dfrac{2}{k+3} = \dfrac{4}{k+13}$ 57. $\dfrac{6}{2a-3} = \dfrac{3}{a+5}$ 58. $\dfrac{10}{5x-4} = \dfrac{2}{x}$ 59. $\dfrac{6n-12}{n-2} = 6$

60. $\dfrac{4a+2}{4} = \dfrac{2a+1}{2}$ 61. $\dfrac{9}{y^2-36} = \dfrac{3}{y-6}$ 62. $\dfrac{32}{x^2-16} = \dfrac{x}{x-4}$ 63. $\dfrac{50}{x^2-25} = \dfrac{x}{x+5}$

64. $\dfrac{x}{4} + \dfrac{x}{8} = 3$ 65. $\dfrac{y}{5} - \dfrac{y}{6} = 1$ 66. $\dfrac{2x}{3} - \dfrac{x}{6} = -1$ 67. $\dfrac{9}{x} - \dfrac{3}{x} = \dfrac{1}{2}$

68. $\dfrac{1}{y} - \dfrac{5}{y} = -14$ 69. $\dfrac{5}{x+2} - \dfrac{2}{x} = \dfrac{1}{3}$ 70. $\dfrac{2c-5}{5} + \dfrac{c}{3} = \dfrac{8}{3}$ 71. $\dfrac{3}{a-3} + 2 = \dfrac{a}{a-3}$

72. $\dfrac{14}{x} - \dfrac{2x-5}{2x} = \dfrac{5}{6}$ 73. $\dfrac{5-a}{a-3} - \dfrac{1}{3-a} = 0$ 74. $\dfrac{8}{x-1} - \dfrac{16}{x^2-1} = 3$ 75. $\dfrac{x^2-9}{x+3} + x^2 = 3$

76. Two fifths of a number is 6 more than one tenth of the number. Find the number.

77. The sum of a number and its reciprocal is $\frac{58}{21}$. Find the number.

78. The sum of the reciprocals of two positive numbers is $\frac{1}{9}$, and one of the numbers is 3 times the other. Find the numbers.

79. If five pounds of beef cost \$12.60, how much do 8 pounds cost ?

80. David can finish a job in 7 days. Coby can finish the same job in 5 days. How many days would it take them to finish the job if they work together ?

81. David can finish a job in 7 days. Coby can finish the same job in 5 days. After working together for 2 days, Coby leaves. How many days will it take David to complete the job ?

82. A 10-gallon alcohol-water solution contains 30% pure alcohol. How much water should be added to produce the solution to 20% alcohol ?

83. A 10-gallon alcohol-water solution contains 30% pure alcohol. How much pure alcohol should be added to produce the solution to 34% alcohol ?

84. If y is inversely proportional to x, and if $y = 6$ when $x = 1.5$, find y when $x = 1.2$.

85. If (5, 14) and (7, y_2) are ordered pairs of the same inverse variation, find y_2.

86. The number of days needed to build a tower varies inversely as the number of workers working on the job. It takes 120 days for 8 workers to finish the jobs. If the job has to be finished in 80 days, how many workers are needed ?

87. If z varies directly as x and inversely as y^2, and if $z = 6$ when $x = 2$ and $y = 4$, find z when $x = 4$ and $y = 2$.

88. Write the equation if f varies directly as the product of m_1 and m_2, and inversely as r^2.

-----Continued-----

Solving the following Rational (fractional) inequalities.

89. $\dfrac{6}{x} > 2$

90. $\dfrac{x-1}{x} + 3 > 0$

91. $\dfrac{x+2}{x} \le \dfrac{x+4}{x+6}$

92. $\left|\dfrac{x-2}{4}\right| < 7$

93. $\left|\dfrac{x-2}{4}\right| \ge 7$

94. $\left|\dfrac{2x-1}{x-2}\right| \ge 1$

Simplify each complex fraction.

95. $\dfrac{\frac{1}{2}+3}{\frac{1}{3}+2}$

96. $\dfrac{\frac{x}{3}-2}{x-6}$

97. $\dfrac{\frac{1}{x-1}}{\frac{1}{x^2-1}}$

98. $\dfrac{\frac{x}{2}+\frac{3}{4}}{4-\frac{3}{x}}$

99. $\dfrac{\frac{x+h}{(x+h)+2}-\frac{2}{x+2}}{h}$

100. $\dfrac{5x^{\frac{2}{3}} - x^{-\frac{1}{3}}}{2x^{-\frac{1}{3}}}$

Decompose into its partial fractions for each expression.

101. $\dfrac{1}{x^2 - x}$

102. $\dfrac{x-6}{x^2-2x}$

103. $\dfrac{x+13}{x^2-x-20}$

104. $\dfrac{6x-26}{(x-3)^2}$

105. $\dfrac{x^2-7x-10}{x^2+2x-3}$

106. $\dfrac{x^2+x-5}{(x-1)(x^2-2)}$

Graph each of the following rational (fractional) functions. Use a graphing calculator to verify your graphs. State the domain and range of each function.

107. $y = \dfrac{1}{x-1}$

108. $y = \dfrac{x-2}{x+1}$

109. $y = \dfrac{1}{(x-4)^2}$

110. $y = \dfrac{x}{x+2}$

111. $y = \dfrac{1}{|x+2|}$

112. $y = \dfrac{x^2-1}{x-1}$

113. A closed right circular cylinder that has a radius r and a volume of 400 cubic feet is designed. a) Find a function $S(x)$ for the surface area of the material needed.

  b) Use a graphing calculator to find the value of r that minimize the surface area.

Space for Taking Notes

Answers

CHAPTER 6 – 1   Simplify Rational Expressions   page 196

1. $x = 0$ **2.** $x = 1$ **3.** $a = 5$ **4.** $y = -4$ **5.** $x = 3$ **6.** $n = 2, -1$ **7.** $n = -5, 1$ **8.** $t = 0, -2$

9. None **10.** None **11.** $y = 5, -1$ **12.** $y = 0, 2, -1$ **13.** $3, x \neq 0$ **14.** $3x^3, x \neq 0$ **15.** $\dfrac{3}{2x^3}, x \neq 0$

16. $\dfrac{3b^3}{5a^4}, a \neq 0$ **17.** $2x + y$ **18.** $\dfrac{1}{2x + y}, x \neq -\dfrac{y}{2}$ **19.** $\dfrac{2}{y - 4}, y \neq 4$ **20.** $\dfrac{y - 4}{2}$ **21.** $\dfrac{x + y}{2(x - y)}, x \neq y$

22. $-1, x \neq 5$ **23.** $-2, a \neq 3$ **24.** $-(a + 2), a \neq 2$ **25.** $-\dfrac{1}{2 + a}, a \neq -2$ **26.** $\dfrac{1}{x + 8}, x \neq -8, 8$

27. $\dfrac{2 + x}{x}, x \neq 0$ **28.** $3, y \neq -3$ **29.** $\dfrac{2}{x - y}, x \neq -2y, y$ **30.** $\dfrac{x - y}{3}, x \neq -y$ **31.** $a - 2, a \neq 2$

32. $\dfrac{1}{3(a - 1)}, a \neq 1$ **33.** $\dfrac{a - b}{a + b}, a \neq -b, b$ **34.** $x - 4, x \neq 4$ **35.** $x - 3, x \neq 5$ **36.** $-\dfrac{1}{n + 1}, n \neq 3, -1$

37. $\dfrac{2}{a + 2}, a \neq 6, -2$ **38.** $\dfrac{k - 3}{5}, k \neq -4$ **39.** $-\dfrac{2}{p + 1}, p \neq 5, -1$ **40.** $\dfrac{x - 2}{x - 4}, x \neq -7, 4$

41. It is undefined when $x = 6$ or -3 **42.** It equals zero when $x = 0$ or -4.
43. $x = a - b, a \neq -b$ **44.** The width is given by $x + 3$.

CHAPTER 6 – 2   Adding and Subtracting Rational Expressions   Page 198

1. 15 **2.** 36 **3.** 6 **4.** 15 **5.** 30 **6.** 24 **7.** $10x$ **8.** $12x$ **9.** $12x$ **10.** $40x$ **11.** $8ab$ **12.** $8ab$

13. $6x^2$ **14.** xy **15.** $28y^3$ **16.** $(x + 3)(x - 3)$ **17.** $48x^2y^2$ **18.** $a^2 - 9$ **19.** $30(a - 2)^2$ **20.** $x(x^2 - 1)$

21. 15 **22.** 15 **23.** $10x$ **24.** $40x$ **25.** $8ab$ **26.** $6x^2$ **27.** $(x + 3)(x - 3)$ **28.** $a^2 - 9$ **29.** x

30. a **31.** xy **32.** x^2y **33.** $x(x - 5)$ **34.** $6c(c - 2)$ **35.** $(t + 1)^2$ **36.** $(x - 2)(x + 1)$ **37.** $\dfrac{8x}{15}$

38. $\dfrac{11a}{15}$ **39.** $-\dfrac{1}{10x}, x \neq 0$ **40.** $\dfrac{41}{40x}, x \neq 0$ **41.** $\dfrac{6b - 1}{8ab}, a \neq 0, b \neq 0$ **42.** $\dfrac{4x - 5}{6x^2}, x \neq 0$

43. $\dfrac{7x - 3}{(x + 3)(x - 3)}, x \neq -3, 3$ **44.** $\dfrac{4}{a + 3}, a \neq -3, 3$ **45.** $\dfrac{x^2 + 2}{x}, x \neq 0$ **46.** $\dfrac{1 - 5a}{a}, a \neq 0$

47. $\dfrac{2y + 5x}{xy}, x \neq 0, y \neq 0$ **48.** $\dfrac{2xy + 9}{x^2y}, x \neq 0, y \neq 0$ **49.** $\dfrac{x - 10}{x(x - 5)}, x \neq 0, 5$

50. $-\dfrac{11c + 8}{6c(c - 2)}, c \neq 0, 2$ **51.** $\dfrac{3}{(t + 1)^2}, t \neq -1$ **52.** $\dfrac{5}{2(x + 1)}, x \neq -1$ **53.** $\dfrac{10x - 11}{(x - 2)(x + 1)}, x \neq 2, -1$

54. $\dfrac{6x - 1}{3(x - 4)}, x \neq 4$ **55.** $-\dfrac{1}{2(x + 3)}, x \neq -3, 3$ **56.** $\dfrac{x - 1}{x - 2}, x \neq 2$ **57.** $2, n \neq \dfrac{1}{2}$

58. $\dfrac{-3}{(a - 1)(a + 2)}, a \neq 1, a \neq -2$ **59.** $\dfrac{1}{x - 1}, x \neq -3, 1$ **60.** $\dfrac{-y}{(y - 3)^2}, y \neq 3$

61. $\dfrac{2b - 4}{(b + 1)(b + 3)(b - 5)}, b \neq -3, -1, 5$ **62.** $\dfrac{13 - 6c}{9 - c^2}, c \neq -3, 3$ **63.** $\dfrac{-6x + 10}{x(x + 5)(x - 5)}, x \neq 0, -5, 5$

Roger was absent yesterday.
Teacher: Roger, why were you absent ?
Roger: My tooth was aching.
   I went to see my dentist.
Teacher: Is your tooth still aching ?
Roger: I don't know. My dentist has it.

CHAPTER 6 – 3   Multiplying and Dividing Rational Expressions   Page 200

1. $\dfrac{1}{2x}, x \neq 0$ 2. $4x^2, x \neq 0$ 3. $\dfrac{9}{y}, y \neq 0$ 4. $\dfrac{5y^3}{2}, y \neq 0$ 5. $\dfrac{3y^3}{x^3}, x \neq 0, y \neq 0$

6. $\dfrac{9}{a^4 b^2}, a \neq 0, b \neq 0$ 7. $\dfrac{3a^3 b^4}{2}, a \neq 0, b \neq 0$ 8. $\dfrac{6}{d^5}, c \neq 0, d \neq 0$ 9. $\dfrac{4x^2}{y^2}, x \neq 0, y \neq 0$

10. $\dfrac{8x}{y}, x \neq 0, y \neq 0$ 11. $\dfrac{x+2}{x-2}, x \neq 2, 5$ 12. $4y, y \neq 0, x \neq 2y$ 13. $-1, a \neq 3$ 14. $-5, a \neq 2b$

15. $\dfrac{n-4}{2}, n \neq 0, -4$ 16. $\dfrac{n+1}{n+5}, n \neq 5, -5$ 17. $\dfrac{3}{2x}, x \neq 0$ 18. $\dfrac{x^2}{y^2}, x \neq 0, y \neq 0$

19. $\dfrac{10p^2}{n^2}, n \neq 0, p \neq 0$ 20. $\dfrac{2b^2}{a^2}, a \neq 0, b \neq 0$ 21. $\dfrac{9}{x}, x \neq 0$ 22. $\dfrac{9y}{40}, y \neq 0$ 23. $\dfrac{x+2}{x-2}, x \neq 2, 5$

24. $4y, y \neq 0, x \neq 2y$ 25. $\dfrac{3a}{2}, a \neq 0, -1$ 26. $3(x+2), x \neq 2$ 27. $-\dfrac{1}{2}, x \neq y$ 28. $\dfrac{4}{3}, a \neq 5$

29. $\dfrac{4(x+3)}{x+4}$ 30. $\dfrac{2(x-2)}{x+3}$ 31. $\dfrac{x-3}{5}$ 32. $\dfrac{1}{(x-3)(x-1)}$ 33. $\dfrac{(m-4)(m-5)}{m-2}$ 34. $\dfrac{x+y}{4x^2}$

35. $\dfrac{a+3}{a+2}$ 36. $\dfrac{9(a-6)}{5(a-4)}$ 37. $x - 7 + \dfrac{20}{x+2}$ 38. $x + 7 + \dfrac{8}{x-2}$ 39. $2a + 3 - \dfrac{9}{2a-1}$

40. $2a - 3 + \dfrac{16}{3a+2}$ 41. $x^2 - 2x + 3 - \dfrac{11}{2x+4}$ 42. $x^2 + 2x + 5 + \dfrac{20}{2x-5}$

43. $a^3 + 4a^2 + 16a + 64 + \dfrac{252}{a-4}$ 44. $2n^2 - 5n + 8 - \dfrac{13n+3}{n^2+2n+1}$

CHAPTER 6 – 4   Solving Rational Equations   page 202

1. $x = 2\frac{2}{5}$ 2. $x = 3\frac{3}{5}$ 3. $x = 5\frac{5}{6}$ 4. $x = 1\frac{11}{24}$ 5. $y = 1\frac{3}{4}$ 6. $y = 5\frac{2}{5}$ 7. $x = 5$ 8. $x = -14$

9. $a = 9$ 10. $a = 7$ 11. $n = 53$ 12. $n = 16$ 13. $x = 1\frac{1}{2}$ 14. $n = 2\frac{4}{5}$ 15. $x = 1$ 16. $t = 8$

17. $a = \frac{4}{7}$ 18. No solution   19. $y = -\frac{2}{5}$ 20. $x = -\frac{2}{3}, 1$ 21. No solution

22. All real numbers except $x = \frac{3}{2}$ 23. $y = 6\frac{1}{2}$ 24. $y = -1\frac{5}{9}$ 25. $x = 4\frac{4}{9}$ 26. $x = -42$

27. $x = \frac{7}{18}$ 28. $x = \frac{23}{42}$ 29. $x = 8\frac{1}{2}$ 30. $x = 1$ 31. $y = \frac{3}{4}$ 32. $y = -1\frac{1}{3}$ 33. $n = 1$ (a double root)

34. $n = \frac{1}{3}, 4$ 35. $a = -6$ 36. $x = -4$ 37. $c = 5$ 38. $c = 3$ 39. $m = 20$ 40. No solution

41. All real numbers except 2   42. $a = 4$ 43. $\dfrac{x}{y} = -5$ 44. $\dfrac{x}{y} = \dfrac{2}{7}$

45. Slope: $m = -\frac{2}{3}$ (Hint: Simplify the equation in the form $y = mx + b$.)

CHAPTER 6 – 5   Solving Word Problems by Rational Equations   page 204

1. 12   2. 12   3. 12 and 32   4. 4 and 6   5. $\frac{3}{4}$, or $\frac{4}{3}$ 6. 5 and 7   7. $16.4   8. $8.28   9. 10 gallons

10. 272 miles   11. 5000 voters   12. 300 meters by 200 meters   13. 1.5 hours   14. 2.4 hours

15. $2\frac{3}{4}$ hours   16. $3\frac{1}{3}$ miles/h   17. 60 miles/h   18. 50 km/h   19. 5 miles/h   20. 8 gal. of water

21. 4.5 gal. of water   22. 1.125 gal. of pure salt

CHAPTER 6 – 6   Inverse Variations   page 206

1. $k = 20; xy = 20$ 2. $k = 50; xy = 50$ 3. $k = 4000; mn = 4000$ 4. $k = 2700; mn = 2700$

5. $k = 1500; (x+8)y = 1500$ 6. $y = \frac{5}{3}$ or $1\frac{2}{3}$ 7. $x = \frac{5}{3}$ or $1\frac{2}{3}$ 8. $y = \frac{5}{3}$ or $1\frac{2}{3}$ 9. $n = 45$

10. $m = 45$ 11. $y_1 = 36$ 12. $y_1 = 106\frac{2}{3}$ 13. $y_1 = 12.8$ 14. $y_2 = 2\frac{2}{3}$ 15. $x_1 = 90$

16. $x_2 = 0.4$ 17. $y_1 = \frac{1}{20}$ 18. $y_2 = \frac{3}{100}$ 19. $y_1 = \frac{4}{49}$ 20. $y_2 = \frac{1}{5}$

21. Yes, If $xy = 5$, then $y = \frac{5}{x}$. y is decreased when x is increased.

22. Yes. $6 \times 4 = 24$, $3 \times 8 = 24$, $2 \times 12 = 24$ 23. 8 workers   24. 4.8 hours   25. $36m$

26. 40 miles/h   27. $z = 40.5$ (Hint: $z = k\frac{x}{y}$)

CHAPTER 6 – 7   Fractional Inequalities and Absolute Values   page 208

1. $x < 0$ or $x > \frac{1}{2}$ **2.** $0 \le x \le \frac{1}{2}$ **3.** $x < 0$ or $x > \frac{1}{2}$ **4.** $0 < x < \frac{1}{2}$ **5.** $x < -1$ or $0 < x < 1$

6. $-1 < x < 1, x \ne 0$ or $x > 1$ **7.** $x \le -1$ or $0 < x \le 1$ **8.** $x < 0$ or $x > \frac{1}{2}$ **9.** $x < -5$ or $x > -2$

10. $-3 < x < 0$ **11.** $x < 2$ or $x \ge 5$ **12.** $-2 < x < \frac{11}{5}$ **13.** $-14 < x < -2$ or $x > 6$

14. $0 < x < 3$ **15.** $y < -2$ or $y \ge 8$ **16.** $x < -14$ or $x > 14$ **17.** $-14 < x < 14$ **18.** $x \le -10$ or $x \ge 18$

19. $-10 \le x \le 18$ **20.** $-13 < x < 17$ **21.** $x < -13$ or $x > 17$ **22.** $x \ge -\frac{1}{2}$ **23.** $x < \frac{23}{5}$ or $x > 13$

24. $x < \frac{29}{6}$ or $x > \frac{19}{2}$ **25.** $x \le -1$ or $x \ge \frac{1}{5}$ **26.** $-1 < x < \frac{1}{5}, x \ne -\frac{1}{4}$ **27.** $\frac{29}{6} \le x \le \frac{19}{2}, x \ne 6$

CHAPTER 6 – 8   Complex Fractions   page 210

1. $\frac{2}{3}$ **2.** $-\frac{7}{11}$ **3.** $1\frac{1}{8}$ **4.** $3\frac{2}{3}$ **5.** 12 **6.** $2\frac{3}{11}$ **7.** $\frac{5+x}{4-x}, x \ne 0, 4$ **8.** $2, x \ne 0$ **9.** $2(x-1), x \ne -1$

10. $\frac{b}{a}, a \ne 0, b \ne 0$ **11.** $-\frac{b}{a}, a \ne 0, b \ne 0$ **12.** $b + a, a \ne 0, b \ne 0$ **13.** $\frac{x}{2}, x \ne 0, -6$ **14.** $\frac{2(x-4)}{x+4}, x \ne -4$

15. $a(a-b), a \ne 0, b \ne 0$ **16.** $\frac{x^2+xy+y^2}{x+y}, x \ne 0, y \ne 0, x \ne y$ **17.** $x-1, x \ne 1$ **18.** $\frac{x-1}{x+1}, x \ne 0, -1$

19. $-\frac{1}{x-4}, x \ne 3, 4$ **20.** $\frac{x^2-3x-3}{x-4}, x \ne 3, 4$ **21.** $-\frac{1}{x^2+hx}, h \ne 0$ **22.** $\frac{-2x-h}{x^4+2hx^3+x^2h^2}, h \ne 0$ **23.** $\frac{a^2+a+1}{a}, a \ne 0, 1$

24. $\frac{y-x}{xy}, x \ne 0, y \ne 0, x \ne -y$ **25.** $\frac{xy}{(x+y)^2}, x \ne 0, y \ne 0, x \ne -y$ **26.** $\frac{a^2+a+1}{a}, a \ne 0, 1$

27. $ab, a \ne 0, b \ne 0, a \ne -b$ **28.** $2\frac{4}{5}$ **29.** $2\frac{5}{7}$ **30.** $\frac{x^5+4x^3+3x}{x^4+3x^2+1}, x \ne 0$

CHAPTER 6 – 9   Decomposing Rational Expressions (Partial Fractions)   page 214

1. $\frac{1}{x} - \frac{1}{x+1}$ **2.** $\frac{1}{x-2} - \frac{1}{x}$ **3.** $\frac{1}{2}\left(\frac{1}{x-1} - \frac{1}{x+1}\right)$ **4.** $\frac{2}{x-1} - \frac{2}{x+2}$ **5.** $\frac{3}{x-3} - \frac{3}{x+4}$ **6.** $\frac{1}{3}\left(\frac{2}{2x-3} - \frac{2}{2x+3}\right)$ **7.** $\frac{2}{x+1} - \frac{3}{x-1}$

8. $\frac{2}{x+4} + \frac{7}{x-2}$ **9.** $\frac{3}{x-3} - \frac{2}{x+2}$ **10.** $\frac{1}{4}\left(\frac{5}{x-4} - \frac{1}{x}\right)$ **11.** $1 + \frac{2}{x-1} - \frac{1}{x}$ (Hint: must divide before decomposing)

12. $1 + \frac{3}{x-1} - \frac{2}{x}$ (Hint: must divide before decomposing)   **13.** $1 - \frac{2}{x+2}$ (Hint: must divide before de-

composing)   **14.** $1 - \frac{3x+1}{x^2+x+1}$ (Hint: $x^2 + x + 1$ can't be factored)   **15.** $\frac{2}{x} + \frac{4}{x+1} - \frac{3}{x-1}$ **16.** $\frac{5}{2x-1} - \frac{3}{x+1}$

17. $\frac{1}{x-1} + \frac{6}{(x-1)^2}$ **18.** $\frac{2}{x-3} + \frac{6}{(x-3)^2}$ **19.** $\frac{2}{x-1} - \frac{3}{(x-1)^2}$ **20.** $\frac{3}{x-1} + \frac{2}{(x-1)^2} + \frac{1}{(x-1)^3}$ **21.** $\frac{2}{x+2} - \frac{5}{x-4} + \frac{6}{x-3}$

22. $\frac{3}{x} + \frac{2}{x+2} - \frac{1}{(x+2)^2}$ **23.** $1 + \frac{3}{x-1} + \frac{3}{(x-1)^2} + \frac{1}{(x-1)^3}$ (Hint: must divide before decomposing)

24. $x + 3 + \frac{6}{x-1} + \frac{4}{(x-1)^2} + \frac{1}{(x-1)^3}$ (Hint: must divide before decomposing)   **25.** $\frac{2}{x-1} - \frac{2x-2}{x^2+1}$

26. $\frac{2}{x-1} - \frac{x+3}{x^2+x+1}$ **27.** $\frac{4}{x-3} + \frac{3x-2}{x^2+5}$ **28.** $x + 2 + \frac{1}{x+1} + \frac{1}{x-1}$ (Hint: must divide before decomposing)

29. $x - 4 + \frac{1}{x+3} - \frac{2}{x-2}$ (Hint: must divide before decomposing)   **30.** $\frac{2}{x} + \frac{x+2}{x^2+4}$ **31.** $\frac{5x}{x^2+2} - \frac{6x}{(x^2+2)^2}$

32. $x + \frac{5}{x} - \frac{1}{x+1} + \frac{3}{(x+1)^2}$

CHAPTER 6 –10   Graphing Rational Functions   page 219

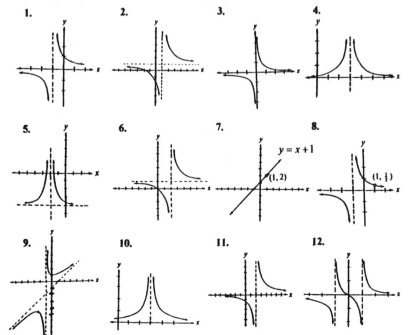

CHAPTER 6 – 10　Graphing Rational Functions　page 219 (Continued)

1. Domain: All real numbers except -1
　Range: All nonzero real numbers

3. Domain: All nonzero real numbers
　Range: All nonzero real numbers

5. Domain: All real numbers except -2
　Range: All real numbers of $y > -4$

7. Domain: All real number except 1
　Range: All real numbers except 2

9. Domain: All real number except -1
　Range: All real numbers of $y \geq 1$ or $y \leq -5$

11. Domain: All real numbers except 2
　Range: All real numbers of $y \neq 0$

13. H: $y = 0$　14. H: $y = 0$　15. H: $y = 0$　16. H: $y = 0$　17. H: $y = 4$　18. H: $y = -4$　19. H: $y = 1$
　V: $x = 0$　　V: $x = 0$　　V: $x = -1$　V: $x = 2$　　V: $x = 0$　　V: $x = -3$　　V: $x = -5$

20. H: $y = 1$　21. H: $y = \frac{2}{3}$　22. H: $y = 3$　23. H: $y = 5$　24. H: $y = \frac{3}{2}$　25. H: $y = \frac{1}{2}$　26. H: $y = 10$
　V: $x = 1$　　V: $x = \frac{1}{3}$　V: $x = 5$　　V: $x = -4$　V: $x = 2$　　V: $x = 3$　　V: $x = 9$

27. H: $y = 0$　28. H: $y = \frac{3}{2}$　29. H: $y = -4$　30. H: $y = 1$　31. H: $y = 0$　32. H: $y = 0$
　V: $x = 3$　　V: $x = -\frac{1}{2}$　V: $x = -2$　　V: $x = 2$　　V: $x = -1$　　V: $x = 1, x = -1$

33. H: none　34. H: none　35. H: none　36. H: none　37. H: none　38. H: none
　V: none　　V: $x = 6$　　V: $x = 1$　　V: $x = -1$　　V: $x = 0$　　V: $x = 0, x = 1$
　O: $y = x$　O: $y = x + 6$　O: none　　O: $y = x - 1$　O: $y = 3x$　O: $y = -x - 1$

39. H: $y = 2$
　V: $x = \frac{1}{3}, x = -2$
　O: none

40. H: none
　V: $x = \frac{1}{3}, x = -2$
　O: none

41. a) $S(x) = 2x^2 + \frac{16000}{x}$
　b) $x = 15.42 \sim 15.95$ inches
　(Hint: $x = 15.87$ inches by using calculus.)

CHAPTER 6　Exercises　Rational Equations and Functions　page 221 ~ 223

1. $\frac{2}{x}, x \neq 0$　2. $3x^2, x \neq 0$　3. $\frac{3a^2}{4b^3}, a \neq 0, b \neq 0$　4. $2x - 3$　5. $-2, x \neq \frac{1}{2}$　6. $n + 3, n \neq 3$

7. $\frac{8(a+3)^2}{3}, a \neq -3$　8. $\frac{1}{x+5}, x \neq 4, -5$　9. $\frac{a+5}{a+2}, a \neq 5, -2$　10. $\frac{t-2}{t}, t \neq 0, -2$

11. $\frac{y+6}{y+7}, y \neq 4, -7$　12. $\frac{x-1}{x(x+2)}, x \neq 0, -3, -2$　13. $\frac{9x}{14}$　14. $\frac{5y}{14}$　15. $\frac{5}{14x}$　16. $\frac{9}{14y}$　17. $\frac{2}{a^3}$

18. $\frac{3}{x}$　19. $\frac{10}{n^3}$　20. $4n^3$　21. $\frac{x+y}{x^2y^2}$　22. $\frac{2(a+2)}{a^2b}$　23. $\frac{a}{2}$　24. $\frac{y}{2x}$　25. $\frac{3a+7}{(a+2)(a+3)}$　6. $\frac{-n+12}{(n-2)(n+3)}$

27. $-\frac{1}{2}$　28. $-\frac{4}{3}$, or $-1\frac{1}{3}$　29. $\frac{7a+6}{a(a+2)}$　30. $\frac{-a+14}{a^2-4}$　31. $2(a+2)$　32. $2(a-2)$　33. $\frac{25-4x}{x^2-25}$

34. $\frac{2(x+8)}{x^2-25}$　35. $\frac{6}{x+5}$　36. $\frac{x+4}{3}$　37. x　38. $x + 6$　39. $\frac{a^3(a+b)}{2}$　40. $m(m+3)$　41. $\frac{3}{c+1}$

42. $-\frac{1}{x+3}$　43. $\frac{1}{2}$　44. $x - 6 + \frac{7}{x+2}$　45. $x^2 + 2x + 12 + \frac{42}{x-4}$　46. $2x^2 - x - 1 + \frac{5}{2x+1}$

47. $x^2 - ax + a^2 - \frac{2a^3}{x+a}$　48. $x = 3\frac{3}{4}$　49. $x = 22\frac{1}{2}$　50. $y = 6$　51. $y = 6$　52. $n = 4\frac{6}{7}$　53. $n = -\frac{3}{4}$

54. $x = 5$　55. $x = 5$　56. $k = 7$　57. ϕ (no solution)　58. ϕ (no solution)　59. All real numbers

60. All real numbers　61. $y = -3$　62. $x = -8$　63. $x = 10$　64. $x = 8$　65. $y = 30$　66. $x = -2$

67. $x = 12$　68. $y = \frac{2}{7}$　69. $x = 3$ and 4　70. $c = 5$　71. ϕ (no solution)　72. $x = 9$　73. $a = 6$

74. $x = 1\frac{2}{3}$　75. $x = 2$　76. 20　77. $\frac{3}{7}$, or $\frac{7}{3}$　78. 12 and 36　79. \$20.16　80. $2\frac{11}{12}$ days　81. $2\frac{1}{5}$ days

82. 5 gallons of water　83. 0.61 gallons of pure alcohol　84. $y = 7.5$　85. $y_2 = 10$　86. 12 workers

87. $z = 48$　88. $f = k\frac{m_1 \cdot m_2}{r^2}$ (Hint: It is the Law of Universal Gravitation.)　89. $0 < x < 3$

90. $x < 0$ or $x > \frac{1}{4}$　91. $x < -6$ or $-3 \leq x < 0$　92. $-26 \leq x \leq 30$　93. $x \leq -26$ or ≥ 30

94. $x \leq -1$ or $1 \leq x < 2$ or $x > 2$　95. $1\frac{1}{2}$　96. $\frac{1}{3}, x \neq 6$　97. $x + 1, x \neq \pm 1$　98. $\frac{x(2x+3)}{4(4x-3)}; x \neq 0, \frac{3}{4}$

99. $\frac{2}{(x+h+2)(x+2)}, h \neq 0$　100. $\frac{5x-1}{2}, x \neq 0$　101. $-\frac{1}{x} + \frac{1}{x-1}$　102. $\frac{3}{x} - \frac{2}{x-2}$　103. $\frac{2}{x-5} - \frac{1}{x+4}$

104. $\frac{6}{x-3} - \frac{8}{(x-3)^2}$　105. $1 - \frac{5}{x+3} - \frac{4}{x-1}$　106. $\frac{3}{x-3} - \frac{2x+1}{x^2-2}$　107 ~ 112 are left to the students.

113. a) $r = 3.936 \sim 4.043$ feet (Hint: $x = 3.99$ feet by using calculus)

Graphing a rational function
by a graphing calculator

Examples

1: Graph $y = \dfrac{3}{x-2}$.

Solution:

Press: **Y=** $\to$ 3 $\to$ ÷ $\to$ ($\to$ X,T,θ, n $\to$ – $\to$ 2 $\to$)
$\to$ **GRAPH**

To graph a function, we must select **FUNC**
and **connected-mode** from mode settings
before we enter the function.

Press: **MODE** $\to$ **Func** $\to$ **Connected** $\to$ **ENTER**
The screen shows:

2: Graph $y = \dfrac{x^2}{x+1}$.

Solution:

Press: **Y=** $\to$ X,T,θ,n $\to$ ∧ $\to$ 2 $\to$ ÷ ($\to$ X,T,θ,n $\to$
+ $\to$ 1) $\to$ **GRAPH**

Press: **MODE** $\to$ **Func** $\to$ **Connected** $\to$ **ENTER**
The screen shows:

A middleaged single lady told her friend.
  Lady: I prefer to marry an archaeologist rather than a
        doctor or a lawyer.
Friend: Why do you want to marry an archaeologist ?
  Lady: If my husband is an archaeologist, the more I
        get older, the more he is interested in me.

Space for Taking Notes

Exponential and Logarithmic Functions

7-1 Exponential and Logarithmic Functions

The function $y = 2^x$ is called an **exponential function** with base 2.

The function $x = 2^y$ is the inverse function of $y = 2^x$ by interchanging variables x and y.

To write the function $x = 2^y$ by y in terms of x, we use the form of logarithm $y = \log_2 x$.

The function $y = \log_2 x$ is called a **logarithmic function** with base 2.

For example, we want to find the value y of the equation $2^y = 8$, we write it in the form of logarithm: $y = \log_2 8$ → Read "y is the logarithm of 8 to the base 2.".

We may simply say that $\log_2 8$ represents **" 2 to what power gives 8. "**.

The following comparison gives us the value of logarithm:

Example: $2^3 = 8$ → $\log_2 8 = 3$ It represents " 2 to third power gives 8 ".

We can use a calculator to evaluate exponential and logarithmic expressions.

When we write a logarithm with base 10, the subscript 10 is omitted. $\log_{10} 5.2 = \log 5.2$

Logarithms with base 10 are called **common logarithms**.

Relationships between exponential and logarithmic functions:

1) If $a > 0$ and $a \neq 1$, the equation $y = a^x$ is called the **exponential function** with base a.
   The constant base a is a positive real number and the exponent x is any real number.
   If $a > 1$, $y = a^x$ is an exponential growth function. y increases as x increases.
   If $0 < a < 1$, $y = a^x$ is an exponential decay function. y decreases as x increases.
   The line $y = 0$ (the $x-axis$) is an asymptote.

2) $x = a^y$ is the inverse function of $y = a^x$. The **logarithmic function** of $x = a^y$ is
   written by $y = \log_a x$. $y = \log_a x$ and $x = a^y$ are equivalent.

3) $y = \log_a x$ and $y = a^x$ are inverses each other. The graphs of $y = \log_a x$ and
   $y = a^x$ are reflection of each other about the line $y = x$.

4) The domains and ranges of the two functions are interchanged.
$$y = a^x; \quad x \in R, \; y > 0.$$
$$y = \log_a x \; \text{ or } \; x = a^y; \quad y \in R, \; x > 0.$$

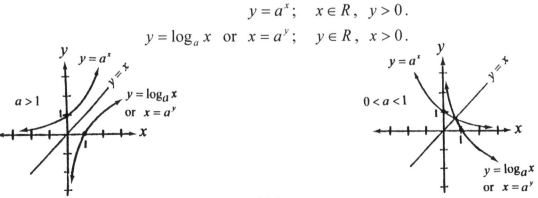

EXERCISES

Use a calculator to evaluate each expression. Round your answer to two decimal places. (Hint: $\log 20$ is equivalent to $\log_{10} 20$. It is called **common logarithm**.)

1. $5^{2.42}$ **2.** $5^{4.35}$ **3.** $2^{1.35}$ **4.** $16^{0.25}$ **5.** $5^{-2.42}$

6. $32^{0.2}$ **7.** $2^{2.718}$ **8.** $2^{\sqrt{5}}$ **9.** $2.4^{-1.5}$ **10.** $2.4^{0.92}$

11. $2.4^{-4.5}$ **12.** $3.14^{2.718}$ **13.** $3.141^{2.71}$ **14.** $\pi^{3.14}$ **15.** $\pi^{\sqrt{3}}$

16. $25^{-1.5}$ **17.** $25^{-4/7}$ **18.** $\pi^{-0.5}$ **19.** $\pi^{-\sqrt{2}}$ **20.** $2.718^{-1.4}$

21. $\log 20$ **22.** $\log 54$ **23.** $\log 35$ **24.** $\log 89$ **25.** $\log 20.6$

26. $\log 42.4$ **27.** $\log 3.141$ **28.** $\log 2.718$ **29.** $\log 0.364$ **30.** $\log 0.0364$

Write each exponential expression to an equivalent logarithmic form.

31. $2^4 = 16$ **32.** $2^5 = 32$ **33.** $3^2 = 9$ **34.** $3^4 = 81$ **35.** $10^3 = 1000$

36. $8^{1/3} = 2$ **37.** $16^{1/4} = 2$ **38.** $8^{-2} = \frac{1}{64}$ **39.** $8^{-1/3} = \frac{1}{2}$ **40.** $10^{-3} = 0.001$

41. $9^{3/2} = 27$ **42.** $a^5 = 3.54$ **43.** $3^x = 8.6$ **44.** $x^{\sqrt{3}} = \pi$ **45.** $2.71^x = 12$

Write each logarithmic expression to an equivalent exponential form.

46. $\log_4 16 = 2$ **47.** $\log_3 27 = 3$ **48.** $\log_5 125 = 3$ **49.** $\log_7 49 = 2$

50. $\log_7 \frac{1}{49} = -2$ **51.** $\log_{10} 10 = 1$ **52.** $\log_{10} \frac{1}{10} = -1$ **53.** $\log_{10} \frac{1}{1000} = -3$

54. $\log_{16} 8 = \frac{3}{4}$ **55.** $\log_{2.71} 1 = 0$ **56.** $\log_{10} 0.955 = -0.02$ **57.** $\log_a 5 = 12$

58. $\log_2 9 = x$ **59.** $\log_5 x = 3.97$ **60.** $\log_{2.71} \pi = x$

Identify each exponential function as a growth or decay function.

61. $y = 2^x$ **62.** $y = \left(\frac{1}{2}\right)^x$ **63.** $y = \left(\frac{2}{5}\right)^x$ **64.** $y = \left(\frac{5}{2}\right)^x$

65. $y = 3^x$ **66.** $y = 3^{-x}$ **67.** $y = -3^x$ **68.** $y = -3^{-x}$

69. $y = (2.71)^x$ **70.** $y = (2.71)^{-x}$

7-2 Exponential Functions and their Graphs

Definition of an exponential function:

An exponential function $f(x)$ with a constant base a is a function of the form

$$f(x) = a^x \quad \text{or} \quad y = a^x$$

where $a > 0$, $a \neq 1$, and x is any real number. It is a one-to-one function.

Graphs of $y = a^x$ and $y = a^{-x}$:

1. The graph of $y = a^x$ is an increasing function. y increases as x increases.
   The graph lies above the x – axis and passes through the point (0, 1) and (1, a).
   The x – axis ($y = 0$) is a horizontal asymptote to the graph as $x \to -\infty$.

2. The graph of $y = a^{-x}$ is a decreasing function. y decreases as x increases.
   The graph lies above the x – axis and passes through the point (0, 1) and $(1, \frac{1}{a})$.
   The x – axis ($y = 0$) is a horizontal asymptote to the graph as $x \to \infty$.

3. The graph of $y = a^{-x}$ is the reflection of the graph of $y = a^x$ through the y – axis.

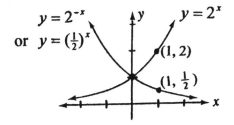

Domain: All real numbers ($x \in R$)

Range: All positive numbers ($y > 0$)

Graphing exponential functions using transformations (see Section 5~7) :

1. The graph of $y = 3^{x-1}$ shifts the graph of $y = 3^x$, 1 unit to the right.

2. The graph of $y = 3^x - 1$ shifts the graph of $y = 3^x$, 1 unit downward.

3. The graph of $y = -3^x$ reflects the graph of $y = 3^x$ over the x – axis.

4. The graph of $y = 3^{-x}$ reflects the graph of $y = 3^x$ over the y – axis.

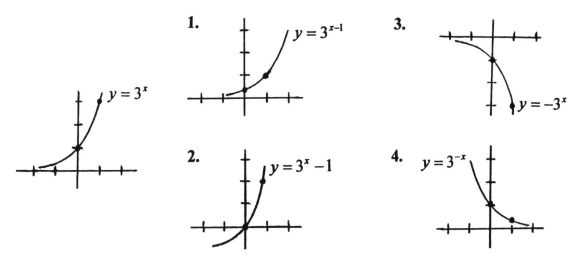

EXERCISES

Use the graph of $y = 2^x$ to write each function that is graphed after the indicated transformations.

1. Shift up 5 units   **2.** Shift down 5 units   **3.** Shift right 5 units

4. Shift left 5 units   **5.** Reflect about $x-$axis   **6.** Reflect about $y-$axis

7. Shift left 5 units   **8.** Shift up 3 units   **9.** Reflect about $y-$axis

    Reflect about $x-$axis   Reflect about $y-$axis   shift down 3 units

10. Shift right 3 units   **11.** Reflect about $y-$axis   **12.** Shift up 2 units

    Reflect about $x-$axis   shift down 3 units   Reflect about $y-$axis

                                              Shift right 1 unit

13. The graph of a function is made after the following three transformations applied to the graph of $y = 3^x$. Write the function.

        1. Reflect about the $x-$axis   2. Shift up 4 units   3. Shift right 6 units

14. The graph of a function is made after the following three transformations applied to the graph of $y = (\frac{1}{3})^x$ (or $y = 3^{-x}$). Write the function.

        1. Shift left 6 units   2. Reflect about the $y-$axis   3. Shift down 4 units

Match each of the function with its graph. Write the domains and range.

15. $y = 2^x$ **16.** $y = 2^{-x}$ **17.** $y = -2^x$ **18.** $y = -2^{-x}$

19. $y = 2^x - 3$ **20.** $y = 2^x + 3$ **21.** $y = 2^{x+3}$ **22.** $y = 2^{x-3}$

23. $y = 2^{3-x}$ **24.** $y = 2^{-x} + 3$ **25.** $y = 2^{-x} - 3$ **26.** $y = 2^{-x-3}$

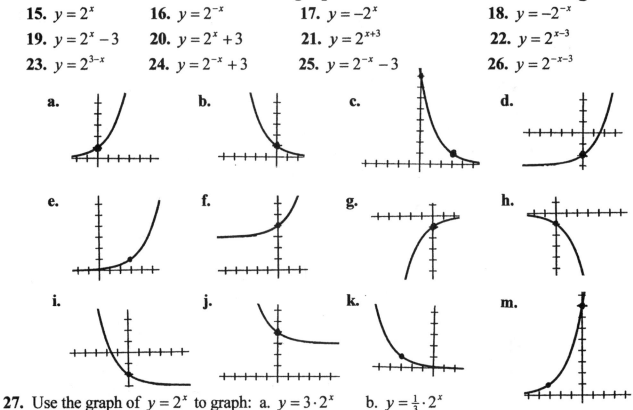

27. Use the graph of $y = 2^x$ to graph: a. $y = 3 \cdot 2^x$ b. $y = \frac{1}{3} \cdot 2^x$

28. A total of $500 is deposited in a bank at an annual interest rate of 6%. Find the balance after 3 years if it is compounded monthly.

7-3 Logarithmic Functions and their Graphs

We have learned in the preceding section that the exponential function $y = a^x$, for $a > 0$ and $a \neq 1$, is a one-to-one function. Since every one-to-one function has an inverse function, $y = a^x$ has the inverse function by the equation:

$$x = a^y, \ a > 0 \text{ and } a \neq 1$$

In order to solve the equation $x = a^y$ for the exponent y in terms of x, we create the word **logarithm** for the exponent y:

$$y = \log_a x \qquad \text{Read " } y \text{ is the logarithm of } x \text{ to the base } a. \text{ ".}$$

We simply say that $\log_2 8$ represents " 2 to what power gives 8. ". Therefore, $\log_2 8 = 3$.

$y = \log_a x$ is the equivalent logarithmic form of $x = a^y$.

$y = \log_a x$ and $y = a^x$ are inverse functions of each other (see graphs on section 7~1).

The domains and ranges of $y = \log_a x$ and $y = a^x$ are interchanged.

Definition of a logarithmic function:

A logarithmic function $f(x)$ to a constant base a is a function of the form

$$f(x) = \log_a x \quad \text{or} \quad y = \log_a x$$

where $a > 0$, $a \neq 1$, and x is any real number. It is a one-to-one function.

Graphs of $y = \log_a x$:

1. The function is an increasing function if $a > 1$, y increases as x increases. The $y-$axis ($x = 0$) is a vertical asymptote to the graph as $x \to 0$.

2. The function is a decreasing function if $0 < a < 1$, y decreases as x increases. The $y-$axis ($x = 0$) is a vertical asymptote to the graph as $x \to 0$.

3. The graph lies to the right of $y-$axis and passes through the point (1, 0)

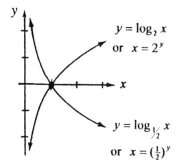

$y = \log_2 x$
or $x = 2^y$

$y = \log_{1/2} x$
or $x = (\frac{1}{2})^y$

Domain: All positive real numbers ($x > 0$)
Range: All real numbers ($y \in R$)

Graphing logarithmic functions using transformations (see Section 5~7):

1. The graph of $y = \log_3 (x-1)$ shifts the graph of $y = \log_3 x$, 1 unit to the right.

2. The graph of $y = \log_3 x - 1$ shifts the graph of $y = \log_3 x$, 1 unit downward.

3. The graph of $y = -\log_3 x$ reflects the graph of $y = \log_3 x$ over the $x-$axis.

4. The graph of $y = \log_3 (-x)$ reflects the graph of $y = \log_3 x$ over the $y-$axis.

(The graphs are shown in the exercise in this section.)

EXERCISES

Use the graph of $y = \log_3 x$ to write each function that is graphed after the indicated transformations.

1. Shift up 5 units

2. Shift down 5 units

3. Shift right 5 units

4. Shift left 5 units

5. Reflect about x – axis

6. Reflect about y – axis

7. Shift left 5 units
Reflect about x – axis

8. Shift up 3 units
Reflect about y – axis

9. Reflect about y – axis
shift down 3 units

10. Shift right 3 units
Reflect about x – axis

11. Reflect about y – axis
shift down 1 units

12. Shift up 2 units
Reflect about y – axis
Shift right 1 unit

13. The graph of a function is made after the following three transformations applied to the graph of $y = \log_2 x$. Write the function.

1. Reflect about the x – axis   2. Shift up 4 units   3. Shift right 6 units

14. The graph of a function is made after the following three transformations applied to the graph of $y = \log_{1/2} x$. Write the function.

1. Shift left 6 units   2. Reflect about the y – axis   3. Shift down 4 units

Match each of the function with its graph.

15. $y = \log_3 x$ **16.** $y = 4 + \log_3 x$ **17.** $y = \log_3(-x)$ **18.** $y = \log_3(x-1)$

19. $y = \log_3 x - 1$ **20.** $y = 1 - \log_3 x$ **21.** $y = 2\log_3 x$ **22.** $y = \log_3(1-x)$

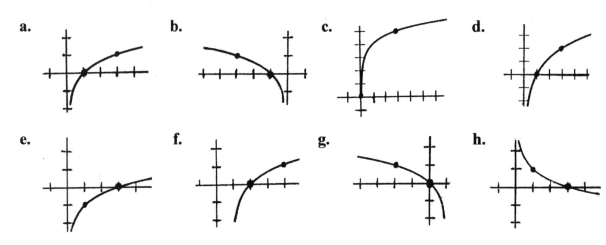

a.   b.   c.   d.

e.   f.   g.   h.

23. Use the graph of $y = \log_3 x$ to graph: a. $y = 3\log_3 x$ b. $y = \frac{1}{3}\log_3 x$

24. Graph $y = \log_3 |x|$ **25.** Graph $y = \log_3 x^2$

26. The loudness L in number of decibels of a sound is related to its intensity I in watts per square meter. The relationship between L and I is given by the formula below. Find L if I is 1,000 times as great as I_0. ($I_0 = 10^{-12}$ watt per square meter)

$$L = 10 \log \frac{I}{I_0} \text{, Where } I_0 = \text{is the weakest sound that can be heard by human ear.}$$

7-4 The Law of Logarithms

The Law of Logarithms

For all positive real numbers a, p, q, with $a \neq 1$, and n is any real number:

1) $\log_a 1 = 0$ **2)** $\log_a a = 1$ **3)** $\log_a a^n = n$

4) $\log_a pq = \log_a p + \log_a q$ **5)** $\log_a \frac{p}{q} = \log_a p - \log_a q$ **6)** $\log_a p^n = n \log_a p$

7) $a^{\log_a n} = n$ **8)** Formula for Changing Base: If $b \neq 1$, then $\log_a n = \dfrac{\log_b n}{\log_b a}$.

In practice, we use $b = 10$. Therefore, $\log_a n = \dfrac{\log_{10} n}{\log_{10} a}$.

9) Formula for Antilogarithms: If $\log x = a$, then $x = anti\log a$.

Also: $anti\log_a p = a^p$ and $anti\log p = 10^p$

$$anti\log_a(p+q) = a^{p+q} = a^p \cdot a^q = (anti\log_a p) \cdot (anti\log_a q)$$

Examples

Given $\log_{10} 2 = 0.301$, $\log_{10} 3 = 0.477$, $\log_{10} 5 = 0.699$. Find:

(Hint: we may omit the subscript 10)

1. $\log_{10} 20 = \log_{10} 4 \times 5 = \log_{10} 4 + \log_{10} 5 = \log_{10} 2^2 + \log_{10} 5 = 2\log_{10} 2 + \log_{10} 5$
     $= 2(0.301) + 0.699 = 1.301$.

2. $\log_{10} 0.2 = \log_{10} \frac{2}{10} = \log_{10} 2 - \log_{10} 10 = 0.301 - 1 = -0.699$.

3. $\log_{10} 81 = \log_{10} 3^4 = 4\log_{10} 3 = 4(0.477) = 1.908$.   **4.** $\log_{0.1} 3 = \dfrac{\log_{10} 3}{\log_{10} 0.1} = \dfrac{0.477}{-1} = -0.477$.

Examples

1. $\log 1 = 0$ **2.** $\log 100 = 2$ **3.** $\log 1000 = 3$ **4.** $\log 1000 = 4$
5. $\log 0.1 = -1$ **6.** $\log 0.01 = -2$ **7.** $\log 0.001 = -3$ **8.** $\log 0.0001 = -4$

9. $\log_2 8 = \dfrac{\log 8}{\log 2} = \dfrac{0.9031}{0.3010} = 3.0003$ **10.** $5^{\log_5 -9} = -9$ **11.** $4^{\log_4 \pi} = \pi$

12. Expand: $\log_a x^2 \sqrt{x+1} = \log_a x^2 + \log_a (x+1)^{1/2} = 2\log_a x + \frac{1}{2}\log_a(x+1)$

13. Condense: $4\log_a x + \frac{1}{2}\log_a(x-1) - 2\log_a(x+2) = \log_a x^4 + \log_a \sqrt{x-1} - \log_a(x+2)^2$

$$= \log_a(x^4 \sqrt{x-1}) - \log_a(x+2)^2 = \log_a \frac{x^4 \sqrt{x-1}}{(x+2)^2}$$

14. $anti\log 4 = 10^4 = 10{,}000$ **15.** $anti\log 0.9445 = 10^{0.9445} = 8.8$

16. $anti\log(-1.8763) = 10^{-1.8763} = \dfrac{1}{10^{1.8763}} = \dfrac{1}{75.2142} = 0.0133$

EXERCISES

Evaluate each expression.

1. $\log_2 1$ **2.** $\log_2 2$ **3.** $\log 1$ **4.** $\log_7 7$ **5.** $\log_3 3$

6. $\log_2 2^5$ **7.** $\log_5 5^2$ **8.** $\log_2 16$ **9.** $\log_2 64$ **10.** $\log_5 125$

11. $\log_3 9^5$ **12.** $\log_4 16^{10}$ **13.** $\log(0.1)^3$ **14.** $\log(0.01)^3$ **15.** $3^{\log_3 5}$

16. $2^{\log_2 \pi}$ **17.** $5^{\log_5 9}$ **18.** $\log_{2.7} 2.7^t$ **19.** $\log_{0.2} 0.2^{-\sqrt{5}}$ **20.** $\log_\pi \pi^{2.5}$

Use $\log_2 3 = 1.585$ and $\log_2 7 = 2.807$ to evaluate each expression. Round your answer to three decimal places.

21. $\log_2 \dfrac{3}{7}$ **22.** $\log_2 \dfrac{7}{3}$ **23.** $\log_2 21$ **24.** $\log_2 28$ **25.** $\log_2 1.75$

26. $\log_2 3.5$ **27.** $\log_2 0.75$ **28.** $\log_2 36$ **29.** $\log_2 108$ **30.** $\log_2 10.5$

Use the Change-of-Base formula and a calculator to evaluate each expression. Round your answer to three decimal places.

31. $\log_2 3$ **32.** $\log_2 7$ **33.** $\log_3 20$ **34.** $\log_5 24$ **35.** $\log_{1/2} 50$

36. $\log_{1/3} 68$ **37.** $\log_{\sqrt{2}} 0.9$ **38.** $\log_\pi \sqrt{5}$ **39.** $\log_{2.718} \pi$ **40.** $\log_4 \frac{25}{7}$

Expand each expression as the sum or difference of logarithms.

41. $\log_5 2x$ **42.** $\log_6 3x^4$ **43.** $\log x^2 y^3$ **44.** $\log_4 16x^2$

45. $\log x^{1/2} y^4$ **46.** $\log_a 3x^{2/3} y^4$ **47.** $\log_a (x^3 \sqrt{x-1})$ **48.** $\log (x^5 \sqrt{x^2 + 1})$

49. $\log \dfrac{8x}{7}$ **50.** $\log_2 \dfrac{3x}{y^2}$ **51.** $\log_a \dfrac{3y^5}{x^2}$ **52.** $\log \dfrac{x^3}{x-1}$

53. $\log_2 \dfrac{\sqrt{x}}{(x+1)^3}$ **54.** $\log \dfrac{\sqrt[3]{x+1}}{(x-1)^2}$ **55.** $\log_a \dfrac{x^2}{(x-1)^{1/4}}$ **56.** $\log_a \dfrac{x^2 \sqrt{x^3 + 2}}{(x+3)^3}$

Condense each expression as a simpler logarithm.

57. $\log_a 3 + \log_a 5$ **58.** $\log 36 - \log 6$ **59.** $2\log x + 3\log y$

60. $\frac{1}{2}\log_a 16 - \log_a 5$ **61.** $\frac{2}{3}\log_a 8 - \log_a 15$ **62.** $\frac{2}{5}\log_4 32 - 2\log_4 3$

63. $\log_2 \left(\frac{1}{x}\right) + \log_2 \left(\frac{1}{x}\right)^2$ **64.** $\log_a x + \log_a 20 - \log(x^2 + 1) - \log_a 10$

Evaluate: 65. antilog 5   **66.** antilog 0.5623   **67.** antilog (-0.5623)

68. Chemists use the pH value to describe the acidity of a solution. It is given by the formula $pH = -\log[H^+]$, where $[H^+]$ is measured in moles of hydrogen ions per liter in the solution.

     a) Find the pH of a solution with $[H^+] = 0.0063$ moles per liter.

     b) Find the amount of hydrogen ions $[H^+]$ in moles per liter of a solution with pH = 6.6.

7-5 The Number e and Natural Logarithms

When we graph the exponential function $y = a^x$, where $a > 0$ and $a \neq 1$, we notice the following two facts:

   1. The slope (m) of the tangent to $y = 2^x$ through the point $p(0, 1)$ is less than 1.

   2. The slope (m) of the tangent to $y = 3^x$ through the point $p(0, 1)$ is more than 1.

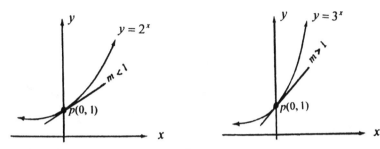

Therefore, we expect that there must be a number a, $2 < a < 3$, so that the slope of the tangent to $y = a^x$ through the point $p(0, 1)$ is exactly equal to 1. This number is designed by the letter e. The number e is called the **Euler number**.

The number e is one of the most famous numbers found in the history of mathematics. It is as famous as π and i. In advanced mathematics, we have found e from the following function in limit notation:

$$e = \lim_{n \to \infty}(1 + \tfrac{1}{n})^n. \qquad e = 2.71828\cdots\cdots. \text{ It is an irrational number.}$$

$x = e^y$ is the inverse function of $y = e^x$. The **natural logarithmic function** of $x = e^y$ is written by $y = \log_e x$. (Note that : $y = \log_e x$ and $x = e^y$ are equivalent.)

Natural logarithms are abbreviated by $\ln$, with the base understood to be e.

     $y = \ln x \rightarrow$ read as " **y is the logarithm of x to the base e** ".

$y = \ln x$ and $y = e^x$ are inverses each other. The graph of $y = \ln x$ and $y = e^x$ are reflection of each other about the line $y = x$.

The domains and ranges of the two functions are interchanged.

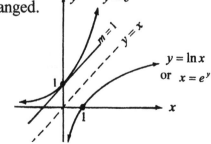

$$y = e^x \rightarrow x \in R, \ y > 0.$$
$$y = \ln x \text{ or } x = e^y \rightarrow y \in R, \ x > 0.$$

Natural logarithms are widely used in calculus and science. All laws of logarithms and the formulas for the antilogarithms that we have learned apply to the natural logarithms as well. (See Section 7-4)

The function $y = e^x$ is called the **natural exponential function**. Its graph is shown below.

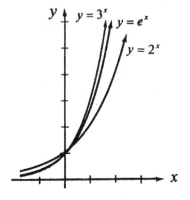

Most calculators have the key " e^x " to evaluate the exponential function for a given x-value and the key " LN " to evaluate the natural logarithms. The symbol for the antilogarithm of x to the base e is $anti \ln x$.

If $\ln x = 0.6931$, then $x = anti \ln 0.6931 = e^{0.6931} \approx 2$.

Examples

1. $e^2 = (2.7182\cdots)^2 = 7.389$

2. $e^3 = (2.7182\cdots)^3 = 20.086$

3. $e^{1.2} = (2.7182\cdots)^{1.2} = 3.320$

4. $e^{5.8} = (2.7282\cdots)^{5.8} = 330.30$

5. $e^{-1.2} = \dfrac{1}{e^{1.2}} = \dfrac{1}{3.320} = 0.301$

6. $e^{-0.04} = \dfrac{1}{e^{0.04}} = \dfrac{1}{1.041} = 0.961$

7. $e^8 \cdot e^5 = e^{8+5} = e^{13}$

8. $3e^5 \cdot 4e^2 = 12e^{5+2} = 12e^7$

9. $\dfrac{12e^{10}}{2e^7} = 6e^{10-7} = 6e^3$

10. $\dfrac{12e^7}{2e^{10}} = 6e^{7-10} = 6e^{-3} = \dfrac{6}{e^3}$

11. $(3e^5)^2 = 3^2 e^{5(2)} = 9e^{10}$

12. $(2e^{-4x})^3 = 2^3 e^{-4x \cdot 3} = 8e^{-12x} = \dfrac{8}{e^{12x}}$

13. Given $\ln 2 = 0.693$, $\ln 3 = 1.099$, and $\ln 5 = 1.609$, Find:

$\ln 20 = \ln(4 \times 5) = \ln 4 + \ln 5 = \ln 2^2 + \ln 5 = 2\ln 2 + \ln 5 = 2(0.693) + 1.609 = 2.995$.

$\ln 0.2 = \ln \frac{1}{5} = \ln 1 - \ln 5 = 0 - 1.609 = -1.609$.

$\ln 81 = \ln 3^4 = 4\ln 3 = 4(1.099) = 4.396$.

$\log_{0.1} 3 = \dfrac{\ln 3}{\ln 0.1} = \dfrac{\ln 3}{\ln 1 - \ln 10} = \dfrac{\ln 3}{0 - (\ln 2 + \ln 5)} = \dfrac{1.099}{-(0.693 + 1.609)} = -0.477$.

$\log_{\sqrt{2}} \sqrt{5} = \dfrac{\ln \sqrt{5}}{\ln \sqrt{2}} = \dfrac{\frac{1}{2}\ln 5}{\frac{1}{2}\ln 2} = \dfrac{\ln 5}{\ln 2} = \dfrac{1.609}{0.693} = 2.322$.

14. If $\ln N = 0.231$, find N.
Solution:

$N = anti \ln(0.231) = e^{0.231} = 1.26$.

15. If $\ln x = 0.492$, find x.
Solution:

$x = anti \ln(0.492) = e^{0.492} = 1.636$

Formulas for compound interest

The amount A after t years in an account with principal P, annual interest rate r is given by the following formulas:

1. For n compoundings per year: $A = P\left(1 + \dfrac{r}{n}\right)^{nt}$

2. For continuous compoundings:

We rewrite the formula as

$$A = P\left(1 + \frac{r}{n}\right)^{nt} = P\left(\left[1 + \frac{1}{n/r}\right]^{n/r}\right)^{rt}$$

When n increases without bound, n/r increases without bound.

It is denoted by $n/r \to \infty$ as $n \to \infty$.

We use the fact $\lim\limits_{k \to \infty}\left(1 + \dfrac{1}{k}\right)^{k} = e$. Therefore, we have $A = P\, e^{rt}$.

Examples

1: \$5,000 is deposited in a bank at an annual interest rate 6%. Find the new principal after 5 years if it is compounded monthly.

Solution:

$$A = P\left(1 + \frac{r}{n}\right)^{nt} = 5000\left(1 + \frac{0.06}{12}\right)^{12(5)} = 5000(1.005)^{60} = 5000(1.3489) = \$6,744.50. \text{ Ans.}$$

2. \$5,000 is deposited in a bank at an annual interest rate 6%. Find the new principal after 5 years if it is compounded continuously.

Solution:

$$A = P\, e^{rt} = 5000 \cdot e^{0.06(5)} = 5000 \cdot e^{0.3} = 5000(1.3499) = \$6,749.50. \text{ Ans.}$$

3. If you need \$8,000 after 10 years at an annual interest rate 8%, compounded quarterly, find the principal (present value) needed to invest now.

Solution:

$$8,000 = P\left(1 + \frac{0.08}{4}\right)^{4(10)} = P(1.02)^{40} = P(2.208) \quad \therefore P = \frac{8000}{2.208} = \$3,623.20. \text{ Ans.}$$

4. If you need \$8,000 after 10 years at an annual interest rate 8%, compounded continuously, find the principal (present value) needed to invest now.

Solution: $8,000 = P\, e^{0.08(10)} = P\, e^{0.8} \quad \therefore P = \dfrac{8000}{e^{0.8}} = \dfrac{8000}{2.226} = \$3,593.89. \text{ Ans.}$

Other Application

5. The population of a certain type of bacteria increases at time t hours is given by the exponential equation $n(t) = 20\, e^{0.05t}$. Find the population of the bacteria after 60 hours.

Solution:

$$n(60) = 20\, e^{0.05(60)} = 20\, e^{3} = 20(20.086) = 402. \text{ Ans.}$$

242   A-Plus Notes for Algebra

EXERCISES

Use a calculator to evaluate each expression. Round your answer to three decimal places.

1. $e^{1.5}$
2. e^5
3. $e^{2.8}$
4. $e^{3.5}$

5. $e^{8.2}$
6. $e^{-1.5}$
7. $e^{-0.2}$
8. $e^{-0.45}$

9. e^{π}
10. π^e
11. $\ln 8.25$
12. $\ln 0.274$

13. $\ln 0.005$
14. $\ln 1.008$
15. $\ln 1,000$
16. $\ln 2.564$

17. $anti \ln(0.45)$
18. $anti \ln(0.387)$
19. $anti \ln(-0.124)$
20. $anti \ln(-2.536)$

Simplify each expression.

21. $e^5 \cdot e^9$
22. $4e^3 \cdot 6e^9$
23. $(-4e^3)^2$
24. $\dfrac{27e^{12}}{3e^3}$

25. $\dfrac{3e^3}{27e^{12}}$
26. $\dfrac{15e^4}{18e^8}$
27. $\dfrac{3e^{6x}}{2e^{2x}}$
28. $\dfrac{10e^{4x}}{15e^{7x}}$

29. $\sqrt{9e^4x^2}$
30. $\sqrt{16e^8x^4}$
31. $(5e^{-3x})^{-2}$
32. $(4^{-2}e^3)^{-3x}$

Given $\ln 3 = 1.099$, $\ln 4 = 1.386$, and $\ln 6 = 1.792$, evaluate each expression. Round your answer to three decimal places.

33. $\ln 12$
34. $\ln 27$
35. $\ln 54$
36. $\ln 0.25$

37. $\ln 216$
38. $\ln \sqrt{12}$
39. $\log_3 4$
40. $\log_{\sqrt{3}} 16$

41. $5,000 is deposited in a bank at an annual interest rate 5%. Find the new principal after 10 years if it is compounded quarterly.
42. $5,000 is deposited in a bank at an annual interest rate 5%. Find the new principal after 10 years if it is compounded continuously.
43. If you need $50,000 after 5 years to marry your girl friend, find the principal (present value) needed now to deposit in an account that pays 5.8% annual interest rate, compounded continuously.
44. If you want to double your investment in 10 years, what annual rate compounded annually you should need ?
45. The population of a certain type of bacteria increases at time t hours is given by the exponential equation $n(t) = 32\,e^{0.08t}$. Find the population of the bacteria after 50 hours.
46. The value of a new car that costs $25,000 decreases after t years is given by the exponential equation $V(t) = 25,000\,(0.85)^t$. Find the value of the car after 5 years.

7-6 Exponential and Logarithmic Equations

Equations that contains the terms a^x, $a > 0$, $a \neq 1$ are called **exponential equations**. Equations that contains the terms $\log_a x$, where a is a positive real number, with $a \neq 1$, are called **logarithmic equations**. There are different ways to solve various types of exponential and logarithmic equations.

To solve **exponential equations**, we apply the laws of exponents (see Section 5~1).
     1. Rewrite the equation using the same base on both sides.
     2. Take log on both sides.
     3. Make a substitution.

To solve **logarithmic equations**, we apply the laws of logarithms (see Section 7~4).
     1. Change the equation to exponential form if it has only a single logarithm.
     2. Combine it as a single logarithm if it has multiple logarithms.
To solve an exponential or logarithmic equation, we must check each solution in the original equation and discard solutions that are not permissible.

Examples

1. Solve $3^{x+1} = 9^{2x-1}$.
Solution:
$$3^{x+1} = (3^2)^{2x-1}$$
$$3^{x+1} = 3^{4x-2}$$
$$x + 1 = 4x - 2$$
$$3 = 3x$$
$$\therefore x = 1. \text{ Ans.}$$

2. Solve $3^x = 20$.
Solution:
$$\log 3^x = \log 20$$
$$x \log 3 = \log 20$$
$$\therefore x = {\log 20}/{\log 3} = {1.301}/{0.477} = 2.727. \text{ Ans.}$$

3. Solve $4^x - 3 \cdot 2^x - 4 = 0$
Solution:
$$2^{2x} - 3 \cdot 2^x - 4 = 0$$
Let $a = 2^x$ (make a substitution):
$$a^2 - 3a - 4 = 0$$
$$(a-4)(a+1) = 0$$
$$a - 4 = 0 \quad \text{or} \quad a + 1 = 0$$

$a = 4$	$a = -1$
$2^x = 4$	$2^x = -1$
$\therefore x = 2.$	Not permissible
Ans.	$(2^x > 0)$

4. Solve $\log_a 16 = 4$.
Solution:
$$a^4 = 16, \quad a = 2 \text{ or } -2$$
$$(a = -2 \text{ is not permissible, } a > 0)$$
$$\therefore a = 2. \text{ Ans.}$$

5. Solve $\log_4 x + \log_4 (x-6) = 2$
Solution:
$$\log_4 x(x-6) = 2$$
$$x(x-6) = 4^2$$
$$x^2 - 6x - 16 = 0, \quad (x-8)(x+2) = 0$$
$$x = 8 \quad \text{or} \quad -2 \text{ (not permissible, } x > 0)$$
$$\therefore x = 8. \text{ Ans.}$$

EXERCISES

Solve each equation. Round your answer to three decimal places.

1. $3^x = 81$

2. $3^{2x} = 81$

3. $3^{4x} = 81$

4. $3^{x+1} = 81$

5. $3^{x-1} = 3^{4x+1}$

6. $3^{4x+1} = 9^{x-1}$

7. $3^{4x+1} = \left(\frac{1}{9}\right)^{x-1}$

8. $x^{2/3} = 16$

9. $x^{3/2} = 27$

10. $3^x = 40$

11. $3^{x-1} = 40$

12. $2^{2x+1} = 8$

13. $4^x = 15$

14. $\left(\frac{1}{4}\right)^x = 15$

15. $3^{1-2x} = 9^x$

16. $3^{1-2x} = 4^x$

17. $4^{1-2x} = \frac{1}{4}$

18. $2^{x+1} = 5^{1-2x}$

19. $2.5^x = 4^{-x}$

20. $e^{x+2} = 4$

21. $e^{x+3} = \pi$

22. $e^{x+5} = \pi^x$

23. $e^x = \pi^{2-x}$

24. $8^{x^2-2x} = \frac{1}{2}$

25. $3^{2x} + 3^x - 2 = 0$

26. $3^{2x} - 2 \cdot 3^x - 3 = 0$

27. $x^{1/2} - 6x^{-1/2} - 5 = 0$

28. $e^{\ln(2x-1)} = 4$

29. $e^{2x-1} = 4$

30. $\log_x 5 = -\frac{1}{2}$

31. $\log_{\frac{1}{4}} x = -2$

32. $\log_{\frac{1}{2}} 16 = x$

33. $\log_2(3x-1) = 3$

34. $\log_5(4x-1) = 2$

35. $\log_4(x^2 + 4x + 11) = 2$

36. $\log_5(x^2 + 1) = 2$

37. $\log_4(x+4)^3 = 3$

38. $\log_3(x^2 + x + 4) = 2$

39. $e^{3x+5} = 40$

40. $e^x - 3e^{-x} - 2 = 0$

41. $e^x - e^{-x} - 10 = 0$

42. $\ln(x+1) = 0$

43. $e^{\ln(1+x)} = 3x$

44. $\ln e^{\sqrt{x+1}} = 8$

45. $\ln x^2 = (\ln x)^2$

46. $\ln e^{x^2} = 4$

47. $\log(x-6) + \log(x+6) = 1$

48. $\log_2 x = 2\log_2 5 + \log_2 6$

49. $\log_5 3x - \log_5(x+4) = 0$ **50.** $\log 5 - \log x = 1$

51. $\log_{11}(x-5) + \log_{11}(x-5) = 2$

52. $\log x + \log(x-3) = 1$ **53.** $\log(x+21) + \log x = 2$ **54.** $\log_4 x - 1 = -\log_4(x-3)$

55. $\log_4(x^2 - 9) - \log_4(x+3) = 2$ **56.** $3\log_3 x + \log_9 x = 7$ **57.** $\log_2(x^2 + 1) - \log_4 x^2 = 1$

58. $\log(x^2 - 4) - \log(x-2) = \log x^2$ **59.** $\log(x+5) = \log(x-1) - \log(x+1)$

60. $\ln(x+2) = 1 + \ln x$ **61.** $\ln(2x+2) + \ln(x-3) = 2\ln x$ **62.** $\ln(x-2) + \ln(2x-3) = 2\ln x$

7-7 Exponential Growth and Decay

In physics, exponential and logarithmic functions are useful in studying the growth of individuals (such as bacteria) and the decay of substances (such as radioactive elements). The physicists have proved that the growth and the decay of certain individuals and substances, under ideal conditions, follow the **general exponential form** of changes.

1) Formula for Exponential Growth

$$n(t) = n_0 \cdot e^{kt}, \quad k > 0.$$

$n(t) \to$ the population at time t .

$n_0 \to$ the population at $t = 0$.

$k \to$ a constant depends on a
     particular individual.

2) Formula for Exponential Decay

$$n(t) = n_0 \cdot e^{kt}, \quad k < 0 \quad \text{or} \quad n(t) = n_0 \cdot 2^{-t/h}$$

$n(t) \to$ the amount of the substance
     remaining at time t .

$n_0 \to$ the amount at $t = 0$.

Half-life time $(h) \to$ the time it needs for only half of the amount remains unchanged.
(See proof on page 250)

Examples

1. If the population of a bacteria grew double every 20 hours, how long does it take to triple ? Assuming exponential growth.

   Solution: $n(t) = n_0 \cdot e^{kt}$, $2n_0 = n_0 \cdot e^{kt}$, $2 = e^{k(20)}$, $\ln 2 = 20k$, $\therefore k = \frac{\ln 2}{20} = \frac{0.693}{20} = 0.035$

   We have: $n(t) = n_0 \cdot e^{0.035t}$, $3n_0 = n_0 \cdot e^{0.035t}$, $3 = e^{0.035t}$, $\ln 3 = 0.035t$

   $$\therefore t = \frac{\ln 3}{0.035} = \frac{1.0986}{0.035} = 31.39 \text{ hrs. Ans.}$$

2. A radioactive element has a half-life of 2.8 years. How long would it take 1 gram amount to decay to 0.2 gram ? Assuming exponential decay.

   Solution: $n(t) = n_0 \cdot e^{kt}$, $0.5 = 1 \cdot e^{k(2.8)}$

   $\ln 0.5 = 2.8k$, $k = \frac{\ln 0.5}{2.8} = \frac{-0.693}{2.8} = -0.2475$

   We have: $0.2 = 1 \cdot e^{-0.2475t}$, $\ln 0.2 = -0.2475t$

   $\therefore t = \frac{\ln 0.2}{-0.2475} = \frac{-1.609}{-0.2475} = 6.50$ years. Ans.

   OR: $n(t) = n_0 \cdot 2^{-t/h}$, $0.2 = 1 \cdot 2^{-t/28}$

   $\ln 0.2 = -\frac{t}{2.8} \ln 2$

   $\therefore t = \frac{-2.8(\ln 0.2)}{\ln 2} = \frac{-2.8(-1.609)}{0.693}$

   $= \frac{4.5052}{0.693} = 6.50$ years. Ans.

3. If the students in Hawthorne high school increases by 8% per year, how long does it take to double ?

   Solution: $n = n_0(1 + 0.08)^t$

   $$2n_0 = n_0(1.08)^t$$

   $$2 = (1.08)^t$$

   $$\log 2 = t \log 1.08$$

   $$\therefore t = \frac{\log 2}{\log 1.08} = \frac{0.301}{0.033} = 9.12 \text{ years. Ans.}$$

4. If a new car has a value of $25,000 and decreases by 12% per year, how many years will it have a value of $10,000 ?

   Solution: $p = a(1 - 0.12)^t$

   $$10000 = 25000(0.88)^t$$

   $$0.4 = (0.88)^t$$

   $$\log 0.4 = t \log 0.88$$

   $$\therefore t = \frac{\log 0.4}{\log 0.88} = \frac{-0.398}{-0.056} = 7.11 \text{ years. Ans.}$$

EXERCISES

1. The students in a high school increases by 6% per year. How long does it take to double ?

2. The students in a high school increases by 6% per year. How long does it take to triple ?

3. A new car has a value of $28,000 and decreased by 15% per year. How many years will it have a value of $5,000 ?

4. A new car has a value of $28,000 and decreased by 15% per year. How many years will it have a half-value ?

5. The population of a bacteria grew double every 10 hours. How long does it take to triple ? Assuming an exponential growth.

6. A radioative material has a half-life of 3.5 years. How long would it take 5 grams to decay to 1 gram ? Assuming an exponential decay.

7. The population (p) of a certain type of bacteria increases at time t (in hours) follows the equation $p = 800 \cdot e^{0.05t}$. How long will the population reach 1,600 ?

8. The population (p) of a certain type of bacteria increases at time t (in hours) follows the equation $p = 800 \cdot e^{0.05t}$. How long will the population reach 2,000 ?

9. The amount (n) of a certain type of radioactive material decays at time t (in years) follows the equation $n(t) = n_0 e^{-0.03t}$, where n_0 is the initial amount. How long does it take 50 grams to decay to 10 grams ?

10. The amount (n) of a certain type of radioactive material decays at time t (in years) follows the equation $n(t) = n_0 e^{-0.03t}$, where n_0 is the initial amount. What is the half-life time of this material ?

11. The population (p) (in thousands) of a city increases at time t (in years) follows the exponential growth equation $p = 432 \cdot e^{0.025t}$. How long does it take the population reach 650,000 ?

12. The loudness $L(x)$ in number of decibels of a sound is related to its intensity (x) in watts per square meter. The relationship between L and x is given by $L(x) = 10\log_{10}(x/I_0)$, where I_0 is the weakest intensity of sound that can be heard by human ear ($I_0 = 10^{-12}$ watt per square meter). Find the level of loudness in decibels of a machine that at a sound intensity of 10^{-4} watt per square meter.

13. The magnitude (R) of an earthquake is given by the Richter scale formula $R = \log_{10}(I/I_0)$, where I_0 is the zero-level intensity of an earthquake whose seismographic reading is 0.001 millimeter ($I_0 = 10^{-3}$ millimeter). Find the magnitude R of an earthquake whose seismographic reading is 10 millimeters.

14. The population (p) of a certain type of insect grows rapidly initially. Its growth rate decreases under constraints and reach the maximum population after a limit of time t (in days) by the equation of the form

$$p(t) = \frac{210}{1 + 25 \cdot e^{-0.15t}} \ .$$ Its graph is called **Logistic Growth curve.**

   a) What is the maximum population as $t \to \infty$?

   b) When will the population reach 116 ?

   c) Use a graphing calculator to graph the function.

CHAPTER 7   EXERCISES

Apply the laws of exponents in Section 5~1 to evaluate each expression.

1. $49^{\frac{1}{2}}$

2. $49^{-\frac{1}{2}}$

3. $49^{\frac{3}{2}}$

4. $49^{-\frac{3}{2}}$

5. $81^{\frac{1}{4}}$

6. $81^{\frac{3}{4}}$

7. $81^{-\frac{3}{4}}$

8. $16^{\frac{3}{4}}$

9. $16^{-\frac{3}{4}}$

10. $125^{\frac{2}{3}}$

11. $25^{\frac{3}{2}}$

12. $25^{-1.5}$

13. $125^{0.\overline{3}}$

14. $8^{1.5}$

15. $8^{1.\overline{3}}$

16. $16^{0.25}$

17. $(\frac{8}{27})^{\frac{2}{3}}$

18. $(\frac{27}{8})^{-\frac{2}{3}}$

19. $\sqrt{x^4 y^2}$

20. $\sqrt[3]{8x^3 y^{-9}}$

21. $(9^{\frac{1}{2}} + 4^{\frac{1}{2}})^2$

22. $(4^{-\frac{1}{2}} - 9^{-\frac{1}{2}})^{-2}$

23. $4^{\frac{1}{2}} \cdot 16^{\frac{3}{2}}$

24. $3^{0.5} \cdot 3^{0.7}$

25. $a^{-\frac{1}{2}} \cdot a^{\frac{1}{2}} \div a^{\frac{5}{6}}$

26. $8^{-2} \div 16^{-4} \cdot 2^3$

27. $(9a^{\frac{2}{3}} b^{-2})^{\frac{1}{2}}$

28. $(27a^{-6} b^9)^{\frac{1}{3}}$

29. $27^{\pi} \cdot 3^{3-3\pi}$

30. $3^{1+\sqrt{2}} \cdot 9^{1-\sqrt{2}}$

31. $3^{1+\sqrt{2}} \div 3^{1-\sqrt{2}}$

32. $(3^{\sqrt{2}})^{-\sqrt{2}}$

33. $\dfrac{16^{\sqrt{2}}}{4^{\sqrt{2}}}$

34. $\dfrac{16^{\sqrt{3}-\pi}}{2^{\sqrt{3}+\pi}}$

35. $\dfrac{9^{3+\pi}}{3^{3-\pi}} \cdot 27^{1-\pi}$

36. $\sqrt{\sqrt{a} \cdot \dfrac{a}{\sqrt[3]{a}}}$

37. $5e^3 \cdot 2e^{-7}$

38. $(3^{-2} e^2)^{-4x}$

39. $4e^{5x} \Big/ 6e^{8x}$

40. $\sqrt{8e^6 x^8}$

Solve each exponential equation. Round your answer to three decimal places.

41. $x^{-\frac{2}{3}} = 4$

42. $3^x = 243$

43. $81^{x+3} = 3^{3x+1}$

44. $(x+2)^{\frac{2}{3}} = 9$

45. $x^{\frac{1}{2}} - 8x^{-\frac{1}{2}} = 2$

46. $2^{7-x} = 16$

47. $(\frac{1}{2})^x = 4^{x-1}$

48. $(\frac{1}{125})^x = 25^{1-x}$

49. $(\frac{2}{3})^x = (\frac{3}{2})^{2x-3}$

50. $2^{2x} + 3 \cdot 2^x - 10 = 0$

51. $3^{2x} - 7 \cdot 3^x - 18 = 0$

52. $2^{2x} - 8 = 2^{x+1}$

53. $\dfrac{2^{x^2+1}}{2^{x-1}} = 16$

54. $\dfrac{2^{x^2+1}}{4^{x-1}} = 16$

55. $\dfrac{3^{x^2-1}}{9^{x+2}} = \dfrac{1}{9}$

56. $4^x = 2$

57. $(\frac{1}{4})^x = 2$

58. $4^{1-2x} = 3^x$

59. $e^{x-2} = 5$

60. $e^{x-3} = \pi$

61. $e^x = \pi^{x+3}$

62. $e^{\ln 2x} = 8$

63. $e^{2x} + 4 \cdot e^x - 5 = 0$

64. $e^x - 4 \cdot e^{-x} - 3 = 0$

-----Continued-----

Apply the laws of logarithms in Section 7~4 to evaluate each expressions.

65. $\log_3 3$ **66.** $\log_3 1$ **67.** $\log_{12} 12$ **68.** $\log_{12} 1$

69. $\log_b b$ **70.** $\log_x 1$ **71.** $\log_2 32$ **72.** $\log_3 27$

73. $\log_{\frac{1}{3}} 27$ **74.** $\log_{10} \frac{1}{10}$ **75.** $\log_{10} \frac{1}{1000}$ **76.** $\log_6 \frac{1}{36}$

77. $\log_{\frac{1}{6}} 36$ **78.** $\log_8 0.125$ **79.** $\log_{\frac{1}{2}} 16$ **80.** $\log_{\frac{1}{2}} 2$

81. $\log_2 \frac{1}{8}$ **82.** $\log_{1.5} \frac{27}{8}$ **83.** $2^{\log_2 7}$ **84.** $9^{\log_3 2}$

85. $\log_{\frac{1}{7}} 49$ **86.** $\log_{\sqrt{3}} 9$ **87.** $\log_{\frac{1}{2}} \frac{1}{4}$ **88.** $\log_4 8$

89. $\log_{\sqrt{4}} 16$ **90.** antilog 0.9395 **91.** antilog (-2.0938) **92.** $\log(\log 10^{10})$

93. $\ln e^2$ **94.** $\ln e^4$ **95.** $\ln e^{-3}$ **96.** $\ln \frac{1}{e}$

97. $\ln \frac{1}{e^2}$ **98.** $\ln \sqrt{e}$ **99.** $\ln \frac{1}{\sqrt{e}}$ **100.** $e^{\ln 7}$

101. antiln (0.56) **102.** antiln (-4.51) **103.** antiln (-0.24) **104.** $\ln (\ln e^e)$

Solve each equation. Round your answer to three decimal places.

105. $\log_5 x = 4$ **106.** $\log_x 16 = 4$ **107.** $\log_{\sqrt{2}} x = 6$

108. $\log_4 x = -2$ **109.** $\log_x 27 = \frac{3}{2}$ **110.** $\log_3 (x^2 - 1) = 2$

111. $\log_{\sqrt{5}} 25 = y$ **112.** $\log_x 9\sqrt{3} = 5$ **113.** $10^{\log x} = \sqrt{1000}$

114. $\log_3 x = \log_3 12 + \log_3 x^2$ **115.** $\log_5 x^3 = \frac{3}{2}\log_5 4$ **116.** $x^{\log x} = \frac{1}{100} x^3$

117. $\log(3x - 1) + \log(x - 1) = 0$ **118.** $x^{\log x} = 1000x^2$ **119.** $3^{\log x} = 2^{\log 3}$

120. $\log_2 x = \log_4 2x$ **121.** $\log_4 x = \log_{16}(x + 2)$ **122.** $2\log_5 x + \log_{25} x = 10$

123. $\ln x = 4$ **124.** $\ln x = -4$ **125.** $\ln(x - 1) = 1$

126. $\ln(x - 2) = 1 + \ln x$ **127.** $\ln(2x - 1) + \ln x = 2\ln x$ **128.** $\ln(x - 1) + \ln(2x - 4) = 2\ln x$

Use a graphing calculator to graph each function.

129. $y = 3^x + 4$ **130.** $y = 3^{-x} - 4$ **131.** $y = e^{x+2}$

132. $y = \log_{10}(2x)$ **133.** $y = \ln(1 + 100x)$ **134.** $y = \dfrac{50}{1 + e^{-x}}$

-----Continued-----

135. If you deposit $1,000 in the bank with a compound annual interest rate of 8%, how many years later will it take for your money to reach $5,000 ?

136. A $8,000 investment earns interest at the annual rate of 6.5% compound monthly. What is the investment worth after 10 years ?

137. If the students in a high school increases by 5% per year, how long does it take to double ?

138. If a new car has a value of $25,000 and decreases by 15% per year, how many years will it have a value of $10,000 ?

139. If the population of a bacteria grew double every 15 years, how long does it take to triple ? Assuming an exponential growth.

140. If a radioactive substance has a half-life of 10 years, how long would it take for 2 grams to decay to 0.01 gram ? Assuming an exponential decay.

141. A radioactive element is decaying according to the formula $y = y_0 (2.718)^{-0.2x}$, where y_0 is the initial amount of the element and y is the amount of the element remaining after x years. Find the half-life of this element if the initial amount is 40 grams.

142. Carbon-14 is a radioactive element with a half-life of 5750 years. A human skeleton is found to contain one-fifth of its original amount of Carbon-14. How old is the skaleton ?

143. The population (p) of a certain type of bacteria increases at time t (in hours) follows the equation $p = 400 \cdot e^{0.08t}$. How long will the population reach 900 ?

144. The amount (n) of a certain type of radioactive material decays at time t (in years) follows the equation $n(t) = n_0 \cdot e^{-0.05t}$, where n_0 is the initial amount. What is the half-life time of this material ?

145. $12,000 is invested in a mutual fund at an annual interest rate 7%, compound continuously. How long will it take for the original investment to double in value ?

146. $12,000 is invested in a mutual fund at an annual interest rate 7%, compound continuously. What interest rate is required for the investment to double in 10 years ?

147. The unit price (p) of a product depends on its demand (x) in the market. The unit price of a certain product is given by the equation $p = 250 - 1.5 \cdot e^{0.003x}$. Find the unit price if the demand $x = 1,000$ units.

148. The unit price (p) of a product depends on its demand (x) in the market. The unit price of a certain product is given by the equation $p = 250 - 1.5 \cdot e^{0.003x}$. Find the demand if the unit price is $190.

149. A student was given an exam and then retested monthly with an equivalent exam. His average scores decreases according the formula
$$S(t) = 95 - 16 \cdot \log(t+1).$$ It is called **Human Memory formula.**
    a) What was the average score after 5 months ?
    b) How long will it take for the average score to drop to 75 ?

150. The bell-shaped of a normal probability curve is given by the function
$$y = e^{-x^2}.$$ It is also called **Gaussian normal curve.**
    a) What is the y-value when $x = 1.5$?   b) What is the y-value when $x = -1.5$?
    c) Use a graphing calculator to graph the function and find the maximum y-value.

Proof of the Exponential Decay Formula
if the Half-life time is given

Prove: The exponential decay formula $n(t) = n_0 \cdot e^{k \cdot t}$, $k < 0$ is equivalent to $n(t) = n_0 \cdot 2^{-\frac{t}{h}}$, where h is the half-life time.

Proof:

$$n(t) = n_0 \cdot e^{k \cdot t}$$

$$\frac{1}{2} n_0 = n_0 \cdot e^{k \cdot h}$$

$$\therefore \frac{1}{2} = e^{k \cdot h}$$

$$\ln\left(\frac{1}{2}\right) = kh \qquad \therefore k = \frac{\ln\left(\frac{1}{2}\right)}{h}$$

We have:

$$n(t) = n_0 \cdot e^{\frac{\ln\left(\frac{1}{2}\right)}{h} \cdot t} = n_0 \cdot e^{\frac{t}{h}\ln\left(\frac{1}{2}\right)} = n_0 \cdot e^{\ln\left(\frac{1}{2}\right)^{\frac{t}{h}}} = n_0 \cdot \left(\frac{1}{2}\right)^{\frac{t}{h}} = n_0 \cdot 2^{-\frac{t}{h}}$$

$$\therefore n(t) = n_0 \cdot 2^{-\frac{t}{h}}. \text{ The proof is complete.}$$

Space for Taking Notes

Answers

CHAPTER 7 – 1   Exponential and Logarithmic Functions   page 232
1. 49.15 **2.** 1097.79 **3.** 2.55 **4.** 2 **5.** 0.02 **6.** 2 **7.** 6.58 **8.** 4.71 **9.** 0.27 **10.** 2.24 **11.** 0.02
12. 22.42 **13.** 22.24 **14.** 36.40 **15.** 7.26 **16.** 0.01 **17.** 0.16 **18.** 0.56 **19.** 0.20 **20.** 0.25 **21.** 1.30
22. 1.73 **23.** 1.54 **24.** 1.95 **25.** 1.31 **26.** 1.63 **27.** 0.50 **28.** 0.43 **29.** −0.44 **30.** −1.44
31. $\log_2 16 = 4$ **32.** $\log_2 32 = 5$ **33.** $\log_3 9 = 2$ **34.** $\log_3 81 = 4$ **35.** $\log 1000 = 3$ **36.** $\log_8 2 = \frac{1}{3}$
37. $\log_{16} 2 = \frac{1}{4}$ **38.** $\log_8 \frac{1}{64} = -2$ **39.** $\log_8 \frac{1}{2} = -\frac{1}{3}$ **40.** $\log 0.001 = -3$ **41.** $\log_9 27 = \frac{3}{2}$
42. $\log_a 3.54 = 5$ **43.** $\log_3 8.6 = x$ **44.** $\log_x \pi = \sqrt{3}$ **45.** $\log_{2.71} 12 = x$ **46.** $4^2 = 16$ **47.** $3^3 = 27$
48. $5^3 = 125$ **49.** $7^2 = 49$ **50.** $7^{-2} = \frac{1}{49}$ **51.** $10^1 = 10$ **52.** $10^{-1} = \frac{1}{10}$ **53.** $10^{-3} = \frac{1}{1000}$ **54.** $16^{3/4} = 8$
55. $2.71^0 = 1$ **56.** $10^{-0.02} = 0.955$ **57.** $a^{12} = 5$ **58.** $2^x = 9$ **59.** $5^{3.97} = x$ **60.** $2.71^x = \pi$
61. Growth **62.** Decay **63.** Decay **64.** Growth **65.** Growth **66.** Decay **67.** Decay **68.** Growth
69. Decay

CHAPTER 7 – 2   Exponential Functions and their Graphs   page 234
1. $y = 2^x + 5$ **2.** $y = 2^x - 5$ **3.** $y = 2^{x-5}$ **4.** $y = 2^{x+5}$ **5.** $y = -2^x$ **6.** $y = 2^{-x}$ **7.** $y = -2^{x+5}$
8. $y = 2^{-x} + 3$ **9.** $y = 2^{-x} - 3$ **10.** $y = -2^{x-3}$ **11.** $y = 2^{-x} - 3$ **12.** $y = 2^{-(x-1)} + 2$ or $y = 2^{1-x} + 2$
13. $y = -3^{x-6} + 4$ **14.** $y = \left(\frac{1}{3}\right)^{-x+6} - 4$ or $y = 3^{x-6} - 4$ **15.** a **16.** b **17.** h **18.** g **19.** d **20.** f
21. m **22.** e **23.** c **24.** j **25.** i **26.** k (15 ~ 26 Domains and Ranges are left to the students.)
27. a) It stretches the graph of $y = 2^x$ vertically by a factor of 3.   a.   b.
   b) It compresses the graph of $y = 2^x$ vertically by a factor of 3.
28. $598.34 (Hint: $A = P(1 + \frac{r}{n})^{nt}$)

CHAPTER 7 – 3   Logarithmic Functions and their Graphs   page 236
1. $y = \log_3 x + 5$ **2.** $y = \log_3 x - 5$ **3.** $y = \log_3 (x - 5)$ **4.** $y = \log_3 (x + 5)$ **5.** $y = -\log_3 x$
6. $y = \log_3 (-x)$ **7.** $y = -\log(x + 5)$ **8.** $y = \log_3 (-x) + 3$ **9.** $\log_3 (-x) - 3$ **10.** $y = -\log_3 (x - 3)$
11. $y = \log_3 (-x) - 1$ **12.** $y = \log_3 -(x-1) + 2$ or $y = \log_3 (1-x) + 2$ **13.** $y = -\log_2 (x-6) + 4$
14. $y = \log_{1/2} (-x + 6) - 4$ **15.** a **16.** c **17.** b **18.** f **19.** e **20.** h **21.** d **22.** g

23. a. It stretches the graph of $y = \log_3 x$ vertically by a factor of 3.
   b. It compresses the graph of $y = \log_3 x$ vertically by a factor of 3.

a.   b.   **24.**   **25.**   **26.** 30 decibels

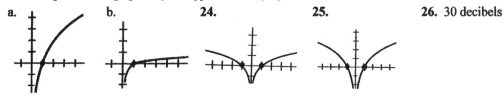

CHAPTER 7 – 4   The Law of Logarithms   page 238
1. 0 **2.** 0 **3.** 0 **4.** 1 **5.** 1 **6.** 5 **7.** 2 **8.** 4 **9.** 6 **10.** 3 **11.** 10 **12.** 20 **13.** −3 **14.** −6 **15.** 5 **16.** π
17. 9 **18.** t **19.** $-\sqrt{5}$ **20.** 2.5 **21.** −1.222 **22.** 1.222 **23.** 4.392 **24.** 4.807 **25.** 0.807 **26.** 1.807
27. −0.415 **28.** 5.170 **29.** 6.755 **30.** 3.392 **31.** 1.585 **33.** 2.807 **33.** 2.727 **34.** 1.974 **35.** −5.645
36. −3.843 **37.** −0.305 **38.** 0.702 **39.** 1.145 **40.** 0.919 **41.** $\log_5 2 + \log_5 x$ **42.** $\log_6 3 + 4\log_6 x$
43. $2\log x + 3\log y$ **44.** $2 + 2\log_4 x$ **45.** $\frac{1}{2}\log x + 4\log y$ **46.** $\log_a 3 + \frac{2}{3}\log_a x + 4\log_a y$
47. $3\log_a x + \frac{1}{2}\log_a (x-1)$ **48.** $5\log x + \frac{1}{2}\log(x^2 + 1)$ **49.** $\log 8 + \log x - \log 7$
50. $\log_2 3 + \log_2 x - 2\log_2 y$ **51.** $\log_a 3 + 5\log_a y - 2\log_a x$ **52.** $3\log x - \log(x-1)$
53. $\frac{1}{2}\log_2 x - 3\log_2 (x+1)$ **54.** $\frac{1}{3}\log(x+1) - 2\log(x-1)$ **55.** $2\log_a x - \frac{1}{4}\log_a (x-1)$
56. $2\log_a x + \frac{1}{2}\log_a (x^3 + 2) - 3\log_a (x+3)$ **57.** $\log_a 15$ **58.** $\log 6$ **59.** $\log(x^2 y^3)$ **60.** $\log_a \frac{4}{5}$
61. $\log_a \frac{4}{15}$ **62.** $\log_4 \frac{4}{9}$ **63.** $-3\log_2 x$ **64.** $\log_a \frac{2x}{x^2 + 1}$ **65.** 100,000 **66.** 3.65 **67.** 0.274
68. a. pH = 2.2   b. $[H^+] = 2.512 \times 10^{-7}$ moles per liter

CHAPTER 7 – 5   The Number e and Natural Logarithms   page 242

1. 4.482 **2.** 48.413 **3.** 16.445 **4.** 33.115 **5.** 3640.950 **6.** 0.223 **7.** 0.819 **8.** 0.638 **9.** 23.141
10. 22.459 **11.** 2.110 **12.** −1.295 **13.** −5.298 **14.** 0.008 **15.** 6.908 **16.** 0.942 **17.** 1.568 **18.** 1.473
19. 0.883 **20.** 0.079 **21.** e^{14} **22.** $24e^{12}$ **23.** $16e^6$ **24.** $9e^9$ **25.** $\frac{1}{9}e^{-9}$ **26.** $\frac{5}{6}e^{-4}$ **27.** $\frac{3}{2}e^{4x}$ **28.** $\frac{2}{3}e^{-3x}$
29. $3e^2|x|$ **30.** $4e^4x^2$ **31.** $\frac{1}{25}e^{6x}$ **32.** $4^{6x}e^{-9x}$ **33.** 2.485 **34.** 3.297 **35.** 3.990 **36.** −1.386
37. 5.376 **38.** 1.243 **39.** 1.261 **40.** 5.045 **41.** \$8,218.10 **42.** \$8,243.61 **43.** \$37,413.18
44. 7.18% **45.** \$1,747.14 **46.** \$11,092.63

CHAPTER 7 – 6   Exponential and Logarithmic Functions   244

1. $x = 4$ **2.** $x = 2$ **3.** $x = 1$ **4.** $x = 3$ **5.** $x = -\frac{2}{3}$ **6.** $x = -1\frac{1}{2}$ **7.** $x = \frac{1}{6}$ **8.** $x = 64$ **9.** $x = 9$
10. $x = 1.085$ **11.** $x = 2.085$ **12.** $x = 1$ **13.** 1.953 **14.** −1.953 **15.** $x = \frac{1}{4}$ **16.** 0.307 **17.** $x = 1$
18. $x = 0.234$ **19.** $x = 0$ **20.** $x = -0.614$ **21.** $x = -1.855$ **22.** $x = 34.483$ **23.** $x = 1.068$
24. $x = \frac{1}{3}(3 \pm \sqrt{6})$ **25.** $x = 0$ **26.** $x = 1$ **27.** $x = 36$ **28.** $x = 2\frac{1}{2}$ **29.** $x = 1.193$ **30.** $x = \frac{1}{25}$ **31.** $x = 16$
32. $x = -4$ **33.** $x = 3$ **34.** $x = 6\frac{1}{2}$ **35.** $x = 1$ or $x = -5$ **36.** $x = \pm2\sqrt{6}$ **37.** $x = 0$
38. $x = -\frac{1}{2}(1 \pm \sqrt{21})$ **39.** $x = -0.437$ **40.** $x = \ln 3$ or 1.099 **41.** $x = \ln(5 + \sqrt{26})$ or 2.312 **42.** $x = 0$
43. $x = \frac{1}{2}$ **44.** $x = 63$ **45.** $x = 1$ or $x = e^2$ **46.** $x = \pm2$ **47.** $x = \sqrt{46}$ or 6.702 **48.** $x = 150$
49. $x = 2$ **50.** $x = \frac{1}{2}$ **51.** $x = 16$ **52.** $x = 5$ **53.** $x = 4$ **54.** $x = 4$ **55.** $x = 19$
56. $x = 3$ (Hint: Change $\log_9 x$ to base 3) **57.** $x = \pm1$ (Hint: Change $\log_4 x^2$ to base 2) **58.** $x = 2$
59. No solution **60.** $x = \frac{2}{e-1}$ or 1.164 **61.** $x = 2 + \sqrt{10}$ or 5.162 **62.** $x = 6$

CHAPTER 7 – 7   Exponential Growth and Decay   page 246

1. 12.04 years **2.** 19.08 years **3.** 10.6 years **4.** 4.26 years
5. 15.85 hours **6.** 8.13 years **7.** 13.86 hours **8.** 18.33 hours
9. 53.65 years **10.** 23.1 years **11.** 16.35 years **12.** 80 decibels
13. $R = 4$

14. a. 210
b. 22.86 days
c.

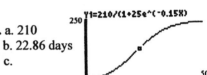

CHAPTER 7   Exercises   page 247 ~ 249

1. 7 **2.** $\frac{1}{7}$ **3.** 343 **4.** $\frac{1}{343}$ **5.** 3 **6.** 27 **7.** $\frac{1}{27}$ **8.** 9 **9.** $\frac{1}{8}$ **10.** 25 **11.** 125 **12.** $\frac{1}{125}$ **13.** 5 **14.** $16\sqrt{2}$ **15.** 16
16. 2 **17.** $\frac{4}{9}$ **18.** $\frac{4}{9}$ **19.** $x^2|y|$ **20.** $\frac{2x}{y^3}$ **21.** 25 **22.** 36 **23.** 128 **24.** $3\sqrt[5]{3}$ **25.** $a^{-\frac{5}{6}}$ **26.** 2^{13} **27.** $\frac{3\sqrt{a}}{b}$
28. $\frac{3b^3}{a^2}$ **29.** 27 **30.** $3^{3-\sqrt{2}}$ **31.** $9^{\sqrt{2}}$ **32.** $\frac{1}{9}$ **33.** $4^{\sqrt{2}}$ **34.** $2^{3\sqrt{3}-5\pi}$ **35.** 729 **36.** $a^{\frac{7}{12}}$ **37.** $10e^{-4}$ **38.** $\left(\frac{3}{e}\right)^{8x}$
39. $\frac{2}{3}e^{-3x}$ **40.** $2\sqrt{2}e^3x^4$ **41.** $x = \frac{1}{8}$ **42.** $x = 5$ **43.** $x = -11$ **44.** $x = 25$ **45.** $x = 16$ **46.** $x = 3$
47. $x = \frac{2}{3}$ **48.** $x = -2$ **49.** $x = 1$ **50.** $x = 1$ **51.** $x = 2$ **52.** $x = 2$ **53.** $x = 2$ or $x = -1$ **54.** $x = 1 \pm \sqrt{2}$
55. $x = 3$ or $x = -1$ **56.** $x = \frac{1}{2}$ **57.** $x = -\frac{1}{2}$ **58.** $x = \frac{\log 4}{\log 3 + 2\log 4}$ or 0.358 **59.** $x = 2 + \ln 5$ or 3.609
60. $x = 3 + \ln \pi$ or 4.145 **61.** $x = \frac{3\ln \pi}{1 - \ln \pi}$ or −23.690 **62.** $x = 4$ **63.** $x = 0$ **64.** $x = \ln 4$ or 1.386
65. 1 **66.** 0 **67.** 1 **68.** 0 **69.** 1 **70.** 0 **71.** 5 **72.** 3 **73.** −3 **74.** −1 **75.** −3 **76.** −2 **77.** −2 **78.** −1
79. −4 **80.** −1 **81.** −3 **82.** 3 **83.** 7 **84.** 4 **85.** −2 **86.** 4 **87.** 2 **88.** $1\frac{1}{2}$ **89.** 4 **90.** 8.700 **91.** 0.008
92. 1 **93.** 2 **94.** 4 **95.** −3 **96.** −1 **97.** −2 **98.** $\frac{1}{2}$ **99.** $-\frac{1}{2}$ **100.** 7 **101.** 1.751 **102.** 0.011 **103.** 0.787
104. 1 **105.** $x = 625$ **106.** $x = 2$ **107.** $x = 8$ **108.** $x = \frac{1}{16}$ **109.** $x = 9$ **110.** $x = \pm\sqrt{10}$ **111.** $y = 4$
112. $x = \sqrt{3}$ **113.** $x = 10\sqrt{10}$ **114.** $x = \frac{1}{12}$ **115.** $x = 2$ **116.** $x = 100$ or $x = 10$ **117.** $x = \frac{4}{3}$
118. $x = 1,000$ or $x = \frac{1}{10}$ **119.** $x = 2$ **120.** $x = 2$ (Hint: Change $\log_4 2x$ to base 2)
121. $x = 2$ (Hint: Change $\log_{16}(x + 2)$ to base 4) **122.** $x = 625$ (Hint: Change $\log_{25} x$ to base 5)
123. $x = e^4$ or 54.598 **124.** $x = e^{-4}$ or 0.018 **125.** $x = 1 + e$ or 3.718 **126.** $x = \frac{2}{1-e}$ or −1.164
127. $x = 1$ **128.** $x = 3 \pm \sqrt{5}$
129 ~ 134 Graphs are left to the students. (See page 253)
135. 20.9 years **136.** \$15,267.20 **137.** 14.33 years
138. 5.61 years **139.** 23.78 years **140.** 76.45 years
141. 3.47 years **142.** 13,353 years **143.** 10.14 hours
144. 13.86 years **145.** 9.9 years **146.** 6.93%
147. \$219.87 **148.** 1,230 units **149.** a. 82.60 b. 16.78 months

150. a. $y(1.5) = 0.1054$
b. $y(-1.5) = 0.1054$
c. $y_{max} = 1$

Graphing an exponential function
by a graphing calculator

Examples

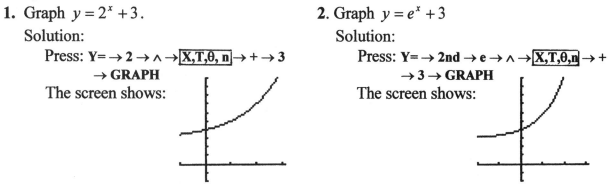

1. Graph $y = 2^x + 3$.

   Solution:

   Press: **Y= → 2 → ∧ → X,T,θ, n → + → 3**
   → GRAPH

   The screen shows:

2. Graph $y = e^x + 3$

   Solution:

   Press: **Y= → 2nd → e → ∧ → X,T,θ,n → +**
   → 3 → GRAPH

   The screen shows:

Graphing a logarithmic function
by a graphing calculator

3. Graph $y = \log_3(x - 1)$.

   Solution:

   $$y = \frac{\log_{10}(x - 1)}{\log_{10} 3}$$

   Press: **Y= → LOG (→ X,T,θ,n → − → 1 →)**
   → ÷ → LOG (→ 3 →) → GRAPH

   The screen shows:

4. Graph $y = \ln(x - 1)$

   Solution:

   Press: **Y= → LN (→ X,T,θ,n → − 1 →)**

   → GRAPH

   The screen shows:

Solving an exponential equation
by a graphing calculator

5. Solve $3^{1-2x} = 9^x$.

   Solution:

   $$y = 3^{1-2x} - 9^x$$

   Press: **Y= → 3 → ∧ → (→ 1 → − → 2 → X,T,θ,n →) → −**
   → 9 → ∧ → X,T,θ,n → GRAPH

   Press: **ZOOM → 6:Zstandard → ENTER**

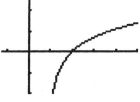

X=.24468085   Y=-.0645161

   Adjust the viewing window: [−1, 2] for x and [−1, 10] for y.

   Move the cursor ◄ and ► to estimate the x-intercept.

   The screen shows: ($x = 0.24468085$)

Finding a scatter plot and the best-fitting curve of an exponential function (a regression curve) by a graphing calculator

Example: Use a graphing calculator to find a scatter plot and the best-fitting curve of the following data.

x	0	1	2	3	4	5	6	7	8	9
y	15	12.6	10.6	8.9	7.5	6.3	5.3	4.4	3.7	3.1

Solution:

1. To enter the data:
Press: **STAT → 1:Edit → ENTER**
Move the cursor and enter the data
on the screen:

L1	L2
0	15
1	12.6
2	10.6
3	8.9
4	7.5
5	6.3
6	5.3
7	4.4
8	3.7
9	3.1

2. To adjust the size of the graph:
Press: **WINDOW**
Change the viewing window to:
$$x_{min} = 0 \ , \quad x_{max} = 10$$
$$y_{min} = 0 \ , \quad y_{max} = 18$$
Set the scale factors to:
$$x_{scl} = 1 \ , \ y_{scl} = 1$$

Press: **2nd → STAT PLOT → ENTER**
Select the scatter plot on the screen,
move cursor on it and press **ENTER**
to highlight it:

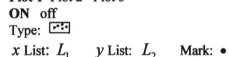

Plot 1 Plot 2   Plot 3
ON   off
Type:
x List: L_1 y List: L_2 Mark: •

3. To draw the scatter plot on the screen:
Press: **GRAPH**
The scatter plot shows on the screen.
It must be a polynomial or exponential
curve.

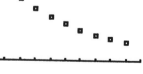

4. To graph the exponential function of best-fitting:
Press: **STAT → CALC → 0:ExpReg → ENTER**
Select L1 as the x list and L2 as the y list:

Press: **2nd → L1 → , → 2nd → L2 → ,**
Press: **VARS → Y–VARS → ENTER**
    → FUNCTION → ENTER
The screen shows: **ExpReg L1, L2, Y1 → ENTER**
The screen shows: **ExpReg**
    $y = a \times b^{\wedge}x$, $a = 15.04892315$, $b = 0.8393653732$
Press: **GRAPH** (the screen shows the best-fitting curve.)

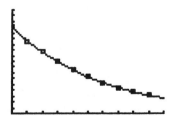

Additional   Example

Using three 2's to write any integer

Formula: a number $n = -\log_2 (\log_2 \sqrt{\sqrt{\sqrt{\cdots \sqrt{2}}}} \,)$

There are n square roots of 2 in the radicals.

Proof:

For n square roots of 2 in the expression:

$$-\log_2 (\log_2 \sqrt{\sqrt{\sqrt{\cdots \sqrt{2}}}} \,) = -\log_2 \cdot \log_2 [((2^{1/2})^{1/2})^{1/2} \cdots]$$

$$= -\log_2 \log_2 (2)^{\left(\frac{1}{2}\right)^n} = -\log_2 \left(\tfrac{1}{2}\right)^n = -\log_2 2^{-n} = -(-n) = n .$$

The proof is complete.

Examples: 1. If $n = 1$, $-\log_2 \log_2 \sqrt{2} = -\log_2 \log_2 2^{1/2} = -\log_2 (\tfrac{1}{2}) = -\log_2 2^{-1} = -(-1) = 1$.

2. If $n = 2$, $-\log_2 \log_2 \sqrt{\sqrt{2}} = -\log_2 \log_2 2^{1/4} = -\log_2 (\tfrac{1}{4}) = -\log_2 2^{-2} = -(-2) = 2$.

3. If $n = 3$, $-\log_2 \log_2 \sqrt{\sqrt{\sqrt{2}}} = -\log_2 \log_2 2^{1/8} = -\log_2 (\tfrac{1}{8}) = -\log_2 2^{-3} = -(-3) = 3$.

Before a test, a student asks his math teacher.
  Student: Could you please tell me the answers
              to the test questions ?
  Teacher: Do you promise to keep the answers
              secret ?
  Student: Yes, I do.
  Teacher: So do I.

Puzzle
A farmer wants to give his 17 lambs to his three
sons. Each son will have $\frac{1}{2}$, $\frac{1}{3}$, and $\frac{1}{9}$ of the
lambs. How could the farmer give the lambs to
his sons perfectly.

Space for Taking Notes

CHAPTER EIGHT

Analytic Geometry and Conic Sections

8-1 Distance and Midpoint Formulas

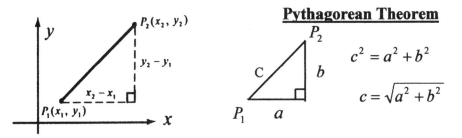

Pythagorean Theorem

$$c^2 = a^2 + b^2$$

$$c = \sqrt{a^2 + b^2}$$

Distance Formula: The distance between two points P_1 and P_2 :

$$\overline{P_1 P_2} = \sqrt{(x_2 - x_1)^2 + (y_2 - y_1)^2}$$

Midpoint Formula: The midpoint between two points P_1 and P_2 :

$$M = (\frac{x_1 + x_2}{2}, \frac{y_1 + y_2}{2})$$

Examples

1. Find the distance and midpoint between $P_1(-2, 1)$ and $P_2(5, 4)$.

Solution:

$$\overline{P_1 P_2} = \sqrt{(x_2 - x_1)^2 + (y_2 - y_1)^2} = \sqrt{[5 - (-2)]^2 + (4 - 1)^2} = \sqrt{49 + 9} = \sqrt{58} \ .$$

$$M = (\frac{x_1 + x_2}{2}, \frac{y_1 + y_2}{2}) = (\frac{-2 + 5}{2}, \frac{1 + 4}{2}) = (1.5 \ , \ 2.5).$$

2. Find the distance and midpoint between $(4, 3\sqrt{3})$ and $(2, -\sqrt{3})$.

Solution:

$$\overline{P_1 P_2} = \sqrt{(4 - 2)^2 + [3\sqrt{3} - (-\sqrt{3})]^2} = \sqrt{2^2 + (4\sqrt{3})^2} = \sqrt{4 + 48} = \sqrt{52} = 2\sqrt{13} \ .$$

$$M = (\frac{4 + 2}{2}, \frac{3\sqrt{3} + (-\sqrt{3})}{2}) = (3, \sqrt{3}).$$

3. If $M(1, 0)$ is the midpoint of the segment $\overline{PQ}$ and $Q(-1, -2)$ is the coordinates of Q. Find the coordinates of P.

Solution: $M(1, 0)$, $P(x, y)$, $Q(-1, -2)$

$$\frac{x + (-1)}{2} = 1 \quad \therefore \ x - 1 = 2 \ , \ x = 3.$$

$$\frac{y + (-2)}{2} = 0 \quad \therefore \ y - 2 = 0 \ , \ y = 2. \quad \text{Ans: The coordinates of } P \text{ is } P(3, 2).$$

EXERCISES

Find the distance between two points with the given coordinates.

1. $(1, 3)$, $(2, 6)$ **2.** $(3, 1)$, $(6, 2)$ **3.** $(-1, 3)$, $(2, -6)$

4. $(-1, -3)$, $(-2, 6)$ **5.** $(1, -3)$, $(-2, -6)$ **6.** $(-1, -3)$, $(-2, -6)$

7. $(5, 0)$, $(0, 7)$ **8.** $(5, 0)$, $(7, 0)$ **9.** $(5, 7)$, $(0, 0)$

10. $(0.5, -3)$, $(0.8, 2)$ **11.** $(1, \sqrt{3})$, $(3, 2\sqrt{3})$ **12.** $(0, 2\sqrt{3})$, $(\sqrt{3}, 0)$

13. $(2\sqrt{5}, -\sqrt{3})$, $(-\sqrt{5}, 2\sqrt{3})$ **14.** $(\sqrt{2}, 2)$, $(-2, \sqrt{2})$ **15.** $(\frac{1}{2}, 3)$, $(1, 4)$

16. $\left(\dfrac{3\sqrt{2}}{2}, \dfrac{\sqrt{3}}{3} \right)$, $\left(\dfrac{6\sqrt{2}}{2}, \dfrac{\sqrt{3}}{6} \right)$ **17.** $\left(\dfrac{\sqrt{3}}{3}, \dfrac{\sqrt{2}}{4} \right)$, $\left(\dfrac{-2\sqrt{3}}{3}, \dfrac{\sqrt{2}}{2} \right)$ **18.** $(2, \frac{2}{3})$, $(4, -\frac{1}{2})$

19. $(\sqrt{3}, \sqrt{2}+1)$, $(4\sqrt{3}, \sqrt{2}-5)$ **20.** $(\sqrt{5}-1, \sqrt{6})$, $(\sqrt{5}+3, 4\sqrt{6})$

Find the midpoint of the line segment between two points.

21. $(4, 6)$, $(2, 8)$ **22.** $(7, 5)$, $(11, 7)$ **23.** $(-3, 2)$, $(4, 6)$

24. $(3, -9)$, $(-7, 3)$ **25.** $(-5, -8)$, $(-7, -4)$ **26.** $(8, 0)$, $(12, 0)$

27. $(1.5, 15)$, $(2.5, 7)$ **28.** $(-2.5, 4.8)$, $(-4.7, -5.4)$ **29.** $(3, \sqrt{5})$, $(5, 3\sqrt{5})$

30. $(\sqrt{5}, 4)$, $(-3\sqrt{5}, -6)$ **31.** $(4, \frac{1}{2})$, $(5, \frac{3}{2})$ **32.** $(\frac{1}{2}, -\frac{2}{3})$, $(\frac{1}{2}, 2\frac{2}{3})$

33. $(\sqrt{3}, \sqrt{2}+5)$, $(3\sqrt{3}, \sqrt{2}-5)$ **34.** $(\sqrt{2}+3, 4)$, $(2\sqrt{2}-3, -6)$

35. $(\sqrt{7}+11, \sqrt{5}-12)$, $(3\sqrt{7}-11, 5\sqrt{5}+12)$

36. If $M(3, 2)$ is the midpoint of the line segment $\overline{PQ}$ and $Q(-5, 4)$ is the coordinates of one endpoint, find the coordinates of the other endpoint.

37. If $M(\frac{2}{3}, \frac{1}{4})$ is the midpoint of the line segment $\overline{AB}$ and $A(\frac{5}{6}, 6)$ is the coordinates of one endpoint, find the coordinates of the other endpoint.

38. The line segment $\overline{AB}$ is the diameter of a circle. If the center is at $(4, -2)$ and A is at $(-1, 5)$, Find the coordinates of B.

39. Find the equation of the line that passes $(3, -2)$ and is perpendicular to the line $2x + y = 1$.

40. Find the equation of the line that is perpendicular bisector of the line segment between two points $(5, -1)$ and $(3, 7)$.

41. The vertices of a triangle are $(3, 3)$, $(-3, 3)$ and $(0, -1)$. Identify the triangle as scalene, isosceles, or equilateral.

8-2 Conic Sections and Parabolas

A conic section is curve that can be formed by slicing a right circular cone with a plane. The figure of the intersection of the cone and the plane is a conic section.

By changing the position of the plane (horizontally or slant) , we obtain four types of conic sections, **a circle**, **an ellipse**, **a parabola**, or part of **a hyperbola**.

In this chapter, we will learn to obtain an equation for each of the conic sections by applying its geometric definition.

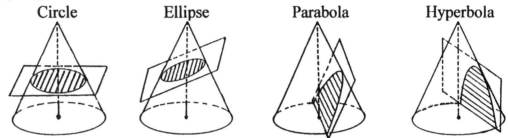

Circle   Ellipse   Parabola   Hyperbola

If the intersection of the cone and plane is a point, a line, or two lines that lie on the plane, these are called **degenerate conics**. (See page 284)

In Section 5~3, we have learned the basic equation of a parabola $y = ax^2 + bx + c$ $(a \neq 0)$.

In this section, we will learn to derive equations for parabolas by applying a general definition of a parabola and learn how to graph a parabola.

Definition of a parabola: A parabola is the set of all points equidistant from a fixed line (called the directrix) and a fixed point (called the focus).

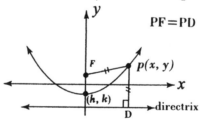

Examples

1. Find the equation of a parabola whose focus is the point $(0, 2)$ and whose directrix is the line $y = -4$.

Solution:

$$PF = PD$$

$$\sqrt{(x-0)^2 + (y-2)^2} = \sqrt{(x-x)^2 + (y+4)^2}$$

$$x^2 + (y-2)^2 = (y+4)^2$$

$$x^2 + y^2 - 4y + 4 = y^2 + 8y + 16$$

$$x^2 = 12y + 12$$

$$x^2 = 12(y+1) \checkmark$$

$$\therefore y + 1 = \tfrac{1}{12}x^2. \text{ Ans.}$$

Vertex $(0, -1)$.

$a = \tfrac{1}{12} > 0$ open upward.

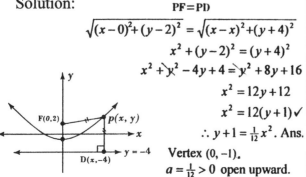

2. Find the equation of a parabola whose focus is the point $(0, 3)$ and whose directrix is the line $x = 4$.

Solution:

$$PF = PD$$

$$\sqrt{(x-0)^2 + (y-3)^2} = \sqrt{(x-4)^2 + (y-y)^2}$$

$$x^2 + (y-3)^2 = (x-4)^2$$

$$x^2 + (y-3)^2 = x^2 - 8x + 16$$

$$(y-3)^2 = -8x + 16$$

$$(y-3)^2 = -8(x-2) \checkmark$$

$$\therefore x - 2 = -\tfrac{1}{8}(y-3)^2. \text{ Ans.}$$

Vertex $(2, 3)$,

$a = -\tfrac{1}{8} < 0$, open to the left.

Parabola is commonly applied to the design of a car's headlight. A small light bulb is placed at the focus of the headlight mirror which is shaped like a parabola. When the light rays from the small bulb hit the mirror, the rays are collected and reflected as parallel lines of light creating an amplified light beam.

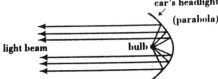

In Section 5~3, we have learned the parabolas that have a vertical axis and open upward or downward. Now, we will also learn the parabola that have a horizontal axis and open to the right or to the left. The standard form of each parabola shown below are useful to locate the vertex, the focus, the directrix, and latus rectum.

Standard Forms of Parabolas

Latus Rectum: The line segment passing through the focus and perpendicular to the axis.

1. $y^2 = 4ax$

   vertex: (0, 0), open to the right
   focus : $(a, 0)$
   directrix: $x = -a$
   latus rectum: $4a$

2. $(y - k)^2 = 4a(x - h)$

   vertex: (h, k), open to the right
   focus: $(h + a, k)$
   directrix: $x = h - a$
   latus rectum: $4a$

3. $x^2 = 4ay$

   vertex: (0, 0), open upward
   focus: $(0, a)$
   diretrix: $y = -a$
   latus rectum: $4a$

4. $(x - h)^2 = 4a(y - k)$

   vertex: (h, k), open upward
   focus: $(h, k + a)$
   diretrix: $y = k - a$
   latus rectum: $4a$

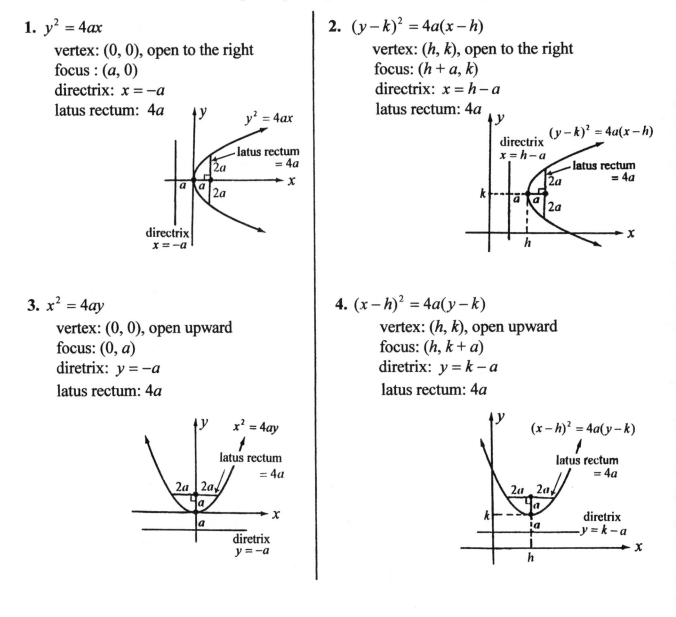

5. $y^2 = -4ax$
 	
   vertex: $(0, 0)$, open to the left
   focus: $(-a, 0)$
   diretrix: $x = a$
   latus rectum: $4a$
	6. $(y - k)^2 = -4a(x - h)$
	
	   vertex: (h, k), open to the left
	   focus: $(h - a, k)$
	   diretrix: $x = h + a$
	   latus rectum: $4a$

7. $x^2 = -4ay$
 	
   vertex: $(0, 0)$, open downward
   focus: $(0, -a)$
   diretrix: $y = a$
   latus rectum: $4a$
	8. $(x - h)^2 = -4a(y - k)$
	
	   vertex: (h, k), open downward
	   focus: $(h, k - a)$
	   diretrix: $y = k + a$
	   latus rectum: $4a$

Examples

1. Find an equation of the parabola with vertex $(0, 0)$ and focus $(2, 0)$.

Solution: Sketch and Find:

It is open to the right.

$y^2 = 4ax$ and $a = 2$

$y^2 = 4(2)x = 8x$

$\therefore y^2 = 8x$ is the equation.

2. Find an equation of the parabola with vertex $(0, 0)$ and focus $(-2, 0)$.

Solution: Sketch and Find:

It is open to the left.

$y^2 = -4ax$ and $a = 2$

$y^2 = -4(2)x = -8x$

$\therefore y^2 = -8x$ is the equation.

3. Find the equation of a parabola whose focus is the point $(0, 4)$ and whose directrix is the line $y = -4$.

Solution: Sketch and find:

It is open upward. vertex $(0, 0)$

$x^2 = 4ay$ and $a = 4$

$x^2 = 4(4)y = 16y$

$\therefore x^2 = 16y$ is the equation.

4. Find the equation of a parabola whose focus is the point $(0, 2)$ and whose directrix is the line $y = -4$.

Solution: Sketch and find:

It is open upward. vertex $(0, -1)$

$(x - h)^2 = 4a(y - k)$ and $a = 3$

$(x - 0)^2 = 4(3)(y + 1)$

$\therefore x^2 = 12((y + 1)$ is the equation.

5. Find the vertex, focus, diretrix, and latus rectum of the parabola $x^2 + 2x + 2y - 4 = 0$.

Solution: Completing the square

$(x^2 + 2x + 4) - 4 + 2y - 4 = 0$, $(x + 2)^2 + 2y - 8 = 0$

$(x + 2)^2 = -2(y - 4)$, open downward

$-4a = -2 \therefore a = \frac{1}{2}$

vertex: $(h, k) = (-2, 4)$

focus: $(h, k - a) = (-2, \frac{7}{2})$

directrix: $y = k + a$, $y = \frac{9}{2}$

latus rectum: $4a = 2$

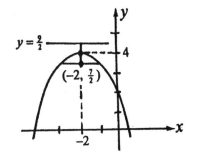

EXERCISES

The coordinates of the vertex and a point of a parabola are given. Write an equation in standard form for each parabola.

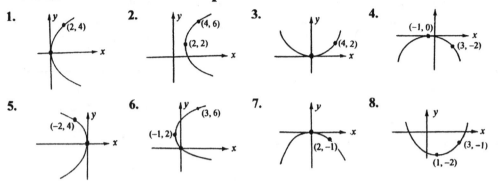

Write the equation of each parabola having the given conditions. (Hint: Sketch it.)

9. Vertex $(0, 0)$; Focus $(3. 0)$ **10.** Vertex $(0, 0)$; Focus $(-3, 0)$

11. Vertex $(0, 0)$; Focus $(0, 3)$ **12.** Vertex $(0, 0)$; Focus $(0, -3)$

13. Vertex $(2, 3)$; Focus $(4, 3)$ **14.** Vertex $(2, 3)$; Focus $(0, 3)$

15. Vertex $(2, 3)$; Focus $(2, 5)$ **16.** Vertex $(2, 3)$; Focus $(2, 1)$

17. Vertex $(2, 1)$; Focus $(2, 4)$ **18.** Vertex $(-2, 3)$; Focus $(0, 3)$

19. Vertex $(2, -3)$; Focus $(2, -5)$ **20.** Vertex $(4, -2)$; Focus $(2, -2)$

21. Vertex $(0, 0)$; Focus $(0, -\frac{3}{2})$ **22.** Vertex $(0, 0)$; Focus $(\frac{5}{8}, 0)$

23. Vertex $(0, 0)$; Directrix $x = -4$ **24.** Vertex $(0, 0)$; Directrix $x = 4$

25. Vertex $(0, 0)$; Directrix $y = -4$ **26.** Focus $(0, 0)$; Directrix $y = 4$

27. Vertex $(0, 0)$; Directrix $y = 8$ **28.** Focus $(2, 4)$; Directrix $x = -4$

29. Vertex $(0, -1)$; Directrix $y = 1$ **30.** Focus $(-3, 4)$; Directrix $y = 2$

Find the vertex, focus, directrix, latus rectum, and the direction of opening of each parabola.

31. $y^2 = 32x$ **32.** $x^2 = -6y$ **33.** $x^2 = \frac{5}{2}y$

34. $y^2 = -\frac{9}{4}x$ **35.** $(y-4)^2 = 16(x-1)$ **36.** $(x-3)^2 = 12(y+6)$

37. $(y+2)^2 = -4(x-5)$ **38.** $(x+1)^2 = -(y+7)$ **39.** $x^2 + 8x = 4y - 8$

40. $y^2 - 4y = -4x - 4$ **41.** $x^2 + 6x - 4y + 21 = 0$ **42.** $y^2 - 8y - 12 + 4 = 0$

43. The satellite TV dish antenna is shaped like a parabola. The dish is 12 feet in diameter at its opening and is 4 feet deep at its center. Find the position of the receiver from the base (vertex) of the dish to the focus.

44. The equation of the parabola is $y^2 = 4ax$. Prove that the equation of the line tangent to the point $p(x_0, y_0)$ on the parabola is $y_0 y = 2a(x_0 + x)$. (Hint: $D = b^2 - 4ac = 0$)

45. Use the formula described in Problem 44 to find the equation of the tangent line to the point $(2, 4)$ on the parabola $y^2 = 8x$. (See Chart of Equations of Tangent lines on page 301)

8-3 Circles

Definition of a circle: A circle is the set of all points in a plane that are a fixed distance r (called the radius) from a fixed point (called the center).

Standard Forms of Circles

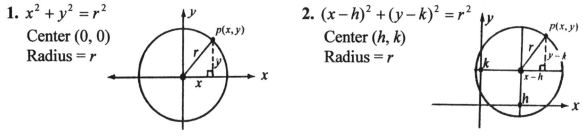

1. $x^2 + y^2 = r^2$

   Center $(0, 0)$

   Radius $= r$

2. $(x-h)^2 + (y-k)^2 = r^2$

   Center (h, k)

   Radius $= r$

When the equation of a circle is given in the form $x^2 + y^2 + ax + by + c = 0$, we rewrite it to match the standard form by the method of completing the square. Then find its center and radius.

Examples

1. Find the center and radius of $x^2 + y^2 = 25$.
   Solution:

   $$x^2 + y^2 = 25$$
   $$x^2 + y^2 = 5^2$$
   $$\therefore \text{ Center } (0, 0)$$
   $$\text{Radius } r = 5$$

2. Find the equation of the circle with center $(0, 0)$ and radius $3\sqrt{2}$.
   Solution:

   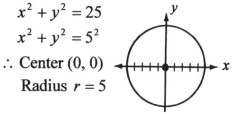

   $$x^2 + y^2 = r^2$$
   $$x^2 + y^2 = (3\sqrt{2})^2$$
   $$\therefore x^2 + y^2 = 18. \text{ Ans.}$$

3. Find the equation of the circle with center $(-2, 1)$ and radius 3.
   Solution:

   $$(x-h)^2 + (y-k)^2 = r^2$$
   $$(x+2)^2 + (y-1)^2 = 3^2 \text{ (Standard form)}$$
   $$x^2 + 4x + 4 + y^2 - 2y + 1 = 9$$
   $$\therefore x^2 + y^2 + 4x - 2y - 4 = 0 \text{ (General form)}$$

4. Graph $x^2 + y^2 + 4x - 8y + 11 = 0$
   Solution: Completing the square

   $$x^2 + y^2 + 4x - 8y + 11 = 0$$
   $$(x^2 + 4x) + (y^2 - 8y) + 11 = 0$$
   $$(x^2 + 4x + 4) - 4 + (y^2 - 8y + 16)$$
   $$-16 + 11 = 0$$
   $$(x+2)^2 + (y-4)^2 = 9$$
   $$(x+2)^2 + (y-4)^2 = 3^2$$
   $$\therefore \text{Center } (-2, 4), \text{ Radius} = 3$$

5. Write the equation of a circle if the endpoints of the diameter are at $(5, 4)$ and $(-1, -2)$.
   Solution:

   $$\text{Center} = \left(\tfrac{5+(-1)}{2}, \tfrac{4+(-2)}{2}\right) = (2, 1)$$
   $$\text{Radius } r = \sqrt{(5-2)^2 + (4-1)^2} = \sqrt{18}$$
   $$\therefore (x-2)^2 + (y-1)^2 = 18. \text{ Ans.}$$

EXERCISES

Write the equation in standard form of each circle.

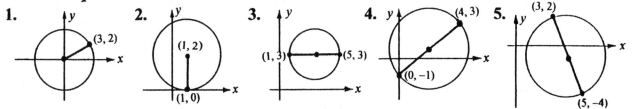

1. (3, 2)
2. (1, 2) ; (1, 0)
3. (1, 3) ; (5, 3)
4. (4, 3) ; (0, −1)
5. (3, 2) ; (5, −4)

Write the standard form of the equation of each circle.

6. $r = 1$; Center $(0, 0)$ **7.** $r = 1$; Center $(0, 5)$ **8.** $r = 1$; Center $(-5, 0)$

9. $r = 1$; Center $(3, 5)$ **10.** $r = 3$; Center $(5, -3)$ **11.** $r = 5$; Center $(-3, 5)$

12. $r = 8$; Center $(-1, -6)$ **13.** $r = \frac{1}{2}$; Center $(4, -2)$ **14.** $r = \frac{1}{2}$; Center $(\frac{1}{2}, 0)$

Find the center and radius of each circle.

15. $x^2 + y^2 = 16$ **16.** $x^2 + y^2 = 18$ **17.** $5x^2 + 5y^2 = 40$

18. $(x-4)^2 + y^2 = 36$ **19.** $x^2 + (y+6)^2 = 36$ **20.** $(x+4)^2 + (y-6)^2 = 18$

21. $3(x-4)^2 + 3y^2 = 54$ **22.** $4x^2 + 4(y+6)^2 = 40$ **23.** $9(x+4)^2 + 9(y-6)^2 = 90$

24. $x^2 + y^2 - 6x + 5 = 0$ **25.** $x^2 + y^2 + 4y = 0$ **26.** $x^2 + y^2 - 2x + 4y - 4 = 0$

27. $x^2 + y^2 + 2x + 4y = 9$ **28.** $x^2 + y^2 - 8x + 12y = -44$ **29.** $x^2 + 2\sqrt{5}x + y^2 = 4$

30. Find the equation in general form of the circle with center $(3, 5)$ and tangent to the $x-$axis. (Hint: tangent line to a circle is the line that intersects the circle at only a single point.)

31. Find the equation in general form of the circle with center $(3, 5)$ and tangent to the $y-$axis.

32. Find the equation of the line that is tangent to the circle $x^2 + y^2 = 34$ at point $(3, 5)$.

33. Find the equation of the line that is tangent to the circle $x^2 + y^2 = 34$ at point $(3, -5)$.

34. Find the equation of the line that is tangent to the circle $x^2 + y^2 = 16$ at point $(3, \sqrt{7})$.

35. Find the equation of the line that is tangent to the circle $x^2 + y^2 - 8x + 12y + 42 = 0$ at $(5, -3)$.

36. Write the equation of the circle if the endpoints of the diameter are at $(4, 5)$ and $(-2, -1)$.

37. If the equation of the circle is $x^2 + y^2 = r^2$ and the point of tangent is $p(x_0, y_0)$, prove that the equation of the tangent line is $x_0 x + y_0 y = r^2$.

38. Use the formula described in Problem 37 to find the equation of the tangent line to the circle $x^2 + y^2 = 34$ and the point of tangent is $(3, 5)$.

39. If the equation of the circle is $x^2 + y^2 = r^2$ and the slope of the tangent line is m, prove that: the equation of the tangent line is $y = mx \pm r\sqrt{1+m^2}$. (Hint: $D = b^2 - 4ac = 0$)

40. Use the formula described in Problem 39 to find the equation of the tangent line to the circle $x^2 + y^2 = 4$ from the point $(0, 5)$ not on the circle.(See Chart of Tangent lines on page 301)

8-4 Ellipses

Definition of an ellipse: An ellipse is the set of all points in a plane such that for each point, the sum of the distances (called the focal radii) from two fixed points (called the foci) is a constant.

Standard Forms of Ellipses

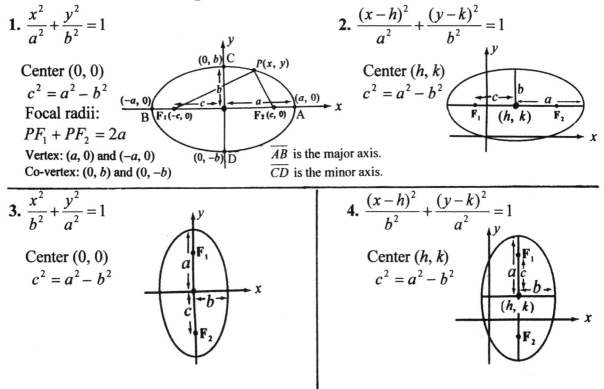

1. $\dfrac{x^2}{a^2} + \dfrac{y^2}{b^2} = 1$

Center $(0, 0)$

$c^2 = a^2 - b^2$

Focal radii:

$PF_1 + PF_2 = 2a$

Vertex: $(a, 0)$ and $(-a, 0)$

Co-vertex: $(0, b)$ and $(0, -b)$

$\overline{AB}$ is the major axis.

$\overline{CD}$ is the minor axis.

2. $\dfrac{(x-h)^2}{a^2} + \dfrac{(y-k)^2}{b^2} = 1$

Center (h, k)

$c^2 = a^2 - b^2$

3. $\dfrac{x^2}{b^2} + \dfrac{y^2}{a^2} = 1$

Center $(0, 0)$

$c^2 = a^2 - b^2$

4. $\dfrac{(x-h)^2}{b^2} + \dfrac{(y-k)^2}{a^2} = 1$

Center (h, k)

$c^2 = a^2 - b^2$

The ellipse is commonly applied to the movement of the earth. The earth's orbit around the sun is shaped as an ellipse in which the sun is located and fixed at one of its two foci.

Examples

1. Find the equation of an ellipse having foci $(0, 3)$ and $(0, -3)$ and the sum of its focal radii is 10.

Solution:

$$PF_1 + PF_2 = 10$$
$$\sqrt{(x-0)^2 + (y-3)^2} + \sqrt{(x-0)^2 + (y+3)^2} = 10$$
$$\sqrt{x^2 + (y-3)^2} = 10 - \sqrt{x^2 + (y+3)^2}$$
$$x^2 + (y-3)^2 = 100 - 20\sqrt{x^2 + (y+3)^2} + x^2 + (y+3)^2$$
$$y^2 - 6y + 9 = 100 - 20\sqrt{x^2 + (y+3)^2} + y^2 + 6y + 9$$
$$12y + 100 = 20\sqrt{x^2 + (y+3)^2}$$
$$3y + 25 = 5\sqrt{x^2 + y^2 + 6y + 9}$$
$$9y^2 + 150y + 625 = 25(x^2 + y^2 + 6y + 9)$$
$$9y^2 + 150y + 625 = 25x^2 + 25y^2 + 150y + 225$$
$$25x^2 + 16y^2 = 400$$
$$\frac{25x^2}{400} + \frac{16y^2}{400} = 1 \quad \therefore \quad \frac{x^2}{16} + \frac{y^2}{25} = 1 \text{ is the equation. Ans.}$$

2. Graph $9x^2 + 25y^2 = 225$.

Solution:

$9x^2 + 25y^2 = 225$

Divided both sides by 225:

$\dfrac{9x^2}{225} + \dfrac{25y^2}{225} = 1$

$\dfrac{x^2}{25} + \dfrac{y^2}{9} = 1$

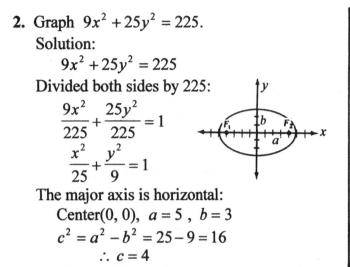

The major axis is horizontal:

Center(0, 0), $a = 5$, $b = 3$

$c^2 = a^2 - b^2 = 25 - 9 = 16$

$\therefore c = 4$

3. Graph $25x^2 + 9y^2 = 225$.

Solution:

$25x^2 + 9y^2 = 225$

Divided both sides by 225:

$\dfrac{25x^2}{225} + \dfrac{9y^2}{225} = 1$

$\dfrac{x^2}{9} + \dfrac{y^2}{25} = 1$

The major axis is vertical:

Center(0, 0), $a = 5$, $b = 3$

$c^2 = a^2 - b^2 = 25 - 9 = 16$

$\therefore c = 4$

4. Graph $4x^2 + y^2 - 16x + 10y + 25 = 0$

Solution:

Completing the square:

$4(x^2 - 4x) + (y^2 + 10y) + 25 = 0$

$4[(x^2 - 4x + 4) - 4] + (y^2 + 10y + 25) - 25 + 25 = 0$

$4[(x - 2)^2 - 4] + (y + 5)^2 = 0$

$4(x - 2)^2 + (y + 5)^2 = 16$

Divided both sides by 16:

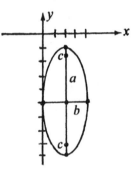

$\dfrac{(x - 2)^2}{4} + \dfrac{(y + 5)^2}{16} = 1$, The major axis is vertical.

Center $(2, -5)$; $a = 4$; $b = 2$

$c^2 = a^2 - b^2 = 16 - 4 = 12$ $\therefore c = \sqrt{12} = 3.36$

5. Find the equation of an ellipse having foci $(0, 3)$ and $(0, -3)$ and the sum of its focal radii is 10.

Solution: (See Example 1)

Center $(0, 0)$ and $c = 3$

$2a = 10$ $\therefore a = 5$

$c^2 = a^2 - b^2$

$3^2 = 5^2 - b^2$

$b^2 = 16$ $\therefore b = 4$

The major axis is vertical:

$\dfrac{x^2}{b^2} + \dfrac{y^2}{a^2} = 1$ $\therefore \dfrac{x^2}{16} + \dfrac{y^2}{25} = 1$. Ans.

(See graph on Example 1)

6. Find the equation of an ellipse having vertices at $(0, 5)$ and $(0, -5)$ and foci at $(0, 3)$ and $(0, -3)$.

Solution: (See Example 1)

Center $(0, 0)$ and $a = 5$, $c = 3$

We find b:

$c^2 = a^2 - b^2$

$3^2 = 5^2 - b^2$

$b^2 = 16$ $\therefore b = 4$

The major axis is vertical:

$\dfrac{x^2}{b^2} + \dfrac{y^2}{a^2} = 1$ $\therefore \dfrac{x^2}{16} + \dfrac{y^2}{25} = 1$

(See graph on Example 1)

7. Find the equation of an ellipse passing through the point (3, 5) and having the foci at (3, 2) and (−1, 2).

Solution:

We graph and find: Center (1, 2) and $c = 2$

The major axis is horizontal.

$$\frac{(x-1)^2}{a^2} + \frac{(y-2)^2}{b^2} = 1$$

Passing the point (3, 5), we apply the definition of an ellipse:

$$\sqrt{(5-2)^2 + (3-3)^2} + \sqrt{(5-2)^2 + (3+1)^2} = 2a$$

$$\sqrt{9} + \sqrt{25} = 2a$$

$$3 + 5 = 2a \quad \therefore a = 4$$

Since $c^2 = a^2 - b^2$, $c = 2$ we have $4 = 16 - b^2$ $\therefore b = \sqrt{12}$

$$\therefore \frac{(x-1)^2}{16} + \frac{(y-2)^2}{12} = 1. \text{ Ans.}$$

8. The earth's orbit around the sun is shaped as an ellipse in which the sun is located and fixed at one of its two foci. At its closest distance (called the perihelion), the earth is 91.45 million miles from the center of the sun. At its farthest distance (called the aphelion), the earth is 94.55 million miles from the center of the sun. Write the equation of the orbit.

Solution:

$$2a = 91.45 + 94.55 = 186 \text{ million miles}$$

$$\therefore a = 93 \text{ million miles}$$

We have: $c = 93 - 91.45 = 1.55 \text{ million miles}$

$$c^2 = a^2 - b^2$$

$$(1.55)^2 = (93)^2 - b^2$$

$$b^2 = (93)^2 - (1.55)^2 = 8646.60$$

$$\frac{x^2}{a^2} + \frac{y^2}{b^2} = 1, \quad \frac{x^2}{(93)^2} + \frac{y^2}{8646.60} = 1 \quad \therefore \frac{x^2}{8649} + \frac{y^2}{8646.60} = 1 \text{ (in million miles).}$$

It is nearly circular.

On graduation day, a high school graduate said to his math teacher.
Student: Thank you very much for giving me a "D" in your class.
        For your kindness, do you want me to do anything for you ?
Teacher: Yes. Please do me a favor.
        Just don't tell anyone that I was your math teacher.

EXERCISES

Find the center, vertices on the major axis, and foci of each ellipse. Graph each equation.

1. $\dfrac{x^2}{25}+\dfrac{y^2}{9}=1$

2. $\dfrac{x^2}{9}+\dfrac{y^2}{25}=1$

3. $\dfrac{(x-1)^2}{25}+\dfrac{(y+4)^2}{9}=1$

4. $\dfrac{x^2}{4}+\dfrac{y^2}{16}=1$

5. $\dfrac{x^2}{16}+\dfrac{y^2}{4}=1$

6. $\dfrac{(x+1)^2}{4}+\dfrac{(y-4)^2}{16}=1$

7. $x^2+\dfrac{y^2}{4}=1$

8. $\dfrac{x^2}{4}+y^2=1$

9. $(x-3)^2+\dfrac{y^2}{4}=1$

10. $9x^2+4y^2=36$

11. $25x^2+16y^2=400$

12. $4(x+3)^2+16(y+2)^2=64$

13. $9x^2+y^2-18x=0$

14. $x^2+4y^2+8y-12=0$

15. $x^2+4y^2-2x+16y+1=0$

Write the equation in standard form for each ellipse having the given conditions. (Hint: Sketch and Find)

16. Center: (0, 0)
    Vertex: (5, 0)
    Co-vertex: (0, 2)

17. Center: (0, 0)
    Vertex: (0, −5)
    Co-vertex: (2, 0)

18. Center: (0, 0)
    Vertex: (0, 8)
    Focus: (0, 5)

19. Center: (0, 0)
    Co-vertex: (0, 1)
    Focus: (2, 0)

20. Center: (2, −1)
    Vertex: (8, −1)
    Focus: (6, −1)

21. Length of major axis: 10
    Foci: (6, 1) and (−2, 1)

22. Center: (1, 2)
    Vertex: (1, 4)
    Passing through the point (2, 2)

Using a graphing calculator to graph each ellipse.

23. $4x^2+16y^2=64$

24. $10x^2+4y^2=40$

25. $4x^2+y^2=16$

26. $4y^2+x^2=8$

27. $x^2+4y^2-2x+16y+13=0$

28. $9x^2+4y^2+18x-8y-23=0$

29. The moon's orbit around the earth is shaped as an ellipse in which the earth is located and fixed at one of its two foci. At its closest distance (called the perigee), the moon is 364.20 million meters from the center of the earth. At its farthest distance (called the apogee), the moon is 404.60 million meters from the center of the earth. Write the equation of the orbit.

30. The equation of the ellipse is $\dfrac{x^2}{a^2}+\dfrac{y^2}{b^2}=1$ and the slope of the tangent line is m. Prove that the equation of the tangent line $y=mx\pm\sqrt{a^2m^2+b^2}$. (Hint: $D=b^2-4ac=0$)

31. The equation of the ellipse is $\dfrac{x^2}{a^2}+\dfrac{y^2}{b^2}=1$. Prove that the equation of the line tangent to the point $p(x_0,\ y_0)$ on the ellipse is $\dfrac{x_0 x}{a^2}+\dfrac{y_0 y}{b^2}=1$.

32. Find the equation of tangent line tangent to the point $(5, \frac{9}{5})$ on the ellipse $9x^2+25y^2=225$. (See Chart of the Equations of Tangent line on page 301)

8-5 Hyperbolas

Definition of a hyperbola: A hyperbola is the set of all points in a plane such that for each point., the difference of the distances (called the focal radii) from two fixed points (called the foci) is a constant.

Standard Forms of Hyperbolas

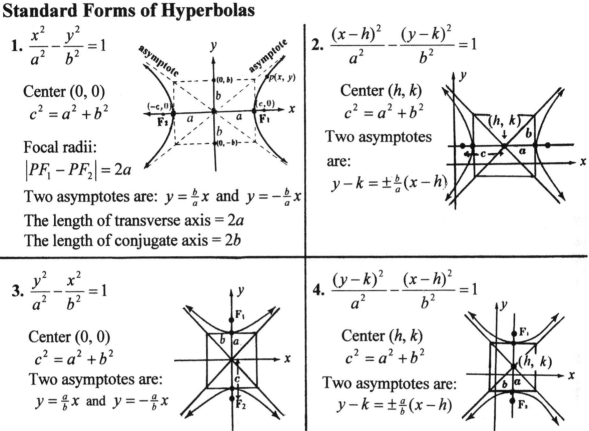

1. $\dfrac{x^2}{a^2} - \dfrac{y^2}{b^2} = 1$

Center $(0, 0)$

$c^2 = a^2 + b^2$

Focal radii:

$|PF_1 - PF_2| = 2a$

Two asymptotes are: $y = \frac{b}{a}x$ and $y = -\frac{b}{a}x$

The length of transverse axis $= 2a$
The length of conjugate axis $= 2b$

2. $\dfrac{(x-h)^2}{a^2} - \dfrac{(y-k)^2}{b^2} = 1$

Center (h, k)

$c^2 = a^2 + b^2$

Two asymptotes are:

$y - k = \pm\frac{b}{a}(x-h)$

3. $\dfrac{y^2}{a^2} - \dfrac{x^2}{b^2} = 1$

Center $(0, 0)$

$c^2 = a^2 + b^2$

Two asymptotes are:

$y = \frac{a}{b}x$ and $y = -\frac{a}{b}x$

4. $\dfrac{(y-k)^2}{a^2} - \dfrac{(x-h)^2}{b^2} = 1$

Center (h, k)

$c^2 = a^2 + b^2$

Two asymptotes are:

$y - k = \pm\frac{a}{b}(x-h)$

Examples

1. Find the equation of a hyperbola having foci $(0, 3)$ and $(0, -3)$ and the difference of its focal radii is 4.

Solution:

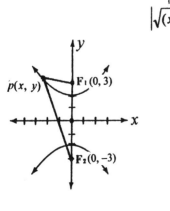

$$|PF_1 - PF_2| = 4$$

$$\left|\sqrt{(x-0)^2 + (y-3)^2} - \sqrt{(x-0)^2 + (y+3)^2}\right| = 4$$

$$\sqrt{x^2 + (y-3)^2} - \sqrt{x^2 + (y+3)^2} = \pm 4$$

$$\sqrt{x^2 + (y-3)^2} = \sqrt{x^2 + (y+3)^2} \pm 4$$

$$x^2 + (y-3)^2 = x^2 + (y+3)^2 \pm 8\sqrt{x^2 + (y+3)^2} + 16$$

$$y^2 - 6y + 9 = y^2 + 6y + 9 \pm 8\sqrt{x^2 + (y+3)^2} + 16$$

$$12y + 16 = \pm 8\sqrt{x^2 + (y+3)^2}$$

$$3y + 4 = \pm 2\sqrt{x^2 + (y+3)^2}$$

$$9y^2 + 24y + 16 = 4(x^2 + y^2 + 6y + 9)$$

$$9y^2 + 24y + 16 = 4x^2 + 4y^2 + 24y + 36$$

$$5y^2 - 4x^2 = 20$$

$$\frac{5y^2}{20} - \frac{4x^2}{20} = 1 \qquad \therefore \ \frac{y^2}{4} - \frac{x^2}{5} = 1 \text{ is the equation. Ans.}$$

2. Graph $16x^2 - 9y^2 = 144$.

Solution:

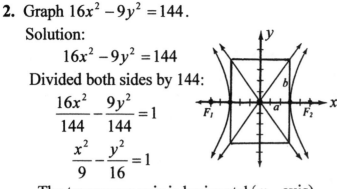

$$16x^2 - 9y^2 = 144$$

Divided both sides by 144:

$$\frac{16x^2}{144} - \frac{9y^2}{144} = 1$$

$$\frac{x^2}{9} - \frac{y^2}{16} = 1$$

The transverse axis is horizontal (x – axis).

Center $(0, 0)$, $a = 3$, $b = 4$

$$c^2 = a^2 + b^2 = 9 + 16 = 25 \quad \therefore c = 5$$

3. Graph $4y^2 - 9x^2 = 36$.

Solution:

$$4y^2 - 9x^2 = 36$$

Divided both sides by 36:

$$\frac{4y^2}{36} - \frac{9x^2}{36} = 1$$

$$\frac{y^2}{9} - \frac{x^2}{4} = 1$$

The transverse axis is vertical (y – axis).

Center $(0, 0)$, $a = 3$, $b = 2$

$$c^2 = a^2 + b^2 = 9 + 4 = 13 \quad \therefore c = \sqrt{13}$$

4. Graph $y^2 - 2x^2 + 6y + 8x - 3 = 0$

Solution:

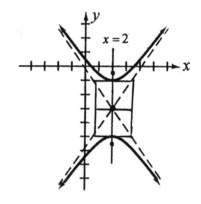

Completing the square:

$$(y^2 + 6y) - 2(x^2 - 4x) - 3 = 0$$

$$(y^2 + 6y + 9) - 9 - 2[(x^2 - 4x + 4) - 4] - 3 = 0$$

$$(y + 3)^2 - 9 - 2(x - 2)^2 + 8 - 3 = 0$$

$$(y + 3)^2 - 2(x - 2)^2 = 4$$

Divided both sides by 4:

$$\frac{(y + 3)^2}{4} - \frac{(x - 2)^2}{2} = 1$$

The transverse axis is vertical ($x = 2$).

Center $(2, -3)$, $a = 2$, $b = \sqrt{2}$

$$c^2 = a^2 + b^2 = 4 + 2 = 6 \quad \therefore c = \sqrt{6}$$

5. Find the equation of a hyperbola having foci $(0, 3)$ and $(0, -3)$ and the difference of its focal radii is 4.

Solution:

Center $(0, 0)$, $c = 3$

$$2a = 4 \quad \therefore a = 2$$

We find b: $c^2 = a^2 + b^2$

$$3^2 = 2^2 + b^2$$

$$b^2 = 5 \quad \therefore b = \sqrt{5}$$

The transverse axis is vertical (y – axis).

$$\frac{y^2}{a^2} - \frac{x^2}{b^2} = 1 \quad \therefore \frac{y^2}{4} - \frac{x^2}{5} = 1. \text{ Ans.}$$

(See graph on Example 1)

6. Find the equation of a hyperbola having vertices at $(0, 2)$ and $(0, -2)$ and foci $(0. 3)$ and $(0, -3)$.

Solution:

Center $(0, 0)$ and $a = 2$, $c = 3$

We find b: $c^2 = a^2 + b^2$

$$3^2 = 2^2 + b^2$$

$$b^2 = 5$$

$$\therefore b = \sqrt{5}$$

The transverse axis is vertical (y – axis).

$$\frac{y^2}{a^2} - \frac{x^2}{b^2} = 1 \quad \therefore \frac{y^2}{4} - \frac{x^2}{5} = 1. \text{ Ans.}$$

(See graph on Example 1)

7. Find the equation of a hyperbola passing through the point $(1, -5)$ and having the foci at $(-2, -1)$ and $(-2, -5)$.

Solution:

We graph and find: Center $(-2, -3)$ and $c = 2$

The transverse axis is vertical.

We have the equation: $\dfrac{(y+3)^2}{a^2} - \dfrac{(x+2)^2}{b^2} = 1$

Passing the point $(1, -5)$, we apply the definition of a hyperbola:

$$\left| \sqrt{(1+2)^2 + (-5+1)^2} - \sqrt{(1+2)^2 + (-5+5)^2} \right| = 2a$$

$$\left| \sqrt{25} - \sqrt{9} \right| = 2a$$

$$|5 - 3| = 2a \quad \therefore a = 1$$

Since $c^2 = a^2 + b^2$, $c = 2$ we have $2^2 = 1^2 + b^2 \quad \therefore b = \sqrt{3}$

$$\therefore (y+3)^2 - \frac{(x+2)^2}{3} = 1. \text{ Ans.}$$

---

8. Find the equations of the asymptotes of the hyperbola $x^2 - 4y^2 = 4$.

Solution:

$$x^2 - 4y^2 = 4$$

Divided both sides by 4:

$$\frac{x^2}{4} - \frac{y^2}{1} = 1 \quad \therefore a = 2,\ b = 1$$

The equations of the asymptotes are:

$$y = \pm \tfrac{b}{a} x$$

$$\therefore y = \pm \tfrac{1}{2} x. \text{ Ans.}$$

9. Find the equations of the asymptotes of the hyperbola $3y^2 - 4x^2 = 36$.

Solution:

$$3y^2 - 4x^2 = 36$$

Divided both sides by 35:

$$\frac{y^2}{12} - \frac{x^2}{9} = 1 \quad \therefore a = \sqrt{12},\ b = 3$$

The equations of the asymptotes are:

$$y = \pm \tfrac{a}{b} x$$

$$\therefore y = \pm \tfrac{2\sqrt{3}}{3} x. \text{ Ans.}$$

---

10. Find the equations of the asymptotes of the hyperbola $9x^2 - 4y^2 - 36x - 24y - 36 = 0$.

Solution:

$$9x^2 - 4y^2 - 36x - 24y - 36 = 0$$

Completing the square, we have:

$$\frac{(x-2)^2}{4} - \frac{(y+3)^2}{9} = 1 \quad \therefore a = 2,\ b = 3$$

The equations of the asymptotes are:

$$y - k = \pm \tfrac{b}{a}(x - h)$$

$$\therefore y + 3 = \pm \tfrac{3}{2}(x - 2). \text{ Ans.}$$

11. Find the equations of the asymptotes of the hyperbola $y^2 - x^2 - 6y - 4x + 1 = 0$.

Solution:

$$y^2 - x^2 - 6y - 4x + 1 = 0$$

Completing the square, we have:

$$\frac{(y-3)^2}{4} - \frac{(x+2)^2}{4} = 1 \quad \therefore a = b = 2$$

(Hint: It is an **equilateral hyperbola**)

The equations of the asymptotes are:

$$y - k = \pm \tfrac{a}{b}(x - h)$$

$$\therefore y - 3 = \pm(x + 2). \text{ Ans.}$$

EXERCISES

Find the center, vertices, and foci of each hyperbola. Graph each equation.

1. $\dfrac{x^2}{25} - \dfrac{y^2}{9} = 1$

2. $\dfrac{x^2}{9} - \dfrac{y^2}{25} = 1$

3. $\dfrac{(x-1)^2}{25} - \dfrac{(y+4)^2}{9} = 1$

4. $\dfrac{y^2}{4} - \dfrac{x^2}{16} = 1$

5. $\dfrac{y^2}{16} - \dfrac{x^2}{4} = 1$

6. $\dfrac{(y+1)^2}{4} - \dfrac{(x-4)^2}{16} = 1$

7. $x^2 - \dfrac{y^2}{4} = 1$

8. $\dfrac{y^2}{4} - x^2 = 1$

9. $(x-3)^2 - \dfrac{y^2}{4} = 1$

10. $9x^2 - 4y^2 = 36$

11. $25y^2 - 16x^2 = 400$

12. $4(x+3)^2 - 16(y+2)^2 = 64$

13. $9y^2 - x^2 - 18y = 0$

14. $x^2 - 4y^2 + 8y = 0$

15. $2x^2 - y^2 + 4x + 4y - 4 = 0$

Write the equation in standard form for each hyperbola having the given conditions.

16. Center: (0, 0)
    Vertex: (5, 0)
    Focus: (6, 0)

17. Center: (0, 0)
    Vertex: (0, –5)
    Focus: (0, 6)

18. Center: (0, 0)
    Vertex: (0, 5)
    Focus: (0, 8)

19. Center: (0, 0)
    Vertex: (0, 1)
    Focus: (0, 2)

20. Center: (2, –1)
    Vertex: (8, –1)
    Focus: (9, –1)

21. Vertex: (–2, 1)
    Foci: (7, 1), (–3, 1)

22. Vertex: (0, 4), (0, –4)
    Asymptote: $4x - 3y = 0$

23. Vertices: (1, –3), (1, 1)
    Asymptote: $3x - 2y - 5 = 0$

24. Foci: (3, 0), (–3, 0)
    Passing the point: (5, 4)

Find the equations of the asymptotes of each hyperbola.

25. $9x^2 - 4y^2 = 36$

26. $4(x+3)^2 - 16(y+2)^2 = 64$

27. $25y^2 - 16x^2 = 400$

28. $9y^2 - x^2 - 18y = 0$

29. $y^2 - 9x^2 = 9$

30. $9x^2 - y^2 = 9$

Using a graphing calculator to graph each hyperbola.

31. $4x^2 - 16y^2 = 64$

32. $10x^2 - 4y^2 = 40$

33. $4y^2 - x^2 = 16$

34. $4x^2 - y^2 = 16$

35. $x^2 - 4y^2 + 8y = 0$

36. $9x^2 - y^2 - 18x = 0$

37. Prove that the hyperbola $x^2/a^2 - y^2/b^2 = 1$ has the two asymptotes $y = \pm(b/a)x$.

38. If the equations of asymptotes are $bx + ay = 0$ and $bx - ay = 0$, prove that the equation of the hyperbola is $(bx + ay)(bx - ay) = k$, where k is a constant.

39. Use the formula described in Problem 38 to find the equation of the hyperbola having the length of horizontal transverse axis equals 6 and the equations of asymptotes are $2x - y = 0$ and $2x + y = 0$. (Hint: See distance Formulas on page 304)

40. The equation of the hyperbola is $x^2/a^2 - y^2/b^2 = 1$ and the slope of the tangent line is m.

    Prove that the equation of the tangent line is $y = mx \pm \sqrt{a^2m^2 - b^2}$. (Hint: $D = b^2 - 4ac = 0$)
    (See Chart of Equations of Tangent Lines on Page 301)

8-6 Solving Quadratic Systems

We have learned how to solve a system of linear equations in basic algebra. We use the same methods to solve a system of quadratic equations.

      1. Substitution Method

      2. Linear-combination Methods (Addition, Subtraction, Multiplication)

1) The system of one quadratic equation and one linear equation may have 2, 1, or no real solutions (intersections).

2) The system of two quadratic equations may have 4, 3, 2, 1 or no real solutions (intersections).

Examples

1.

Solve $\begin{cases} x^2 - y = 2 \text{------①} \\ \\ x + y = 0 \text{------②} \end{cases}$

Solution:

  From ②: $x = -y$

  Substitute $x = -y$ in ①

    $(-y)^2 - y = 2$

    $y^2 - y - 2 = 0$

    $(y - 2)(y + 1) = 0$

  $y - 2 = 0 \quad | \quad y + 1 = 0$

  $\therefore\ y = 2 \quad | \quad \therefore y = -1$

Substitute in ② | Substitute in ②

  $x + 2 = 0 \quad | \quad x + (-1) = 0$

  $\therefore\ x = -2 \quad | \quad \therefore\ x = 1$

Ans: $(-2, 2)$ and $(1, -1)$.

2.

Solve $\begin{cases} x^2 + 2y^2 = 41 \text{------①} \\ \\ 2x^2 - y^2 = 2 \text{------②} \end{cases}$

Solution:

  Eliminate x:

    $① \times 2 \rightarrow\ \ 2x^2 + 4y^2 = 82$

    $-\ ② \rightarrow\ \ 2x^2 - y^2 = 2$

                $5y^2 = 80$

                $y^2 = 16$

            $\therefore\ y = \pm 4$

Substitute $y = 4$ in ② | $y = -4$ in ②

  $2x^2 - 4^2 = 2 \quad | \quad 2x^2 - (-4)^2 = 2$

    $2x^2 = 18 \quad | \quad\quad 2x^2 = 18$

    $x^2 = 9 \quad | \quad\quad\quad x^2 = 9$

    $x = \pm 3 \quad | \quad\quad\quad x = \pm 3$

Ans: $(3, 4)$, $(-3, 4)$, $(3, -4)$, $(-3, -4)$.

3.

Solve $\begin{cases} x^2 - y = 2 \text{------①} \\ \\ x + y = -3 \text{------②} \end{cases}$

Solution:

  Eliminate y:

  $① + ② \quad\quad x^2 + x = -1$

            $x^2 + x + 1 = 0$

         $\therefore x = \dfrac{-1 \pm \sqrt{3}\,i}{2}$

Substitute in ②

    $\dfrac{-1 \pm \sqrt{3}\,i}{2} + y = -3$

    $y = -3 - \dfrac{-1 \pm \sqrt{3}\,i}{2}$

    $\therefore\ y = \dfrac{-5 \pm \sqrt{3}\,i}{2}$

Ans: The system has no real root.
     (no intersection)

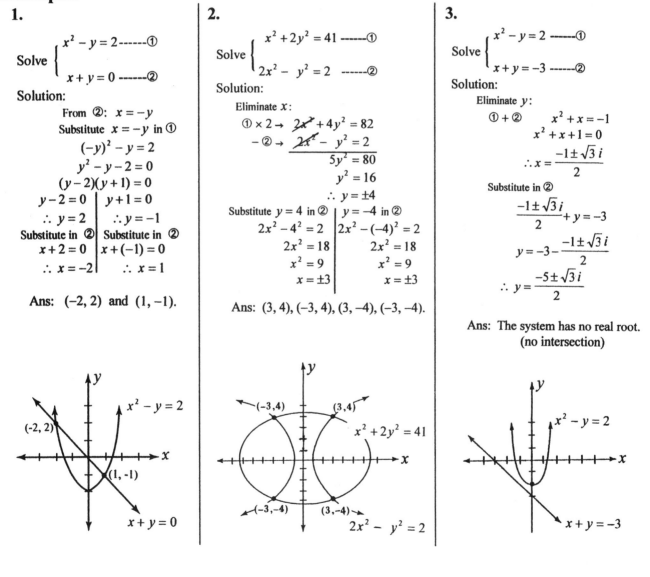

EXERCISES

Solve each system of linear equations by using any algebraic method.

1. $\begin{cases} 3x - 5y = 37 \\ 2x + y = 16 \end{cases}$

2. $\begin{cases} x - 3y = -29 \\ 4x + 10y = 38 \end{cases}$

3. $\begin{cases} 6x - y = -3 \\ 3x + 5y = 37 \end{cases}$

4. $\begin{cases} 2x - 6y = -54 \\ 4x - 5y = -17 \end{cases}$

5. $\begin{cases} 5x - 8y = -37 \\ -3x + 4y = 23 \end{cases}$

6. $\begin{cases} 6a + 3b = 4 \\ 8a - 3b = 3 \end{cases}$

7. $\begin{cases} 0.5x - 0.5y = -0.1 \\ 0.1x + 0.4y = 0.38 \end{cases}$

8. $\begin{cases} 0.4x + 0.6y = 0.22 \\ 1.6x - 1.8y = -1.22 \end{cases}$

9. $\begin{cases} 4x + 5y = 4 \\ 5x + 4y = 4.1 \end{cases}$

10. $\begin{cases} a = \frac{1}{2}b \\ 8a - 3b = 10 \end{cases}$

11. $\begin{cases} \frac{1}{2}x - \frac{1}{2}y = 1 \\ \frac{1}{4}x + y = 3 \end{cases}$

12. $\begin{cases} \frac{2}{3}x - \frac{1}{2}y = -3 \\ \frac{1}{3}x - \frac{3}{2}y = -4 \end{cases}$

13. $\begin{cases} x + 2y - z = 3 \\ x + y + 2z = 12 \\ -x + 3y - z = 4 \end{cases}$

14. $\begin{cases} 2x + y - 2z = 3 \\ x + 2y + 4z = 16 \\ 4x - 2y - 2z = -2 \end{cases}$

15. $\begin{cases} \frac{6}{x} - \frac{2}{y} + \frac{2}{z} = 4 \\ \frac{3}{x} - \frac{1}{y} - \frac{3}{z} = -1 \\ \frac{6}{x} - \frac{4}{y} - \frac{2}{z} = 6 \end{cases}$

Solve each system of quadratic equations by using any algebraic method.

16. $\begin{cases} x^2 - y = -2 \\ x + y = 4 \end{cases}$

17. $\begin{cases} x^2 - y = -2 \\ x - y = -4 \end{cases}$

18. $\begin{cases} x^2 + y^2 = 25 \\ y + 4 = 0 \end{cases}$

19. $\begin{cases} x^2 + y^2 = 13 \\ x - y = -1 \end{cases}$

20. $\begin{cases} 3x^2 - y^2 = -9 \\ 2x - y = 0 \end{cases}$

21. $\begin{cases} x^2 - y^2 = 21 \\ x + y = 7 \end{cases}$

22. $\begin{cases} x^2 + y^2 = 13 \\ x^2 - y = 7 \end{cases}$

23. $\begin{cases} x^2 - y^2 = 2 \\ x^2 - y = 0 \end{cases}$

24. $\begin{cases} x^2 + y^2 = 16 \\ x^2 - 2y = 8 \end{cases}$

25. $\begin{cases} x^2 + y^2 = 10 \\ x^2 - 8y = 35 \end{cases}$

26. $\begin{cases} x^2 - 4y^2 = -7 \\ 3x^2 + y^2 = 31 \end{cases}$

27. $\begin{cases} x^2 + 2y^2 = 16 \\ 2x^2 + y^2 = 17 \end{cases}$

28. $\begin{cases} x^2 + 4x - y = 14 \\ 2x - y = -1 \end{cases}$

29. $\begin{cases} x^2 + y = 0 \\ x^2 - 4x - y = 0 \end{cases}$

30. $\begin{cases} x^2 - y^2 = 9 \\ x^2 + y^2 - 16x = -33 \end{cases}$

31. $\begin{cases} x^3 - y = 0 \\ x - y = 0 \end{cases}$

32. $\begin{cases} x^2 + y^2 + x - y - 2 = 0 \\ x^2 + x + y - 2 = 0 \end{cases}$

33. $\begin{cases} \frac{2}{x^2} - \frac{4}{y^2} = -1 \\ \frac{6}{x^2} - \frac{11}{y^2} = -4 \end{cases}$

34. $\begin{cases} y = e^x \\ x - y + 1 = 0 \end{cases}$

35. $\begin{cases} y = \ln x \\ x + y - 1 = 0 \end{cases}$

36. $\begin{cases} \ln x - 2\ln y = 0 \\ \log_2 x - \log_2 y = 3 \end{cases}$

8-7 Eccentricity and Latus Rectum

The **eccentricity** e of a quadratic equation (ellipse or hyperbola) is defined as: $e = c/a$.
c is the distance from its focus to its center. a is the distance from its center to the endpoint of its major axis.
The eccentricity of a quadratic equation represents the **flatness** of its graph. It means that if c is close to a, then e is near 1 and the graph is quite flat.

The **eccentricity** of an ellipse is between 0 and 1. The eccentricity of a hyperbola is larger than 1. The eccentricity of a circle is 0 ($c = 0$). The eccentricity of a parabola is 1 ($c = a$).

The **latus rectum** of a quadratic equation is the chord passing through its focus and perpendicular to its axis.
The length of the latus rectum of a parabola is $4a$. The length of the latus rectum of an ellipse or a hyperbola is $2b^2/a$.

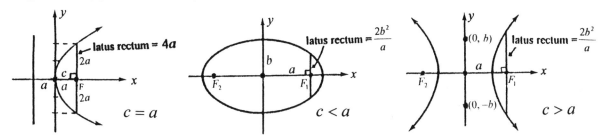

Examples

1. Find the eccentricity and latus rectum of the parabola $(y-2)^2 = 4(x+1)$.
Solution: $(y-k)^2 = 4a(x-h)$

$$a = 1, \ c = a = 1$$

Eccentricity: $e = \frac{c}{a} = \frac{1}{1} = 1$

Latus Rectum: $4a = 4(1) = 4$.

3. Find the eccentricity and latus rectum of the equation $4x^2 - y^2 + 8x + 4y + 4 = 0$.
Solution:

Completing the square, we have:
$\frac{(y-2)^2}{4} - \frac{(x+1)^2}{1} = 1$, It is a hyperbola.

$$a = 2, \ b = 1$$
$$c^2 = a^2 + b^2 = 4 + 1 = 5 \ \therefore c = \sqrt{5}$$

Eccentricity: $e = \frac{c}{a} = \frac{\sqrt{5}}{2}$.

Latus Rectum: $2b^2/a = \frac{2(1)}{2} = 1$.

2. Find the eccentricity and latus rectum of the ellipse $16x^2 + 9y^2 = 144$.
Solution: Divided both sides by 144:

$$\frac{x^2}{9} + \frac{y^2}{16} = 1 \ \therefore a = 4, b = 3$$
$$c^2 = a^2 - b^2 = 16 - 9 = 7 \ \therefore c = \sqrt{7}$$

Eccentricity: $e = \frac{c}{a} = \sqrt{7}/4$.

Latus Rectum: $2b^2/a = \frac{2(9)}{4} = \frac{9}{2}$.

4. Find the equation of the hyperbola with center $(2, -3)$, vertex $(6, -3)$, and $e = 2$.
Solution:

Graph and find: $a = 4$

$$e = \frac{c}{a} = 2 \ \therefore c = 2a = 2(4) = 8$$
$$c^2 = a^2 + b^2$$
$$b^2 = c^2 - a^2 = 64 - 16 = 48$$
$$\frac{(x-2)^2}{16} - \frac{(y+3)^2}{48} = 1, \quad \text{Ans.}$$

EXERCISES

Find the eccentricity (e) and latus rectum of each conic section.

1. $y^2 = 8x$

2. $x^2 = 16y$

3. $(x-3)^2 = 8(x+1)$

4. $(y+4)^2 = -12(x-5)$

5. $4x^2 - 9y^2 = 36$

6. $4x^2 + 5y^2 = 20$

7. $25x^2 + 4y^2 = 100$

8. $25x^2 - 4y^2 = 100$

9. $4(x-1)^2 - 36(y+1)^2 = 144$

10. $36(x-4)^2 + 4y^2 = 144$

11. $x^2 + y^2 = 10$

12. $x^2 - y^2 = 10$

13. $3x^2 + 3y^2 = 39$

14. $x^2 + 9(y-2)^2 = 9$

15. $(x+5)^2 - (y-7)^2 = 1$

16. Given the parabola $y^2 = -16x$, find the vertex, focus, equation of its axis, directrix, eccentricity, and latus rectum.

17. Given the parabola $x^2 - 4x - 8y - 4 = 0$, find the vertex, focus, equation of its axis, directrix, eccentricity, and latus rectum.

18. Find the equation of a parabola having vertex $(2, -1)$, latus rectum is 8, and the axis is parallel to the $x-$axis.

19. Find the equation of the hyperbola having center $(2, -3)$, eccentricity is 2, the length of conjugate axis is 10, and the transverse axis is parallel to $y-$axis.

20. Find the equation of the hyperbola having center $(2, -4)$, vertex $(5, -4)$, and eccentricity $\frac{4}{3}$.

21. Find the equation of the ellipse having vertices $(-5, -1)$ and $(5, -1)$, and eccentricity 0.6.

22. Find the equation of the ellipse having foci $(3, -3)$ and $(3, 5)$, and eccentricity 0.8.

23. Given the ellipse $4x^2 + 9y^2 = 36$, find the center, foci, equations of its major and minor axes, eccentricity, and latus rectum.

24. Given the hyperbola $25y^2 - 9x^2 = 225$, find the center, foci, equations of its transverse and conjugate axes, eccentricity, and latus rectum.

25. Given the equation $x^2 + 4y^2 - 2x + 16y + 1 = 0$, find the center, foci, equations of its axes, eccentricity, eccentricity, and latus rectum.

26. Given the equation $9x^2 - 4y^2 - 18x - 27 = 0$, find the center, foci, equations of its axes, eccentricity, and latus rectum.

27. The earth's orbit around the sun is an ellipse with the sun at one focus. The orbit is nearly circular. The eccentricity of the orbit is $e = 0.0167$ and the length of the major axis is about 186 million miles. Find the equation of the earth's orbit. (See Example 8, page 267)

28. Halley comet's orbit around the sun is an ellipse with the sun at one focus. At its closest distance (called the perihelion), the comet is 54.60 million miles from the center of the sun. The eccentricity of the orbit is $e = 0.967$. Find the farthest distance (called the aphelion) of the comet from the sun. (See Example 8, page 267)

8-8 General Quadratic Form and Rotation of Axes

Rotation of the axes in a coordinate plane is an important tool to simplify and graph a general quadratic equation $ax^2 + bxy + cy^2 + dx + ey + f = 0$ (when $b \neq 0$).
A general quadratic equation represents a circle, a parabola, an ellipse or a hyperbola.
If $b = 0$, the axis of its graph is parallel to either the $x-axis$ or the $y-axis$.
If $b \neq 0$, the axis of its graph is not parallel to either the $x-axis$ or the $y-axis$.

To graph a general quadratic equation (when $b \neq 0$), we transfer the coordinates of its graph from an xy-system to an $x'y'$-system by rotating the axes of the coordinate plane through an angle θ about the origin. The original graph is kept fixed.
The original graph has coordinates (x, y) with respect to the original system and coordinates (x', y') with respect to the new system.

To find the formulas of the relationships between (x, y) and (x', y'), we draw the following figure:

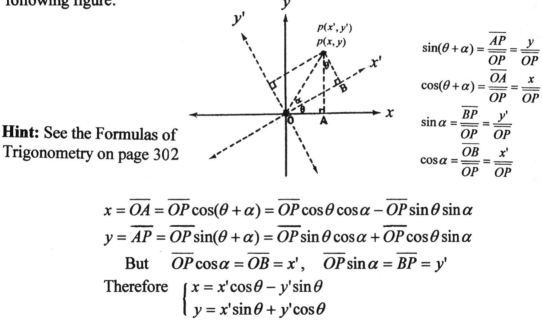

Hint: See the Formulas of Trigonometry on page 302

$$\sin(\theta + \alpha) = \frac{\overline{AP}}{\overline{OP}} = \frac{y}{\overline{OP}}$$

$$\cos(\theta + \alpha) = \frac{\overline{OA}}{\overline{OP}} = \frac{x}{\overline{OP}}$$

$$\sin \alpha = \frac{\overline{BP}}{\overline{OP}} = \frac{y'}{\overline{OP}}$$

$$\cos \alpha = \frac{\overline{OB}}{\overline{OP}} = \frac{x'}{\overline{OP}}$$

$$x = \overline{OA} = \overline{OP}\cos(\theta + \alpha) = \overline{OP}\cos\theta\cos\alpha - \overline{OP}\sin\theta\sin\alpha$$

$$y = \overline{AP} = \overline{OP}\sin(\theta + \alpha) = \overline{OP}\sin\theta\cos\alpha + \overline{OP}\cos\theta\sin\alpha$$

But $\overline{OP}\cos\alpha = \overline{OB} = x'$, $\overline{OP}\sin\alpha = \overline{BP} = y'$

Therefore $\begin{cases} x = x'\cos\theta - y'\sin\theta \\ y = x'\sin\theta + y'\cos\theta \end{cases}$

To express x' and y' in terms of x and y, we solve the above two equations for x' and y'. We have the following **formulas for the rotation of axes**:

$$\begin{cases} x' = x\cos\theta + y\sin\theta \\ y' = -x\sin\theta + y\cos\theta \end{cases}$$

Examples

1. A Cartesian coordinate system is rotated $60°$. Find the coordinates of the point $(3, -2)$ in the rotated system.

    Solution:

    $$x' = x\cos\theta + y\sin\theta = 3\cos 60° + (-2)\sin 60° = 3(\tfrac{1}{2}) + (-2)(\tfrac{\sqrt{3}}{2}) = \tfrac{3}{2} - \sqrt{3}.$$

    $$y' = -x\sin\theta + y\cos\theta = -3\sin 60° + (-2)\cos 60° = -3(\tfrac{\sqrt{3}}{2}) + (-2)(\tfrac{1}{2}) = -\tfrac{3\sqrt{3}}{2} - 1.$$

    The coordinates of the point $(3, -2)$ in the $x'y'-$ system are:

    $$(\tfrac{3}{2} - \sqrt{3}, \ -\tfrac{3\sqrt{3}}{2} - 1). \ \text{Ans.}$$

2. The coordinates of a point $(\tfrac{3}{2} - \sqrt{3}, \ -\tfrac{3\sqrt{3}}{2} - 1)$ is made by rotating $60°$ about the origin from the Cartesian coordinate system. Find the original coordinates of the point (x, y).

    Solution:

$x = x'\cos\theta - y'\sin\theta$	$y = x'\sin\theta + y'\cos\theta$
$= (\tfrac{3}{2} - \sqrt{3})\cos 60° - (-\tfrac{3\sqrt{3}}{2} - 1)\sin 60°$	$= (\tfrac{3}{2} - \sqrt{3})\sin 60° + (-\tfrac{3\sqrt{3}}{2} - 1)\cos 60°$
$= (\tfrac{3}{2} - \sqrt{3}) \cdot \tfrac{1}{2} - (-\tfrac{3\sqrt{3}}{2} - 1) \cdot \tfrac{\sqrt{3}}{2}$	$= (\tfrac{3}{2} - \sqrt{3}) \cdot \tfrac{\sqrt{3}}{2} + (-\tfrac{3\sqrt{3}}{2} - 1) \cdot \tfrac{1}{2}$
$= \tfrac{3}{4} - \tfrac{\sqrt{3}}{2} + \tfrac{9}{4} + \tfrac{\sqrt{3}}{2}$	$= \tfrac{3\sqrt{3}}{4} - \tfrac{3}{2} - \tfrac{3\sqrt{3}}{4} - \tfrac{1}{2}$
$= 3$	$= -2$

    The coordinates of the point in the $xy-$ system are $(3, -2)$. Ans.

3. Transform the equation $3x - y + 1 = 0$ by rotating the axes through an angle $30°$.

    Solution:

    $$x = x'\cos\theta - y'\sin\theta = x'\cos 30° - y'\sin 30° = \tfrac{\sqrt{3}}{2}x' - \tfrac{1}{2}y'$$

    $$y = x'\sin\theta + y'\cos\theta = x'\sin 30° + y'\cos 30° = \tfrac{1}{2}x' + \tfrac{\sqrt{3}}{2}y'$$

    Substituting into the equation:

    $$3(\tfrac{\sqrt{3}}{2}x' - \tfrac{1}{2}y') - (\tfrac{1}{2}x' + \tfrac{\sqrt{3}}{2}y') + 1 = 0$$

    $$\tfrac{3\sqrt{3}}{2}x' - \tfrac{3}{2}y' - \tfrac{1}{2}x' - \tfrac{\sqrt{3}}{2}y' + 1 = 0$$

    $$(\tfrac{3\sqrt{3}}{2} - \tfrac{1}{2})x' - (\tfrac{3}{2} + \tfrac{\sqrt{3}}{2})y' + 1 = 0$$

    The equation in the $x'y'-$ system is:

    $$\tfrac{3\sqrt{3} - 1}{2}x' - \tfrac{3 + \sqrt{3}}{2}y' + 1 = 0. \ \text{Ans.}$$

    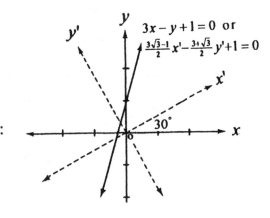

Graphing a General Quadratic Equation

A general quadratic equation has the form $ax^2 + bxy + cy^2 + dx + ey + f = 0$.

The axis of its graph (when $b \neq 0$) is not parallel to either the $x-axis$ or the $y-axis$.

To graph a general quadratic equation (when $b \neq 0$), we transfer its coordinates by means of the **rotation of axes** counterclockwise through an acute angle θ into a new $x'y'-$ system that has no $x'y'-$ term.

In other words, we select an acute angle θ of rotation to obtain $b' = 0$ so that the $x'y'-$ term is missing in the new coordinates (x', y').

Substituting the following **formulas for the rotation of axes** into the general quadratic equation:

$$x = x'\cos\theta - y'\sin\theta, \quad y = x'\sin\theta + y'\cos\theta$$

$$ax^2 + bxy + cy^2 + dx + ey + f = 0$$

$$a(x'\cos\theta - y'\sin\theta)^2 + b(x'\cos\theta - y'\sin\theta)(x'\sin\theta + y'\cos\theta) + c(x'\sin\theta + y'\cos\theta)^2$$
$$+ d(x'\cos\theta - y'\sin\theta) + e(x'\sin\theta + y'\cos\theta) + f = 0$$

Simplifying and adding the resulting $x'y'-$ term, we obtain the following :

$$[2(c-a)\sin\theta\cos\theta + b(\cos^2\theta - \sin^2\theta)] \; x'y'$$

Setting the coefficient (b') of $x'y'-$ term equal to zero to obtain the angle θ :

$$2(c-a)\sin\theta\cos\theta + b(\cos^2\theta - \sin^2\theta) = 0$$

$$(c-a)\sin 2\theta + b\cos 2\theta = 0 \quad \textbf{(Apply the formulas on page 302)}$$

$$(c-a)\sin 2\theta = -b\cos 2\theta$$

$$\frac{\sin 2\theta}{\cos 2\theta} = \frac{-b}{c-a} = \frac{b}{a-c} \quad \therefore \tan 2\theta = \frac{\sin 2\theta}{\cos 2\theta} = \frac{b}{a-c}$$

We have the following formula.

Formula: $\tan 2\theta = \dfrac{b}{a-c}$, $a \neq c$, ($\theta = 45°$ if $a = c$). Or $\cot 2\theta = \dfrac{a-c}{b}$.

If $\tan 2\theta \geq 0$ (1st quadrant), then $0° < 2\theta \leq 90°$, so that $0 < \theta \leq 45°$.

If $\tan 2\theta < 0$ (2nd quadrant), then $90° < 2\theta < 180°$, so that $45° < \theta < 90°$.

Rotating the axes counterclockwise through the acute angle θ obtained from the above formula, we transfer a general quadratic equation with $xy-$ term into an equation in $x'y'-$ system that has no $x'y'-$ term.

Depending on the angle chosen (clockwise or counterclockwise), the rotated equation may be different, but the graphs are equivalent.

Examples

4. Simplify and graph the equation $2x^2 + 3xy - 2y^2 - 10 = 0$ by rotating it into an
$x'y'-$ system having no $x'y'-term$.

Solution:

$$a = 2, \ b = 3, \ c = -2. \quad \tan 2\theta = \frac{b}{a-c} = \frac{3}{2-(-2)} = \frac{3}{4}, \quad \cos 2\theta = \frac{4}{5}$$

Since $\cos 2\theta = 2\cos^2 \theta - 1$, we find :

$$\cos\theta = \sqrt{\frac{1+\cos 2\theta}{2}} = \sqrt{\frac{1+\frac{4}{5}}{2}} = \sqrt{\frac{9}{10}} = \frac{3}{\sqrt{10}}$$

$$\sin\theta = \sqrt{1-\cos^2\theta} = \sqrt{1-\frac{9}{10}} = \frac{1}{\sqrt{10}}$$

The equations of rotation are:

$$x = x'\cos\theta - y'\sin\theta = \frac{3}{\sqrt{10}}x' - \frac{1}{\sqrt{10}}y' = \frac{3x'-y'}{\sqrt{10}}$$

$$y = x'\sin\theta + y'\cos\theta = \frac{1}{\sqrt{10}}x' + \frac{3}{\sqrt{10}}y' = \frac{x'+3y'}{\sqrt{10}}$$

Substituting into the given equation, we have:

$$2\left(\frac{3x'-y'}{\sqrt{10}}\right)^2 + 3\left(\frac{3x'-y'}{\sqrt{10}}\right)\left(\frac{x'+3y'}{\sqrt{10}}\right) - 2\left(\frac{x'+3y'}{\sqrt{10}}\right)^2 - 10 = 0$$

$$\frac{9(x')^2 - 6x'y' + (y')^2}{5} + \frac{3[3(x')^2 + 8x'y' - 3(y')^2]}{10} - \frac{(x')^2 + 6x'y' + 9(y')^2}{5} - 10 = 0$$

Multiplying by 10:

$$18(x')^2 - 12x'y' + 2(y')^2 + 9(x')^2 + 24x'y' - 9(y')^2 - 2(x')^2 - 12x'y' - 18(y')^2 - 100 = 0$$

$$25(x')^2 - 25(y')^2 = 100 \quad \therefore \quad \frac{(x')^2}{4} - \frac{(y')^2}{4} = 1. \text{ Ans.}$$

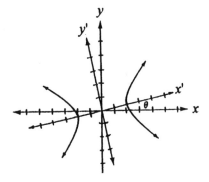

$\tan 2\theta = \frac{1}{4}$, $\quad \tan 2\theta = 0.75$, $\quad 2\theta = 36.8°$, $\quad \theta = 18.4°$
The vertex in the $x'y'-$ system is $(2, 0)$ and $(-2, 0)$.
The vertex in the $xy -$ system is:

$$(2\cos\theta, \ 2\sin\theta) \text{ and } (-2\cos\theta, \ -2\sin\theta)$$
$$= (2\cdot\tfrac{3}{\sqrt{10}}, \ 2\cdot\tfrac{1}{\sqrt{10}}) \text{ and } (-2\cdot\tfrac{3}{\sqrt{10}}, \ -2\sin\cdot\tfrac{1}{\sqrt{10}})$$
$$= (1.9, \ 0.64) \text{ and } (-1.9, \ -0.64).$$

Examples

5. Simplify and graph the equation $x^2 + \sqrt{3}xy + 2y^2 - 10 = 0$ by rotating it into an $x'y'$– system having no $x'y'$–term.

Solution:

$$a = 1, \quad b = \sqrt{3}, \quad c = 2.$$

$$\tan 2\theta = \frac{b}{a-c} = \frac{\sqrt{3}}{1-2} = -\sqrt{3}. \quad \therefore \ 2\theta = 120, \quad \theta = 60°$$

$$\cos\theta = \cos 60° = \frac{1}{2}, \quad \sin\theta = \sin 60° = \frac{\sqrt{3}}{2}$$

The equations of rotations are:

$$x = x'\cos\theta - y'\sin\theta = \frac{1}{2}x' - \frac{\sqrt{3}}{2}y' = \frac{x' - \sqrt{3}y'}{2}$$

$$y = x'\sin\theta + y'\cos\theta = \frac{\sqrt{3}}{2}x' + \frac{1}{2}y' = \frac{\sqrt{3}x' + y'}{2}$$

Substituting into the given equation, we have:

$$\left(\frac{x' - \sqrt{3}y'}{2}\right)^2 + \sqrt{3}\left(\frac{x' - \sqrt{3}y'}{2}\right)\left(\frac{\sqrt{3}x' + y'}{2}\right) + 2\left(\frac{\sqrt{3}x' + y'}{2}\right)^2 - 10 = 0$$

$$\left(\frac{(x')^2 - 2\sqrt{3}x'y' + 3(y')^2}{4}\right) + \sqrt{3} \cdot \left(\frac{\sqrt{3}(x')^2 - 2x'y' - \sqrt{3}(y')^2}{4}\right)$$

$$+ \left(\frac{3(x')^2 + 2\sqrt{3}x'y' + (y')^2}{2}\right) - 10 = 0$$

Multiplying by 4:

$$(x')^2 - 2\sqrt{3}x'y' + 3(y')^2 + 3(x')^2 - 2\sqrt{3}x'y' - 3(y')^2 + 6(x')^2 + 4\sqrt{3}x'y' + 2(y')^2 - 10 = 0$$

$$10(x')^2 + 2(y')^2 = 40 \quad \therefore \ \frac{(x')^2}{4} + \frac{(y')^2}{20} = 1. \text{ Ans.}$$

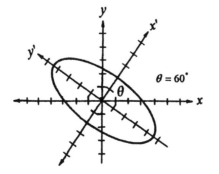

Identifying the graph without a rotation of axes

The graph of the quadratic equation $ax^2 + cy^2 + dx + ey + f = 0$ is one of the following cases (except for the degenerate cases):

    1. If $a = c$, it is a circle.
    2. If $ac = 0$, it is a parabola. ($a = 0$ or $c = 0$, but not both.)
    3. If $ac > 0$ and $a \ne c$, it is an ellipse. (a and c have like signs.)
    4. If $ac < 0$, it is a hyperbola. (a and c have unlike signs.)

The general quadratic equation $ax^2 + bxy + cy^2 + dx + ey + f = 0$ and its rotated equation $a'(x')^2 + b'x'y' + c'(y')^2 + d'x' + e'y' + f' = 0$ through an angle θ have the following **rotation invariants** (the same in both equations). In other words, they are invariant under rotation of axes (see **Example 6**, at the next page).

    1. $f = f'$.   **2.** $a + c = a' + c'$.   **3.** $b^2 - 4ac = (b')^2 - 4a'c'$.

We can prove above rotation **invariants** no matter what angle θ of rotation is chosen.

If we choose the angle θ of rotation satisfies $\tan 2\theta = \dfrac{b}{a-c}$, then $b' = 0$.

We have the **discriminant of the general quadratic equation:**

$$b^2 - 4ac = -4a'c'$$

The value of $a'c'$ determines the type of graph for the rotated equation:

$$a'(x')^2 + c'(y')^2 + d'x' + e'y' + f' = 0$$

Therefore, the value of $b^2 - 4ac$ determines the type of graph for the original equation.

To identify the graph of a general quadratic equation without a rotation of axes, we use the following rules (except for the degenerate cases):

    1. If $b^2 - 4ac = 0$, it is a parabola. ($\because a'c' = 0$.)
    2. If $b^2 - 4ac < 0$, it is an ellipse or a circle. ($\because a'c' > 0$.)
        If $b^2 - 4ac < 0$, $b \ne 0$ **or** $a \ne c$, it is an ellipse.
        If $b^2 - 4ac < 0$, $b = 0$ **and** $a = c$, it is a circle.
    3. If $b^2 - 4ac > 0$, it is a hyperbola. ($\because a'c' < 0$.)

We can also prove that a general quadratic equation with $xy -$ term in it is not a circle, and a general quadratic equation without $x -$ term and $y -$ term in it is not a parabola.
(See proofs **Examples 7 and 8**, at the next page.)

Examples

6. Prove that the equation $2x^2 + 3xy - 2y^2 - 10 = 0$ and its rotated equation

$(x')^2 - (y')^2 - 4 = 0$ are invariant under rotation of axes. (See **Example 4**, page 280)

Proof:

Rewrite the rotated equation to satisfy the invariant $f = f'$ by multiplying each

side of the rotated equation by 2.5: $\quad 2.5(x')^2 - 2.5(y')^2 - 10 = 0$

$$a + c = 2 + (-2) = 0$$
$$a' + c' = 2.5 + (-2.5) = 0 \qquad \therefore a + c = a' + c'.$$
$$b^2 - 4ac = 3^2 - 4(2)(-2) = 25$$
$$(b')^2 - 4a'c' = 0^2 - 4(2.5)(-2.5) = 25 \quad \therefore b^2 - 4ac = (b')^2 - 4a'c'.$$

7. Prove that the equation of the circle $x^2 + y^2 = r^2$ is invariant under rotation of axes.

Proof:

$$x^2 + y^2 = (x'\cos\theta - y'\sin\theta)^2 + (x'\sin\theta + y'\cos\theta)^2$$
$$= (x')^2 \cos^2\theta - 2x'y'\cos\theta\sin\theta + (y')^2 \sin^2\theta$$
$$+ (x')^2 \sin^2\theta + 2x'y'\cos\theta\sin\theta + (y')^2 \cos^2\theta$$
$$= (x')^2 (\cos^2\theta + \sin^2\theta) + (y')^2 (\sin^2\theta + \cos^2\theta)$$

We have: $x^2 + y^2 = (x')^2 + (y')^2 = r^2$.

(It also proves that a general quadratic equation with xy-term in it is not a circle.)

8. Prove that a general quadratic equation without x-term and y-term in it is not a parabola.

Proof: $\quad ax^2 + bxy + cy^2 - f = 0$. If it is a parabola, we have $b^2 - 4ac = 0$.

Therefore $b = 2\sqrt{ac}$, we have $ax^2 + 2\sqrt{ac}\,xy + cy^2 = f$.

It has a perfect square on the left side. $(\sqrt{a}x + \sqrt{c}y)^2 = f \quad \therefore \sqrt{a}x + \sqrt{c}y = \pm\sqrt{f}$

It represents two straight lines, not a parabola.

(It also proves that " If $b^2 - 4ac = 0$, then $ax^2 + bxy + cy^2$ is a perfect square ".)

9. Using the discriminant to identify the graph of each of the following equations.

a. $xy + 1 = 0$.   **b.** $4x^2 + 10xy + 4y^2 - 36 = 0$.   **c.** $8x^2 + 4xy + 5y^2 - 36 = 0$.

d. $4x^2 - 4xy + y^2 - y + 2 = 0$.   **e.** $5x^2 - 2xy + 5y^2 - 12 = 0$.   **f.** $5x^2 + 5y^2 - 12 = 0$.

Solution: **a)** $b^2 - 4ac = 1^2 - 4(0)(0) = 1 > 0$. It is a hyperbola.

     b) $b^2 - 4ac = 10^2 - 4(4)(4) = 36 > 0$. It is a hyperbola.

     c) $b^2 - 4ac = 4^2 - 4(8)(5) = -144 < 0$. $b \neq 0$. It is an ellipse.

     d) $b^2 - 4ac = (-4)^2 - 4(4)(1) = 0$. It is a parabola.

     e) $b^2 - 4ac = (-2)^2 - 4(5)(5) = -96 < 0$. $b \neq 0$. It is an ellipse.

     f) $b^2 - 4ac = 0 - 4(5)(5) = -100 < 0$. $b = 0$ and $a = c$. It is a circle.

Degenerate Cases of Quadratic Equations

Degenerate case of a quadratic equation occurs when the graph is a point, a line, two parallel lines, two intersecting lines, or no graph.

One way to identify a degenerate case is to check whether the equation is factorable or not.

Examples

1. What is the graph of $xy = 0$?

Solution:

$$xy = 0 \text{ if and only if } x = 0 \text{ or } y = 0.$$

The graphs are two intersecting lines:

$$x = 0 \text{ (the } y - axis \text{) and } y = 0 \text{ (the } x - axis \text{).}$$

It is a degenerate case.

2. What is the graph of $x^2 + y^2 + 6x - 2y + 10 = 0$?

Solution:

Completing the square:

$$x^2 + y^2 + 6x - 2y + 10 = 0$$
$$(x^2 + 6x + 9) - 9 + (y^2 - 2y + 1) - 1 + 10 = 0$$
$$(x^2 + 6x + 9) + (y^2 - 2y + 1) = 0$$
$$(x + 3)^2 + (y - 1)^2 = 0$$

It is a circle with center (-3, 1) and radius = 0. The graph is a point (-3, 1).
It is a degenerate case.

3. What is the graph of $4x^2 + 9y^2 + 12xy + 12x + 18y + 9 = 0$?

Solution:

$$4x^2 + 9y^2 + 12xy + 12x + 18y + 9 = 0$$
$$(4x^2 + 12xy + 9y^2) + 6(2x + 3y) + 9 = 0$$
$$(2x + 3y)^2 + 6(2x + 3y) + 9 = 0$$
$$(2x + 3y + 3)^2 = 0$$

The graph is a line $2x + 3y + 3 = 0$. It is a degenerate case.

4. What is the graph of $x^2 + y^2 + 9 = 0$?

Solution:

$$x^2 + y^2 + 9 = 0 \quad \therefore x^2 + y^2 = -9$$

The left side of the equation is a positive number. The right side of the equation is no way can equal a negative number. Therefore, the graph does not exit.
It is a degenerate case.

EXERCISES

Transform by rotating the axes through an angel described.

1. Transform the point $(3, -1)$ by rotating the axes through an angle $30°$.

2. Transform the equation $2x + 3y - 1 = 0$ by rotating the axes through an angle $60°$.

3. Transform the equation $x^2 - y^2 = 2$ by rotating the axes through an angle $30°$.

4. Transform the equation $x^2 - y^2 = 2$ by rotating the axes through an angle $45°$.

5. The equation $x'y' = -1$ is made by rotating $45°$ about the origin from the Cartesian coordinate system. Find the original equation.

Simplify and graph each of the following equations by rotating it into an $x'y'$– system having no $x'y'$– term.

6. $xy + 1 = 0$

7. $2xy - 1 = 0$

8. $4x^2 + 10xy + 4y^2 - 36 = 0$

9. $4x^2 - 10xy + 4y^2 - 36 = 0$

10. $4x^2 - 10xy + 4y^2 - 36 = 0$ if choose $\theta = -45°$

11. $8x^2 + 4xy + 5y^2 - 36 = 0$

12. $3x^2 + 2\sqrt{3}xy + y^2 + 2x - 2\sqrt{3}y - 8 = 0$

13. $x^2 + 4xy + 4y^2 - 16 = 0$

14. $7x^2 - 6\sqrt{3}xy + 13y^2 - 16 = 0$

15. $5x^2 - 2xy + 5y^2 - 12 = 0$

Use the discriminant to identify whether the graph of each equation is a parabola, a circle, an ellipse, or a hyperbola without a rotation of axes.

16. $2xy - 1 = 0$

17. $2x^2 + \sqrt{3}xy + y^2 - 10 = 0$

18. $2x^2 - 4xy - y^2 - 10 = 0$

19. $x^2 - 4xy + 4y^2 + 5\sqrt{5}y + 1 = 0$

20. $4x^2 - 10xy + 4y^2 - 36 = 0$

21. $x^2 + 4xy + 4y^2 - 16 = 0$

22. $3x^2 + 2\sqrt{3}xy + y^2 + 2x - 2\sqrt{3}y - 8 = 0$

23. $x^2 - xy + y^2 - 3\sqrt{2}x = 0$

24. $9x^2 + 24xy + 16y^2 + 90x - 130y = 0$

25. $13x^2 - 8xy + 7y^2 - 45 = 0$

26. $x^2 + \sqrt{3}xy + 2y^2 - 10 = 0.$

27. $2x^2 + 3xy - 2y^2 - 10 = 0$

28. $x^2 + 4xy + 4y^2 + 16 = 0$

29. $x^2 + y^2 - 4x + 2y - 11 = 0.$

30. $x^2 + 4y^2 - 2x + 16y + 13 = 0.$

31. $3x^2 - 8x + 7y^2 - 10 = 0$

32. $2x^2 - 5y^2 - 8xy + x - 5 = 0$

-----Continued-----

Identify the graph of each equation of the following degenerate cases.

33. $x^2 + y^2 = 0$ **34.** $x^2 - y^2 = 0$

35. $x^2 + x - 2 = 0$ **36.** $x^2 - 4x + 4 = 0$

37. $4x^2 - 9y^2 + 24x + 18y + 27 = 0$ **38.** $4x^2 + 9y^2 + 24x - 18y + 45 = 0$

39. $4x^2 + 9y^2 + 10 = 0$ **40.** $4x^2 + 9y^2 + 12xy - 6x - 18 = 0$

42. Prove that the equation $x^2 + \sqrt{3}xy + 2y^2 - 10 = 0$ and its rotated equation $20(x')^2 + 4(y')^2 - 80 = 0$ are invariant under rotation of axes. (See Example 5, page 281)

43. Prove that the equation $4x^2 + 10xy + 4y^2 - 36 = 0$ and its rotated equation $9(x')^2 - (y')^2 - 36 = 0$ are invariant under rotation of axes. (See Problem 8, page 285)

Space for Taking Notes

8-9 Parametric Equations and Graphs

Parametric equations are the equations to express a relationship between x and y in terms of a third variable (called a **parameter**). We write:

$$x = f(t) \quad \text{and} \quad y = f(t)$$

We say that the two equations form a set of parametric equations. Each value of t determines a point (x, y) in the plane. The graph of the parametric equations is the set of points in the plane when t takes on all possible real numbers ($t \in R$).

Parametric equation is useful to trace the movement or orientation of the graph of the equation. Each point (x, y) on the graph is determined from a value chosen for t.

The parametric equations of a line:

$$\begin{cases} x = x_1 + at \\ y = y_1 + bt \end{cases}$$. If $a = 0$, the line is vertical. If $b = 0$, the line is horizontal.

Example 1. Find the parametric equation of the line passing the points $(-2, 5)$ and $(4, 8)$.

Solution:

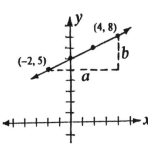

We select $(-2, 5)$ as (x_1, y_1) and trace the movement of the line by finding its slope:

$$m = \frac{8-5}{4-(-2)} = \frac{3}{6} = \frac{1}{2} = \frac{b}{a}$$

a indicates the horizontal movement. b indicates the vertical movement. Therefore, we have the parametric equation:

$$\begin{cases} x = -2 + 2t \\ y = 5 + t; \ t \in R. \end{cases} \quad \text{or} \quad \begin{cases} x = -2 + 6t \\ y = 5 + 3t; \ t \in R. \end{cases} \text{ Ans.}$$

When $t = 0$, we have $(-2, 5)$; $t = 1$, we have $(0, 6)$; $t = 2$, we have $(2, 7)$; $t = 3$, we have $(4, 8)$, and so on. Plotting and connecting these points, we have the graph of the line. There are many parametric equations for this line for $\frac{b}{a} = \frac{1}{2} = \frac{2}{4} = \frac{3}{6} = \frac{4}{8} = \cdots$.

To graph a pair of parametric equations, we eliminate the parameter to obtain the rectangular equation by solving for it in one of the equations and then substituting in the other. We may need to adjust the domain of the resulting rectangular equations to match the restriction of the original parametric equations.

Example 2. Graph the parametric equations $\begin{cases} x = t^2 \\ y = 2t^2 + 3; \ t \in R \end{cases}$

Solution: Since $t^2 \geq 0$, we have $x \geq 0$ and $y \geq 3$.

Substituting $x = t^2$ in to the second equation, we have the rectangular equation:

$$y = 2x + 3, \ x \geq 0. \text{ Ans.}$$

Orientation of the graph of a parametric equation

Parametric equation is very useful to provide the information about the orientation (the direction) of the graph. The graph is oriented by the increasing values of the parameter.
We indicate an orientation by placing arrowheads on or alongside the graph.

Example 3: Graph the parametric equations $x = t^2 - 4$, $y = 2t^2 - 2$; $t \in R$. Indicate the orientation.

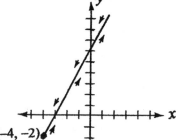

Solution: $t^2 = x + 4$, $x \geq -4$

Substitute into the second equation:

$y = 2(x + 4) - 2$

We have the rectangular equation:

$y = 2x + 6$, $x \geq -4$

If $t = -3$, we have (5, 16). If $t = 0$, we have (−4, −2). If $t = 3$, we have (5, 16).

As t increases from $-\infty$ to 0, $p(x, y)$ slants downward.

As t increases from 0 to ∞, $p(x, y)$ slants upward.

(Hint: The infinite symbol, ∞, indicates the value continues without end.)

The parametric equations of a parabola

1) $y^2 = 4ax$

$$\begin{cases} x = at^2 \\ y = 2at, \ t \in R \end{cases}$$

2) $(y - k)^2 = 4a(x - h)$

$$\begin{cases} x = h + at^2 \\ y = k + 2at, \ t \in R \end{cases}$$

The parametric equations of a circle: $0 \leq \theta < 2\pi$

1) $x^2 + y^2 = r^2$

$$\begin{cases} x = r \cos\theta \\ y = r \sin\theta \end{cases}$$

2) $(x - h)^2 + (y - k)^2 = r^2$

$$\begin{cases} x = h + r \cos\theta \\ y = k + r \sin\theta \end{cases}$$

The parametric equations of an ellipse: $0 \leq \theta < 2\pi$

1) $\dfrac{x^2}{a^2} + \dfrac{y^2}{b^2} = 1$

$$\begin{cases} x = a \cos\theta \\ y = b \sin\theta \end{cases}$$

2) $\dfrac{(x - h)^2}{a^2} + \dfrac{(y - k)^2}{b^2} = 1$

$$\begin{cases} x = h + a \cos\theta \\ y = k + b \sin\theta \end{cases}$$

The parametric equations of a hyperbola: $0 \leq \theta < 2\pi$, $\theta \neq n\pi + \frac{\pi}{2}$

1) $\dfrac{x^2}{a^2} - \dfrac{y^2}{b^2} = 1$

$$\begin{cases} x = a \sec\theta \\ y = b \tan\theta \end{cases}$$

2) $\dfrac{(x - h)^2}{a^2} - \dfrac{(y - k)^2}{b^2} = 1$

$$\begin{cases} x = h + a \sec\theta \\ y = k + b \tan\theta \end{cases}$$

Examples

4. Graph the parametric equations $x = 4t^2 - 5$, $y = 2t + 3$; $t \in R$. Indicate the orientation.

Solution:

$y = 2t + 3$, we have: $t = \frac{y-3}{2}$

Substitute into the first equation: $x = 4(\frac{y-3}{2})^2 - 5$ $(-5, 3)$

We have: $(y-3)^2 = x + 5$; $x \geq -5$

or $y = \pm\sqrt{x+5} + 3$; $x \geq -5$

The graph is the entire parabola. The arrowheads on the graph indicate the direction of the curve as t increases from $-\infty$ to $+\infty$.

5. The position of point $p(x, y)$ at an angle θ is given by the parametric equations $x = a\cos\theta$, $y = a\sin\theta$, $0 \leq \theta \leq \pi$. Find the rectangular equation and indicate the orientation.

Solution:

We have: $\cos\theta = \frac{x}{a}$ and $\sin\theta = \frac{y}{a}$

$$\frac{x^2}{a^2} + \frac{y^2}{a^2} = \cos^2\theta + \sin^2\theta = 1$$

$$\therefore x^2 + y^2 = a^2; \ -a \leq x \leq a$$

or $y = a\sqrt{1-(x/a)^2}$; $-a \leq x \leq a$

The graph is a semicircle. The arrowheads on the graph indicate a counterclockwise direction of the curve as θ increases from 0 to π.

6. Find the equation in x and y of the parametric equations $x = 4\cos\theta$, $y = 9\sin\theta$; $\theta \in R$.

Solution: We have: $\cos\theta = \frac{x}{4}$ and $\sin\theta = \frac{y}{9}$

$$\frac{x^2}{4^2} + \frac{y^2}{9^2} = \cos^2\theta + \sin^2\theta = 1$$

$$\therefore \frac{x^2}{16} + \frac{y^2}{81} = 1 \ (\text{It is an ellipse}). \ \text{Ans.}$$

7. The path of the projectile fired at an angle θ to the horizontal is given by the formula $x = (v_0\cos\theta)t$, $y = -16t^2 + (v_0\sin\theta)t + h_0$, where v_0 is the initial speed, h_0 is the initial height. Find the rectangular equation in x and y if $\theta = 45°$. (See Example 5, Page 122)

Solution: (Hint: $\sin 45° = \cos 45° = \frac{\sqrt{2}}{2}$)

We have: $t = \frac{x}{v_0\cos\theta}$, Substitute t into y:

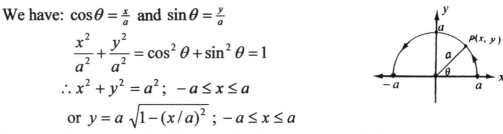

$$y = -16\left(\frac{x}{v_0\cos\theta}\right)^2 + (v_0\sin\theta)\cdot\left(\frac{x}{v_0\cos\theta}\right) + h_0 = -16\left(\frac{x^2}{v_0^2\cos^2 45°} + x + h_0\right)$$

$$= -16\left(\frac{x^2}{v_0(\frac{\sqrt{2}}{2})^2}\right) + x + h_0 = -16\left(\frac{x^2}{v_0^2 \cdot \frac{1}{2}}\right) + x + h_0 \quad \therefore y = \frac{-32x^2}{v_0^2} + x + h_0. \ \text{Ans.}$$

EXERCISES

Find the parametric equations of the line passing the two points if we choose the first point as (x_1, y_1) in the parametric equations $x = x_1 + at$, $y = y_1 + bt$. Use $\frac{b}{a}$ in lowest terms.

1. $(4, 0)$ and $(0, 3)$ 2. $(1, 2)$ and $(11, 8)$ 3. $(-7, 4)$ and $(-7, -4)$

4. $(7, -4)$ and $(-7, -4)$ 5. $(1, \sqrt{2})$ and $(3, 3\sqrt{2})$ 6. $(3, -\sqrt{3})$ and $(5, -2\sqrt{3})$

Find the rectangular equation in x and y for each parametric equation. State the domain for the resulting rectangular equation.

7. $\begin{cases} x = 4 - 4t \\ y = 3t \quad ; t \in R \end{cases}$ 8. $\begin{cases} x = 7 - 14t \\ y = -4 \quad ; t \in R \end{cases}$ 9. $\begin{cases} x = 3 - 2t \\ y = 2 + 3t ; t \in R \end{cases}$

10. $\begin{cases} x = \sqrt{t} \\ y = 2 - t ; t \geq 0 \end{cases}$ 11. $\begin{cases} x = t + 4 \\ y = t^2 ; -1 \leq t \leq 2 \end{cases}$ 12. $\begin{cases} x = t - 2 \\ y = \frac{t}{t-2} ; t > 2 \end{cases}$

13. $\begin{cases} x = -t \\ y = \frac{2}{t} ; t \neq 0 \end{cases}$ 14. $\begin{cases} x = 3t + 2 \\ y = t + 1 ; 0 \leq t \leq 3 \end{cases}$ 15. $\begin{cases} x = 2 - t \\ y = \sqrt{t} ; t \geq 0 \end{cases}$

16. $\begin{cases} x = 2\cos^2 \theta \\ y = 3\sin^2 \theta ; \theta \in R \end{cases}$ 17. $\begin{cases} x = 3\cos t \\ y = 3\sin t ; 0 \leq t \leq \pi \end{cases}$ 18. $\begin{cases} x = 2\cos t \\ y = 4\sin t ; 0 \leq t \leq 2\pi \end{cases}$

19. $\begin{cases} x = \sec \theta \\ y = \tan \theta ; 0 \leq \theta \leq \frac{\pi}{4} \end{cases}$ 20. $\begin{cases} x = 2 + 4\cos \theta \\ y = 3 + 3\sin \theta ; \theta \in R \end{cases}$ 21. $\begin{cases} x = e^{2t} \\ y = e^t ; t \geq 0 \end{cases}$

Graph each parametric equations. Indicate the orientation of each curve.

22. $x = 2t + 3$, $y = t + 1$; $0 \leq t \leq 3$ 23. $x = \sqrt{t} + 2$, $y = \sqrt{t} - 5$; $t \geq 0$

24. $x = 2\sin^2 \theta$, $y = 2\cos^2 \theta$; $0 \leq \theta \leq \frac{\pi}{2}$ 25. $x = \sqrt{t}$, $y = \sqrt{4-t}$; $0 \leq t \leq 4$

26. $x = t$, $y = \sqrt{4-t^2}$; $-2 \leq t \leq 2$ 27. $x = 3\sin \theta$, $y = 3\cos \theta$; $0 \leq \theta \leq 2\pi$

28. $x = 3\cos \theta$, $y = 2\sin \theta$; $\theta \in R$ 29. $x = \sin^2 t$, $y = \cos t$; $0 \leq t \leq \pi$

30. $x = t^3$, $y = \ln t$; $t > 0$ 31. $x = 3e^t$, $y = 1 + e^t$; $t \geq 0$

32. Find the equation in x and y of the parametric equations $x = t^2 + t$, $y = t - 3$.

33. Find the equation in x and y of the parametric equations $x = t^2 + 2t - 2$, $y = t^2 + t - 1$.

34. Find the equation in x and y of the parametric equations $x = 2^t - 2^{-t}$, $y = 2^t + 2^{-t}$.

35. The path of a projectile fired at an angle θ to the horizontal is given by the formula $x = (v_0 \cos \theta)t$, $y = -16t^2 + (v_0 \sin \theta)t + h_0$, where v_0 is the initial speed in feet per second, h_0 is the initial height in feet, t is the time in seconds. a) Find the parametric equations of the projectile as a function of t if $v_0 = 80$, $\theta° = 45°$, $h_0 = 96$. b) How far will it reach the ground again? c) What is the maximum height of the projectile? (See Example 5, P.122)

8-10 Polar Coordinates and Graphs

A point $p(x, y)$ in the rectangular coordinates in x and y can be represented in the **polar coordinates** $p(r, \theta)$:

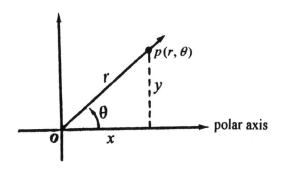

o : the **pole** or **origin**.

r : directed distance from o to p.

θ : directed angle, counterclockwise from polar axis to $\overline{op}$.

The relationship between polar and retangular coordinates are:

$$x = r\cos\theta, \quad y = r\sin\theta$$

$$\tan\theta = \frac{y}{x}, \quad r = \sqrt{x^2 + y^2}$$

In the polar coordinates, the coordinates of the pole is $p(0, \theta)$, where θ can be any angle.

In the polar coordinates, a point $p(r, \theta)$ has infinitely many representations.

The point $p(r, \theta)$ can be represented as $p(r, \theta \pm 2n\pi)$ or $p(-r, \theta \pm (2n\pi + \pi))$, where n is an integer.

Example: $(5, \frac{\pi}{6}) = (5, \frac{\pi}{6} + 2\pi) = (5, \frac{13\pi}{6})$.

$(5, \frac{\pi}{6}) = (5, \frac{\pi}{6} - 2\pi) = (5, -\frac{11\pi}{6})$.

$(5, \frac{\pi}{6}) = (-5, \frac{\pi}{6} + \pi) = (-5, \frac{7\pi}{6})$.

$(5, \frac{\pi}{6}) = (-5, \frac{\pi}{6} - \pi) = (-5, -\frac{5\pi}{6})$.

To locate the point $(-5, \frac{7\pi}{6})$, we use the ray in the direction opposite the terminal side of $\frac{7\pi}{6}$ at a distance 5 units from the pole.

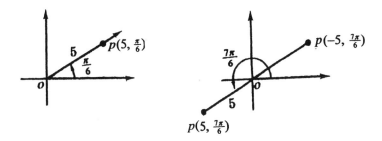

1. Express the point $(\sqrt{3}, 1)$ in terms of polar coordinates.
    Solution:
    $$r = \sqrt{x^2 + y^2} = \sqrt{(\sqrt{3})^2 + 1^2} = 2, \quad \tan\theta = \frac{y}{x} = \frac{1}{\sqrt{3}} \quad \therefore \theta = \frac{\pi}{6}.$$
    The polar coordinates of p are $(2, \frac{\pi}{6})$. Ans.

2. Find the polar coordinates of the point $(0, -5)$.
    Solution:
    The point lies on the negative $y-axis$.
    $r = 5$, $\theta = \frac{3\pi}{2}$. The polar coordinates of p are $(5, \frac{3\pi}{2})$. Ans.

3. Find the polar coordinates of $p(-1, -\sqrt{3})$.
    Solution:
    $$r = \sqrt{x^2 + y^2} = \sqrt{(-1)^2 + (-\sqrt{3})^2} = 2 . \quad \tan\alpha = \frac{-\sqrt{3}}{-1} = \sqrt{3}, \quad \alpha = \frac{\pi}{3}.$$
    The point lies on the 3rd quadrant. $\therefore \theta = \frac{\pi}{3} + \pi = \frac{4\pi}{3}$.
    The polar coordinates of p are $(2, \frac{4\pi}{3})$. Ans.

4. Find the rectangular coordinates of the point $(2, \frac{\pi}{6})$.
    Solution:
    $$x = r\cos\theta = 2\cos\frac{\pi}{6} = 2(\frac{\sqrt{3}}{2}) = \sqrt{3}$$
    $$y = r\sin\theta = 2\sin\frac{\pi}{6} = 2(\frac{1}{2}) = 1$$
    The rectangular coordinates of the point are $(\sqrt{3}, 1)$. Ans.

5. Transform the equation $xy = 2$ to polar form.
    Solution:
    $$xy = 2, \quad (r\cos\theta)(r\sin\theta) = 2, \quad r^2\cos\theta\sin\theta = 2, \quad r^2(\tfrac{1}{2}\sin 2\theta) = 2$$
    The polar form of the equation is $r^2\sin 2\theta = 4$. Ans.

6. Transform the equation $r^2\sin 2\theta = 4$ to rectangular coordinates.
    Solution:
    $$r^2\sin 2\theta = 4, \quad r^2(2\sin\theta\cos\theta) = 4, \quad r^2\sin\theta\cos\theta = 2$$
    $$(r\cos\theta)(r\sin\theta) = 2$$
    The rectangular coordinates of the equation is $xy = 2$. Ans.

Graphing Polar Equations

One method to graph a polar equation is to convert the equation to rectangular form. However, it is not always easy or helpful to graph a polar equation by converting it to rectangular form. Usually, we graph a polar equation by **checking for symmetry and plotting points**.

Checking for symmetry: (See Patterns of Special Polar Graphs on Page 303)

1. If we replace (r, θ) by $(r, -\theta)$ and the equation is unchanged, it is symmetry with respect to the polar axis.

2. If we replace (r, θ) by $(-r, \theta)$ and the equation is unchanged, it is symmetry with respect to the pole (the origin).

3. If we replace (r, θ) by $(r, \pi - \theta)$ and the equation is unchanged, it is symmetry with respect to the line $\theta = \frac{\pi}{2}$.

However, the above rules are not necessary conditions for symmetry. It is possible for a graph to have certain symmetry status which the above rules fail to test.

Examples:

1. Graph the polar equation $\theta = \dfrac{3\pi}{4}$.

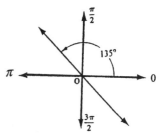

   Solution:

   Converting it to rectangular form.

   $\tan\theta = \frac{y}{x}$, $\tan\frac{3\pi}{4} = -\tan 45° = -1$, $\frac{y}{x} = -1$,

   $\therefore x + y = 0$

   It is a line passing through the pole at an angle of 135° with the polar axis (regardless of the value of r .)

2. Graph the polar equation $r = -4$.

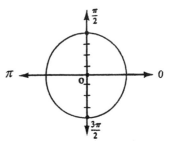

   Solution:

   Converting it to rectangular form.

   $r = \sqrt{x^2 + y^2} = -4$, $\therefore x^2 + y^2 = 16$

   It is a circle with center (0, 0) and radius 4.

3. Graph the polar equation $r = -4\cos\theta$.

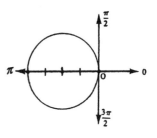

   Solution:

   Converting it to rectangular form.

   $r = -4\cos\theta$, $r^2 = -4r\cos\theta$, $x^2 + y^2 = -4x$

   $x^2 + y^2 + 4x = 0$

   Completing the square: $(x + 2)^2 + y^2 = 4$.

   It is a circle with center (−2, 0) and radius 2.

Examples

4. Graph the polar equation $r = 1 - \cos\theta$.

Solution:

Replacing θ by $(-\theta)$, $r = 1 - \cos(-\theta) = 1 - \cos\theta$.

The graph is symmetric with respect to the polar axis.
Plotting the points, we only need to select values of θ
from 0 to π .

θ	0	$\frac{\pi}{6}$	$\frac{\pi}{3}$	$\frac{\pi}{2}$	$\frac{2\pi}{3}$	$\frac{5\pi}{6}$	π
$\cos\theta$	1	$\frac{\sqrt{3}}{2}$	$\frac{1}{2}$	0	$-\frac{1}{2}$	$-\frac{\sqrt{3}}{2}$	-1
$r = 1 - \cos\theta$	0	0.13	0.5	1	1.5	1.87	2

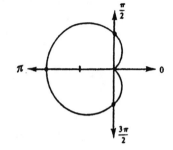

It is a **Cardioid**.
(A heart-shaped curve).

5. Graph the polar equation $r = 3 + 2\sin\theta$.

Solution:

Replacing θ by $(\pi - \theta)$, $r = 3 + 2\sin(\pi - \theta) = 3 + 2\sin\theta$.

The graph is symmetric with respect to the line $\theta = \frac{\pi}{2}$.

Plotting the points, we only need to select values of θ
from $-\frac{\pi}{2}$ to $\frac{\pi}{2}$.

θ	$-\frac{\pi}{2}$	$-\frac{\pi}{3}$	$-\frac{\pi}{6}$	0	$\frac{\pi}{6}$	$\frac{\pi}{3}$	$\frac{\pi}{2}$
$\sin\theta$	-1	$-\frac{\sqrt{3}}{2}$	$-\frac{1}{2}$	0	$\frac{1}{2}$	$\frac{\sqrt{3}}{2}$	1
$r = 3 + 2\sin\theta$	1	1.27	2	3	4	4.73	5

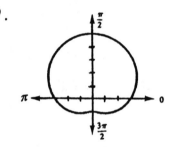

It is a **Limaçon**.

6. Graph the polar equation $r = 4\cos 2\theta$.

Solution:

Replacing θ by $(-\theta)$, $r = 4\cos 2(-\theta) = 4\cos 2\theta$.

Replacing θ by $(\pi - \theta)$,

$r = 4\cos 2(\pi - \theta) = 4\cos(2\pi - 2\theta) = 4\cos 2\theta$.

The graph is symmetric with respect to the polar axis
and the line $\theta = \frac{\pi}{2}$.

Plotting the points, we only need to select values of θ
from 0 to $\frac{\pi}{2}$.

θ	0	$\frac{\pi}{6}$	$\frac{\pi}{4}$	$\frac{\pi}{3}$	$\frac{\pi}{2}$
$\cos 2\theta$	1	$\frac{1}{2}$	0	$-\frac{1}{2}$	-1
$r = 4\cos 2\theta$	4	2	0	-2	-4

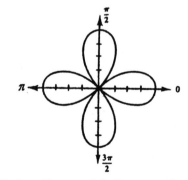

It is a **Rose** with four petals.

Example 7. Graph the polar equation $r = 1 + 2\sin\theta$.

Solution: Replacing θ by $(\pi - \theta)$: $r = 1 + 2\sin(\pi - \theta) = 1 + 2\sin\theta$

The graph is symmetric with respect to the line $\theta = \frac{\pi}{2}$.

Plotting the points, we only need to select values of θ from $-\frac{\pi}{2}$ to $\frac{\pi}{2}$.

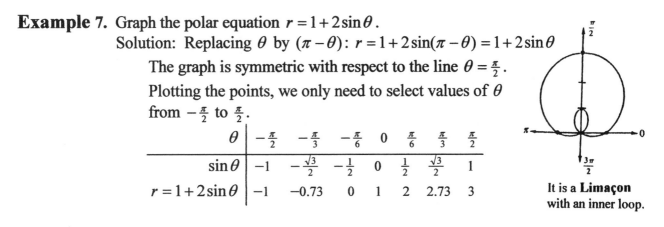

θ	$-\frac{\pi}{2}$	$-\frac{\pi}{3}$	$-\frac{\pi}{6}$	0	$\frac{\pi}{6}$	$\frac{\pi}{3}$	$\frac{\pi}{2}$
$\sin\theta$	-1	$-\frac{\sqrt{3}}{2}$	$-\frac{1}{2}$	0	$\frac{1}{2}$	$\frac{\sqrt{3}}{2}$	1
$r = 1 + 2\sin\theta$	-1	-0.73	0	1	2	2.73	3

It is a **Limaçon** with an inner loop.

Polar Forms of Conics: In order to find a convenient form to represent a conic, we locate a focus at the pole and apply the definition of a conic.

Definition of a conic: A conic is the set of all points such that the distance from a fixed point (called the focus) is in a constant ratio to its distance from a fixed line (called the directrix). The constant ratio is the eccentricity of the conic and is denoted by e.

$\dfrac{\overline{PF}}{\overline{PD}} = e \quad \therefore \overline{PF} = e \cdot \overline{PD}$

$r = e(p + r\cos\theta)$

$r = ep + er\cos\theta$

$r - er\cos\theta = ep$

$r(1 - e\cos\theta) = ep$

$\therefore r = \dfrac{ep}{1 - e\cos\theta}$ (Directrix is located at the left of the pole.)

Similarly, we can obtain the following polar forms of conics:

$r = \dfrac{ep}{1 + e\cos\theta}$ (Directrix is located at the right of the pole.)

$r = \dfrac{ep}{1 - e\sin\theta}$ (Directrix is located below the pole.)

$r = \dfrac{ep}{1 + e\sin\theta}$ (Directrix is located above the pole.)

Hint: $e = 1$ (parabola), $e < 1$ (ellipse), $e > 1$ (hyperbola)

Example 8. Identify and graph the equation $r = \dfrac{9}{3 - 3\sin\theta}$.

Solution: $r = \dfrac{3}{1 - \sin\theta}$ and $e = 1$

It is a parabola with focus at the pole.

$ep = 3 \quad \therefore p = 3$

Directrix is 3 units below the pole.

We plot two addition points $(3, 0°)$ and $(3, \pi)$ to assist in graphing.

EXERCISES

In the polar coordinates, a single point $p(r, \theta)$ has infinitely many represent-tations. Find other polar coordinates of each point.

1. The point $(3, \frac{2\pi}{3})$ for which $r > 0$, $-2\pi \leq \theta < 0$.

2. The point $(3, \frac{2\pi}{3})$ for which $r > 0$, $2\pi \leq \theta < 4\pi$.

3. The point $(3, \frac{2\pi}{3})$ for which $r < 0$, $0 \leq \theta < 2\pi$.

4. The point $(5, \frac{\pi}{4})$ for which $r > 0$, $-2\pi \leq \theta < 4\pi$.

5. The point $(5, \frac{\pi}{4})$ for which $r < 0$, $0 \leq \theta \leq 2\pi$.

Find the polar coordinates for each point.

6. $(-3, 0)$	**7.** $(0, -3)$	**8.** $(2, -2)$	**9.** $(\frac{3}{2}, \frac{\sqrt{3}}{2})$
10. $(-2\sqrt{2}, 2\sqrt{2})$	**11.** $(-1, -\sqrt{3})$	**12.** $(1.2, -3.4)$	**13.** $(-0.8, -2.4)$

Find the rectangular coordinates for each point.

14. $(-5, 0°)$	**15.** $(8, \frac{\pi}{6})$	**16.** $(-2, -\frac{\pi}{4})$	**17.** $(5, \pi)$

Transform each equation to rectangular form.

18. $r = 5$	**19.** $r = \frac{4}{\cos\theta}$	**20.** $r = \frac{-5}{\sin\theta}$	**21.** $r = \sin\theta$
22. $\sin\theta = 4r\cos^2\theta$	**23.** $r^2 = \sin\theta$	**24.** $r = \sin\theta + 2$	**25.** $r = \frac{2}{1-\sin\theta}$
26. $r = 4\sin\theta$	**27.** $\theta = \frac{\pi}{3}$	**28.** $\theta = -\frac{\pi}{4}$	**29.** $r\cos\theta = -5$

Transform each equation to polar form.

30. $y = 4x^2$	**31.** $4xy = 3$	**32.** $x^2 + y^2 = 2y$	**33.** $y = -5$
34. $4x^2 + 4y^2 = 1$	**35.** $y = x$	**36.** $y = \sqrt{3}\,x$	**37.** $x + 5 = 0$

Identify each polar equation by converting it to rectangular form.

38. $r = 5$	**39.** $r = -6$	**40.** $\theta = \frac{\pi}{6}$	**41.** $\theta = -\frac{\pi}{3}$
42. $r\sin\theta = 5$	**43.** $r\cos\theta = 5$	**44.** $r = 3\cos\theta$	**45.** $r = 4\sin\theta$
46. $r\sec\theta = 4$	**47.** $r = \frac{2}{1+\cos\theta}$		

Graph each polar equation.

48. $r = 1 - \sin\theta$	**49.** $r = 2(1+\cos\theta)$	**50.** $r = 3 + 2\cos\theta$	**51.** $r = 2\sin 2\theta$
52. $r = 2\cos 2\theta$	**53.** $r^2 = 4\cos 2\theta$	**54.** $r^2 = 4\sin 2\theta$	**55.** $r = 4\sin 3\theta$
56. $r = 2\cos 3\theta$	**57.** $r = 1 - 2\cos\theta$	**58.** $r = \theta$, $\theta \geq 0$	**59.** $r = e^{\theta/2}$

Identify each graph. Use a graphing calculator to graph each polar equation.

60. $r = \dfrac{4}{1+\sin\theta}$	**61.** $r = \dfrac{2}{1+2\cos\theta}$	**62.** $r = \dfrac{4}{4-2\sin\theta}$	**63.** $r = \dfrac{2}{4-5\cos\theta}$

8-11 Polar Coordinates involving Complex Numbers

A **complex number** $z = a + bi$ is formed when we solve an algebraic equation having an imaginary root.

A complex number $z = a + bi$ can be considered as a point in the **polar coordinates**. The polar coordinates used to represent the complex number is called the **complex plane.**

In a complex plane, the horizontal axis is called the **real axis**, and the vertical axis is called the **imaginary axis.**

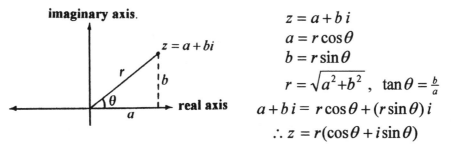

$$z = a + bi$$
$$a = r\cos\theta$$
$$b = r\sin\theta$$
$$r = \sqrt{a^2 + b^2} \ , \quad \tan\theta = \tfrac{b}{a}$$
$$a + bi = r\cos\theta + (r\sin\theta)\,i$$
$$\therefore z = r(\cos\theta + i\sin\theta)$$

The number r is called the **modulus** (or **amplitude**) of z. The angle θ is called an **argument** of z. Since there are infinitely many choices for θ, the polar form of a complex number is not unique. We choice the values of θ for $0 \le \theta < 2\pi$, or $\theta < 0$. The complex number i is represented by the point $(0, 1)$ and $i = 1(\cos 90° + i\sin 90°)$.

A complex number represented in polar form will give us an efficient and easy way to do some applications and computations (add, subtract, multiply, and divide).

To add or subtract two complex numbers, we use the **parallelogram rule.**

Example 1. Find the sum and difference of $z_1 = 2 + 6i$ and $z_2 = 4 - 2i$ by using the polar coordinates.

Solution:

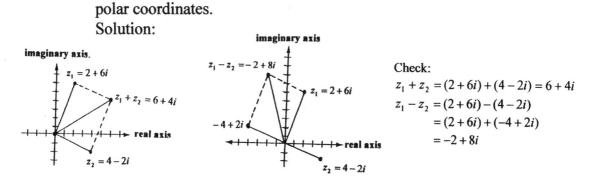

Check:
$$z_1 + z_2 = (2 + 6i) + (4 - 2i) = 6 + 4i$$
$$z_1 - z_2 = (2 + 6i) - (4 - 2i)$$
$$= (2 + 6i) + (-4 + 2i)$$
$$= -2 + 8i$$

To multiply or divide two complex numbers in polar forms, we use the following two formulas:　　$z_1 = r_1(\cos\theta_1 + i\sin\theta_1)$, 　$z_2 = r_2(\cos\theta_2 + i\sin\theta_2)$

1) $z_1 \cdot z_2 = r_1 r_2[\cos(\theta_1 + \theta_2) + i\sin(\theta_1 + \theta_2)]$

2) $\dfrac{z_1}{z_2} = \dfrac{r_1}{r_2}[\cos(\theta_1 - \theta_2) + i\sin(\theta_1 - \theta_2)]$, 　$z_2 \neq 0$

(See proofs on next page)

Given: $z_1 = r_1(\cos\theta_1 + i\sin\theta_1)$, $z_2 = r_2(\cos\theta_2 + i\sin\theta_2)$, find: **1.** $z_1 z_2$ **2.** $\dfrac{z_1}{z_2}$.

Solution: (See formulas of trigonometry on page 302)

1. $z_1 z_2 = r_1(\cos\theta_1 + i\sin\theta_1)\cdot r_2(\cos\theta_2 + i\sin\theta_2)$

$\qquad = r_1 r_2[(\cos\theta_1\cos\theta_2 - \sin\theta_1\sin\theta_2) + i(\sin\theta_1\cos\theta_2 + \cos\theta_1\sin\theta_2)]$

$\qquad = r_1 r_2[\cos(\theta_1 + \theta_2) + i\sin(\theta_1 + \theta_2)]$. Ans.

2. $\dfrac{z_1}{z_2} = \dfrac{r_1(\cos\theta_1 + i\sin\theta_1)}{r_2(\cos\theta_2 + i\sin\theta_2)} = \dfrac{r_1(\cos\theta_1 + i\sin\theta_1)}{r_2(\cos\theta_2 + i\sin\theta_2)} \cdot \dfrac{\cos\theta_2 - i\sin\theta_2}{\cos\theta_2 - i\sin\theta_2}$

$\qquad = \dfrac{r_1}{r_2} \cdot \dfrac{\cos\theta_1\cos\theta_2 + \sin\theta_1\sin\theta_2 + i(\sin\theta_1\cos\theta_2 - \cos\theta_1\sin\theta_2)}{\cos^2\theta_2 + \sin^2\theta_2}$

$\qquad = \dfrac{r_1}{r_2}\left[\dfrac{\cos(\theta_1 - \theta_2) + i\sin(\theta_1 - \theta_2)}{1}\right] = \dfrac{r_1}{r_2}[\cos(\theta_1 - \theta_2) + i\sin(\theta_1 - \theta_2)]$. Ans.

Example 2: Given $z_1 = 2\sqrt{2}(\cos 45° + i\sin 45°)$ and $z_2 = 3\sqrt{2}(\cos 45° + i\sin 45°)$.

Find $z_1 z_2$ and $\dfrac{z_1}{z_2}$.

Solution:

$\qquad z_1 z_2 = 2\sqrt{2}(\cos 45° + i\sin 45°) \cdot 3\sqrt{2}(\cos 45° + i\sin 45°)$

$\qquad\qquad = 2\sqrt{2}\cdot 3\sqrt{2}(\cos 90° + i\sin 90°) = 12(0 + i) = 12i$. Ans.

$\qquad \dfrac{z_1}{z_2} = \dfrac{2\sqrt{2}(\cos 45° + i\sin 45°)}{3\sqrt{2}(\cos 45° + i\sin 45°)} = \dfrac{2\sqrt{2}}{3\sqrt{2}}(\cos 0° + i\sin 0°) = \dfrac{2}{3}(1 + 0) = \dfrac{2}{3}$. Ans.

$\qquad\qquad$ Check: $z_1 z_2 = (2 + 2i)(3 + 3i) = 6 + 6i + 6i - 6 = 12i$

$\qquad\qquad\qquad \dfrac{z_1}{z_2} = \dfrac{2 + 2i}{3 + 3i} \cdot \dfrac{3 - 3i}{3 - 3i} = \dfrac{6 - 6i + 6i + 6}{9 + 9} = \dfrac{12}{18} = \dfrac{2}{3}$

To find the nth power of a complex number, we use De Moivre's theorem.

De Moivre's Theorem
For any positive integer n and $z = r(\cos\theta + i\sin\theta)$, then

$$z^n = r^n(\cos n\theta + i\sin n\theta)$$

To find the nth roots of a complex number, we use the Nth-root formula.

Nth root formula
For any positive integer n and $z = r(\cos\theta + i\sin\theta)$, then

$$z^{1/n} = r^{1/n}\left(\cos\frac{\theta + 2k\pi}{n} + i\sin\frac{\theta + 2k\pi}{n}\right), \quad k = 0, 1, 2, \cdots, n-1.$$

Examples

3. Express the complex number $z = 2\sqrt{3} - 2i$ in polar form.

Solution:

$$r = \sqrt{(2\sqrt{3})^2 + (-2)^2} = \sqrt{16} = 4, \quad \tan\theta = \frac{b}{a} = \frac{-2}{2\sqrt{3}} = -\frac{\sqrt{3}}{3}.$$

It lies in 4th quadrant, we have $\theta = 360° - 30° = 330°$.

$\therefore \ z = r(\cos\theta + i\sin\theta) = 4(\cos 330° + i\sin 330°)$. Ans.

4. Express $z = 4(\cos 330° + i\sin 330°)$ in standard form $a + bi$.

Solution:

$$z = 4(\cos 330° + i\sin 330°) = 4[\frac{\sqrt{3}}{2} + i(-\frac{1}{2})] = 2\sqrt{3} - 2i. \text{ Ans.}$$

5. Find z^3 of $z = 2(\cos 45° + i\sin 45°)$ in polar form.

Solution: $z^3 = 2^3[\cos 3(45°) + i\sin 3(45°) = 8(\cos 135° + i\sin 135°)$. Ans.

6. Find the cubic roots of $2 + 2i$ in standard form $a + bi$.

Solution:

$$r = \sqrt{2^2 + 2^2} = \sqrt{8}, \quad \tan\theta = \frac{b}{a} = \frac{2}{2} = 1, \quad \therefore \theta = 45°$$

$$(2 + 2i)^{\frac{1}{3}} = [\sqrt{8}(\cos 45° + i\sin 45°)]^{\frac{1}{3}} = (\sqrt{8})^{\frac{1}{3}}(\cos\frac{45° + 2k\pi}{3} + i\sin\frac{45° + 2k\pi}{3})$$

$$= [(2)^{\frac{3}{2}}]^{\frac{1}{3}}(\cos\frac{45° + 2k\pi}{3} + i\sin\frac{45° + 2k\pi}{3})$$

$$= \sqrt{2}(\cos\frac{45° + 2k\pi}{3} + i\sin\frac{45° + 2k\pi}{3})$$

Therefore, for $k = 0, 1, 2$, we have three cubic roots:

$k = 0$, $z_0 = \sqrt{2}(\cos 15° + i\sin 15°) = \sqrt{2}(0.9659 + i \cdot 0.2588) = 1.3658 + 0.3659i$

$k = 1$, $z_1 = \sqrt{2}(\cos 135° + i\sin 135°) = \sqrt{2}(-0.707 + i \cdot 0.707) = -1 + i$

$k = 2$, $z_2 = \sqrt{2}(\cos 255° + i\sin 255°) = \sqrt{2}(-0.2588 + i \cdot -0.9659) = -0.3659 - 1.3658i$

7. Find all solutions to the equation $x^2 + 4 = 0$ in standard form $a + bi$.

Solution: $x^2 = -4$, $x = (-4)^{\frac{1}{2}}$

Convert -4 into polar form: $-4 = 4(\cos\pi + i\sin\pi)$. It is the 3rd quadrant.

$$x = (-4)^{\frac{1}{2}} = 4^{\frac{1}{2}}[\cos\frac{\pi + 2k\pi}{2} + i\sin\frac{\pi + 2k\pi}{2}]$$

Therefore, for $k = 0, 1$, we have two square roots:

$k = 0$, $z_0 = 2[\cos 90° + i\sin 90°) = 2(0 + i) = 2i$

$k = 1$, $z_1 = 2[\cos 270° + i\sin 270°) = 2(0 - i) = -2i$

EXERCISES

Express each complex number in polar form.

1. $z = \sqrt{3} + i$ 2. $z = -2 - 2\sqrt{3}\,i$ 3. $z = -2 + 2\sqrt{3}\,i$

4. $z = 5\,i$ 5. $z = 5$ 6. $z = -4\,i$

Find $z_1 + z_2$ and $z_1 - z_2$ using parallelogram rule in the polar coordinates.

7. $z_1 = 4 + 2\,i$, $z_2 = 3 - 4\,i$ 8. $z_1 = 2 + 3\,i$, $z_2 = 2 - 3\,i$

9. $z_1 = -3 + 2\,i$, $z_2 = -2 + 3\,i$ 10. $z_1 = 3 + 5\,i$, $z_2 = 5 + 3\,i$

11. $z_1 = 3 - 5\,i$, $z_2 = 5 + 3\,i$ 12. $z_1 = 2 + 3\,i$, $z_2 = 3 - 4\,i$

Express each complex number in rectangular (standard) form $a + b\,i$.

13. $z = 2(\cos 30^\circ + i \sin 30^\circ)$ 14. $z = 4(\cos 45^\circ + i \sin 45^\circ)$

15. $z = 4(\cos 120^\circ + i \sin 120^\circ)$ 16. $z = 6(\cos 210^\circ + i \sin 210^\circ)$

17. $z = 5(\cos \frac{\pi}{2} + i \sin \frac{\pi}{2})$ 18. $z = 8(\cos \frac{7\pi}{4} + i \sin \frac{7\pi}{4})$

19. $z = 0.2(\cos 110^\circ + i \sin 110^\circ)$ 20. $z = 0.4(\cos 220^\circ + i \sin 220^\circ)$

Find $z_1 z_2$ and z_1/z_2 by using polar form.

21. $z_1 = 4(\cos 60^\circ + i \sin 60^\circ)$, $z_2 = 2(\cos 30^\circ + i \sin 30^\circ)$

22. $z_1 = 3(\cos \frac{2\pi}{3} + i \sin \frac{2\pi}{3})$, $z_2 = \frac{1}{2}(\cos \frac{\pi}{3} + i \sin \frac{\pi}{3})$

23. $z_1 = 3(\cos \frac{\pi}{9} + i \sin \frac{\pi}{9})$, $z_2 = 7(\cos \frac{5\pi}{9} + i \sin \frac{5\pi}{9})$

24. $z_1 = 6(\cos 60^\circ + i \sin 60^\circ)$, $z_2 = 3(\cos 240^\circ + i \sin 240^\circ)$

25. $z_1 = 1 - i$, $z_2 = 1 - \sqrt{3}\,i$

Evaluate each expression in rectangular (standard) form.

26. $z = (1 + i)^4$ 27. $z = [2(\cos \frac{\pi}{8} + i \sin \frac{\pi}{8})]^4$ 28. $z = [2(\cos 45^\circ + i \sin 45^\circ)]^3$

29. $z = (1 - i)^8$ 30. $z = (3 + 4i)^4$ 31. $z = (1 + i)^{20}$

32. Find the three cubic roots of i in polar form.

33. Find the three cubic roots of $27i$ in standard form.

34. Find all the fourth roots of $\sqrt{3} - i$ in polar form.

35. Solve $x^4 + 16 = 0$.

36. Solve $x^5 - i = 0$.

37. Find the three cubic roots of 1.

38. Find the three cubic roots of -1.

Equations of the Tangent Lines
on Conic Sections

Conic Sections	The slope (m) is known	$P(a_0, b_0)$ on the curve is known
Parabolas		
$y^2 = 4ax$	$y = mx + \frac{a}{m}$	$y_0 y = 2a(x_0 + x)$
$x^2 = 4ay$	$y = mx - am^2$	$x_0 x = 2a(y_0 + y)$
$(y-k)^2 = 4a(x-h)$	$y - k = m(x-h) + \frac{a}{m}$	$(y_0 - k)(y-k) = 2a[(x_0 - h) + (x-h)]$
$(x-h)^2 = 4a(y-k)$	$y - k = m(x-h) - am^2$	$(x_0 - h)(x-h) = 2a[(y_0 - k) + (y-k)]$
Circles		
$x^2 + y^2 = r^2$	$y = mx \pm r\sqrt{1+m^2}$	$x_0 x + y_0 y = r^2$
$(x-h)^2 + (y-k)^2 = r^2$	$y - k = m(x-h) \pm r\sqrt{1+m^2}$	$(x_0 - h)(x-h) + (y_0 - k)(y-k) = r^2$
Ellipses		
$\dfrac{x^2}{a^2} + \dfrac{y^2}{b^2} = 1$	$y = mx \pm \sqrt{a^2 m^2 + b^2}$	$\dfrac{x_0 x}{a^2} + \dfrac{y_0 y}{b^2} = 1$
$\dfrac{y^2}{a^2} + \dfrac{x^2}{b^2} = 1$	$y = mx \pm \sqrt{b^2 m^2 + a^2}$	$\dfrac{y_0 y}{a^2} + \dfrac{x_0 x}{b^2} = 1$
$\dfrac{(x-h)^2}{a^2} + \dfrac{(y-k)^2}{b^2} = 1$	$y - k = m(x-h) \pm \sqrt{a^2 m^2 + b^2}$	$\dfrac{(x_0 - h)(x-h)}{a^2} + \dfrac{(y_0 - k)(y-k)}{b^2} = 1$
$\dfrac{(y-k)^2}{a^2} + \dfrac{(x-h)^2}{b^2} = 1$	$y - k = m(x-h) \pm \sqrt{b^2 m^2 + a^2}$	$\dfrac{(y_0 - k)(y-k)}{a^2} + \dfrac{(x_0 - h)(x-h)}{b^2} = 1$
Hyperbolas		
$\dfrac{x^2}{a^2} - \dfrac{y^2}{b^2} = 1$	$y = mx \pm \sqrt{a^2 m^2 - b^2}$	$\dfrac{x_0 x}{a^2} - \dfrac{y_0 y}{b^2} = 1$
$\dfrac{y^2}{a^2} - \dfrac{x^2}{b^2} = 1$	$y = mx \pm \sqrt{-b^2 m^2 + a^2}$	$\dfrac{y_0 y}{a^2} - \dfrac{x_0 x}{b^2} = 1$
$\dfrac{(x-h)^2}{a^2} - \dfrac{(y-k)^2}{b^2} = 1$	$y - k = m(x-h) \pm \sqrt{a^2 m^2 - b^2}$	$\dfrac{(x_0 - h)(x-h)}{a^2} - \dfrac{(y_0 - k)(y-k)}{b^2} = 1$
$\dfrac{(y-k)^2}{a^2} - \dfrac{(x-h)^2}{b^2} = 1$	$y - k = m(x-h) \pm \sqrt{-b^2 m^2 + a^2}$	$\dfrac{(y_0 - k)(y-k)}{a^2} - \dfrac{(x_0 - h)(x-h)}{b^2} = 1$

Formulas in Basic Trigonometry

1. Reciprocal Relationships

$$\sin\theta = \frac{1}{\csc\theta}, \quad \cos\theta = \frac{1}{\sec\theta}, \quad \tan\theta = \frac{1}{\cot\theta}$$

$$\csc\theta = \frac{1}{\sin\theta}, \quad \sec\theta = \frac{1}{\cos\theta}, \quad \cot\theta = \frac{1}{\tan\theta}$$

2. Basic Relationships

$$\tan\theta = \frac{\sin\theta}{\cos\theta}$$

$$\cot\theta = \frac{\cos\theta}{\sin\theta}$$

3. The signs of the function in the various quadrants

	Quadrants			
Functions	1st.	2nd.	3rd.	4th.
$\sin\theta$ and $\csc\theta$	+	+	−	−
$\cos\theta$ and $\sec\theta$	+	−	−	+
$\tan\theta$ and $\cot\theta$	+	−	+	−

4. The values of special angles $30°, 45°, 60°$

θ	$\sin\theta$	$\cos\theta$	$\tan\theta$	$\cot\theta$	$\sec\theta$	$\csc\theta$
$30°, \dfrac{\pi}{6}$	$\dfrac{1}{2}$	$\dfrac{\sqrt{3}}{2}$	$\dfrac{\sqrt{3}}{3}$	$\sqrt{3}$	$\dfrac{2\sqrt{3}}{3}$	2
$45°, \dfrac{\pi}{4}$	$\dfrac{\sqrt{2}}{2}$	$\dfrac{\sqrt{2}}{2}$	1	1	$\sqrt{2}$	$\sqrt{2}$
$60°, \dfrac{\pi}{3}$	$\dfrac{\sqrt{3}}{2}$	$\dfrac{1}{2}$	$\sqrt{3}$	$\dfrac{\sqrt{3}}{3}$	2	$\dfrac{2\sqrt{3}}{3}$

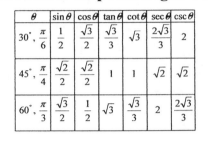

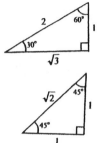

5. Pythagorean Relationships

$$\sin^2\theta + \cos^2\theta = 1$$
$$1 + \tan^2\theta = \sec^2\theta$$
$$1 + \cot^2\theta = \csc^2\theta$$

6. Triangular Relationships

The Law of Sines: $\dfrac{\sin A}{a} = \dfrac{\sin B}{b} = \dfrac{\sin C}{c}$

The Law of Cosines: $a^2 = b^2 + c^2 - 2bc\cos A$
$$b^2 = a^2 + c^2 - 2ac\cos B$$
$$c^2 = a^2 + b^2 - 2ab\cos C$$

7. Double-angle Formulas

$$\sin 2\theta = 2\sin\theta\cos\theta$$
$$\cos 2\theta = \cos^2\theta - \sin^2\theta$$
$$= 1 - 2\sin^2\theta$$
$$= 2\cos^2\theta - 1$$

$$\tan 2\theta = \frac{2\tan\theta}{1 - \tan^2\theta}$$

8. Half-angle Formulas

$$\sin\frac{A}{2} = \pm\sqrt{\frac{1 - \cos A}{2}}, \quad \cos\frac{A}{2} = \pm\sqrt{\frac{1 + \cos A}{2}}$$

$$\tan\frac{A}{2} = \pm\sqrt{\frac{1 - \cos A}{1 + \cos A}} = \frac{\sin A}{1 + \cos A} = \frac{1 - \cos A}{\sin A}$$

9. Sum and Difference Formulas

$$\sin(A + B) = \sin A\cos B + \cos A\sin B$$
$$\sin(A - B) = \sin A\cos B - \cos A\sin B$$
$$\tan(A + B) = \frac{\tan A + \tan B}{1 - \tan A\tan B}$$

$$\cos(A + B) = \cos A\cos B - \sin A\sin B$$
$$\cos(A - B) = \cos A\cos B + \sin A\sin B$$
$$\tan(A - B) = \frac{\tan A - \tan B}{1 + \tan A\tan B}$$

10. Formulas for Transformation

$\sin(90° - \theta) = \cos\theta$	$\sin(90° + \theta) = \cos\theta$	$\sin(180° - \theta) = \sin\theta$	$\sin(180° + \theta) = -\sin\theta$
$\cos(90° - \theta) = \sin\theta$	$\cos(90° + \theta) = -\sin\theta$	$\cos(180° - \theta) = -\cos\theta$	$\cos(180° + \theta) = -\cos\theta$
$\tan(90° - \theta) = \cot\theta$	$\tan(90° + \theta) = -\cot\theta$	$\tan(180° - \theta) = -\tan\theta$	$\tan(180° + \theta) = \tan\theta$
$\cot(90° - \theta) = \tan\theta$	$\cot(90° + \theta) = -\tan\theta$	$\cot(180° - \theta) = -\cot\theta$	$\cot(180° + \theta) = \cot\theta$
$\sec(90° - \theta) = \csc\theta$	$\sec(90° + \theta) = -\csc\theta$	$\sec(180° - \theta) = -\sec\theta$	$\sec(180° + \theta) = -\sec\theta$
$\csc(90° - \theta) = \sec\theta$	$\csc(90° + \theta) = \sec\theta$	$\csc(180° - \theta) = \csc\theta$	$\csc(180° + \theta) = -\csc\theta$
$\sin(270° - \theta) = -\cos\theta$	$\sin(270° + \theta) = -\cos\theta$	$\sin(360° - \theta) = -\sin\theta$	$\sin(360° + \theta) = \sin\theta$
$\cos(270° - \theta) = -\sin\theta$	$\cos(270° + \theta) = \sin\theta$	$\cos(360° - \theta) = \cos\theta$	$\cos(360° + \theta) = \cos\theta$
$\tan(270° - \theta) = \cot\theta$	$\tan(270° + \theta) = -\cot\theta$	$\tan(360° - \theta) = -\tan\theta$	$\tan(360° + \theta) = \tan\theta$
$\cot(270° - \theta) = \tan\theta$	$\cot(270° + \theta) = -\tan\theta$	$\cot(360° - \theta) = -\cot\theta$	$\cot(360° + \theta) = \cot\theta$
$\sec(270° - \theta) = -\csc\theta$	$\sec(270° + \theta) = \csc\theta$	$\sec(360° - \theta) = \sec\theta$	$\sec(360° + \theta) = \sec\theta$
$\csc(270° - \theta) = -\sec\theta$	$\csc(270° + \theta) = -\sec\theta$	$\csc(360° - \theta) = -\csc\theta$	$\csc(360° + \theta) = \csc\theta$

Patterns of Special Polar Graphs

1. **Cardioid** (a heart-shaped curve).

   $r = a(1 \pm \sin\theta)$, $a > 0$. It is symmetric with respect to the line $\theta = \frac{\pi}{2}$.

   $r = a(1 \pm \cos\theta)$, $a > 0$. It is symmetric with respect to the polar axis.

2. **Rose**.

   $r = a \sin n\theta$, $a > 0$.

   $r = a \cos n\theta$, $a > 0$.

   There are n petals if n is odd, and $2n$ petals if n is even.

   If $n = 1$, there is one petal and it is a circle.

3. **a) Limaçon** without inner loop.

   $r = a \pm b \sin\theta$, $a > b > 0$.

   $r = a \pm b \cos\theta$, $a > b > 0$.

   b) Limaçon with inner loop.

   $r = a \pm b \sin\theta$, $b > a > 0$.

   $r = a \pm b \cos\theta$, $b > a > 0$.

4. **Lemniscate** (a propeller-shaped curve). See Problems 53 and 54 on page 296.

   $r^2 = a^2 \sin 2\theta$, $a \neq 0$.

   $r^2 = a^2 \cos 2\theta$, $a \neq 0$.

5. **Spiral**. See Problems 58 and 59 on page 296.

   $r = k\theta$. It is called **spiral of Archimedes**.

   $r = a \cdot b^{k\theta}$. It is called a **logarithmic spiral**.

A patient said to his dentist.
  Patient: Why are you charging me $300 ?
              Last time, you charged me only $100.
  Dentist: Two patients in the waiting room ran
              away when they heard you screaming.

Distance Formulas between Lines and Planes

1. The distance between a point (x_1, y_1) and a straight line $ax + by + c = 0$ is given by the formula:

$$d = \overline{PQ} = \frac{|ax_1 + by_1 + c|}{\sqrt{a^2 + b^2}}$$

2. The distance between two parallel lines $ax + by + c_1 = 0$ and $ax + by + c_2 = 0$ is given by the formula:

$$d = \overline{PQ} = \frac{|c_1 - c_2|}{\sqrt{a^2 + b^2}}$$

3. The distance between a point (x_1, y_1, z_1) and a plane $ax + by + cz + d = 0$ is given by the formula:

$$D = \overline{PQ} = \frac{|ax_1 + by_1 + cz_1 + d|}{\sqrt{a^2 + b^2 + c^2}}$$

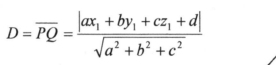

Examples

1. Find the distance between the line $2x + 4y - 5 = 0$ and the point $(3, -1)$.
   Solution:

$$d = \frac{|ax_1 + by_1 + c|}{\sqrt{a^2 + b^2}} = \frac{|2 \cdot 3 + 4(-1) + (-5)|}{\sqrt{2^2 + 4^2}} = \frac{|-3|}{\sqrt{20}} = \frac{3}{4.472} = 0.67.$$

2. Find the distance between the two parallel lines $x - 2y + 3 = 0$ and $x - 2y - 5 = 0$.
   Solution:

$$d = \frac{|c_1 - c_2|}{\sqrt{a^2 + b^2}} = \frac{|3 - (-5)|}{\sqrt{1^2 + (-2)^2}} = \frac{8}{\sqrt{5}} = \frac{8}{2.236} = 3.58.$$

3. Find the distance between the point $(1, 2, 1)$ and the plane $2x + y - z - 1 = 0$.
   Solution:

$$d = \frac{|ax_1 + by_1 + cz_1 + d|}{\sqrt{a^2 + b^2 + c^2}} = \frac{|2 \cdot 1 + 1 \cdot 2 - 1 \cdot 1 - 1|}{\sqrt{2^2 + 1^2 + (-1)^2}} = \frac{2}{\sqrt{6}} = \frac{2}{2.449} = 0.82.$$

4. Two parallel planes are $2x - 3y + z - 1 = 0$ and $2x - 3y + z - 5 = 0$. Find the distance between the planes.
   Solution:

   Find one point on the first plane $(0, 0, 1)$

$$d = \frac{|ax_1 + by_1 + cz_1 + d|}{\sqrt{a^2 + b^2 + c^2}} = \frac{|0 + 0 + 1 - 5|}{\sqrt{4 + 9 + 1}} = \frac{4}{\sqrt{14}} = \frac{4}{3.742} = 1.07.$$

Kepler's Three Laws of Planetary Motion

Kepler's Three Laws of Planetary Motion were first introduced on the basis of observation by Johannes Kepler (1571-1630), a German mathematician and physicist. These Laws were later proved by Isaac Newton (1642-1727) and are now playing a crucial role in modern science and astronomy. Kepler's Laws show that they apply to any object that is driven by a force that obeys the inverse square law (The Law of Universal Gravitation). The Law of University Gravitation: See the book " A-Plus Notes for Beginning Algebra" on Page 41.

Kepler's Three Laws of Planetary Motion

1. **Kepler's First Law:** Each planet moves in an elliptical orbit with the sun as a focus.

2. **Kepler's Second Law:** A ray from the sun to the planet sweeps out equal areas in equal times.

3. **Kepler's Thirs Law:** The square of the orbital period (T) is proportional to the cube of the mean distance (a) from the sun.

$$T^2 = K \, a^3$$

T : The time (in years) that take a planet to go around the sun once.

a: The mean distance (in AU) is the semimajor axis of the planet's elliptical orbit.

AU (Astronomical Unit): It is the distance from the earth to the sun.

1 AU = 150,000,000 meters.

K: The constant of proportionality. It is the same ($K = 1$) for every planet in the solar system.

Examples: 1. For the Earth:

$T = 1$ year, $a = 1$ AU, $K = 1$

2. For the Mars:

$T = 1.88$ years, $K = 1$

$T^2 = a^3$

$(1.88)^2 = a^3$, $3.5344 = a^3$

$\therefore a = \sqrt[3]{3.5344} = 1.523$ AU $= 1.523 \times 150,000,000$

$= 228.45$ million meters.

Space for Taking Notes

CHAPTER 8   EXERCISES

Find the distance and midpoint between the given points.

1. (2, 6) and (4, 3)   **2.** (–2, 6) and (4, –3)   **3.** (–2, –6) and (–4, –3)

4. (2, –6) and (–4, 3)   **5.** (2, 0) and (0, 3)   **6.** (–2, 0) and (0, –3)

7. (0, –6) and (0, 0)   **8.** $(2, \sqrt{3})$ and $(4, -2\sqrt{3})$ **9.** $(\sqrt{3}, 1)$ and $(-1, \sqrt{3})$

10. $(2, -\sqrt{3})$ and $(6, \sqrt{3})$ **11.** $(3\sqrt{3}, -\sqrt{2})$ and $(-\sqrt{3}, 3\sqrt{2})$ **12.** $(1, \frac{2}{3})$ and (4, 1)

13. $(1, \frac{2}{3})$ and $(4, -\frac{1}{2})$ **14.** $(1, -\frac{2}{3})$ and $(\frac{1}{2}, \frac{2}{3})$ **15.** $(\sqrt{2}, \sqrt{3}+2)$ and $(3\sqrt{2}, \sqrt{3}-2)$

If M is the midpoint of $\overline{PQ}$, find the coordinates of Q .

16. $M(-2, 3)$, $P(-1, 4)$ **17.** $M(-1, -1.5)$, $P(0, -3)$ **18.** $M(\sqrt{3}, \sqrt{2})$, $P(3\sqrt{3}, -\sqrt{2})$

Graph each equation.

19. $x^2 + y^2 = 121$ **20.** $x^2 + y^2 - 144 = 0$ **21.** $x^2 + y^2 - 6 = 0$

22. $4x^2 + 9y^2 = 36$ **23.** $4x^2 + 25y^2 = 100$ **24.** $25x^2 + 9y^2 - 900 = 0$

25. $4x^2 - 9y^2 = 36$ **26.** $4x^2 - 25y^2 = 100$ **27.** $25y^2 - 9x^2 = 900$

28. $36x^2 + 4y^2 = 9$ **29.** $36x^2 - 4y^2 = 9$ **30.** $y^2 - x^2 = 100$

31. $x^2 + 2x + 8y + 1 = 0$ **32.** $x^2 + 6x - 4y + 17 = 0$ **33.** $5x^2 + 9y^2 - 10x - 40 = 0$

34. $x^2 + 4y^2 - 2x + 16y + 13 = 0$ **35.** $4x^2 - y^2 + 8x + 4y + 4 = 0$

36. $x^2 + y^2 - 4x + 2y - 11 = 0$ **37.** $3x^2 - y^2 - 18x - 6y = 0$

38. A flashlight mirror is shaped like a parabola. Its diameter is 6 inches and its depth is 2 inches. Where should the bulb be placed from the vertex so that the emitted rays will be parallel to the axis ?

39. The equation of the tangent line to the parabola $(x-h)^2 = 4a(y-k)$ from a point (x, y) not on the parabola is given by the formula $y - k = m(x-h) - am^2$, where m is the slope of the tangent line. Find the equation of the tangent line to $y = x^2 + 3x$ from the point $(0, -1)$.

40. The equation of the tangent line to the circle $x^2 + y^2 = r^2$ from a point (x, y) not on the circle is given by the formula $y = mx \pm r\sqrt{1+m^2}$, where m is the slope of the tangent line Find the equation of the tangent line to the circle $x^2 + y^2 = 4$ from the point (2, 3).

41. Prove that the equation of the tangent line to the circle $(x-h)^2 + (y-k)^2 = r^2$ at a point $p(x_0, y_0)$ on the circle is given by the formula $(x_0 - h)(x-h) + (y_0 - k)(y-k) = r^2$.

42. Use the formula described in Problem 41 to find the equation of the tangent line that is tangent to the circle $x^2 + y^2 - 8x + 12y + 42 = 0$ at the point (5, –3).

(See Chart of Equations of Tangent Lines on Page 301)   -----Continued-----

Solve each system of quadratic equations.

43. $\begin{cases} x^2 + y = 2 \\ x - y = 0 \end{cases}$

44. $\begin{cases} x^2 - 2y = 2 \\ x - y = 1 \end{cases}$

45. $\begin{cases} x^2 - 2y = 2 \\ x^2 + 2y = 4 \end{cases}$

46. $\begin{cases} x^2 + y^2 = 4 \\ x^2 + y = 2 \end{cases}$

47. $\begin{cases} x^2 - y^2 = 4 \\ x^2 + y = 6 \end{cases}$

48. $\begin{cases} 2x^2 - y^2 = 4 \\ x^2 + y^2 = 8 \end{cases}$

49. $\begin{cases} x^2 + y^2 = 8 \\ xy = 4 \end{cases}$

50. $\begin{cases} x^2 - 4y - 8 = 0 \\ x + 2y = 0 \end{cases}$

51. $\begin{cases} 4x^2 + 9y^2 = 36 \\ y^2 - x = 4 \end{cases}$

52. Find the equation of the line that pass (–3, 2) and is perpendicular to the line $2x - 3y = 4$.
53. Find the equation of the line that pass (–3, 2) and is perpendicular to the line $2x + 3y = 4$.
54. Find the equation of the line that pass (–3, 2) and is parallel to the line $2x - 3y = 4$.
55. Find the equation of the line that pass (0, 0) and is parallel to the line between two points (4, 5) and (2, –7).
56. Find the equation of the line that is perpendicular bisector of the segment between two points (4, 5) and (2, –7).
57. Find the distance from the point (4, 5) to the line $2x - y = 5$.(See formula on Page 304)
58. Find the equation of the circle having center (2, –3) and radius 6.
59. Find the equation of the circle having center (2, –3) and passing through the point (–4, 2).
60. Find the equation of the circle having center on the line $x - y = 5$ and tangent to the x-axis at (8, 0).
61. Find the equation of the circle having center on the line $x = 4$ and tangent to the y-axis at (0, 8).
62. Find the equation of the parabola with focus (2, –3) and directrix $y - 3 = 0$.
63. Find the equation of the parabola with focus (2, –3) and directrix $x + 3 = 0$.
64. Find the equation of the parabola with focus (2, –3) and vertex (–1, –3).
65. Find the equation of the parabola with focus (–2, 3) and directrix $x + 3 = 0$.
66. Find the equation of the parabola with vertex (–1, –3) and directrix $x - 3 = 0$.
67. Find the equation of the ellipse having foci (4, 0) and (–4, 0) and the sum of its focal radii equals to 10.
68. Find the equation of the ellipse having foci (0, 4) and (0, –4) and the sum of its focal radii equals to 12.
69. Find the equation of the hyperbola having foci (0, 6) and (0, –6) and the difference of its focal radii is 4.
70. Find the equation of the hyperbola having foci (5, 0) and (–5, 0) and the difference of its focal radii is 6.
71. Find the equation of the hyperbola having foci (6, 2) and (–4, 2) and the difference of its focal radii is 8.
72. Find the eccentricity and latus rectum of the parabola $(y + 4)^2 = 8(x - 2)$.

-----Continued-----

73. Find the eccentricity and latus rectum of the ellipse $5x^2 + 4y^2 = 20$.

74. Find the eccentricity and latus rectum of the hyperbola $9x^2 - 16y^2 - 18x + 64y - 199 = 0$.

75. Find the equation of the hyperbola with center $(-1, 2)$, vertex $(-1, 4)$, and $e = \frac{\sqrt{5}}{2}$.

76. Find the foci of the ellipse $9(x-3)^2 + 4(y+1)^2 = 36$.

77. Find the foci of the ellipse $4(x+5)^2 + 16(y-4)^2 = 64$.

78. Find the area of the ellipse $4x^2 + 16y^2 = 64$. (Hint: $A = \pi ab$)

79. Find the area of the ellipse $9x^2 + 4y^2 + 36x + 24y + 36 = 0$. (Hint: $A = \pi ab$)

80. The arch over a tunnel for a freeway is to be constructed in the shape of an upper half of a semiellipse. The opening is to have a major axis 24 meters and a height at the center of 8 meters. Using the rectangular coordinate system to write the equation of the elliptical arch of the tunnel.

81. The moon's orbit around the earth is an ellipse with the earth at one focus. The orbit is nearly circular. The eccentricity is $e = 0.0525$ and the length of the major axis is about 768.80 million meters. Find the equation of the moon's orbit.

82. Prove that the hyperbola $\dfrac{y^2}{a^2} - \dfrac{x^2}{b^2} = 1$ has two asymptotes $y = \pm \dfrac{a}{b}x$.

83. Prove that the hyperbola $\dfrac{(x-h)^2}{a^2} - \dfrac{(y-k)^2}{b^2} = 1$ has two asymptotes $y - k = \pm \dfrac{b}{a}(x-h)$.

84. Use the formula described in Problem 38 on page 272 to find the equation of the hyperbola having vertex $(4, 1)$ and the equations of asymptotes $2x + 3y - 5 = 0$ and $2x - 3y + 1 = 0$.

85. Use the formula described in Problem 40 on page 272 to find the equation of the tangent line to the hyperbola $x^2 - 2y^2 = 7$ from the point $(-, 5)$ not on the curve.

86. Transform the point $(-2, 3)$ by rotating the axes through an angle $60°$.

87. Express the equation $xy = 1$ by rotating the axes through a $30°$ angle into an $x'y'$ - system.

88. Express the equation $xy = 1$ by rotating the axes into an $x'y'$ - system having no $x'y'$ - term.

89. Prove that the equation $xy = 1$ and its rotated equation $(x')^2 - (y')^2 = 2$ are invariant under rotation of axes.

90. Use discriminant to identify the graph of $x^2 - 4xy - 2y^2 + 5 = 0$.

91. Use discriminant to identify the graph of $2x^2 - 4xy + 3y^2 - 10 = 0$.

92. Identify the graph of $2x^2 + 3xy - 2y^2 + x - 3y - 1 = 0$. (Hint: A degenerate case)

93. Find the parametric equations of the line passing the two points $(-1, \sqrt{3})$ and $(0, 2\sqrt{3})$.

94. Graph the parametric equation $x = t + 2$, $y = t - 5$; $-3 \le t \le 2$. State the domain and indicate the orientation of the graph.

95. Graph the parametric equation $x = 2\cos t$, $y = \sin t$; $0 \le t \le \pi$. State the domain and indicate the orientation of the curve.

96. Use the formula described in Problem 35 on page 290 to write the parametric equations of a rocket with an initial velocity 240 feet per second at an angle of $60°$ to the horizontal.

-----**Continued**-----

97. Find the three additional polar representations of the point $(15, -\frac{2}{3}\pi)$, $-2\pi < \theta < 2\pi$.

98. Convert the point $(-2, 2)$ to polar coordinates.

99. Convert the point $(-2.1, 4.3)$ to polar coordinates.

100. Convert the point $(\sqrt{3}, \frac{\pi}{3})$ to rectangular coordinates.

101. Convert the polar equation $r = 3$ to rectangular form.

102. Convert the polar equation $r = \sec\theta$ to rectangular form.

103. Convert the polar equation $r = \sin\theta$ to rectangular form.

104. Convert the polar equation $r = \dfrac{1}{2 - 2\cos\theta}$ to rectangular form.

105. Convert the rectangular equation $x^2 + y^2 = 4y$ to polar form.

106. Convert the rectangular equation $x = 10$ to polar form.

107. Convert the rectangular equation $3x - 6y + 8 = 0$ to polar form.

108. Convert the rectangular equation $y^2 - 2x - 1 = 0$ to polar form.

109. Use a graphing calculator to graph the equation $\theta = \dfrac{\pi}{3}$.

110. Use a graphing calculator to graph $r = 5$.

111. Use a graphing calculator to graph $r = 3\cos 2\theta$.

112. Use a graphing calculator to graph $r = 1 + 4\cos\theta$.

113. Use a graphing calculator to graph $r^2 = 16\cos 2\theta$

114. Use a graphing calculator to graph $r = \dfrac{3}{1 - 2\cos\theta}$.

115. Express the complex number $z = -2 - 2\sqrt{3}\,i$ in polar form.

116. Express the complex number $z = 3(\cos 60° + i\sin 60°)$ in rectangular form $a + bi$.

117. Express the complex number $z = 5(\cos 95° + i\sin 95°)$ in rectangular form $a + bi$.

118. Find $z_1 z_2$ if $z_1 = 3(\cos 20° + i\sin 20°)$ and $z_2 = 5(\cos 30° + i\sin 30°)$.

119. Find $\dfrac{z_1}{z_2}$ if $z_1 = 4(\cos\pi + i\sin\pi)$ and $z_2 = 8(\cos\dfrac{\pi}{4} + i\sin\dfrac{\pi}{4})$.

120. Evaluate $(-1 + i)^{10}$ in standard form $a + bi$.

121. Find the three cubic root of $z = -2 + 2i$ in polar form and θ in degrees.

122. Find the three cubic root of $z = -1 + \sqrt{3}\,i$ in standard form $a + bi$.

123. The earth's orbit around the sun is an ellipse with the sun at one focus. The orbit is nearly circular. The eccentricity of the orbit is $e = 0.0167$ and the length of the major axis is about 186 million miles. Find the equation of the earth's orbit in polar form.
(See Problem 27 on page 276)

Answers

CHAPTER 8 – 1　Distance and Midpoint Formulas　page 257

1. $\sqrt{10}$　**2.** $\sqrt{10}$　**3.** $3\sqrt{10}$　**4.** $\sqrt{82}$　**5.** $3\sqrt{2}$　**6.** $\sqrt{10}$　**7.** $\sqrt{74}$　**8.** 2　**9.** $\sqrt{74}$　**10.** 5.009　**11.** $\sqrt{7}$　**12.** $\sqrt{15}$　**13.** $6\sqrt{2}$

14. $2\sqrt{3}$　**15.** $\frac{1}{2}\sqrt{5}$　**16.** $\frac{1}{6}\sqrt{165}$　**17.** $\frac{5}{4}\sqrt{2}$　**18.** $\frac{1}{6}\sqrt{193}$　**19.** $\sqrt{63}$　**20.** $\sqrt{70}$　**21.** $(3, 7)$　**22.** $(9, 6)$　**23.** $(\frac{1}{2}, 4)$

24. $(-2, -3)$　**25.** $(-6, -6)$　**26.** $(10, 0)$　**27.** $(2, 11)$　**28.** $(-3.6, -0.3)$　**29.** $(4, 2\sqrt{5})$　**30.** $((-\sqrt{5}, -1)$　**31.** $(4.5, 1)$

32. $(\frac{1}{2}, 1)$　**33.** $(2\sqrt{3}, \sqrt{2})$　**34.** $(\sqrt{2}, -1)$　**35.** $(2\sqrt{7}, 3\sqrt{5})$　**36.** $P(11, 0)$　**37.** $B(\frac{1}{2}, -\frac{11}{2})$　**38.** $B(9, -9)$　**39.** $x - 2y = 7$

40. $x - 4y = -8$　**41.** Isosceles

CHAPTER 8 – 2　Conic Sections and Parabolas　page 262

1. $y^2 = 8x$　**2.** $(y-2)^2 = 8(x-2)$　**3.** $x^2 = 8y$　**4.** $(x+1)^2 = -8y$　**5.** $y^2 = -8x$　**6.** $(y-2)^2 = 4(x+1)$　**7.** $x^2 = -4y$

8. $(x-1)^2 = 4(y+2)$　**9.** $y^2 = 12x$　**10.** $y^2 = -12x$　**11.** $x^2 = 12y$　**12.** $x^2 = -12y$　**13.** $(y-3)^2 = 8(x-2)$

14. $(y-3)^2 = -8(x-2)$　**15.** $(x-2)^2 = 8(y-3)$　**16.** $(x-2)^2 = -8(y-3)$　**17.** $(x-2)^2 = 12(y-1)$

18. $(y-3)^2 = 8(x+2)$　**19.** $(x-2)^2 = -8(y+3)$　**20.** $(y+2)^2 = -8(x-4)$　**21.** $x^2 = -6y$　**22.** $y^2 = \frac{5}{2}x$　**23.** $y^2 = 16x$

24. $y^2 = -16x$　**25.** $x^2 = 16y$　**26.** $x^2 = -8(y-2)$　**27.** $x^2 = -16(y-4)$　**28.** $(y-4)^2 = 12(x+1)$　**29.** $x^2 = -4y$

30. $(x+3)^2 = 4(y-3)$　**31.** V(0, 0), open to the right, F(8, 0), directrix $x = -8$, latus lactum = 32.

33. V(0, 0), open up, F(0, $\frac{5}{8}$), directrix $y = -\frac{5}{8}$, latus rectum = $\frac{5}{2}$　**35.** V(1, 4), open to the right, F(5, 4),

directrix $x = -3$, latus rectum = 16　**37.** V(5, -2), open to the left, F(4, -2), directrix $x = 6$, latus rectum = 4

39. V(-4, -2), open up, F(-4, -1), directrix $y = -3$, latus rectum = 4　**41.** V(-3, 3), open up, F(-3, 4),

directrix $y = 2$, latus rectum = 4　**43.** 2.25 feet

44. Hint: Eliminate x by substituting the line $y - y_0 = m(x - x_0)$ into $y^2 = 4x$.

We have $my^2 - 4ay + (4ay_0 - 4amx_0) = 0$. Since they intersect at one point, we apply the discriminant $b^2 - 4ac = 0$ to find $m = 2a/y_0$. Substitute m value into the line, we have the equation of the tangent line $y_0 y = 2a(x + x_0)$.　**45.** $x - y + 2 = 0$

CHAPTER 8 – 3　Circles　page 264

1. $x^2 + y^2 = 14$　**2.** $(x-1)^2 + (y-2)^2 = 4$　**3.** $(x-3)^2 + (y-3)^2 = 4$　**4.** $(x-2)^2 + (y-1)^2 = 8$

5. $(x-4)^2 + (y+1)^2 = 10$　**6.** $x^2 + y^2 = 1$　**7.** $x^2 + (y-5)^2 = 1$　**8.** $(x+5)^2 + y^2 = 1$　**9.** $(x-3)^2 + (y-5)^2 = 9$

10. $(x-5)^2 + (y+3)^2 = 9$　**11.** $(x+3)^2 + (y-5)^2 = 25$　**12.** $(x+1)^2 + (y+6)^2 = 64$　**13.** $(x-4)^2 + (y+2)^2 = \frac{1}{4}$

14. $(x-\frac{1}{2})^2 + y^2 = \frac{1}{4}$　**15.** C(0, 0), $r = 4$　**16.** C(0, 0), $r = 3\sqrt{2}$　**17.** C(0, 0), $r = 2\sqrt{2}$　**18.** C(4, 0), $r = 6$

19. (0, -6), $r = 6$　**20.** C(-4, 6), $r = 3\sqrt{2}$　**21.** C(4, 0), $r = 3\sqrt{2}$　**22.** C(0, -6), $r = \sqrt{10}$　**23.** C(-4, 6), $r = \sqrt{10}$

24. C(3, 0), $r = 2$　**25.** C(0, -2), $r = 2$　**26.** C(1, -2), $r = 3$　**27.** C(-1, -2), $r = \sqrt{14}$　**28.** C(4, -6), $r = 2\sqrt{2}$

29. C($-\sqrt{5}$, 1), $r = 3$　**30.** $x^2 + y^2 - 6x - 10y + 9 = 0$　**31.** $x^2 + y^2 - 6x - 10y + 25 = 0$　**32.** $3x + 5y - 34 = 0$

33. $3x - 5y - 34 = 0$　**34.** $3x + \sqrt{7}y - 16 = 0$　**35.** $x + 3y + 4 = 0$　**36.** $(x-1)^2 + (y-2)^2 = 18$

37. Hint: The slope of the line passing (0, 0) and (x_0, y_0) is y_0/x_0. The slope of the tangent line is $-x_0/y_0$.

We have the equation of the tangent line by applying points (x_0, y_0) and (x, y) and slope $-x_0/y_0$.

38. $3x + 5y - 34 = 0$ (See problem 32. Which method is easier ?)

39. Hint: Eliminate y by substituting the line $y = mx + b$ into $x^2 + y^2 = r^2$.

We have $(1 + m^2)x^2 + 2mbx + (b^2 - y^2) = 0$. Since they intersect at one point, we apply the discriminant $b^2 - 4ac = 0$ to find $b = \pm r\sqrt{1 + m^2}$. Substitute b value into the line, we have the equation of the tangent line $y = mx \pm r\sqrt{1 + m^2}$.

40. $y = \pm\dfrac{\sqrt{21}}{2}x + 5$

CHAPTER 8 – 4   Ellipses   page 268

1 ~ 15 graphs are left to the students

1. C(0, 0); V: (5, 0), (–5, 0); F: (4, 0), (–4, 0)   **2.** C(0, 0); V: (0, 5), (0, –5); F: (0, 4), (0, –4)

3. C(1, –4); V: (6, –4), (–4, –4); F: (5, –4), (–3, –4)   **4.** C(0,0); V: (0, 4), (0, –4); F: $(0, 2\sqrt{3})$, $(0, -2\sqrt{3})$

5. C(0, 0); V:(4, 0),(–4, 0); F:$(2\sqrt{3}, 0)$,$(-2\sqrt{3}, 0)$ **6.** C(–1,4); V:(–1,0),(–1, 8); F:$(-1, 4 - 2\sqrt{3})$,$(-1, 4 + 2\sqrt{3})$

7. C(0, 0); V: (0, 2), (0, –2); F: $(0, \sqrt{3})$, $(0, -\sqrt{3})$ **8.** C(0, 0); V: (2, 0), (–2, 0); F: $(\sqrt{3}, 0)$, $(-\sqrt{3}, 0)$

9. C(3, 0); V: (3, 2), (3, –2); F: $(3, \sqrt{3})$, $(3, -\sqrt{3})$ **10.** C(0, 0); V: (0, 3), (0, –3); F: $(0, \sqrt{5})$, $(0, -\sqrt{5})$

11. C(0, 0); V:(0, 5),(0,–5); F:(0, 3),(0,–3)   **12.** C(–3, –2); V:(1,–2),(–7,–2); F:$(2\sqrt{3} - 3, -2)$,$(-2\sqrt{3} - 3, -2)$

13. C(1, 0); V:(1, 3),(1,–3); F:$(1, 2\sqrt{2})$,$(1, -2\sqrt{2})$ **14.** C(0, –1); V: (4, –1), (–4, –1); F: $(2\sqrt{3}, -1)$, $(-2\sqrt{3}, -1)$

15. C(1, –2); V: (5, –2), (–3, –2); F: $(2\sqrt{3} + 1, -2)$, $(-2\sqrt{3} + 1, -2)$

16. $\dfrac{x^2}{25} + \dfrac{y^2}{4} = 1$ **17.** $\dfrac{x^2}{4} + \dfrac{y^2}{25} = 1$ **18.** $\dfrac{x^2}{39} + \dfrac{y^2}{64} = 1$ **19.** $\dfrac{x^2}{5} + \dfrac{y^2}{1} = 1$ **20.** $\dfrac{x^2}{36} + \dfrac{y^2}{20} = 1$ **21.** $\dfrac{(x-2)^2}{25} + \dfrac{(y-1)^2}{9} = 1$

22. $\dfrac{(x-1)^2}{1} + \dfrac{(y-2)^2}{4} = 1$ **23~28 Graphs are left to the students**   **29.** $\dfrac{x^2}{(384.40)^2} + \dfrac{y^2}{(383.87)^2} = 1$ (million meters)

30. Hint: Eliminate y by substituting the line $y = mx + k$ into $x^2 / a^2 + y^2 / b^2 = 1$.

We have $(a^2 m^2 + b^2)x^2 + 2a^2 mkx + (a^2 k^2 - a^2 b^2) = 0$. Since they intersect at one point, we apply the

discriminant $b^2 - 4ac = 0$ to find $k = \pm\sqrt{a^2 m^2 + b^2}$. Substitute k value into the line, we have the

equation of the tangent line $y = mx \pm r\sqrt{a^2 m^2 + b^2}$.

31. Hint: Substitute $P(x_0, y_0)$ and $Q(x, y)$ into the equation of the ellipse respectively. $\dfrac{x^2}{a^2} + \dfrac{y^2}{b^2} = 1$ subtracts

$\dfrac{x_0^2}{a^2} + \dfrac{y_0^2}{b^2} = 1$ to find the equation of $\overline{PQ}$: $y - y_0 = -\dfrac{b^2(x + x_0)}{a^2(y + y_0)}(x - x_0)$. As Q approachs P, we have

$x \to x_0$ and $y \to y_0$ and the tangent line $y - y_0 = -\dfrac{b^2 \cdot 2x_0}{a^2 \cdot 2y_0}(x - x_0)$. Simplify it, we have the equation

of the tangent line $\dfrac{x_0 x}{a^2} + \dfrac{y_0 y}{b^2} = 1$.   **32.** $x + y = 5$

CHAPTER 8 – 5   Hyperbolas   page 272

1 ~ 15 graphs are left to the students

1. C(0,0); V:(5, 0), (–5, 0); F:$(\sqrt{34}, 0)$, $(-\sqrt{34}, 0)$ **2.** C(0, 0); V:(3, 0), (–3, 0); F:$(\sqrt{34}, 0)$, $(-\sqrt{34}, 0)$

3. C(1, –4); V:(6,–4),(–4,–4) F:$(\sqrt{34} + 1, -4)$,$(-\sqrt{34} + 1, -4)$ **4.** C(0, 0); V:(0, 2), (0,–2); F:(0, $2\sqrt{5}$), $(0, -2\sqrt{5})$

5. C(0, 0); V:(0, 4), (0,–4) F:(0, $2\sqrt{5}$), $(0, -2\sqrt{5})$ **6.** C(4,–1); V:(4, 1),(4,–3); F:$(4, 2\sqrt{5} - 1)$,$(4, -2\sqrt{5} - 1)$

7. C(0, 0); V:(1, 0), (–1, 0) F:$(\sqrt{5}, 0)$, $(-\sqrt{5})$ **8.** C(0, 0); V:(0, 2), (0,–2); F:(0, $\sqrt{5}$), $(0, -\sqrt{5})$

9. C(3, 0); V:(2, 0), (4, 0) F:$(3 + \sqrt{5}, 0)$, $(3 - \sqrt{5}, 0)$ **10.** C(0, 0); V:(2, 0), (–2, 0); F:$(\sqrt{13}, 0)$, $(-\sqrt{13}, 0)$

11. C(0, 0); V:(0, 4),(0,–4) F:$(0, \sqrt{41})$,$(0, -\sqrt{41})$ **12.** C(–3,–2); V:(1,–2),(–7,–2); F:$(2\sqrt{5} - 3, -2)$,$(-2\sqrt{5} - 3, -2)$

13. C(0, 1); V:(0,2),(0,0) F:$(0, \sqrt{10} + 1)$,$(0, -\sqrt{10} + 1)$ **14.** C(0, 1); V:(0, 2),(0,0); F:$(0, \sqrt{5} + 1)$,$(0, -\sqrt{5} + 1)$

15. C(–1, 2); V:(0, 2), (–2, 2) F:$(\sqrt{3} - 1, 2)$, $(-\sqrt{3} - 1, 2)$

16. $\dfrac{x^2}{25} - \dfrac{y^2}{11} = 1$ **17.** $\dfrac{y^2}{25} - \dfrac{x^2}{11} = 1$ **18.** $\dfrac{y^2}{25} - \dfrac{x^2}{39} = 1$ **19.** $y^2 - \dfrac{x^2}{3} = 1$ **20.** $\dfrac{(x-2)^2}{36} - \dfrac{(y+1)^2}{13} = 1$

21. $\dfrac{(x-2)^2}{16} - \dfrac{(y-1)^2}{9} = 1$ **22.** $\dfrac{y^2}{16} - \dfrac{x^2}{9} = 1$ **23.** $\dfrac{(y+1)^2}{4} - \dfrac{9(x-1)^2}{16} = 1$ **24.** $\dfrac{x^2}{5} - \dfrac{y^2}{4} = 1$

25. $y = \pm\frac{2}{3}x$ **26.** $y + 2 = \pm\frac{1}{2}(x + 3)$ **27.** $y = \pm\frac{4}{3}x$ **28.** $y - 1 = \pm\frac{1}{3}x$ **29.** $y = \pm 3x$ **30.** $y = \pm 3x$

31 ~ 36 graphs are left to the students

37. Hint: Solve $\dfrac{x^2}{a^2} - \dfrac{y^2}{b^2} = 1$ for y, we have $y = \pm\dfrac{bx}{a}\sqrt{1 - \dfrac{a^2}{x^2}}$. As x approaches ∞, $\dfrac{a^2}{x^2} \to 0$.

Therefore, the lines $y = \pm\dfrac{b}{a}x$ are asymptotes.

38. Hint: (See distance formula on page 304)

Let $p(x, y)$ be a point on the hyperbola. The distance from $p(x, y)$ to $bx + ay = 0$ is $D_1 = \left|\dfrac{bx + ay}{\sqrt{a^2 + b^2}}\right|$.

The distance from $p(x, y)$ to $bx - ay = 0$ is $D_2 = \left|\dfrac{bx - ay}{\sqrt{a^2 + b^2}}\right|$. $D_1 \cdot D_2 = \dfrac{a^2 b^2}{a^2 + b^2} = k$ (It is a constant.)

We have $D_1 = \dfrac{k}{D_2}$ and $D_2 = \dfrac{k}{D_1}$. As $D_2 \to \infty$, $D_1 \to 0$. $bx + ay = 0$ is an asymptote.   **39.** $\dfrac{x^2}{9} - \dfrac{y^2}{36} = 1$

As $D_1 \to \infty$, $D_2 \to 0$. $bx - ay = 0$ is an asymptote.

Therefore, $(bx + ay)(bx - ay) = k$ is the hyperbola.   **40.** Hint: See proof of Problem 30, page 268

CHAPTER 8 – 6   Solving Quadratic Systems   page 274

1. $(9, -2)$ **2.** $(-8, 7)$ **3.** $(\frac{2}{3}, 7)$ **4.** $(12, 13)$ **5.** $(-9, -1)$ **6.** $(\frac{1}{2}, \frac{1}{3})$ **7.** $(0.6, 0.8)$ **8.** $(-0.2, 0.5)$ **9.** $(0.5, 0.4)$

10. $(5, 10)$ **11.** $(4, 2)$ **12.** $(-3, 2)$ **13.** $(1, 3, 4)$ **14.** $(3, 16, -2)$ **15.** $(-2\frac{2}{5}, -\frac{2}{5}, 1\frac{1}{5})$ **16.** $(-2, 6), (1, 3)$

17. $(2, 6), (-1, 3)$ **18.** $(3, -4), (-3, -4)$ **19.** $(-3, -2), (2, 3)$ **20.** $(3, 6), (-3, -6)$ **21.** $(5, 2)$

22. $(2, -3), (-2, -3), (3, 2), (-3, 2)$ **23.** No real number solution (no intersection)   **24.** $(0, -4), (2\sqrt{3}, 2), (-2\sqrt{3}, 2)$

25. No real number solution (no intersection)   **26.** $(3, 2), (-3, 2), (3, -2), (-3, -2)$

27. $(\sqrt{6}, \sqrt{5}), (-\sqrt{6}, \sqrt{5}), (\sqrt{6}, -\sqrt{5}), (-\sqrt{6}, -\sqrt{5})$ **28.** $(-5, -9), (3, 7)$ **29.** $(0, 0), (2, -4)$

30. $(2, \sqrt{13}), (2, -\sqrt{13}), (6, 3\sqrt{5}), (6, -3\sqrt{5})$ **31.** $(0, 0), (1, 1), (-1, -1)$ **32.** $(-2, 0), (1, 0), (0, 2), (-1, 2)$

33. $(\frac{\sqrt{6}}{3}, 1), (-\frac{\sqrt{6}}{3}, 1), (\frac{\sqrt{6}}{3}, -1), (\frac{-\sqrt{6}}{3}, -1)$ **34.** $(0, 1)$ **35.** $(1, 0)$ **36.** $(64, 8)$

CHAPTER 8 – 7   Eccentricity and Latus Rectum   page 276

1. $e = 1, \text{LR} = 8$ **2.** $\text{LR} = 16$ **3.** $e = 1, \text{LR} = 8$ **4.** $e = 1, \text{LR} = 12$ **5.** $e = \frac{\sqrt{13}}{3}, \text{LR} = 2\frac{2}{3}$ **6.** $e = \frac{\sqrt{5}}{5}, \text{LR} = \frac{8\sqrt{5}}{5}$

7. $e = \frac{\sqrt{21}}{5}, \text{LR} = 1\frac{3}{5}$ **8.** $e = \frac{\sqrt{29}}{2}, \text{LR} = 25$ **9.** $e = \frac{\sqrt{10}}{3}, \text{LR} = 1\frac{1}{3}$ **10.** $e = \frac{2\sqrt{2}}{3}, \text{LR} = 2\frac{2}{3}$ **11.** $e = 0, \text{LR} = \text{none}$

12. $e = \sqrt{2}, \text{LR} = 2\sqrt{10}$ **13.** $e = 0, \text{LR} = \text{none}$ **14.** $e = \frac{2\sqrt{2}}{3}, \text{LR} = \frac{2}{3}$ **15.** $e = \sqrt{2}, \text{LR} = 2$

16. $V(0, 0), F(-4, 0)$, Axis: $y = 0$, Directrix: $x = 4$, $e = 1, \text{LR} = 16$

17. $V(2, -1), F(2, 1)$, Axis: $x = 2$, Directrix: $y = -3$, $e = 1, \text{LR} = 8$

18. $(y + 1)^2 = 8(x - 2)$ or $(y + 1)^2 = -8(x - 2)$ **19.** $3(x - 2)^2 - (y + 3)^2 = 25$ **20.** $7(x - 2)^2 - 9(y + 4)^2 = 63$

21. $16x^2 + 25(y + 1)^2 = 400$ **22.** $9(x - 3)^2 + 25(y - 1)^2 = 225$

23. $C(0, 0), F(\pm\sqrt{5}, 0)$, Major axis: $y = 0$, Minor axis: $x = 0$, $e = \frac{\sqrt{5}}{3}, \text{LR} = 2\frac{2}{3}$

24. $C(0, 0), F(0, \pm\sqrt{34})$, Transverse axis: $x = 0$, Conjugate axis: $y = 0$, $e = \frac{\sqrt{34}}{3}, \text{LR} = 16\frac{2}{3}$

25. $C(1, -2), F(1 \pm 2\sqrt{3}, -2)$, Major axis: $y + 2 = 0$, Minor axis: $x - 1 = 0$, $e = \frac{\sqrt{3}}{2}, \text{LR} = 2$

26. $C(1, 0), F(1 \pm \sqrt{13}, 0)$, Transverse axis: $y = 0$, Conjugate axis: $x - 1 = 0$, $e = \frac{\sqrt{13}}{2}, \text{LR} = 9$

27. $\dfrac{x^2}{8649} + \dfrac{y^2}{8646.60} = 1$ (million miles). It is nearly circular.   **28.** $D_{\max} = 3254.60$ million miles

CHAPTER 8 – 8   General Quadratic Form and Rotation of Axes   page 285

1. $(\dfrac{3\sqrt{3} - 1}{2}, -\dfrac{\sqrt{3} + 3}{2})$ **2.** $\dfrac{2 + 3\sqrt{3}}{2} x' - (\dfrac{2\sqrt{3} - 3}{2})y' - 1 = 0$ **3.** $\dfrac{1}{2}(x')^2 - \sqrt{3}x'y' - \dfrac{1}{2}(y')^2 = 2$ **4.** $x'y' = -1$

5. $x^2 - y^2 = 2$ **6.** $\dfrac{(y')^2}{2} - \dfrac{(x')^2}{2} = 1$ **7.** $(x')^2 - (y')^2 = 1$

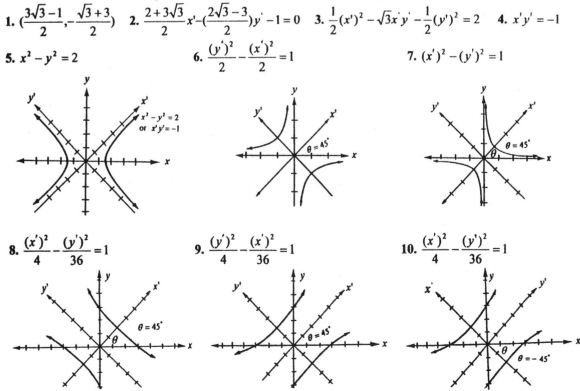

8. $\dfrac{(x')^2}{4} - \dfrac{(y')^2}{36} = 1$ **9.** $\dfrac{(y')^2}{4} - \dfrac{(x')^2}{36} = 1$ **10.** $\dfrac{(x')^2}{4} - \dfrac{(y')^2}{36} = 1$

CHAPTER 8 – 8   General Quadratic Form and Rotation of Axes   page 285 (Continued)

11. $\dfrac{(x')^2}{4}+\dfrac{(y')^2}{9}=1$ 12. $(x')^2=y'+2$ 13. $x'=\pm 1.79$

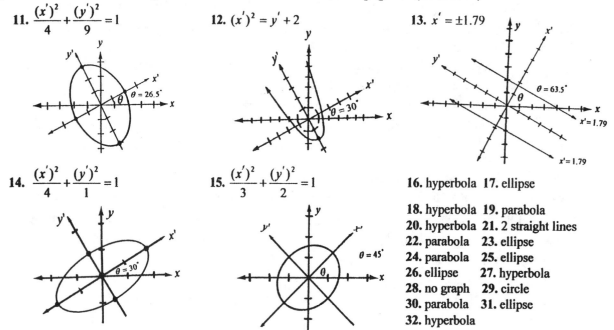

14. $\dfrac{(x')^2}{4}+\dfrac{(y')^2}{1}=1$ 15. $\dfrac{(x')^2}{3}+\dfrac{(y')^2}{2}=1$

16. hyperbola   17. ellipse

18. hyperbola   19. parabola
20. hyperbola   21. 2 straight lines
22. parabola   23. ellipse
24. parabola   25. ellipse
26. ellipse   27. hyperbola
28. no graph   29. circle
30. parabola   31. ellipse
32. hyperbola

33. A point $(0, 0)$ 34. Two intersecting lines $x+y=0$ and $x-y=0$
35. Two parallel lines $x+2=0$ and $x-1=0$ 36. A straight line $x-2=0$
37. Two intersecting lines $2x+3y-3=0$ and $2x-3y+9=0$ 38. A point $(-3, 1)$ 39. No graph
40. Two parallel lines $2x+3y-6=0$ and $2x+3y+3=0$ 42 and 43 proofs are left to the students

CHAPTER 8 – 9   Parametric Equations and Graphs   page 290

1. $\begin{cases} x=4-4t \\ y=3t \ ; \ t\in R \end{cases}$ 2. $\begin{cases} x=1+5t \\ y=2+3t \ ; \ t\in R \end{cases}$ 3. $\begin{cases} x=-7 \\ y=4-8t \ ; \ t\in R \end{cases}$ 4. $\begin{cases} x=7-14t \\ y=-4 \ ; \ t\in R \end{cases}$

5. $\begin{cases} x=1+t \\ y=\sqrt{2}+\sqrt{2}\,t \ ; \ t\in R \end{cases}$ 6. $\begin{cases} x=3+2t \\ y=-\sqrt{3}-\sqrt{3}\,t \ ; t\in R \end{cases}$ 7. $y=-\tfrac{3}{4}x+3 \ ; \ x\in R$ 9. $y=-\tfrac{3}{2}x+\tfrac{13}{2} \ ; \ x\in R$

8. $y=-4 \ ; \ x\in R$ 10. $y=2-x^2 \ ; \ x\geq 0$

11. $y=(x-2)^2 ; 1\leq x\leq 4$ 12. $y=1+\dfrac{1}{x} ; x>0$ 13. $y=\dfrac{2}{x} ; x\neq 0$ 14. $y=\dfrac{x}{3}+\dfrac{1}{3} \ ; \ 2\leq x\leq 11$

15. $y=\sqrt{2-x} \ ; x\leq 2$ 16. $y=-\dfrac{3}{2}x+3 ; 0\leq x\leq 2$ 17. $y=3\sqrt{1-(\tfrac{x}{3})^2} \ ; \ -3\leq x\leq 3$

18. $y=\pm 4\sqrt{1-(\tfrac{x}{2})^2} \ ; \ -2\leq x\leq 2$ 19. $y=\sqrt{x^2-1} ; 1\leq x\leq \sqrt{2}$ 20. $(\dfrac{x-2}{4})^2+(\dfrac{y-3}{3})^2=1 ; \ -2\leq x\leq 6$

21. $y=\sqrt{x} \ ; x\geq 1$ **22 ~ 31** Graphs are left to the students.   22. $y=\dfrac{x}{2}-\dfrac{1}{2} ; 3\leq x\geq 9$ 23. $y=x-7 ; x\geq 0$

24. $y=2-x ; 0\leq x\leq 2$ 25. $x^2+y^2=4$ and $x\geq 0, \ y\geq 0$ 26. $x^2+y^2=4$ and $y\geq 0$

27. $x^2+y^2=9$ 28. $\dfrac{x^2}{9}+\dfrac{y^2}{4}=1$ 29. $y^2=-(x-1)$ and $0\leq x\leq 1$ 30. $y=\dfrac{1}{3}\ln x$ and $x>0$

31. $y=\dfrac{1}{3}x+1$ and $x\geq 3$ 32. $y^2+7y-x+12=0$ 33. $x^2-2xy+y^2+3x-4y+1=0$

34. $y^2-x^2=4$ 35. a) $x=40\sqrt{2}\,t$ and $y=-16t^2+40\sqrt{2}\,t+96$ b) 271 feet   c) $y_{max}=146$ feet

CHAPTER 8 – 10   Polar Coordinates and Graphs   page 296

1. $(3, -\frac{4\pi}{3})$ **2.** $(3, \frac{8\pi}{3})$ **3.** $(-3, \frac{5\pi}{3})$ **4.** $(5, \frac{9\pi}{4})$ and $(5, -\frac{7\pi}{4})$ **5.** $(-5, \frac{5\pi}{4})$ **6.** $(r, \theta) = (3, \pi)$ **7.** $(r, \theta) = (3, \frac{3\pi}{2})$

8. $(r, \theta) = (2\sqrt{2}, \frac{7\pi}{4})$ **9.** $(r, \theta) = (\sqrt{3}, \frac{\pi}{6})$ **10.** $(r, \theta) = (4, \frac{3\pi}{4})$ **11.** $(r, \theta) = (2, \frac{4\pi}{3})$

12. $(r, \theta) = (3.61, -70.56°)$ or $(3.61, -1.23)$ **13.** $(r, \theta) = (2.53, -117.51°)$ or $(2.53, -2.5)$

14. $(-5, 0)$ **15.** $(4\sqrt{3}, 4)$ **16.** $(-\sqrt{2}, \sqrt{2})$ **17.** $(-5, 0)$ **18.** $x^2 + y^2 = 25$ **19.** $x - 4 = 0$ **20.** $y + 5 = 0$

21. $x^2 + y^2 - y = 0$ **22.** $4x^2 - y = 0$ **23.** $(x^2 + y^2)^{3/2} - y = 0$ **24.** $x^2 + y^2 - 2\sqrt{x^2 + y^2} - y = 0$ **25.** $x^2 - 4y - 4 = 0$

26. $x^2 + (y-2)^2 = 4$ **27.** $\sqrt{3}x - y = 0$ **28.** $x + y = 0$ **29.** $x + 5 = 0$ **30.** $4r\cos^2\theta - \sin\theta = 0$ **31.** $2r^2\sin 2\theta = 3$

32. $r = 2\sin\theta$ **33.** $r\sin\theta = -5$ **34.** $r = \frac{1}{2}$ **35.** $\theta = \frac{\pi}{4}$ **36.** $\theta = \frac{\pi}{3}$ **37.** $r\cos\theta = -5$ **38.** $x^2 + y^2 = 25$ (a circle)

39. $x^2 + y^2 = 36$ (a circle) **40.** $\sqrt{3}x - 3y = 0$ (a straight line) **41.** $\sqrt{3}x + y = 0$ (a straight line)

42. $y - 5 = 0$ (a horizontal line) **43.** $x - 5 = 0$ (a vertical line) **44.** $x^2 + y^2 - 3x = 0$ (a circle)

45. $x^2 + y^2 - 4y = 0$ (a circle) **46.** $x^2 + y^2 - 4x = 0$ (a circle) **47.** $y^2 + 4x - 4 = 0$ (a parabola)

48. $r = 1 - \sin\theta$ (Cardioid) **49.** $r = 2(1 + \cos\theta)$ (Cardioid) **50.** $r = 3 + 2\cos\theta$ (Limaçon)

51. $r = 2\sin 2\theta$ (Rose) **52.** $r = 2\cos 2\theta$ (Rose) **53.** $r^2 = 4\cos 2\theta$ (Lemniscate)

54. $r^2 = 4\sin 2\theta$ (Lemnicate) **55.** $r = 4\sin 3\theta$ (Rose) **56.** $r = 2\cos 3\theta$ (Rose)

57. $r = 1 - 2\cos\theta$ (Limaçon) **58.** $r = \theta$, $\theta \geq 0$ (Spiral of Archimedes) **59.** $r = e^{\theta/2}$ (Logarithmic Spiral)

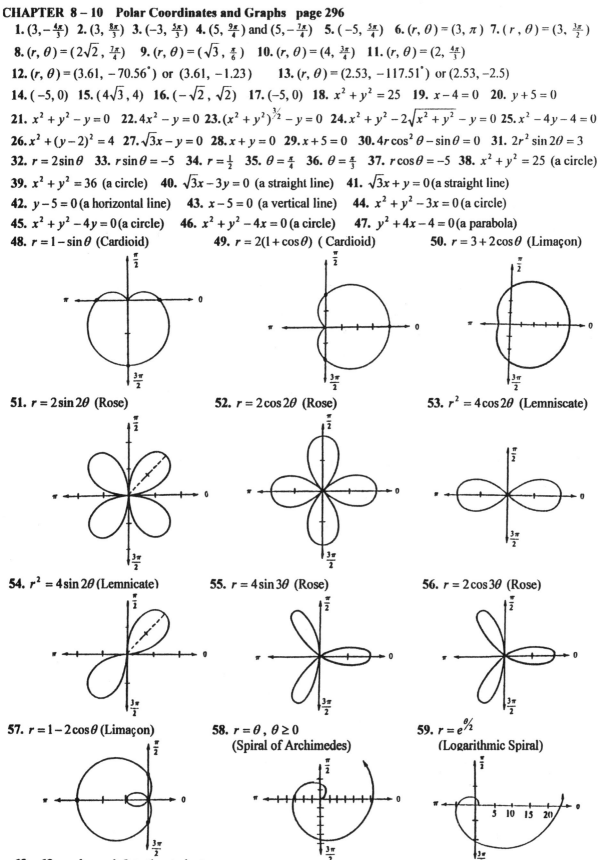

60 ~ 63 graphs are left to the students

60. $e = 1$, a parabola **61.** $e = 2$, a hyperbola **62.** $e = \frac{1}{2}$, an ellipse **63.** $e = \frac{5}{4}$, a hyperbola

CHAPTER 8 – 11 Polar Coordinates involving Complex Numbers page 300

1. $z = 2(\cos 30^\circ + i \sin 30^\circ)$ **2.** $z = 4(\cos 240^\circ + i \sin 240^\circ)$ **3.** $z = 4(\cos 120^\circ + i \sin 120^\circ)$

4. $z = 5(\cos 90^\circ + i \sin 90^\circ)$ **5.** $z = 5(\cos 0^\circ + i \sin 0^\circ)$ **6.** $z = 4(\cos 270^\circ + i \sin 270^\circ)$

7 ~ 12 graphs are left to the students   **7.** $z_1 + z_2 = 7 - 2i,\ z_1 - z_2 = 1 + 6i$ **8.** $z_1 + z_2 = 4,\ z_1 - z_2 = 6i$

9. $z_1 + z_2 = 4,\ z_1 - z_2 = -1 - i$ **10.** $z_1 + z_2 = 8 + 8i,\ z_1 - z_2 = -2 + 2i$ **11.** $z_1 + z_2 = 8 - 2i,\ z_1 - z_2 = -2 - 8i$

12. $z_1 + z_2 = 5 - i,\ z_1 - z_2 = -1 + 7i$ **13.** $z = \sqrt{3} + i$ **14.** $z = 2\sqrt{2} + 2\sqrt{2}\,i$ **15.** $z = -2 + 2\sqrt{3}\,i$

16. $z = -3\sqrt{3} - 3i$ **17.** $z = 5i$ **18.** $z = 4\sqrt{2} - 4\sqrt{2}\,i$ **19.** $z = -0.0684 + 0.1880\,i$ **20.** $z = -0.3064 - 0.2571\,i$

21. $z_1 z_2 = 8(\cos 90^\circ + i \sin 90^\circ),\ z_1/z_2 = 2(\cos 30^\circ + i \sin 30^\circ)$

22. $z_1 z_2 = \frac{3}{2}(\cos \pi + i \sin \pi),\ z_1/z_2 = 6(\cos \frac{\pi}{3} + i \sin \frac{\pi}{3})$ **23.** $z_1 z_2 = 21(\cos \frac{2\pi}{3} + i \sin \frac{2\pi}{3}),\ z_1/z_2 = \frac{3}{7}(\cos \frac{14\pi}{9} + i \sin \frac{14\pi}{9})$

24. $z_1 z_2 = 18(\cos 300^\circ + i \sin 300^\circ),\ z_1/z_2 = 2(\cos 180^\circ + i \sin 180^\circ)$

25. $z_1 z_2 = 2\sqrt{2}(\cos 255^\circ + i \sin 255^\circ),\ z_1/z_2 = \frac{\sqrt{2}}{2}(\cos 15^\circ + i \sin 15^\circ)$ **26.** $z = -4$ **27.** $z = 16i$

28. $z = -4\sqrt{4} + 4\sqrt{2}\,i$ **29.** $z = 16$ **30.** $z = -528 - 555i$ **31.** $z = -1024$

32. $z_0 = \cos 30^\circ + i \sin 30^\circ,\ z_1 = \cos 150^\circ + i \sin 150^\circ,\ z_2 = \cos 270^\circ + i \sin 270^\circ$

33. $z_0 = \frac{3\sqrt{3}}{2} + \frac{3}{2}i,\ z_1 = -\frac{3\sqrt{3}}{2} + \frac{3}{2}i,\ z_2 = -3i$

34. $z_0 = 2^{1/4}(\cos 82.5^\circ + i \sin 82.5^\circ),\ z_1 = 2^{1/4}(\cos 172.5^\circ + i \sin 172.5^\circ),\ z_2 = 2^{1/4}(\cos 262.5^\circ + i \sin 262.5^\circ)$

     $z_3 = 2^{1/4}(\cos 352.5^\circ + i \sin 352.5^\circ)$ **35.** $z_0 = \sqrt{2} + \sqrt{2}\,i,\ z_1 = -\sqrt{2} + \sqrt{2}\,i,\ z_2 = -\sqrt{2} - \sqrt{2}\,i,\ z_3 = \sqrt{2} - \sqrt{2}\,i$

36. $z_0 = 0.9511 + 0.3090\,i,\ z_1 = 1 + i,\ z_2 = -0.9511 + 0.3090\,i,\ z_3 = -0.5878 - 0.8090\,i,\ z_4 = 0.5878 - 0.8090\,i$

37. $z_0 = 1,\ z_1 = -\frac{1}{2} + \frac{\sqrt{3}}{2}i,\ z_2 = -\frac{1}{2} - \frac{\sqrt{3}}{2}i$ (It is called the three cubic roots of unity. They are equispaced on the

                              complex plane by 120°)

38. $z_0 = \frac{1}{2} + \frac{\sqrt{3}}{2}i,\ z_1 = -1,\ z_2 = \frac{1}{2} - \frac{\sqrt{3}}{2}i$ (They are equispaced on the complex plane by 120° .)

CHAPTER 8 Exercises page 305 ~ 308

1. $D = \sqrt{13},\ M = (3,\ 4.5)$ **2.** $D = 3\sqrt{13}, M = (1, \frac{3}{2})$ **3.** $D = \sqrt{13}, M = (-3, -4.5)$ **4.** $D = 3\sqrt{13},\ M = (-1, -\frac{3}{2})$

5. $D = \sqrt{13}, M = (1, \frac{3}{2})$ **6.** $D = \sqrt{13}, M = (-1, -\frac{3}{2})$ **7.** $D = 6,\ M = (0, -3)$ **8.** $D = \sqrt{31},\ M = (3, -\frac{\sqrt{3}}{2})$

9. $D = 2\sqrt{2},\ M = (\frac{\sqrt{3}-1}{2}, \frac{1+\sqrt{3}}{2})$ **10.** $D = 2\sqrt{7}, M = (4, 0)$ **11.** $D = 4\sqrt{5}, M = (\sqrt{3}, \sqrt{2})$ **12.** $D = \frac{\sqrt{82}}{3}, M = (\frac{5}{2}, \frac{5}{6})$

13. $D = \frac{\sqrt{373}}{6}, M = (3, \frac{1}{12})$ **14.** $D = \frac{\sqrt{73}}{6}, M = (\frac{3}{4}, 0)$ **15.** $C = 2\sqrt{6}, M = (2\sqrt{2}, \sqrt{3})$ **16.** $Q(-3, 2)$ **17.** $Q(-2, 0)$

18. $Q(-\sqrt{3}, 3\sqrt{2})$

19 ~ 37 Graphs are left to the students   **19.** A circle, $C(0, 0), r = 11$ **20.** A circle, $C(0, 0), r = 12$

21. A circle, $C(0, 0), r = 2.45$ **22.** An ellipse, $C(0, 0), a = 3, b = 2, c = 2.24$

23. An ellipse, $C(0, 0), a = 5, b = 2, c = 4.58$ **24.** A ellipse, $C(0, 0), a = 10, b = 6, c = 8$

25. A hyperbola, $C(0, 0), a = 3, b = 2, c = 3.61$ **26.** A hyperbola, $C(0, 0), a = 5, b = 2, c = 5.39$

27. A hyperbola, $C(0, 0), a = 6, b = 10, c = 11.67$ **28.** An ellipse, $C(0, 0), a = \frac{3}{2}, b = \frac{1}{2}, c = 1.41$

29. A hyperbola, $C(0, 0), a = \frac{1}{2}, b = \frac{3}{2}, c = 1.58$ **30.** A hyperbola, $C(0, 0), a = 10, b = 10, c = 14.1$

31. A parabola, vertex$(-1, 0), a = -2$, open downward   **32.** A parabola, vertex$(-3, 2), a = 1$, open upward

33. An ellipse, $C(1, 0), a = 3, b = \sqrt{5}, c = 2$ **34.** An ellipse, $C(1, -2), a = 2, b = 1, c = 1.73$

35. A hyperbola, $C(-1, 2), a = 2, b = 1, c = 2.24$ **36.** A circle, $C(2, -1), r = 4$

37. A hyperbola, $C(3, -3), a = 2.45, b = 4.24, c = 4.9$

38. 1.25 inches from the vertex   **39.** $5x - y - 1 = 0$ and $x - y - 1 = 0$ **40.** $5x - 12y + 26 = 0$ and $x - 2 = 0$

41. Hint: The slope of the line passing (h, k) and (x_0, y_0) is $(y_0 - k)/(x_0 - h)$. The slope of the tangent line is

     $-(x_0 - h)/(y_0 - k)$. We find the equation of the tangent line by applying the points (x_0, y_0) and (x, y)

     and its slope. The original formula is $(x_0 - h)(x - x_0) + (y_0 - k)(y - y_0) = 0$. We combine it with the

     equation $(x_0 - h)^2 + (y_0 - k)^2 = r^2$ to find the final formula.   **42.** $x + 3y + 4 = 0$

CHAPTER 8 Exercises page 305 ~ 308 (continued)

43. $(-2, 2), (1, 1)$ **44.** $(0, -1), (2, 1)$ **45.** $(\sqrt{3}, \frac{1}{2}), (-\sqrt{3}, \frac{1}{2})$ **46.** $(0, 2), (\sqrt{3}, -1), (-\sqrt{3}, -1)$

47. $(2\sqrt{2}, -2), (-2\sqrt{2}, -2), (\sqrt{5}, 1), (-\sqrt{5}, 1)$ **48.** $(2, 2), (2, -2), (-2, 2), (-2, -2)$ **49.** $(2, 2), (-2, -2)$ **50.** $(-4, 2), (2, -1)$

51. $(0, 2), (0, -2), (-\frac{9}{4}, \frac{\sqrt{7}}{2}), (-\frac{9}{4}, -\frac{\sqrt{7}}{2})$ **52.** $3x + 2y + 5 = 0$ **53.** $3x - 2y + 13 = 0$ **54.** $2x - 3y + 12 = 0$ **55.** $y = 6x$

56. $x + 6y + 3 = 0$ **57.** $D = \frac{2\sqrt{5}}{5}$ **58.** $(x-2)^2 + (y+3)^2 = 36$ **59.** $(x-2)^2 + (y+3)^2 = 61$ **60.** $(x-8)^2 + (y-3)^2 = 9$

61. $(x-4)^2 + (y-8)^2 = 16$ **62.** $x^2 - 4x + 12y + 4 = 0$ **63.** $y^2 + 6y - 10x + 4 = 0$ **64.** $y^2 + 6y - 12x - 3 = 0$

65. $y^2 - 6y - 2x + 4 = 0$ **66.** $y^2 + 6y + 16x + 25 = 0$ **67.** $9x^2 + 25y^2 = 225$ **68.** $9x^2 + 5y^2 = 180$

69. $32y^2 - 4x^2 = 128$ **70.** $16x^2 - 9y^2 = 144$ **71.** $9x^2 - 16y^2 - 18x + 64y - 199 = 0$ **72.** $e = 1$, LR $= 8$

73. $e = \frac{\sqrt{5}}{5}$, LR $= \frac{8\sqrt{5}}{5}$ **74.** $e = 1\frac{1}{4}$, LR $= 4\frac{1}{2}$ **75.** $(y-2)^2 - 4(x+1)^2 = 4$ **76.** Foci: $(3, -1+\sqrt{5})$, $(3, -1-\sqrt{5})$

77. Foci: $(-5 - 2\sqrt{3}, 4), (-5 + 2\sqrt{3}, 4)$ **78.** Area $= 64\pi$ **79.** Area $= 6\pi$ **80.** $(x^2/144) + (y^2/64) = 1$

81. $(x^2/(384.40)^2) + (y^2/(383.87)^2) = 1$ **82 ~ 83.** See proof of Problem 37 on Page 272

84. Hint: Let $(2x+3y-5)(2x-3y+1) = k$. $k = 36$ by substituting $(4, 1)$. Simplify $(2x+3y-5)(2x-3y+1) = 36$
to obtain the answer $4(x-1)^2 - 9(y-1)^2 = 36$.

85. $3x + 2y - 7 = 0$ and $19x - 6y + 49 = 0$ **86.** $(-\frac{2-3\sqrt{3}}{2}, \frac{3+2\sqrt{3}}{2})$ **87.** $\frac{\sqrt{3}}{4}(x')^2 - x'y' + \frac{\sqrt{3}}{4}(y')^2 = 1$ **88.** $(x')^2 - (y')^2 = 2$

89. $f = f' = 2$, $a + c = a' + c' = 0$, $b^2 - 4ac = (b')^2 - 4a'c' = 4$ **90.** A hyperbola **91.** An ellipse

92. Two perpendicular lines **93.** $x = -1 + \sqrt{3}t$; $y = \sqrt{3} + t$; $t \in R$ **94 ~ 95** Graphs are left to the students

94. $y = x - 7$; $-1 \le x \le 4$ **95.** $y = \frac{1}{2}\sqrt{4 - x^2}$; $-2 \le x \le 2$ **96.** $x = 120t$; $y = -16t^2 + 120\sqrt{3}t$

97. $(15, \frac{4}{3}\pi), (15, \frac{1}{3}\pi), (15, -\frac{5}{3}\pi)$ **98.** $(r, \theta) = (2\sqrt{2}, \frac{3}{4}\pi)$ **99.** $(r, \theta) = (4.79, 154°)$ or $(4.79, 2.69 \text{ radian})$

100. $(x, y) = (\frac{\sqrt{3}}{2}, \frac{3}{2})$ **101.** $x^2 + y^2 = 9$ **102.** $x = 1$ **103.** $x^2 + (y - \frac{1}{2})^2 = \frac{1}{4}$ **104.** $y^2 = x + \frac{1}{4}$ **105.** $r = 4\sin\theta$

106. $r = 10\sec\theta$ **107.** $r = -8/(6\sin\theta - 3\cos\theta)$ **108.** $r = 1/(1 - \cos\theta)$ or $r = -1/(1 + \cos\theta)$

109 ~ 114 Graph are left to the students. **115.** $z = 4(\cos\frac{4\pi}{3} + i\sin\frac{4\pi}{3})$ **116.** $z = \frac{3}{2} + \frac{\sqrt{3}}{2}i$ **117.** $z = (-0.345 + 4.98i)$

118. $z_1 z_2 = 15(\cos 50° + i\sin 50°)$ **119.** $z_1/z_2 = \frac{1}{2}(\cos\frac{3\pi}{4} + i\sin\frac{3\pi}{4})$ **120.** $(-1+i)^{10} = -32i$

121. $z^{1/3} = \sqrt{2}(\cos 45° + i\sin 45°)$, $\sqrt{2}(\cos 165° + i\sin 165°)$, $\sqrt{2}(\cos 285° + i\sin 285°)$

122. $z^{1/3} = 0.398 + 3.341i$, $-4.884 + 1.777i$, $0.904 - 5.118i$ **123.** $r = 94.13/(1 - 0.0167\cos\theta)$ (in million miles)

Roger was absent yesterday.
Teacher: Roger, why were you absent?
Roger: My tooth was aching.
I went to see my dentist.
Teacher: Is your tooth still aching?
Roger: I don't know. My dentist has it.

Graphing a circle by a graphing calculator

Example: Graph $x^2 + y^2 = 25$.

       Solution:

          Solve the equation for y: $y = \pm\sqrt{25 - x^2}$

          Press: **Y =**

          Enter: **Y1**= → **2nd** → $\sqrt{}$ (→ 25 → − → $\boxed{\text{X,T,}\theta\text{, }n}$ → ∧ → 2 →) → **ENTER**

                Y2= → (−) →**2nd** → $\sqrt{}$ (→ 25 → − → $\boxed{\text{X, T,}\theta\text{,}n}$ → ∧ → 2 →) → **ENTER**

            We can simply enter **Y2** as **−Y1** :

            Press: **Y2**= → (−) → **VARS** → **Y-VARS** → **1:FUNCTION** → **ENTER**

          The screen shows:

$$y_1 = \sqrt{25 - x^2}$$
$$y_2 = -\sqrt{25 - x^2}$$

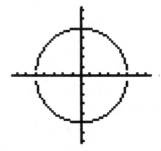

To graph a circle, the window screen must be
set for equal scales on x-axes and y-axes.
Adjust the viewing window:

$$x_{min} = -7, \ x_{max} = 7 \ , \ x_{scl} = 1$$
$$y_{min} = -7, \ y_{max} = 7 \ , \ y_{scl} = 1$$

          Press: **ZOOM** → **5:Zsquare** → **ENTER**
          The screen shows:

Graphing an ellipse by a graphing calculator

Example: Graph $9x^2 + 4y^2 = 36$.

       Solution:

          Solve the equation for y: $\ \ y = \pm\sqrt{\dfrac{36 - 9x^2}{4}}$

          Press: **Y=**

          Enter: **Y1**= → **2nd** → $\sqrt{}$ (→ (→ 36 → − 9 → $\boxed{\text{X,T,}\theta\text{,}n}$ → ∧ → 2 →) → ÷ → 4 →) → **ENTER**

          Simply enter **Y2** as **−Y1**:

            Y2= → (−) → **VARS** → **Y-VARS** → **1:FUNCTION** → **ENTER** → **1:Y1** → **ENTER**

          The screen shows:

$$y_1 = \sqrt{((36 - 9x^2)/4)}$$
$$y_2 = -y_1$$

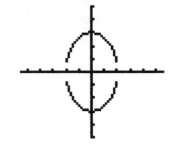

          Press: **ZOOM** → **5:Zsquare** → **ENTER**
          The screen shows:

Graphing a hyperbola by a graphing calculator

Example: Graph $y^2 - 2x^2 + 6y + 18x - 3 = 0$.

Solution:

Complete the square and solve for y : $y = -3 \pm \sqrt{4 + 2(x-2)^2}$

Let $y_1 = \sqrt{4 + 2(x-2)^2}$, $y_2 = -3 + y_1$, $y_3 = -3 - y_1$

Press: **Y=**

Enter: **Y1**= $\to$ **2ⁿᵈ** $\to$ $\sqrt{\ }$ ($\to$ 4 $\to$ + $\to$ 2 $\to$ ($\to$ **X,T,θ,n** $\to$ –2 $\to$) $\to$ ∧ $\to$ 2 $\to$) $\to$ **ENTER**

   Y2= $\to$ –3 $\to$ + $\to$ **VARS** $\to$ **Y-VARS** $\to$ **1:FUNCTION**
        $\to$ **ENTER** $\to$ **1:Y1** $\to$**ENTER**
   Y3= $\to$ –3 $\to$ – $\to$ **VARS** $\to$ **Y-VARS** $\to$ **1:FUNCTION**
        $\to$ **ENTER** $\to$ **1:Y1** $\to$**ENTER**

To turn off the graph of **Y1**, highlight the equal sign
next to **Y1**. Then press **ENTER**

Press: **ZOOM** $\to$ **5:Zsquare** $\to$ **ENTER**
The screen shows:

Solving qudratic systems by a graphing calculator

Example: Graph $\begin{cases} x^2 + 2y^2 = 41 \\ 2x^2 - y^2 = 2 \end{cases}$

Solution:

Solve each equation for y: $y = \pm\sqrt{\dfrac{41 - x^2}{2}}$ and $y = \pm\sqrt{2x^2 - 2}$

Let $y_1 = \sqrt{\dfrac{41 - x^2}{2}}$, $y_2 = -y_1$ and $y_3 = \sqrt{2x^2 - 2}$, $y_4 = -y_3$

Press: **Y=**

Enter: **Y1**= $\to$ **2nd** $\to$ $\sqrt{\ }$ ($\to$ ($\to$ 41 $\to$ – $\to$ **X,T,θ,n**
      $\to$ ∧ $\to$ 2 $\to$) $\to$ ÷ 2 $\to$) $\to$ **ENTER**
   Y2= $\to$ **VARS** $\to$ **Y-VARS** $\to$ **1:FUNCTION**
        $\to$ **ENTER** $\to$ **1:Y1** $\to$ **ENTER**
   Y3= $\to$ **2nd** $\to$ $\sqrt{\ }$ ($\to$ ($\to$ 2 $\to$ **X,T,θ,n** $\to$ ∧
        $\to$ 2 $\to$) $\to$ –2 $\to$) $\to$ **ENTER**
   Y4= $\to$ **VARS** $\to$ **Y-VARS** $\to$ **1:FUNCTION**
        $\to$ **ENTER** $\to$ **3:Y3** $\to$ **ENTER**

Press: **ZOOM** $\to$ **5:Zsquare** $\to$ **ENTER**
To find the solutions on the points of intersection, press: **TRACE**
The screen shows the solutions are: (2.979, 4.008), (2.979, –4.008), (–2.979, 4.008),
(–2.979, –4.008). (See Example 2 on Page 273)

Graphing a parametric equation by a graphing calculator

Example: Graph $\begin{cases} x = t + 2 \\ y = t^2 \; ; \quad -3 \le t \le 2 \end{cases}$

Solution:

Press: **MODE $\rightarrow$ Par $\rightarrow$ ENTER**

Enter the equation, press: **Y=**

Type: $X_{1T} = \boxed{\text{X,T,}\theta\text{,n}} \rightarrow + \rightarrow 2$

$Y_{1T} = \boxed{\text{X,T,}\theta\text{, n}} \rightarrow x^2$

The screen shows: $X_{1T} = T + 2$ and $Y_{1T} = T^2$

Select the viewing window for the interval of t
and adjust the suitable viewing window for x and y.

Press: **WINDOW**

Enter: $T_{min} = -3$ $X_{min} = -2$ $Y_{min} = -1$

$T_{max} = 2$ $X_{max} = 5$ $Y_{max} = 5$

$T_{step} = 0.1$ $X_{scl} = 1$ $Y_{scl} = 1$

Press: **GRAPH**   The screen shows the graph:

Hint: T_{step} is the selected increment for t. The smaller the T_{step} is, the more
points the calculator will plot. With $T_{step} = 0.1$, it will evaluate and
plot at $t = -3, -2.9, -2.8, -2.7, \cdots\cdots$.

Graphing a polar equation by a graphing calculator

Example: Graph $r = 4\cos 2\theta$

Solution:

Press: **MODE $\rightarrow$ Pol $\rightarrow$ ENTER**

Enter the equation, press: **Y=**

Type: $r\,1 = 4 \rightarrow \cos \rightarrow 2 \rightarrow \boxed{\text{X,T,}\theta\text{,n}} \rightarrow)$

The screen shows: $r\,1 = 4\cos(2\theta)$

Select the viewing window for the interval of θ
and adjust the suitable viewing window for x and y.

Press: **WINDOW**

Enter: $\theta_{min} = 0$ $X_{min} = -5$ $Y_{min} = -5$

$\theta_{max} = 2\pi$ $X_{max} = 5$ $Y_{max} = 5$

$\theta_{step} = 0.02$ $X_{scl} = 1$ $Y_{scl} = 1$

(Hint: $\theta = 0.02 \approx 1°$)

Press: **ZOON $\rightarrow$ 5:Zsquare $\rightarrow$ ENTER**   The screen shows the graph:

(See Example 6 on Page 294)

<u>**Space for Taking Notes**</u>

Space for Taking Notes

Matrices and Determinants

9-1 Operations with Matrices

Matrix: A rectangular array of numbers enclosed by brackets.

Element: A number in a matrix.

Dimensions of a matrix: The number of rows and the number of columns in a matrix. The number of rows is given first. A 3×4 matrix is a matrix having 3 rows and 4 columns (read " 3 by 4 matrix ").

Row Matrix: A matrix having only one row.

Column Matrix: A matrix having only one column.

Square Matrix: A matrix having the same numbers in rows and columns.

Zero Matrix: A matrix having all elements to be zero.

Identity Matrix: A square matrix whose main diagonal has all elements by 1 and all other elements by 0.

Two matrices are equal if and only if they have the same dimensions and the same elements in all corresponding positions.

Matrix addition: Add the corresponding elements of the matrices being added.

Matrix Subtraction: Subtract the corresponding elements of the matrices being subtracted. Or, add the additive inverse of the second matrix.

Additive Inverse of Matrix: Each element of the additive inverse of a matrix is the opposite of the corresponding element in the matrix.

$$\text{If } A = \begin{bmatrix} 3 & -4 \\ -2 & 6 \end{bmatrix} , \text{ then } -A = \begin{bmatrix} -3 & 4 \\ 2 & -6 \end{bmatrix}$$

Addition and subtraction of matrices of different dimensions (rows and columns) are not allowed.

The product (multiplication) of a real number r and a matrix A is the matrix r A in which the elements is " r times the corresponding element in A ".

Examples:

$$\text{Let } A = \begin{bmatrix} 5 & -4 \\ -1 & 2 \end{bmatrix} , \quad B = \begin{bmatrix} 3 & 6 \\ 2 & -5 \end{bmatrix}$$

1. $A + B = \begin{bmatrix} 8 & 2 \\ 1 & -3 \end{bmatrix}$ **2.** $A - 2B = A + (-2B) = \begin{bmatrix} 5 & -4 \\ -1 & 2 \end{bmatrix} + \begin{bmatrix} -6 & -12 \\ -4 & 10 \end{bmatrix} = \begin{bmatrix} -1 & -16 \\ -5 & 12 \end{bmatrix}$

Matrix Multiplication

The product (multiplication) of two matrices is the matrix in which each element is the sum of the product of corresponding elements in row of the first matrix and in column of the second matrix.

$$A_{m \times n} \cdot B_{n \times p} = X_{m \times p}$$

The dimension of column in the first matrix must equal the dimension of row in the second matrix. Otherwise, they cannot be multiplied.

Examples

1. $\begin{bmatrix} 2 & 3 \end{bmatrix}_{1 \times 2} \cdot \begin{bmatrix} 5 \\ 7 \end{bmatrix}_{2 \times 1} = \begin{bmatrix} 2 \cdot 5 + 3 \cdot 7 \end{bmatrix} = \begin{bmatrix} 31 \end{bmatrix}_{1 \times 1}$

2. $\begin{bmatrix} 5 \\ 7 \end{bmatrix}_{2 \times 1} \cdot \begin{bmatrix} 2 & 3 \end{bmatrix}_{1 \times 2} = \begin{bmatrix} 5 \cdot 2 & 5 \cdot 3 \\ 7 \cdot 2 & 7 \cdot 3 \end{bmatrix} = \begin{bmatrix} 10 & 15 \\ 14 & 21 \end{bmatrix}_{2 \times 2}$

3. $\begin{bmatrix} 3 & -1 \\ 2 & 0 \end{bmatrix} \cdot \begin{bmatrix} -1 \\ 2 \end{bmatrix}_{2 \times 1} = \begin{bmatrix} 3(-1) + (-1) \cdot 2 \\ 2(-1) + 0 \cdot 2 \end{bmatrix}_{2 \times 1} = \begin{bmatrix} -5 \\ -2 \end{bmatrix}_{2 \times 1}$

4. $\begin{bmatrix} -1 \\ 2 \end{bmatrix}_{2 \times 1} \cdot \begin{bmatrix} 3 & -1 \\ 2 & 0 \end{bmatrix}_{2 \times 2}$ They cannot be multiplied.

5. $\begin{bmatrix} -1 & 0 & 5 \\ 3 & 2 & -1 \end{bmatrix}_{2 \times 3} \cdot \begin{bmatrix} -1 & 2 & 0 \\ 3 & -4 & -2 \\ 1 & 1 & 3 \end{bmatrix}_{3 \times 3} = \begin{bmatrix} 6 & 3 & 15 \\ 2 & -3 & -7 \end{bmatrix}_{2 \times 3}$

6. $\begin{bmatrix} 5 & 1 & 0 \\ 4 & 2 & -1 \\ 0 & 1 & 3 \end{bmatrix}_{3 \times 3} \cdot \begin{bmatrix} 1 & 0 \\ 2 & 2 \\ 1 & 1 \end{bmatrix}_{3 \times 2} = \begin{bmatrix} 7 & 2 \\ 7 & 3 \\ 5 & 5 \end{bmatrix}_{3 \times 2}$

7. Let $A = \begin{bmatrix} 3 & 1 \\ 0 & -2 \end{bmatrix}$, $B = \begin{bmatrix} 1 & -2 \\ 2 & 0 \end{bmatrix}$

$$AB = \begin{bmatrix} 3 & 1 \\ 0 & -2 \end{bmatrix} \cdot \begin{bmatrix} 1 & -2 \\ 2 & 0 \end{bmatrix} = \begin{bmatrix} 5 & -6 \\ -4 & 0 \end{bmatrix} \quad , \quad BA = \begin{bmatrix} 1 & -2 \\ 2 & 0 \end{bmatrix} \cdot \begin{bmatrix} 3 & 1 \\ 0 & -2 \end{bmatrix} = \begin{bmatrix} 3 & 5 \\ 6 & 2 \end{bmatrix}$$

It shows that $AB \neq BA$.

8. Solve $2X + 2 \begin{bmatrix} -3 & 1 \\ 0 & 2 \end{bmatrix} = \begin{bmatrix} 5 & 10 \\ 0 & 4 \end{bmatrix}$

Solution:

$$2X = \begin{bmatrix} 5 & 10 \\ 0 & 4 \end{bmatrix} - 2 \begin{bmatrix} -3 & 1 \\ 0 & 2 \end{bmatrix} = \begin{bmatrix} 5 & 10 \\ 0 & 4 \end{bmatrix} + \begin{bmatrix} 6 & -2 \\ 0 & -4 \end{bmatrix} = \begin{bmatrix} 11 & 8 \\ 0 & 0 \end{bmatrix}$$

$$X = \begin{bmatrix} \frac{11}{2} & 4 \\ 0 & 0 \end{bmatrix} . \text{ Ans.}$$

Applications of Matrices

Matrices are useful in solving some word problems in the social and physical sciences, such as marketing survey, Kirchhoff Law in electricity, Leontief Models in economy, transition problems in communication networks and Markov Chains in probability, etc. In this section, one application using Markov Chains is discussed.

Example: There are three supermarkets in a city, supermarket A, supermarket B, and supermarket C. At present, the percentages of customers shopping at three supermarkets are A(30%), B(30%), and C(40%).
Marketing survey shows that the following information is always true within a year:

1) 50% of the customers in supermarket A will continue to shop in A, while 20% will go to B and 30% will go to C.
2) 40% of the customers in supermarket B will continue to shop in B, while 25% will go to A and 35% will go to C.
3) 30% of the customers in supermarket C will continue to shop in C, while 45% will go to A and 25% will go to B.

Find the percentages of the customers each supermarket can have at the end of this year.

Solution:

The transition matrix of customers is
$$T = \begin{array}{c} \\ A \\ B \\ C \end{array} \begin{array}{ccc} A & B & C \\ \left[\begin{array}{ccc} 0.50 & 0.25 & 0.45 \\ 0.20 & 0.40 & 0.25 \\ 0.30 & 0.35 & 0.30 \end{array}\right] \end{array}$$

The percentage matrix at present is
$$P^{(0)} = \begin{array}{c} A \\ B \\ C \end{array}\left[\begin{array}{c} 0.30 \\ 0.30 \\ 0.40 \end{array}\right]$$

The percentage matrix at the end of the first year is:

$$P^{(1)} = T\,P^{(0)} = \begin{bmatrix} 0.50 & 0.25 & 0.45 \\ 0.20 & 0.40 & 0.25 \\ 0.30 & 0.35 & 0.30 \end{bmatrix} \cdot \begin{bmatrix} 0.30 \\ 0.30 \\ 0.40 \end{bmatrix} = \begin{bmatrix} 0.50\times0.30+0.25\times0.30+0.45\times0.40 \\ 0.20\times0.30+0.40\times0.30+0.25\times0.40 \\ 0.30\times0.30+0.35\times0.30+0.30\times0.40 \end{bmatrix} = \begin{array}{c} A \\ B \\ C \end{array}\begin{bmatrix} 0.405 \\ 0.280 \\ 0.315 \end{bmatrix}$$

Ans: A(40.5%), B(28%), C(31.5%).

Hint: After several years, the percentages of customers shopping at three supermarkets will be stable at A(41.69%), B(26.96%), C(31.35%).
(Solve the system in matrix form $AX = X$ and $x + y + z = 1$)

EXERCISES

Perform the indicated operation with matrices.

1. $\begin{bmatrix} 2 & -5 \\ 6 & 4 \end{bmatrix} + \begin{bmatrix} -4 & 3 \\ 1 & -7 \end{bmatrix}$ 2. $\begin{bmatrix} 2 & -5 \\ 6 & 4 \end{bmatrix} - \begin{bmatrix} -4 & 3 \\ 1 & -7 \end{bmatrix}$ 3. $\begin{bmatrix} -4 & 3 \\ 1 & -7 \end{bmatrix} - 2\begin{bmatrix} 2 & -5 \\ 6 & 4 \end{bmatrix}$

4. $\begin{bmatrix} 7 & -2 & 4 \\ -3 & 8 & 1 \end{bmatrix} - 3\begin{bmatrix} 5 & -1 & 0 \\ -10 & 9 & 6 \end{bmatrix}$ 5. $\begin{bmatrix} 5 & -2 \\ -7 & 12 \end{bmatrix} - 6\begin{bmatrix} 15 & 1 & 4 \\ -8 & 0 & 2 \end{bmatrix}$

6. $\begin{bmatrix} 1 & 3 & -4 \\ 2 & -5 & 4 \\ 3 & -3 & 7 \end{bmatrix} + 4\begin{bmatrix} 1 & -2 & 0 \\ 2 & 1 & 7 \\ -1 & 3 & 0 \end{bmatrix}$ 7. $\begin{bmatrix} 2 & 6 & -5 \\ -1 & 0 & 0 \\ 0 & 1 & 3 \end{bmatrix} - \begin{bmatrix} 0 & 3 & 12 \\ -3 & 7 & -8 \\ 5 & -1 & 2 \end{bmatrix} + \begin{bmatrix} 1 & 10 & 7 \\ 0 & -5 & -4 \\ 3 & 0 & 2 \end{bmatrix} \cdot (-2)$

8. $\begin{bmatrix} 2 & -5 \\ 6 & 4 \end{bmatrix} \cdot \begin{bmatrix} -4 & 3 \\ 1 & -7 \end{bmatrix}$ 9. $\begin{bmatrix} 2 & -5 \\ 6 & 4 \end{bmatrix} \cdot \begin{bmatrix} -4 & 12 \\ 1 & -5 \\ 3 & 2 \end{bmatrix}$ 10. $\begin{bmatrix} 1 & -1 & 5 \\ 0 & 3 & 0 \\ 2 & 4 & -2 \end{bmatrix} \cdot \begin{bmatrix} 3 & 2 & -6 \\ 1 & 5 & 7 \\ 4 & -1 & 0 \end{bmatrix}$

Solve each matrix equation for x and y.

11. $\begin{bmatrix} x & -4 \\ 5 & 2y \end{bmatrix} = \begin{bmatrix} -3 & -4 \\ 5 & 22 \end{bmatrix}$ 12. $\begin{bmatrix} 12 & y+2 \\ 2x-1 & -4 \end{bmatrix} = \begin{bmatrix} 12 & 2y-6 \\ 6 & -4 \end{bmatrix}$

13. $\begin{bmatrix} -2 & 1 \\ 3 & 2x \\ 4 & -1 \end{bmatrix} - \begin{bmatrix} 7 & y \\ 4 & 8 \\ -3 & 2 \end{bmatrix} = \begin{bmatrix} -9 & -6 \\ -1 & 4 \\ 7 & -3 \end{bmatrix}$ 14. $3X - 4\begin{bmatrix} -1 & 3 \\ 2 & 7 \end{bmatrix} = -\begin{bmatrix} 5 & 12 \\ 0 & -10 \end{bmatrix}$

15. A lady wants to buy the following fruits for a party. The prices of the two stores in the street are listed in the following table. Use the matrices to write a matrix that shows the total prices of the fruits in each store if she buys all items in one store and the total prices of the fruits if she buys each item having the cheaper price in both stores.

Fruits	Apple	Lemon	Orange	Banana	Pear
Pounds	10	5	12	15	8

Price / lb	Apple	Lemon	Orange	Banana	Pear
Store A	$3.15	2.18	2.43	1.90	4.32
Store B	3.65	1.72	2.14	2.38	3.99

16. A and B are two supermarkets in the same street of a small town. At present, the percentages of customers shopping at two supermarkets are A(30%) and B(70%) . The following Marketing information is true within each year:
   1) 80% of the customers in supermarket A will continue to shop in A, while 20% will go to B.
   2) 60% of the customers in supermarket B will continue to shop in B, while 40% will go to A.
   What is the predicted percentages of customers for supermarkets A and B three years later ?

9-2 Determinants and the Inverse of Matrix

The form of a determinant is similar to a matrix. A matrix is formed with brackets enclosing the entries. A **determinant** is formed with vertical bars enclosing the entries. Each entry is called an **element** of the determinant.

The number of elements in any row or column is called the **order** of the determinant.

Determinant of a 2×2 matrix A is defined as below:

$$\det A = \begin{vmatrix} a & b \\ c & d \end{vmatrix} = ad - bc$$

Determinant of a 3×3 matrix A is defined as below: (Three different ways to evaluate)

1. $\det A = $ 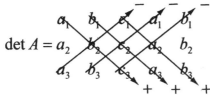 $= a_1b_2c_3 + a_2b_3c_1 + a_3b_1c_2 - a_3b_2c_1 - a_1b_3c_2 - a_2b_1c_3$

The products of elements:
1. run down from left to right are positive.
2. run down from right to left are negative.

2. Expansion by its minors across any row (see next page)

$$\det A = a_1\begin{vmatrix} b_2 & c_2 \\ b_3 & c_3 \end{vmatrix} - b_1\begin{vmatrix} a_2 & c_2 \\ a_3 & c_3 \end{vmatrix} + c_1\begin{vmatrix} a_2 & b_2 \\ a_3 & b_3 \end{vmatrix}$$

3. Copy the first two columns to the right side of the determinant

$$\det A = a_2 \qquad b_2$$

The above formulas are not valid for $n \times n$ if $n \geq 4$ (See next Section).

Examples (Evaluate each determinant)

1. $\begin{vmatrix} 1 & 5 \\ 4 & 8 \end{vmatrix} = 1 \cdot 8 - 5 \cdot 4 = 8 - 20 = -12$ **2.** $\begin{vmatrix} -3 & 6 \\ -2 & 4 \end{vmatrix} = -3 \cdot 4 - 6 \cdot (-2) = -12 + 12 = 0$

3. $\begin{vmatrix} 2 & 1 & -3 \\ 3 & 5 & -4 \\ -4 & 2 & 6 \end{vmatrix} = 60 + (-18) + 16 - 60 - (-16) - 18 = 60 - 18 + 16 - 60 + 16 - 18 = -4$

4. $\begin{vmatrix} \sqrt{2} & -\sqrt{3} \\ \sqrt{15} & \sqrt{6} \end{vmatrix} = \sqrt{2} \cdot \sqrt{6} - (-\sqrt{3} \cdot \sqrt{15}) = \sqrt{12} + \sqrt{45} = 2\sqrt{3} + 3\sqrt{5}$

5. $\begin{vmatrix} \log 5 & \log 2 \\ \log 2 & \log 5 \end{vmatrix} = (\log 5)^2 - (\log 2)^2 = (\log 5 + \log 2)(\log 5 - \log 2) = \log(5 \times 2) \cdot \log\frac{5}{2}$

$$= \log 10 \cdot \log\tfrac{5}{2} = 1 \cdot \log\tfrac{5}{2} = \log\tfrac{5}{2}$$

Expansion of Determinants

Minor of an element : The minor of an element in a determinant is the determinants in which the row and column containing the element are deleted. Multiply by (-1) if the sum of the numbers of rows and columns containing the element is odd. Or, multiplied by $(-1)^{m+n}$.

The determinant of a $n \times n$ matrix can be expanded by minors about any row or column.

The following properties are used to simplify the expansion (evaluation) of a determinant of any order:

1. If all elements in any row or any column are zeros, then the determinant is 0.
2. If all corresponding elements in two rows or two columns are equal, then the determinant is 0.
3. If two rows or two columns of a determinant are interchanged, the resulting determinant is the negative of the original determinant.
4. If each element in one row or one column of a determinant is multiplied by a real number k, then the determinant is multiplied by k.
5. If each element in one row (or one column) of a determinant is multiplied by a real number k and then add to the corresponding element of another row (or another column), the resulting determinant equals the original one.

Examples (Expansion by minors)

1.
$$\begin{vmatrix} 7 & 0 & 0 & 0 \\ 4 & -3 & 0 & 0 \\ 8 & 2 & 2 & 0 \\ -9 & 14 & 3 & 5 \end{vmatrix} = 7\begin{vmatrix} -3 & 0 & 0 \\ 2 & 2 & 0 \\ 14 & 3 & 5 \end{vmatrix} - 0\begin{vmatrix} 4 & 0 & 0 \\ 8 & 2 & 0 \\ -9 & 3 & 5 \end{vmatrix} + 0\begin{vmatrix} 4 & -3 & 0 \\ 8 & 2 & 0 \\ -9 & 14 & 5 \end{vmatrix} - 0\begin{vmatrix} 4 & -3 & 0 \\ 8 & 2 & 2 \\ -9 & 14 & 3 \end{vmatrix}$$

$$= 7\begin{vmatrix} -3 & 0 & 0 \\ 2 & 2 & 0 \\ 14 & 3 & 5 \end{vmatrix} = 7(-30) = -210$$

2.
$$\begin{vmatrix} 1 & 0 & 1 & -1 \\ 2 & -1 & 1 & 2 \\ 2 & ① & 2 & 1 \\ 1 & 2 & 3 & 1 \end{vmatrix} \begin{matrix} r_1 \\ r_2 \\ r_3 \\ r_4 \end{matrix}$$

Hint: Reduce it to 3×3 matrix by "row operations". Choose one column which is the simplest one. Make all its elements to be "0" except one element. Reduce it by its minors. We choose column 2 in this example.)

$$= \begin{matrix} \\ r_2 + r_3 \\ \\ r_4 - 2r_3 \end{matrix} \begin{vmatrix} 1 & 0 & 1 & -1 \\ 4 & 0 & 3 & 3 \\ 2 & 1 & 2 & 1 \\ -3 & 0 & -1 & -1 \end{vmatrix} = (-1)^{3+2} \cdot 1 \cdot \begin{vmatrix} 1 & 1 & -1 \\ 4 & 3 & 3 \\ -3 & -1 & -1 \end{vmatrix} = -1(-10) = 10$$

The Inverse of Matrix

Identity Matrix I : It is a square matrix whose main diagonal has all elements by 1 and all other elements by 0. It is denoted by **I**.

Inverse of Matrix A^{-1} **:** If the product of two square matrices A and B is an identity matrix, B is the inverse of A.

$$\text{If } AB = I, \text{ then } B = A^{-1}$$

The product of a square matrix and its inverse is an identity matrix. $A \cdot A^{-1} = I$.

Not all square matrices have inverses. If a matrix has no inverse, It is called **singular**.

$$A = \begin{bmatrix} 2 & 1 \\ 3 & -1 \end{bmatrix}, \quad B = \begin{bmatrix} \frac{1}{5} & \frac{1}{5} \\ \frac{3}{5} & -\frac{2}{5} \end{bmatrix} \qquad A \cdot B = \begin{bmatrix} 2 & 1 \\ 3 & -1 \end{bmatrix} \cdot \begin{bmatrix} \frac{1}{5} & \frac{1}{5} \\ \frac{3}{5} & -\frac{2}{5} \end{bmatrix} = \begin{bmatrix} 2(\frac{1}{5})+1(\frac{3}{5}) & 2(\frac{1}{5})-1(\frac{2}{5}) \\ 3(\frac{1}{5})-1(\frac{3}{5}) & 3(\frac{1}{5})+1(\frac{2}{5}) \end{bmatrix} = \begin{bmatrix} 1 & 0 \\ 0 & 1 \end{bmatrix}$$

$$\therefore B \text{ is the inverse of matrix } A.$$

We use the following methods to find the inverse of a square matrix:

1. The inverse of a 2×2 matrix can be obtained by the following formula:

If $A = \begin{bmatrix} a & b \\ c & d \end{bmatrix}$, then $A^{-1} = \dfrac{\begin{bmatrix} d & -b \\ -c & a \end{bmatrix}}{\det A}$, $\det A \neq 0$

Example 1: Find the inverse of the matrix:

$$A = \begin{bmatrix} 4 & 2 \\ -3 & -1 \end{bmatrix}$$

Solution:

Method 1: $\det A = -4 - (-6) = 2$

$$A^{-1} = \frac{\begin{bmatrix} -1 & -2 \\ 3 & 4 \end{bmatrix}}{\det A} = \frac{\begin{bmatrix} -1 & -2 \\ 3 & 4 \end{bmatrix}}{2} = \begin{bmatrix} -\frac{1}{2} & -1 \\ \frac{3}{2} & 2 \end{bmatrix}. \text{ Ans.}$$

Method 2:

$$\begin{array}{c} r_1 \\ r_2 \end{array} \begin{bmatrix} 4 & 2 & 1 & 0 \\ -3 & -1 & 0 & 1 \end{bmatrix}$$

$$\begin{array}{c} r_3 \\ r_4 \end{array} \begin{bmatrix} 1 & \frac{1}{2} & \frac{1}{4} & 0 \\ 3 & 1 & 0 & -1 \end{bmatrix} \begin{array}{l} r_1 \div 4 \\ r_2 \div (-1) \end{array}$$

$$\begin{array}{c} r_5 \\ r_6 \end{array} \begin{bmatrix} -1 & 0 & \frac{1}{2} & 1 \\ 0 & -\frac{1}{6} & -\frac{1}{4} & -\frac{1}{3} \end{bmatrix} \begin{array}{l} r_3 \cdot 2 - r_4 \\ r_4 \div 3 - r_3 \end{array}$$

$$\begin{bmatrix} 1 & 0 & -\frac{1}{2} & -1 \\ 0 & 1 & \frac{3}{2} & 2 \end{bmatrix} \begin{array}{l} r_5 \div (-1) \\ r_6 \cdot (-6) \end{array}$$

$$\therefore A^{-1} = \begin{bmatrix} -\frac{1}{2} & -1 \\ \frac{3}{2} & 2 \end{bmatrix}. \text{ Ans.}$$

2. To find the inverse of any square matrix, we adjoin an identity matrix to the right side of the origin matrix, then use the **row operations** to transfer the original matrix into an identity matrix, the new matrix on the right side is the inverse.

Example 2: Find the inverse of the matrix:

$$B = \begin{bmatrix} 0 & 1 & 2 \\ 1 & -1 & 3 \\ 2 & -4 & 1 \end{bmatrix}$$

Solution:

$$\begin{array}{c} r_1 \\ r_2 \\ r_3 \end{array} \begin{bmatrix} 0 & 1 & 2 & 1 & 0 & 0 \\ 1 & -1 & 3 & 0 & 1 & 0 \\ 2 & -4 & 1 & 0 & 0 & 1 \end{bmatrix}$$

$$\begin{array}{c} r_4 \\ r_5 \\ r_6 \end{array} \begin{bmatrix} 1 & 0 & 5 & 1 & 1 & 0 \\ -1 & 1 & -3 & 0 & -1 & 0 \\ 2 & -4 & 1 & 0 & 0 & 1 \end{bmatrix} \begin{array}{l} r_1 + r_2 \\ r_2 (-1) \\ \end{array}$$

$$\begin{array}{c} r_7 \\ r_8 \\ r_9 \end{array} \begin{bmatrix} 1 & 0 & 5 & 1 & 1 & 0 \\ 0 & 1 & 2 & 1 & 0 & 0 \\ 0 & -4 & -9 & -2 & -2 & 1 \end{bmatrix} \begin{array}{l} \\ r_5 + r_4 \\ r_6 - r_4 \cdot 2 \end{array}$$

$$\begin{array}{c} r_{10} \\ r_{11} \\ r_{12} \end{array} \begin{bmatrix} 1 & 0 & 5 & 1 & 1 & 0 \\ 0 & 1 & 2 & 1 & 0 & 0 \\ 0 & 0 & -1 & 2 & -2 & 1 \end{bmatrix} \begin{array}{l} \\ \\ r_9 + r_8 \cdot 4 \end{array}$$

$$\begin{bmatrix} 1 & 0 & 0 & 11 & -9 & 5 \\ 0 & 1 & 0 & 5 & -4 & 2 \\ 0 & 0 & 1 & -2 & 2 & -1 \end{bmatrix} \begin{array}{l} r_{10} + r_{12} \cdot 5 \\ r_{11} + r_{12} \cdot 2 \\ r_{12} (-1) \end{array}$$

$$B^{-1} = \begin{bmatrix} 11 & -9 & 5 \\ 5 & -4 & 2 \\ -2 & 2 & -1 \end{bmatrix}. \text{ Ans.}$$

EXERCISES

Evaluate each determinant.

1. $\begin{vmatrix} 7 & 8 \\ 5 & 3 \end{vmatrix}$
2. $\begin{vmatrix} 7 & 8 \\ -5 & 3 \end{vmatrix}$
3. $\begin{vmatrix} 7 & -8 \\ 5 & -3 \end{vmatrix}$
4. $\begin{vmatrix} \log 25 & \log 4 \\ \log 2 & \log 5 \end{vmatrix}$

5. $\begin{vmatrix} 2 & -5 & 3 \\ 0 & 8 & 1 \\ -5 & 4 & 0 \end{vmatrix}$
6. $\begin{vmatrix} 3 & 2 & -2 \\ 1 & 1 & 2 \\ 2 & -1 & 2 \end{vmatrix}$
7. $\begin{vmatrix} 2 & 2 & -2 \\ 1 & 2 & 1 \\ 2 & 3 & 0 \end{vmatrix}$
8. $\begin{vmatrix} 2 & 1 & 0 \\ 3 & -1 & 0 \\ 4 & 6 & 0 \end{vmatrix}$

9. $\begin{vmatrix} 2 & 4 & -2 \\ 0 & 0 & 0 \\ 0 & 0 & 0 \end{vmatrix}$
10. $\begin{vmatrix} 2 & 5 & 1 \\ -3 & 7 & -8 \\ 11 & -4 & 15 \end{vmatrix}$
11. $\begin{vmatrix} 4 & 0 & -7 \\ 6 & 1 & 11 \\ -6 & 0 & 3 \end{vmatrix}$
12. $\begin{vmatrix} \sqrt{3} & 1 & \sqrt{6} \\ \sqrt{3} & -1 & \sqrt{6} \\ \sqrt{6} & 1 & \sqrt{3} \end{vmatrix}$

13. $\begin{vmatrix} 1 & 0 & -3 \\ 2 & 0 & -1 \\ 4 & 0 & 6 \end{vmatrix}$
14. $\begin{vmatrix} -2 & 3 & 4 \\ -8 & 12 & 16 \\ 5 & 6 & 2 \end{vmatrix}$
15. $\begin{vmatrix} 1 & 1 & 1 \\ 1 & 1+a & 1 \\ 1 & 1 & 1+b \end{vmatrix}$
16. $\begin{vmatrix} 1+a & 1 & 1 \\ 1 & 1+b & 1 \\ 1 & 1 & 1+c \end{vmatrix}$

17. $\begin{vmatrix} 1 & 0 & 1 & 2 \\ 2 & -1 & 2 & 2 \\ 2 & 1 & 3 & 1 \\ 4 & 2 & 1 & 5 \end{vmatrix}$
18. $\begin{vmatrix} 1 & 0 & 0 & 0 \\ 2 & 2 & 0 & 0 \\ 3 & 3 & 3 & 0 \\ 4 & 4 & 4 & 4 \end{vmatrix}$
19. $\begin{vmatrix} 0 & 0 & 0 & k \\ 0 & 0 & k & 6 \\ 0 & k & 4 & 5 \\ k & 1 & 2 & 3 \end{vmatrix}$
20. $\begin{vmatrix} 1 & 1 & 1 & 1 \\ 1 & 1+a & 1 & 1 \\ 1 & 1 & 2+b & 1 \\ 1 & 1 & 1 & 3+c \end{vmatrix}$

Find the inverse of each matrix. If the matrix has no inverse, so state.

21. $\begin{bmatrix} 1 & 2 \\ 1 & 3 \end{bmatrix}$
22. $\begin{bmatrix} 2 & 1 \\ 1 & 3 \end{bmatrix}$
23. $\begin{bmatrix} 2 & 4 \\ 4 & 8 \end{bmatrix}$
24. $\begin{bmatrix} 2 & -3 \\ 3 & 4 \end{bmatrix}$

25. $\begin{bmatrix} 2 & 1 & 1 \\ 0 & 1 & 1 \\ 2 & 0 & 1 \end{bmatrix}$
26. $\begin{bmatrix} 1 & -1 & 0 \\ 1 & 0 & -1 \\ 6 & -2 & -3 \end{bmatrix}$
27. $\begin{bmatrix} 1 & 0 & 1 \\ 0 & 1 & 1 \\ 1 & 1 & 0 \end{bmatrix}$
28. $\begin{bmatrix} 1 & 2 & 3 & 1 \\ 1 & 3 & 3 & 2 \\ 2 & 4 & 3 & 3 \\ 1 & 1 & 1 & 1 \end{bmatrix}$

29. Find the inverse of the matrix.

$$A = \begin{bmatrix} \cos\theta & -\sin\theta \\ \sin\theta & \cos\theta \end{bmatrix}$$

30. If the matrix has no inverse, find x.

$$B = \begin{bmatrix} x^2 & x+3 \\ 2x+1 & 2 \end{bmatrix}$$

9-3 Solving Systems with Matrices and Cramer's Rule

Solving a system with row operations on its augmented matrix

Performing the row operations on the matrix form of the system, we can reduce the system into a system equivalent to original system and make it easy to solve.

The matrices used to represent system of linear equations are called **augmented matrices**.

If the system is $\begin{cases} ax + by = e \\ cx + dy = f \end{cases}$, then the augmented matrix is $\begin{bmatrix} a & b & | & e \\ c & d & | & f \end{bmatrix}$

In writing the augmented matrix of a system, the vertical line represents the equal signs. The coefficient matrix is placed on the left side of the vertical line and the constants on the right side. If any variable is missing, its coefficient is 0.

The method of elimination for solving a system of linear equations in basic algebra can be simplified by the **row operations** on its augmented matrix without writing the variables. The reduced matrix is called the **echelon form** of the original matrix. If a system has no solution, we say that it is **inconsistent**.

Row Operations: Given a matrix of a system of linear equations, we have an equivalent matrix if
1. Interchange any two rows.
2. A row is multiplied or divided by a nonzero real number.
3. A row is multiplied or divided by a nonzero real number and then add to another row.

Echelon form: A matrix in which all elements on the main diagonal are 1's or 0's and all elements below the main diagonal are 0's.

Examples

1. Solve $\begin{cases} 2x + y = 1 \\ 3x - y = 9 \end{cases}$

Solution:

$\begin{matrix} r_1 \\ {} \end{matrix} \begin{bmatrix} 2 & 1 & | & 1 \\ 3 & -1 & | & 9 \end{bmatrix}$

$\begin{matrix} r_3 \\ r_4 \end{matrix} \begin{bmatrix} 1 & \frac{1}{2} & | & \frac{1}{2} \\ 3 & -1 & | & 9 \end{bmatrix} r_1 \div 2$

$\begin{matrix} {} \\ r_6 \end{matrix} \begin{bmatrix} 1 & \frac{1}{2} & | & \frac{1}{2} \\ 0 & -\frac{5}{2} & | & \frac{15}{2} \end{bmatrix} r_4 - r_3(3)$

From the reduced echelon form, we have

$\begin{cases} x + \frac{1}{2}y = \frac{1}{2} \\ y = -3 \end{cases}$

$\begin{bmatrix} 1 & \frac{1}{2} & | & \frac{1}{2} \\ 0 & 1 & | & -3 \end{bmatrix} r_6(-\frac{2}{5})$

Solve the above system, we have

Ans: $x = 2$, $y = -3$

Examples

2. Solve $\begin{cases} x + 2y = 4 \\ x + 2y = -2 \end{cases}$

Solution:

$$\begin{array}{c} r_1 \\ r_2 \end{array}\left[\begin{array}{cc|c} 1 & 2 & 4 \\ 1 & 2 & -2 \end{array}\right]$$

$$\left[\begin{array}{cc|c} 1 & 2 & 4 \\ 0 & 0 & -4 \end{array}\right] r_1 - r_2$$

All elements on the left side of the last row are zeros. The system has no solutions. They are parallel.

(Inconsistent)

4. Solve $\begin{cases} x - y + 3z = 2 \\ y + 2z = 1 \\ 2x - 4y + z = 3 \end{cases}$

Solution:

$$\begin{array}{c} r_1 \\ \\ r_3 \end{array}\left[\begin{array}{ccc|c} 1 & -1 & 3 & 2 \\ 0 & 1 & 2 & 1 \\ 2 & -4 & 1 & 3 \end{array}\right]$$

$$\begin{array}{c} \\ r_5 \\ r_6 \end{array}\left[\begin{array}{ccc|c} 1 & -1 & 3 & 2 \\ 0 & 1 & 2 & 1 \\ 0 & -2 & -5 & -1 \end{array}\right] r_3 - r_1(2)$$

$$\begin{array}{c} \\ \\ r_9 \end{array}\left[\begin{array}{ccc|c} 1 & -1 & 3 & 2 \\ 0 & 1 & 2 & 1 \\ 0 & 0 & -1 & 1 \end{array}\right] r_6 + r_5(2)$$

$$\left[\begin{array}{ccc|c} 1 & -1 & 3 & 2 \\ 0 & 1 & 2 & 1 \\ 0 & 0 & 1 & -1 \end{array}\right] r_9(-1)$$

From the reduced echelon form, we have

$$\begin{cases} x - y + 3z = 2 \\ y + 2z = 1 \\ z = -1 \end{cases}$$

Solve the above system, we have

Ans: $x = 8$, $y = 3$, $z = -1$

3. Solve $\begin{cases} x + 2y = 4 \\ x + 4y = 8 \end{cases}$

Solution:

$$\begin{array}{c} r_1 \\ r_2 \end{array}\left[\begin{array}{cc|c} 1 & 2 & 4 \\ 1 & 4 & 8 \end{array}\right]$$

$$\left[\begin{array}{cc|c} 1 & 2 & 4 \\ 0 & 0 & 0 \end{array}\right] r_1 - r_2$$

All elements in the last row are zeros. The system has unlimited number of solutions. They coincide.

$$x = 4 - 2y, \ y \in R$$

5. Solve $\begin{cases} x + y - z = -2 \\ x + 2y + z = -9 \\ 2x + 3y = -11 \end{cases}$

Solution:

$$\begin{array}{c} r_1 \\ r_2 \\ r_3 \end{array}\left[\begin{array}{ccc|c} 1 & 1 & -1 & -2 \\ 1 & 2 & 1 & -9 \\ 2 & 3 & 0 & -11 \end{array}\right]$$

$$\begin{array}{c} \\ r_5 \\ r_6 \end{array}\left[\begin{array}{ccc|c} 1 & 1 & -1 & -2 \\ 0 & 1 & 2 & -7 \\ 0 & -1 & -2 & 7 \end{array}\right] \begin{array}{c} r_2 - r_1 \\ r_3 - r_2(2) \end{array}$$

$$\left[\begin{array}{ccc|c} 1 & 1 & -1 & -2 \\ 0 & 1 & 2 & -7 \\ 0 & 0 & 0 & 0 \end{array}\right] r_5 + r_6$$

All elements in the last row are zeros. The system has unlimited number of solutions.

The intersection of the first plane and the second plane is a straight line which coincides with the third straight line.

We have $x + y - z = -2$, $y + 2z = -7$
Solve the above system, we have

Ans: $x = 3z + 5$
$y = -2z - 7$, $z \in R$

Solving a system with inverse matrix

The inverse of matrix can be used to solve **system of linear equations** in matrix form.
We can write a system of linear equations in matrix form.

$$\begin{cases} ax + by = e \\ cx + dy = f \end{cases}$$

Performing the multiplication of matrices:

We have: $\begin{bmatrix} a & b \\ c & d \end{bmatrix} \cdot \begin{bmatrix} x \\ y \end{bmatrix} = \begin{bmatrix} e \\ f \end{bmatrix}$ and let $A = \begin{bmatrix} a & b \\ c & d \end{bmatrix}$, $X = \begin{bmatrix} x \\ y \end{bmatrix}$, $C = \begin{bmatrix} e \\ f \end{bmatrix}$

The original system of equations can be written as the equivalent equations in matrix form:

$$A \cdot X = C$$

Multiply both sides by A^{-1}: $A^{-1}A \cdot X = A^{-1} C$

$$I \cdot X = A^{-1} C$$

$$\therefore X = A^{-1} C$$

If there is an inverse A^{-1}, then the system of equations has an unique solution $A^{-1}C$.

If the inverse (A^{-1}) of the coefficient matrix (A) can be easily computed, it is useful in solving a linear system. Otherwise, the inverse method is not beneficial.

If we are using a computational devise, the inverse method is the preferred method.

There are other important uses for the inverse of a matrix in advanced mathematics.

If a square matrix A has an inverse, A is called **invertible** or **nonsingular**. Otherwise, it is called **singular**. If a matrix is not square, it has no inverse.

Examples

1. Solve $\begin{cases} 2x + y = 1 \\ 3x - y = 9 \end{cases}$

    Solution:

    $$\begin{bmatrix} 2 & 1 \\ 3 & -1 \end{bmatrix} \cdot \begin{bmatrix} x \\ y \end{bmatrix} = \begin{bmatrix} 1 \\ 9 \end{bmatrix}$$

    $$A = \begin{bmatrix} 2 & 1 \\ 3 & -1 \end{bmatrix}, \quad X = \begin{bmatrix} x \\ y \end{bmatrix}, \quad C = \begin{bmatrix} 1 \\ 9 \end{bmatrix}$$

    $$A^{-1} = \frac{\begin{bmatrix} -1 & -1 \\ -3 & 2 \end{bmatrix}}{\det A} = \frac{\begin{bmatrix} -1 & -1 \\ -3 & 2 \end{bmatrix}}{-5} = \begin{bmatrix} \frac{1}{5} & \frac{1}{5} \\ \frac{3}{5} & -\frac{2}{5} \end{bmatrix}$$

    $$X = A^{-1}C, \quad \begin{bmatrix} x \\ y \end{bmatrix} = \begin{bmatrix} \frac{1}{5} & \frac{1}{5} \\ \frac{3}{5} & -\frac{2}{5} \end{bmatrix} \cdot \begin{bmatrix} 1 \\ 9 \end{bmatrix} = \begin{bmatrix} 2 \\ -3 \end{bmatrix}$$

    Ans: $x = 2$, $y = -3$

2. Solve $\begin{cases} y + 2z = 1 \\ x - y + 3z = 2 \\ 2x - 4y + z = 3 \end{cases}$

    Solution:

    $$A = \begin{bmatrix} 0 & 1 & 2 \\ 1 & -1 & 3 \\ 2 & -4 & 1 \end{bmatrix}, \quad X = \begin{bmatrix} x \\ y \\ z \end{bmatrix}, \quad C = \begin{bmatrix} 1 \\ 2 \\ 3 \end{bmatrix}$$

    Find A^{-1} (see Example on Page 329):

    $$A^{-1} = \begin{bmatrix} 11 & -9 & 5 \\ 5 & -4 & 2 \\ -2 & 2 & -1 \end{bmatrix}, \quad X = A^{-1}C$$

    $$\begin{bmatrix} x \\ y \\ z \end{bmatrix} = \begin{bmatrix} 11 & -9 & 5 \\ 5 & -4 & 2 \\ -2 & 2 & -1 \end{bmatrix} \cdot \begin{bmatrix} 1 \\ 2 \\ 3 \end{bmatrix} = \begin{bmatrix} 8 \\ 3 \\ -1 \end{bmatrix}$$

    Ans: $x = 8$, $y = 3$, $z = -1$

3. Solve $\begin{cases} 4x - 6y = 3 \\ -2x + 3y = 4 \end{cases}$

Solution:

$$A = \begin{bmatrix} 4 & -6 \\ -2 & 3 \end{bmatrix}$$

det $A = 12 - 12 = 0$.

The coefficient matrix has no inverse.
The system of equations has no solution.
(They are parallel.) (Inconsistent)

4. Solve $\begin{cases} 10x - 6y = -8 \\ -5x + 3y = 4 \end{cases}$

Solution:

$$A = \begin{bmatrix} 10 & -6 \\ -5 & 3 \end{bmatrix}$$

det $A = 30 - 30 = 0$.

The coefficient matrix has no inverse.
The system of equations has unlimited
number of solutions. (they coincide.)

5. Solve $\begin{cases} 2x - y - z = 1 \\ -3x + y + 2z = 2 \\ 5x - y - 3z = 3 \end{cases}$

Solution:

Find the inverse of the coefficient
matrix A. We have: (see Page 329)

$$A = \begin{bmatrix} 2 & -1 & -1 \\ -3 & 1 & 2 \\ 5 & -1 & -3 \end{bmatrix}, \quad A^{-1} = \begin{bmatrix} 1 & 2 & 1 \\ -1 & 1 & 1 \\ 2 & 3 & 1 \end{bmatrix}$$

$$X = A^{-1}C$$

$$\begin{bmatrix} x \\ y \\ z \end{bmatrix} = \begin{bmatrix} 1 & 2 & 1 \\ -1 & 1 & 1 \\ 2 & 3 & 1 \end{bmatrix} \cdot \begin{bmatrix} 1 \\ 2 \\ 3 \end{bmatrix} = \begin{bmatrix} 8 \\ 4 \\ 11 \end{bmatrix}$$

Ans: $x = 8$, $y = 4$, $z = 11$

6. Solve $\begin{cases} x + y - z = -2 \\ x + 2y + z = -9 \\ 2x + 3y = -11 \end{cases}$

Solution: (By row operations)
Find the inverse of the coefficient
matrix.

$$\begin{array}{c} r_1 \\ r_2 \\ r_3 \end{array} \begin{bmatrix} 1 & 1 & -1 \\ 1 & 2 & 1 \\ 2 & 3 & 0 \end{bmatrix} \left| \begin{array}{ccc} 1 & 0 & 0 \\ 0 & 1 & 0 \\ 0 & 0 & 1 \end{array} \right.$$

$$\begin{array}{c} r_4 \\ r_5 \\ r_6 \end{array} \begin{bmatrix} 1 & 1 & -1 \\ 0 & 1 & 2 \\ 0 & -1 & -2 \end{bmatrix} \left| \begin{array}{ccc} 1 & 0 & 0 \\ -1 & 1 & 0 \\ 0 & -2 & 1 \end{array} \right. \begin{array}{c} \\ r_2 - r_1 \\ r_3 - r_2(2) \end{array}$$

$$\begin{bmatrix} 1 & 1 & -1 \\ 0 & 1 & 2 \\ 0 & 0 & 0 \end{bmatrix} \left| \begin{array}{ccc} 1 & 0 & 0 \\ -1 & 1 & 0 \\ -1 & -1 & 1 \end{array} \right. \begin{array}{c} \\ \\ r_6 + r_5 \end{array}$$

All elements in the last row of the left
side are zeros. The coefficient matrix
has no inverse.
The system has unlimited number of
solutions. Ans.

Discussion:
The intersection of the first plane and
the second plane is a straight line which
coincides with the third straight line.
If we solve for x and y in terms of z
from the first and second equations,
we have:
$$x = 3z + 5$$
$$y = -2z - 7$$

The solution set consists of all ordered
triples that satisfy:

$$\begin{array}{l} x = 3t + 5 \\ y = -2t - 7 \\ z = t \ (t \in R) \end{array} \quad \left| \begin{array}{l} \text{Or} \quad x = 3z + 5 \\ \qquad y = -2z - 7 \\ \qquad z \in R \end{array} \right.$$

Cramer's Rule

Cramer's Rule gives us an easy method to solve a system of linear equations in n variables.

Cramer's Rule: If $D \neq 0$, then the system of n linear equations in n variables has the following unique solution:

$$x = \frac{D_x}{D}, \quad y = \frac{D_y}{D}, \quad z = \frac{D_z}{D}, \cdots \cdots$$

$D \rightarrow$ The determinant of the coefficient matrix of the variables.

$D_x \rightarrow$ The determinant of replacing the coefficients of x in D by the constants.

$D_y \rightarrow$ The determinant of replacing the coefficients of y in D by the constants.

$D_z \rightarrow$ The determinant of replacing the coefficients of z in D by the constants.

$\vdots$

A system of n linear equations in n variables has a unique solution if and only if $D \neq 0$.

1) If $D \neq 0$, the system of equations has **consistent and independent** equations. The system has a unique solution.

2) If $D = 0$ and at least one of D_x, D_y, D_z, ... $\neq 0$, the system of equations has **inconsistent** equations. There is no solution (they are parallel).

3) If $D = 0$ and all of D_x, D_y, D_z, ... $= 0$, the system of equations has **consistent and dependent** equations. There are unlimited (infinite) number of solutions (they coincide).

To make a wise and efficient choice of solution method, we need to keep each method (substitution, elimination, inverse matrix, Cramer's Rule, and others) in our memory. Inverse matrix and Cramer's Rule are used to solve a system in which the coefficients matrix must be a square. Cramer's Rule is also not efficient if the system has many equations because many determinants needed to be evaluated.

If the coefficient matrix is not invertible (no inverse), the inverse matrix cannot be used to solve a system. The method of row operations is advisable for all kinds of systems. However, a scientific calculator will simply solve a system.

Examples

1. Solve $\begin{cases} 2x - y = 6 \\ 3x + 5y = 22 \end{cases}$

   Solution:

   $$D = \begin{vmatrix} 2 & -1 \\ 3 & 5 \end{vmatrix} = 10 - (-3) = 13$$

   $$D_x = \begin{vmatrix} 6 & -1 \\ 22 & 5 \end{vmatrix} = 30 - (-22) = 52 , \quad D_y = \begin{vmatrix} 2 & 6 \\ 3 & 22 \end{vmatrix} = 44 - 18 = 26$$

   $$\therefore x = \frac{D_x}{D} = \frac{52}{13} = 4 , \quad y = \frac{D_y}{D} = \frac{26}{13} = 2 . \quad \text{Ans: } x = 4, \ y = 2$$

2. Solve $\begin{cases} 3x + 2y - 2z = 11 \\ x + y + 2y = -1 \\ 2x - y + 2z = -4 \end{cases}$

Solution:

$$D = \begin{vmatrix} 3 & 2 & -2 \\ 1 & 1 & 2 \\ 2 & -1 & 2 \end{vmatrix} = 6 + 2 + 8 - (-4) - (-6) - 4 = 22$$

(Consistent and independent equations)

$$D_x = \begin{vmatrix} 11 & 2 & -2 \\ -1 & 1 & 2 \\ -4 & -1 & 2 \end{vmatrix} = 22 + (-2) + (-16) - 8 - (-22) - (-4) = 22$$

$$D_y = \begin{vmatrix} 3 & 11 & -2 \\ 1 & -1 & 2 \\ 2 & -4 & 2 \end{vmatrix} = -6 + 8 + 44 - 4 - (-24) - 22 = 44$$

$$D_z = \begin{vmatrix} 3 & 2 & 11 \\ 1 & 1 & -1 \\ 2 & -1 & -4 \end{vmatrix} = -12 + (-11) + (-4) - 22 - 3 - (-8) = -44$$

$$x = \frac{D_x}{D} = \frac{22}{22} = 1 \ , \quad y = \frac{D_y}{D} = \frac{44}{22} = 2 \ , \quad z = \frac{D_z}{D} = \frac{-44}{22} = -2$$

Ans: $x = 1$, $y = 2$, $z = -2$.

---

3. Solve $\begin{cases} x + 2y = 4 \\ x + 2y = -2 \end{cases}$

Solution:

$$D = \begin{vmatrix} 1 & 2 \\ 1 & 2 \end{vmatrix} = 2 - 2 = 0$$

$$D_x = \begin{vmatrix} 4 & 2 \\ -2 & 2 \end{vmatrix} = 8 - (-4) = 12 \neq 0$$

$$D_y = \begin{vmatrix} 1 & 4 \\ 1 & -2 \end{vmatrix} = -2 - 4 = -6 \neq 0$$

Inconsistent equation.
They are parallel (no intersection).

Ans: There is no solution.

4. Solve $\begin{cases} x + 2y = 4 \\ 2x + 4y = 8 \end{cases}$

Solution:

$$D = \begin{vmatrix} 1 & 2 \\ 2 & 4 \end{vmatrix} = 4 - 4 = 0$$

$$D_x = \begin{vmatrix} 4 & 2 \\ 8 & 4 \end{vmatrix} = 16 - 16 = 0$$

$$D_y = \begin{vmatrix} 1 & 4 \\ 2 & 8 \end{vmatrix} = 8 - 8 = 0$$

Consistent and dependent equations.
(They coincide.)

Ans: There are unlimited number of solutions.

EXERCISES

Solve each system of equations using matrices (row operations).

1. $\begin{cases} x + y = 6 \\ x - y = 2 \end{cases}$

2. $\begin{cases} x - 2y = 6 \\ x + y = 9 \end{cases}$

3. $\begin{cases} 2x + 3y = 18 \\ x - 2y = -5 \end{cases}$

4. $\begin{cases} 2x - y = 6 \\ 3x + 5y = 22 \end{cases}$

5. $\begin{cases} 3x - y = 2 \\ 6x - 2y = 5 \end{cases}$

6. $\begin{cases} 3x - y = 2 \\ 6x - 2y = 4 \end{cases}$

7. $\begin{cases} x + 2y - z = 3 \\ x + y + 2z = 12 \\ -x + 3y - z = 4 \end{cases}$

8. $\begin{cases} x - y - z = 1 \\ 2x + 3y + z = 3 \\ 3x + 2y = 1 \end{cases}$

9. $\begin{cases} x - y - z = 1 \\ 2x + 3y + z = 3 \\ 3x + 2y = 4 \end{cases}$

10. $\begin{cases} x - y + 3z = 2 \\ 3x + y - z = 2 \\ -x - y + 2z = 3 \end{cases}$

11. $\begin{cases} 2x + y - z = 8 \\ 4x + y - 3z = 14 \end{cases}$

12. $\begin{cases} x - 2y + z = -2 \\ 2x + 2y - z = 11 \\ x - y - z = 2 \\ x + y - 3z = 8 \end{cases}$

13. Show that the following matrix has no inverse.

$$A = \begin{bmatrix} 2 & 6 \\ 3 & 9 \end{bmatrix}$$

14. Show that the following two matrices are inverses of each other.

$$A = \begin{bmatrix} 2 & 1 \\ 3 & -1 \end{bmatrix} \qquad B = \begin{bmatrix} \frac{1}{5} & \frac{1}{5} \\ \frac{3}{5} & -\frac{2}{5} \end{bmatrix}$$

15. Show that the following matrix has no inverse.

$$A = \begin{bmatrix} 1 & 1 & -1 \\ 1 & 3 & 1 \\ 2 & 4 & 0 \end{bmatrix}$$

16. Show that the following two matrices are inverses of each other.

$$A = \begin{bmatrix} 0 & 1 & 2 \\ 1 & -1 & 3 \\ 2 & -4 & 1 \end{bmatrix} \qquad B = \begin{bmatrix} 11 & -9 & 5 \\ 5 & -4 & 2 \\ -2 & 2 & -1 \end{bmatrix}$$

17. Solve the following system of equations using inverse matrix.

$$\begin{cases} x + y = 6 \\ x - y = 2 \end{cases}$$

18. Solve the following system of equations using inverse matrix.

$$\begin{cases} x - y + z = 0 \\ 2y - z = 1 \\ 2x + 3y = 5 \end{cases}$$

19. Solve Problem 4 using Cramer's Rule.   20. Solve Problem 5 using Cramer's Rule.
21. Solve Problem 7 using Cramer's Rule.   22. Solve Problem 9 using Cramer's Rule.
23. Based on Kirchhoff's Laws (see Page 338), the currents I_1, I_2, and I_3 (in amperes) shown in the circuit, are given by the following system of equations. Find the currents I_1, I_2, and I_3.

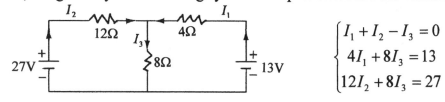

$$\begin{cases} I_1 + I_2 - I_3 = 0 \\ 4I_1 + 8I_3 = 13 \\ 12I_2 + 8I_3 = 27 \end{cases}$$

Ohm's Law and Kirchhoff's Laws

Ohm's Law and Kirchhoff's Laws are the most important theorems in electronics and computer sciences.

DC (direct current) circuit: It is a circuit in which the flow of electricity, or current, that takes places in an unchanging direction.

Ohm's Law: In a DC (direct current) electrical circuit, the voltage (V) at any point is equal to the product of the current (I) and the resistance (R).

Ohm's Law: $V = IR$

V: The electromotive force (in volts)
I : The flow of electrons (in amperes)
R: The resistance that a resistor offers to the flow of electrons (in ohms Ω)

Example: If the series-connected circuit shown below has 3 resistors with the values $R_1 = 400\,\Omega$, $R_2 = 100\,\Omega$, and $R_3 = 300\,\Omega$. The battery has voltage V = 200 volts.
  1. What is the current (I) of the circle.
  2. What are the voltages at the points.

Solution:

  1. $R = R_1 + R_2 + R_3 = 400 + 100 + 300 = 800\ \Omega$

   $\therefore I = \dfrac{V}{R} = \dfrac{200}{800} = 0.25$ ampere

(Figure 1)

  2. $V_1 = IR_1 = 0.25 \times 400 = 100$ volts, $V_2 = IR_2 = 0.25 \times 100 = 25$ volts

   $V_3 = IR_3 = 0.25 \times 300 = 75$ volts

Kirchhoff's Laws:

  1. **Kirchhoff's First Law for Current:** The current going into any point in a DC (direct current) circuit is the same as the current going out of that point, no matter how many branches lead into or out of the point. In other words, the current can never appear from nowhere or disappear into nowhere.

$$I = I_1 + I_2 + I_3$$

  2. **Kirchhoff's Second Law for Voltage:** The sum of all the voltages from a fixed point in a DC circuit all the way around and back to that point from the opposite direction is always zeros. In other words, the voltage can never appear from nowhere or disappear into nowhere.

$$V + V_1 + V_2 + V_3 + \cdots\cdots = 0 \quad \text{(See Figure 1)}$$

CHAPTER 9   EXERCISES

State the dimensions of each matrix.

1. $\begin{bmatrix} 3 & -1 \end{bmatrix}$ **2.** $\begin{bmatrix} 3 \\ -1 \end{bmatrix}$ **3.** $\begin{bmatrix} 3 & 2 \\ -1 & 4 \\ 0 & 3 \end{bmatrix}$ **4.** $\begin{bmatrix} 3 & -1 & 0 \\ 2 & 4 & 3 \end{bmatrix}$ **5.** $\begin{bmatrix} 1 & 2 & 0 & 7 \\ 3 & 0 & -6 & 0 \\ -7 & 9 & 6 & -2 \end{bmatrix}$

Find $A+B$, $A+B-C$, $3A-B$, $A-2B$.

6. $A = \begin{bmatrix} 2 & 3 & 1 \\ 4 & 2 & -1 \end{bmatrix}$, $B = \begin{bmatrix} 0 & 2 & 3 \\ 1 & -2 & 4 \end{bmatrix}$, $C = \begin{bmatrix} -3 & 2 & 1 \\ 2 & 0 & 3 \end{bmatrix}$ **7.** $A = \begin{bmatrix} 1 & 2 \\ 0 & 1 \\ 3 & 1 \end{bmatrix}$, $B = \begin{bmatrix} 2 & 1 \\ 4 & 3 \\ 1 & 0 \end{bmatrix}$, $C = \begin{bmatrix} 4 & -1 \\ 3 & 1 \\ 1 & -3 \end{bmatrix}$

Solve each equation for the matrix X.

8. $X + \begin{bmatrix} 1 & 2 \\ 2 & -1 \end{bmatrix} = \begin{bmatrix} -2 & -4 \\ 3 & 6 \end{bmatrix}$ **9.** $2X - 3\begin{bmatrix} 2 & -3 \\ 1 & 4 \end{bmatrix} = \begin{bmatrix} 4 & 2 \\ 0 & -5 \end{bmatrix}$

Find AB, BA, BC, $(AB)C$. **If the multiplication is not possible, so state.**

10. $A = \begin{bmatrix} 1 & 2 \\ 0 & 1 \end{bmatrix}$, $B = \begin{bmatrix} 1 & 1 & 0 \\ 0 & 2 & 1 \end{bmatrix}$, $C = \begin{bmatrix} 1 \\ 0 \\ 2 \end{bmatrix}$ **11.** $A = \begin{bmatrix} 1 \\ 2 \\ 3 \end{bmatrix}$, $B = \begin{bmatrix} 1 & 2 & 3 \\ 4 & 5 & 6 \end{bmatrix}$, $C = \begin{bmatrix} 1 & 2 \\ 0 & 1 \end{bmatrix}$

12. If $A = \begin{bmatrix} 1 & 1 \\ 0 & 1 \end{bmatrix}$, prove that $A^n = \begin{bmatrix} 1 & n \\ 0 & 1 \end{bmatrix}$ and find A^{20}.

13. If $A = \begin{bmatrix} 1 & 1 & 1 \\ 0 & 1 & 1 \\ 0 & 0 & 1 \end{bmatrix}$, prove that $A^n = \begin{bmatrix} 1 & n & \frac{(1+n)n}{2} \\ 0 & 1 & n \\ 0 & 0 & 1 \end{bmatrix}$ and find A^{20}.

14. A factory makes three products, A, B, and C. Each requires the use of four materials. The quantities (in units) of materials needed are listed in the following table. The amount of each product needed for the next two days are shown below. Use the multiplication of matrices to find the total quantities of each material needed in matrix form for the next two days.

	A	B	C
Material a	8	4	6
Material b	2	9	5
Material c	7	3	1
Material d	2	1	9

	Day 1	Day 2
A	10	6
B	8	15
C	12	10

-----Continued-----

Evaluate each determinant.

15. $\begin{vmatrix} 3 & -4 \\ 2 & 5 \end{vmatrix}$
16. $\begin{vmatrix} -3 & 2 \\ 5 & 1 \end{vmatrix}$
17. $\begin{vmatrix} \sqrt{2} & -\sqrt{3} \\ \sqrt{3} & -\sqrt{6} \end{vmatrix}$
18. $\begin{vmatrix} \log 125 & \log 8 \\ \log 2 & \log 5 \end{vmatrix}$

19. $\begin{vmatrix} 5 & -5 & -5 \\ -18 & 0 & 18 \\ 33 & -44 & 11 \end{vmatrix}$
20. $\begin{vmatrix} 5 & 12 & 4 \\ 10 & 6 & -4 \\ 10 & 18 & 4 \end{vmatrix}$
21. $\begin{vmatrix} 1 & 0 & 1 & -1 \\ 2 & -1 & 1 & 2 \\ 2 & 1 & 2 & 1 \\ 1 & 2 & 3 & 1 \end{vmatrix}$
22. $\begin{vmatrix} 1 & 2 & -1 & 1 \\ 1 & 0 & 1 & -1 \\ 2 & 1 & 1 & -2 \\ 1 & 2 & 2 & -1 \end{vmatrix}$

Find the inverse of each matrix. If the matrix has no inverse, so state.

23. $\begin{bmatrix} 3 & 1 \\ 1 & 2 \end{bmatrix}$
24. $\begin{bmatrix} -3 & -1 \\ 1 & 2 \end{bmatrix}$
25. $\begin{bmatrix} 1 & 1 \\ 0 & 1 \end{bmatrix}$
26. $\begin{bmatrix} 3 & 2 \\ 6 & 4 \end{bmatrix}$

27. $\begin{bmatrix} 2 & 1 & 1 \\ 0 & 1 & 1 \\ 2 & 0 & 1 \end{bmatrix}$
28. $\begin{bmatrix} 1 & 0 & 1 \\ 0 & 2 & -1 \\ -2 & 1 & 3 \end{bmatrix}$
29. $\begin{bmatrix} 1 & 0 & 1 \\ 0 & 1 & 2 \\ 1 & 0 & 1 \end{bmatrix}$
30. $\begin{bmatrix} 1 & 2 & 0 & -1 \\ 0 & 1 & 0 & 1 \\ 1 & 0 & 1 & 0 \\ 0 & 1 & 2 & 1 \end{bmatrix}$

31. $\begin{bmatrix} -10 & 0 & 0 & 0 \\ 0 & 2 & 0 & 0 \\ 0 & 0 & -6 & 0 \\ 0 & 0 & 0 & 8 \end{bmatrix}$
32. $\begin{vmatrix} x^2 & x & 1 \\ y^2 & y & 1 \\ z^2 & z & 1 \end{vmatrix}$
33. $\begin{bmatrix} \cos\theta & \sin\theta \\ \sin\theta & -\cos\theta \end{bmatrix}$
34. $\begin{bmatrix} 1-\cos\theta & \sin\theta \\ \sin\theta & -\cos\theta \end{bmatrix}$

Tell whether the matrices are inverses of each other.

35. $\begin{bmatrix} 5 & 3 \\ 3 & 2 \end{bmatrix}$ and $\begin{bmatrix} 2 & -3 \\ -3 & 5 \end{bmatrix}$
36. $\begin{bmatrix} 3 & 1 \\ 1 & 2 \end{bmatrix}$ and $\begin{bmatrix} 2 & -1 \\ -1 & 3 \end{bmatrix}$
37. $\begin{bmatrix} 1 & 0 & 1 \\ 0 & 1 & 2 \\ 2 & 0 & 1 \end{bmatrix}$ and $\begin{bmatrix} -1 & 0 & 1 \\ -4 & 1 & 2 \\ 2 & 0 & -1 \end{bmatrix}$

38. Prove that the two-point form of the equation of a line containing the two points (x_1, y_1) and (x_2, y_2) can be expressed as the determinant.

$$\begin{vmatrix} x & y & 1 \\ x_1 & y_1 & 1 \\ x_2 & y_2 & 1 \end{vmatrix} = 0$$

39. Three distinct points (x_1, y_1), (x_2, y_2), and (x_3, y_3) lie on the same line. Prove that these points satisfy the determinant.

$$\begin{vmatrix} x_1 & y_1 & 1 \\ x_2 & y_2 & 1 \\ x_3 & y_3 & 1 \end{vmatrix} = 0$$

-----Continued-----

40. The following matrix has no inverse. Find a.

$$A = \begin{bmatrix} 2 & -3 & 1 \\ a & 1 & 5 \\ 1 & 2a & 8 \end{bmatrix}$$

Solve each system of equations with row operations or inverse matrices.

41. $\begin{cases} 2x - y = 4 \\ 2x + 5y = 16 \end{cases}$

42. $\begin{cases} 2x - y = 4 \\ 6x - 2y = 2 \end{cases}$

43. $\begin{cases} 4x - 2y = 7 \\ 10x - 5y = 6 \end{cases}$

44. $\begin{cases} x - y + 3z = 4 \\ y - z = 1 \\ z = 3 \end{cases}$

45. $\begin{cases} x + 2y - z = 3 \\ x + y + 2z = 12 \\ -x + 3y - z = 4 \end{cases}$

46. $\begin{cases} x + 3y - 3z = 0 \\ x - 3y + 7z = 0 \\ 2x + 4z = 0 \end{cases}$

Solve each system of equations using Cramer's Rule.

47. $\begin{cases} 2x - y = 4 \\ 2x + 5y = 16 \end{cases}$

48. $\begin{cases} 3x - 2y = 11 \\ 4x + 3y = 9 \end{cases}$

49. $\begin{cases} 4x - 2y = 6 \\ 10x - 5y = 15 \end{cases}$

50. $\begin{cases} 2x + y + z = 5 \\ x - y + z = 0 \\ 4x - 2y + z = 1 \end{cases}$

51. $\begin{cases} x + 2y + 3z = 2 \\ 2x - y + z = 4 \\ 4x + 3y + 7z = 7 \end{cases}$

52. $\begin{cases} x + 2y - z + u = 6 \\ x + z - u = 0 \\ 2x + y + z - 2u = -1 \\ x + 2y + 2z - u = 7 \end{cases}$

53. Based on Kirchhoff's Laws (see page 338), the currents I_1, I_2, and I_3 (in amperes) shown in the circuit, are given by the following system of equations. Find the currents I_1, I_2, and I_3.

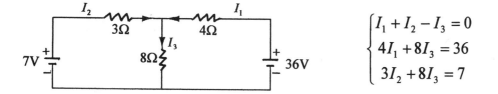

$$\begin{cases} I_1 + I_2 - I_3 = 0 \\ 4I_1 + 8I_3 = 36 \\ 3I_2 + 8I_3 = 7 \end{cases}$$

> A kid said to his mother.
>   Kid: Did you wear contact lenses this morning ?
> Mother: No. Why do you ask ?
>   Kid: I heard you yell to father this morning:
>   "After 20 years of marriage, I have finally
>   seen what kind of man you are !".

Space for Taking Notes

Answers

CHAPTER 9 – 1   Operations with Matrices   page 326

1. $\begin{bmatrix} -2 & -2 \\ 7 & -3 \end{bmatrix}$ **2.** $\begin{bmatrix} 6 & -8 \\ 5 & 11 \end{bmatrix}$ **3.** $\begin{bmatrix} -8 & 13 \\ -11 & -15 \end{bmatrix}$ **4.** $\begin{bmatrix} -8 & 1 & 4 \\ 27 & -19 & -17 \end{bmatrix}$ **5.** Not possible   **6.** $\begin{bmatrix} 5 & -5 & -4 \\ 10 & -1 & 32 \\ -1 & 9 & 7 \end{bmatrix}$

7. $\begin{bmatrix} 0 & -17 & -31 \\ 2 & 3 & 16 \\ -11 & 2 & -3 \end{bmatrix}$ **8.** $\begin{bmatrix} -13 & 41 \\ -20 & -10 \end{bmatrix}$ **9.** Not possible   **10.** $\begin{bmatrix} 22 & -8 & -13 \\ 3 & 15 & 21 \\ 2 & 26 & 16 \end{bmatrix}$ **11.** $x = -3$, $y = 11$

12. $x = 3\frac{1}{2}$, $y = 8$ **13.** $x = 6$, $y = 7$ **14.** $X = \begin{bmatrix} -3 & 0 \\ \frac{8}{3} & \frac{38}{3} \end{bmatrix}$ **15.** $\begin{matrix} \text{Store A} \\ \text{Store B} \\ \text{Cheapest} \end{matrix} \begin{bmatrix} 134.62 \\ 138.40 \\ 126.20 \end{bmatrix}$ **16.** A(64.32%), B(35.68)

CHAPTER 9 – 2   Determinants and the Inverse of Matrix   page 330

1. -19 **2.** 61 **3.** 19 **4.** $2\log\frac{5}{2}$ **5.** 137 **6.** 22 **7.** 0 **8.** 0 **9.** 0 **10.** -134 **11.** -30 **13.** 0 **14.** 0

15. ab Hint: If $a_{11} = 1$ and all other elements above and below the main diagonal are 1's, The value of the determinant is the product of "each element (besides a_{11}) on the diagonal minus 1".

16. $abc(1 + \frac{1}{a} + \frac{1}{b} + \frac{1}{c})$ Hint: Divide the elements in each column by a, b, or c and write abc on the left side of determinant. Write the sum of elements in each row on the first column. Reduce it.

17. 22 **18.** 24 Hint: If all elements above or below either diagonal are zeros, The value of the determinant is the product of elements on the diagonal.

19. k^4 (see Problem 18)   **20.** $a(1+b)(2+c)$ (see Problem 15)   **21.** $\begin{bmatrix} 3 & -2 \\ -1 & 1 \end{bmatrix}$ **22.** $\begin{bmatrix} \frac{3}{5} & -\frac{1}{5} \\ -\frac{1}{5} & \frac{2}{5} \end{bmatrix}$ **23.** No inverse

24. $\begin{bmatrix} \frac{4}{17} & \frac{3}{17} \\ -\frac{3}{17} & \frac{2}{17} \end{bmatrix}$ **25.** $\begin{bmatrix} \frac{1}{2} & -\frac{1}{2} & 0 \\ 1 & 0 & -1 \\ -1 & 1 & 1 \end{bmatrix}$ **26.** $\begin{bmatrix} -2 & -3 & 1 \\ -3 & -3 & 1 \\ -2 & -4 & 1 \end{bmatrix}$ **27.** $\begin{bmatrix} \frac{1}{2} & -\frac{1}{2} & \frac{1}{2} \\ -\frac{1}{2} & \frac{1}{2} & \frac{1}{2} \\ \frac{1}{2} & \frac{1}{2} & -\frac{1}{2} \end{bmatrix}$ **28.** $\begin{bmatrix} 1 & -2 & 1 & 0 \\ 1 & -2 & 2 & -3 \\ 0 & 1 & -1 & 1 \\ -2 & 3 & -2 & 3 \end{bmatrix}$

29. $A^{-1} = \begin{bmatrix} \cos\theta & -\sin\theta \\ \sin\theta & \cos\theta \end{bmatrix}$ **30.** $x = -\frac{3}{7}$

CHAPTER 9 – 3   Solving Systems with Matrices and Cramer's Rule   page 337

1. $x = 4$, $y = 2$ **2.** $x = 8$, $y = 1$ **3.** $x = 3$, $y = 4$ **4.** $x = 4$, $y = 2$ **5.** Inconsistent (no solution)

6. $x = \frac{2}{3} - \frac{1}{3}y$, $y \in R$ (unlimited number of solutions)   **7.** $x = 1$, $y = 3$, $z = 4$ **8.** Inconsistent (no solution)

9. $x = \frac{6}{5} + \frac{2}{5}z$, $y = \frac{1}{5} - \frac{3}{5}z$, $z \in R$ (unlimited number of solutions)   **10.** Inconsistent (no solution)

11. $x = 3 + z$, $y = 2 - z$, $z \in R$ (unlimited number of solutions)   **12.** $x = 3$, $y = 2$, $z = -1$

13. Hint: $D = 0$ **14.** Hint: $AB = I$ **15.** Hint: See Example 6, Page 334   **16.** Hint: $AB = I$

17. $x = 4$, $y = 2$ **18.** $x = -2$, $y = 3$, $z = 5$ **19.** $x = 4$, $y = 2$ **20.** Inconsistent (no solution)

21. $x = 1$, $y = 3$, $z = 4$ **22.** Unlimited number of solution

23. $I_1 = 0.25$ ampere, $I_2 = 1.25$ amperes, $I_3 = 1.5$ amperes

CHAPTER 9   **Exercises**   **page 339 ~ 341**

1. 1×2 matrix (or 1 by 2 matrix)   **2.** 2×1 matrix (or 2 by 1 matrix)   **3.** 3×2 matrix (or 3 by 2 matrix)

4. 2×3 matrix (or 2 by 3 matrix)   **5.** 3×4 matrix (or 3 by 4 matrix)

6. $A + B = \begin{bmatrix} 2 & 5 & 4 \\ 5 & 0 & 3 \end{bmatrix}$, $\quad A + B - C = \begin{bmatrix} 5 & 3 & 3 \\ 3 & 0 & 0 \end{bmatrix}$, $\quad 3A - B = \begin{bmatrix} 6 & 7 & 0 \\ 11 & 8 & -7 \end{bmatrix}$, $\quad A - 2B = \begin{bmatrix} 2 & -1 & -5 \\ 2 & 6 & -9 \end{bmatrix}$

7. $A + B = \begin{bmatrix} 3 & 3 \\ 4 & 4 \\ 4 & 1 \end{bmatrix}$, $\quad A + B - C = \begin{bmatrix} -1 & 4 \\ 1 & 3 \\ 3 & 4 \end{bmatrix}$, $\quad 3A - B = \begin{bmatrix} 1 & 5 \\ -4 & 0 \\ 8 & 3 \end{bmatrix}$, $\quad A - 2B = \begin{bmatrix} -3 & 0 \\ -8 & -5 \\ 1 & 1 \end{bmatrix}$

8. $X = \begin{bmatrix} -3 & -6 \\ 1 & 7 \end{bmatrix}$ **9.** $x = \begin{bmatrix} 5 & -\frac{7}{2} \\ \frac{3}{2} & \frac{7}{2} \end{bmatrix}$ **10.** $AB = \begin{bmatrix} 1 & 5 & 2 \\ 0 & 2 & 1 \end{bmatrix}$, $\;BA$ (Not possible), $BC = \begin{bmatrix} 1 \\ 2 \end{bmatrix}$, $(AB)C = \begin{bmatrix} 5 \\ 2 \end{bmatrix}$

11. AB (Not possible), $\;BA = \begin{bmatrix} 14 \\ 32 \end{bmatrix}$, $\;BC$ (Not possible), $\;(AB)C$ (Not possible)

12. Hint: Find the values of $A^2 = AA$, $A^3 = AA^2$, $A^4 = AA^3$, $\cdots\cdots$ and write the patterns. $\;A^{20} = \begin{bmatrix} 1 & 20 \\ 0 & 1 \end{bmatrix}$

13. Hint: Find the values of $A^2 = AA$, $A^3 = AA^2$, $A^4 = AA^4$, $\cdots\cdots$ and write the patterns. $\;A^{20} = \begin{bmatrix} 1 & 20 & 210 \\ 0 & 1 & 20 \\ 0 & 0 & 1 \end{bmatrix}$

14.
$$\begin{array}{c} & \text{day 1} & \text{day 2} \\ a & 184 & 168 \\ b & 152 & 197 \\ c & 106 & 97 \\ d & 136 & 117 \end{array}$$

15. 23   **16.** -13 **17.** $3 - 2\sqrt{3}$ **18.** $3\log \frac{5}{2}$ **19.** -3960 **20.** 0   **21.** 10   **22.** 6

23. $\begin{bmatrix} \frac{2}{5} & -\frac{1}{5} \\ -\frac{1}{5} & \frac{3}{5} \end{bmatrix}$ **24.** $\begin{bmatrix} -\frac{2}{5} & -\frac{1}{5} \\ \frac{1}{5} & \frac{3}{5} \end{bmatrix}$ **25.** $\begin{bmatrix} 1 & -1 \\ 0 & 0 \end{bmatrix}$ **26.** No inverse (Hint: D = 0)

27. $\begin{bmatrix} \frac{1}{2} & -\frac{1}{2} & 0 \\ 1 & 0 & -1 \\ -1 & 1 & 1 \end{bmatrix}$ **28.** $\begin{bmatrix} \frac{7}{11} & \frac{1}{11} & -\frac{2}{11} \\ \frac{2}{11} & \frac{5}{11} & \frac{1}{11} \\ \frac{4}{11} & -\frac{1}{11} & \frac{2}{11} \end{bmatrix}$ **29.** No inverse   **30.** $\begin{bmatrix} 0 & \frac{1}{2} & 1 & -\frac{1}{2} \\ \frac{1}{3} & \frac{1}{6} & -\frac{1}{3} & \frac{1}{6} \\ 0 & -\frac{1}{2} & 0 & \frac{1}{2} \\ -\frac{1}{3} & \frac{5}{6} & \frac{1}{3} & -\frac{1}{6} \end{bmatrix}$ **31.** $\begin{bmatrix} -\frac{1}{10} & 0 & 0 & 0 \\ 0 & \frac{1}{2} & 0 & 0 \\ 0 & 0 & -\frac{1}{6} & 0 \\ 0 & 0 & 0 & \frac{1}{8} \end{bmatrix}$

32. $(x-y)(x-z)(y-z)$ **33.** $\begin{bmatrix} \cos\theta & \sin\theta \\ \sin\theta & -\cos\theta \end{bmatrix}$ (Hint: $A^{-1} = A$)   **34.** No inverse (Hint: D = 0)   **35.** Yes   **36.** No

37. Yes   **38.** Hint: Expand the determinant and arrange it as the two-point form $\dfrac{y - y_1}{x - x_1} = \dfrac{y_2 - y_1}{x_2 - x_1}$.

39. Hint: Expand the determinant and arrange it as three-point form $\dfrac{y_3 - y_1}{x_3 - x_1} = \dfrac{y_2 - y_1}{x_2 - x_1}$.   **40.** $a = 0$ or -2

41. $x = 3$, $y = 2$ **42.** $x = -3$, $y = -10$ **43.** No solution   **44.** $x = -1$, $y = 4$, $z = 3$ **45.** $x = 1$, $y = 3$, $z = 4$

46. Unlimited number of solutions   **47.** $x = 3$, $y = 2$ **48.** $x = 3$, $y = -1$ **49.** Unlimited number of solutions

50. $x = 1$, $y = 2$, $z = 1$ **51.** No solution   **52.** $x = 1$, $y = 2$, $z = 3$, $u = 4$

53. $I_1 = 5$ amperes, $I_2 = -3$ amperes, $I_3 = 2$ amperes

     (Hint: The direction of I_2 is the opposite of the direction shown in the figure.)

Operations with Matrices
by a Graphing Calculator

Example: If $A = \begin{bmatrix} 5 & -4 \\ -1 & 2 \end{bmatrix}$ and $B = \begin{bmatrix} 3 & 6 \\ 2 & -5 \end{bmatrix}$, find $A + B$ and $A \times B$.

Solution:

Step 1. Define matrix A from **MATRX EDIT** menu

Press: **2nd → MATRX → EDIT** and **1: [A] → ENTER**

Select the dimensions (2×2) of the matrix and display the matrix form.

Press: **2 → ENTER → 2 → ENTER**

Enter the elements of matrix A

Press: **5 → ENTER → (−) 4 → ENTER**
         → (−) 1 → ENTER → 2 → ENTER

The screen shows:

```
MATRIX[A] 2 x2
[ 5        -4    ]
[ -1       ▓▓▓  ]
```

Clear the home screen and name the matrix A
from **MATRX NAME** menu

Press: **2nd → QUIT → 2nd → MATRX**
         → NAME and **1:[A] → ENTER**
         → [A] → ENTER

The screen shows:

```
[A]
        [[5   -4]
         [-1  2 ]]
```

2. Define matrix B from **MATRX EDIT** menu

Press: **2nd → MATRX → EDIT** and **2:[B] → ENTER**

Use the same procedure shown in Step 1(matrix A) to enter
the elements and name the matrix B.

The screen shows:

```
[B]
        [[3  6 ]
         [2  -5]]
```

3. Find $A + B$

Recall matrix A: **2nd → MATRX → NAME** and **1:[A] → ENTER**

Press: **+**

Recall matrix B: **2nd → MATRX → NAME** and **2:[B] → ENTER**

Press: **ENTER**

The screen shows:

```
[A]+[B]
        [[8  2 ]
         [1  -3]]
```

4. Find $A \times B$

Recall matrix A: **2nd → MATRX → NAME** and **1:[A] → ENTER**

Press: **×**

Recall matrix B: **2nd → MATRX → NAME** and **2:[B] → ENTER**

Press: **ENTER**

The screen shows:

```
[A]*[B]
        [[7  50 ]
         [1  -16]]
```

Finding the Determinant and the Inverse of a Matrix
by a Graphing Calculator

Example: Find the determinant and the inverse of the matrix $A = \begin{bmatrix} 1 & -1 & 3 \\ 0 & 1 & 2 \\ 2 & -4 & 1 \end{bmatrix}$.

Solution:

Step 1. Define matrix from **MATRX EDIT** menu

Press: **2nd → MATRX → EDIT** and **1:[A] → ENTER**

Select the dimensions (3 × 3) of the matrix and display the matrix form

Press: **3 → ENTER → 3 → ENTER**

Enter the elements of matrix A

Press: **1 → ENTER → (–) 1 → ENTER → 3 → ENTER → 0 → ENTER → 1**
→ ENTER → 2 → ENTER → 2 → (–) 4 → ENTER → 1 → ENTER

The screen shows:

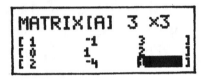

Clear the home screen and name the matrix A from the **MATRX NAME** menu

Press: **2nd → QUIT → 2nd → MATRX → NAME** and **1:[A] → ENTER**
→ [A] → ENTER

The screen shows:

```
[A]
     [[1  -1  3]
      [0   1  2]
      [2  -4  1]]
```

2. Find the determinant

Press: **2nd → MATRX → MATH** and **1:det (→ ENTER**

We have "**det (** " on screen.

Recall matrix A and place [A] on "det ([A])"

Press: **2nd → MATRX → NAME** and **1:[A] → ENTER →)**

Press: **ENTER**

The screen shows:

```
det([A])
                -1
```

3. Find the inverse

Recall matrix A: **2nd → MATRX → NAME** and **1:[A] → ENTER**

Press: x^{-1} **→ ENTER**

The screen shows:

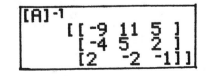

Solving a System with Row Operations
by a Graphing Calculator

Example: Solve $\begin{cases} x - y + 3z = 2 \\ y + 2z = 1 \\ 2x - 4y + z = 3 \end{cases}$

Solution:

Define matrix A from **MATRX EDIT** menu

    Press: **2nd** → **MATRX** → **EDIT** and **1:[A]** → **ENTER**

Select the dimensions (3 × 4) of the matrix and display the matrix form

    Press: **3** → **ENTER** → **4** → **ENTER**

Enter the coefficients as elements in the matrix A

    Press: **1** → **ENTER** → **(−) 1** → **ENTER** → **3** → **ENTER** → **2** → **ENTER**
        → **0** → **ENTER** → **1** → **ENTER** → **2** → **ENTER** → **1** → **ENTER**
        → **2** →**ENTER** → **(−) 4** → **ENTER** → **1** → **ENTER** → **3** → **ENTER**

The screen shows:

```
MATRIX[A] 3 x4
 -1     3     2    1
 1      2     1
 -4     1     3
```

Clear the home screen and name the matrix from **MATRX NAME** menu

    Press: **2nd** → **QUIT** → **2nd** → **MATRX** → **NAME** and **1:[A[** → **ENTER**

The screen shows:

```
[A]
  [[1  -1  3  2]
   [0   1  2  1]
   [2  -4  1  3]]
```

Perform the row operation for reduced row-echelon form from **MATRX**
MATH menu. **B:rref** locates at the end of **MATH** menu.

    Press: **2nd** → **MATRX** → **MATH** and **B:rref (** → **ENTER**

      We have " rref (" on screen.

Recall matrix A and place [A] on " rref ("

    Press: **2nd** → **MATRX** → **NAME** and **1:[A]** → **ENTER** → **)**

    Press: **ENTER**

The screen shows the reduced row-echelon form of the matrix:

```
rref([A])
  [[1  0  0   8 ]
   [0  1  0   3 ]
   [0  0  1  -1]]
```

From the reduced row-echelon form, we have the answer;

    $x = 8$, $y = 3$, $z = -1$

Solving a System with $[A]^{-1} \cdot [B]$
by a Graphing Calculator

Example: Solve $\begin{cases} x - y + 3z = 2 \\ y + 2z = 1 \\ 2x - 4y + z = 3 \end{cases}$

$$A = \begin{bmatrix} 1 & -1 & 3 \\ 0 & 1 & 2 \\ 2 & -4 & 1 \end{bmatrix}, \qquad B = \begin{bmatrix} 2 \\ 1 \\ 3 \end{bmatrix}$$

1. Define matrix A from **MATRX EDIT** menu

   Press: **2nd → MATRX → EDIT and 1:[A] → ENTER**

Select the dimensions (3×3) of matrix A and display the matrix form

   Press: **3 → ENTER → 3 → ENTER**

Enter the elements of matrix A

   Press: **1 → ENTER → (–) 1 → ENTER → 3 → ENTER**
   → 0 → ENTER → 1 → ENTER → 2 → ENTER
   → 2 → ENTER → (–) 4 → ENTER → 1 → ENTER

Clear the home screen and name the matrix A from **MATRIX NAME** menu

   Press: **2nd → QUIT → 2nd → MATRX → NAME and 1:[A] → ENTER**
   → [A] → ENTER

The screen shows:

```
[A]
      [[1  -1  3]
       [0  1   2]
       [2  -4  1]]
```

2. Define matrix B from **MATRX EDIT** menu

   Press: **2nd → MATRX → EDIT and 2:[B] → ENTER**

Select the dimensions (3×1) of matrix B and display the matrix form

   Press: **3 → ENTER → 1 → ENTER**

Enter the elements of matrix B

   Press: **2 → ENTER → 1 → ENTER → 3 → ENTER**

Clear the home screen and name the matrix B from **MATRIX NAME** menu

   Press: **2nd → QUIT → 2nd → MATRX → NAME and 2:[B] → ENTER**
   → [B] → ENTER

The screen shows:

```
[B]
          [[2]
           [1]
           [3]]
```

3. Find $[A]^{-1} \cdot [B]$

   Press: **2nd → MATRX → NAME and 1:[A] → ENTER → [A] → x^{-1}**

   Press: **×**

   Press: **2nd → MATRX → NAME and 2:[B]**
   → ENTER

```
[A]-1*[B]
          [[8 ]
           [3 ]
           [-1]]
```

   Press: **ENTER**

The screen shows the answer:

Space for Taking Notes

Space for Taking Notes

Sequences, Series, Binomial Theorem

10-1 Sequences and Number Patterns

Patterns appear in nature and our daily life. Studying sequences and number patterns helps us to identify, understand patterns and make predictions.

Suppose the chart below shows the seat numbers of economy class in an airplane. There are 42 seats in the economy class. The chart shows six sequences. Each sequence has a pattern with a common difference (d = 6) between any two successive numbers.

1	3	5		6	4	2
7	9	11		12	10	8
13	15	17		18	16	14
.	.	.		.	.	.
.	.	.		.	.	.

1, 7, 13, 19, 25, 31, 37.
3, 9, 15, 21, 27, 33, 39.
5, 11, 17, 23, 29, 35, 41.
6, 12, 18, 24, 30, 36, 42.
4, 10, 16, 22, 28, 34, 40.
2, 8, 14, 20, 26, 32, 38.

Sequence: A consecutive nature numbers in a certain order. Each number is called a term of the sequence.

Arithmetic Sequence: A sequence which has a common difference between any two successive terms.

 1, 2, 3, 4, 5, 6, 7, ······.   Common difference: d = 1
 1, 7, 13, 19, 25, 31, 37, ······.   Common difference: d = 6
 62, 51, 40, 29, 18, 7, ······.   Common difference: d = −11

Fibonacci Sequence: Each number in the sequence is the sum of the two preceding numbers.

 1, 1, 2, 3, 5, 8, 13, ······.
 2, 3, 5, 8, 13, 21, 34, ······.
 3, 4, 7, 11, 18, 29, 47, ······.

Triangular Sequence: Each number in the sequence can be arranged to form a triangle.

 1 , 3 , 6 , 10 , 15 , 21, ······.

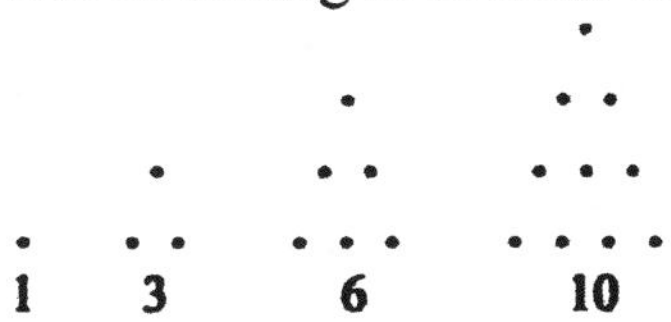

Rectangular Sequence: Each number in the sequence can be arranged to form a rectangle.

 2 , 6 , 12 , 20 . 30 , ······,
1 × 2 2 × 3 3 × 4 4 × 5 5 × 6

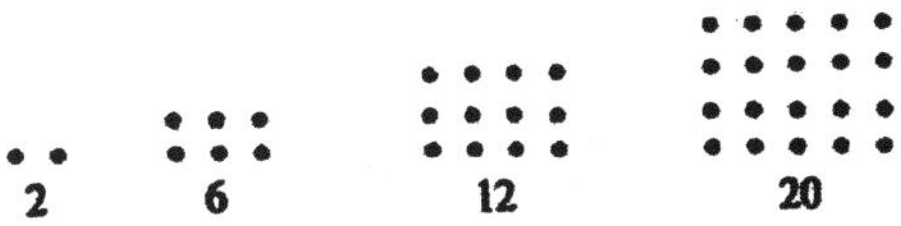

Finite Sequence: A sequence with a last term: 1, 3. 5, 7, 9
Infinite Sequence: A sequence having its terms increases without limit: 1, 3, 5, 7, 9, ·····

Studying and discovering the pattern of a number sequence, we make charts, tables or diagrams to organize the information into a pattern to show how the numbers relate to each other. Then, we continue the pattern to extend the numbers in the sequence.
To find the pattern of a sequence, we find the **differences** between any two successive terms.

Example 1: Find the next five terms of the sequence 2, 4, 6, 8, 10, ·····.
    Solution: The terms increase by 2. The differences are constant.
        The next five terms are 12, 14, 16, 18, and 20.

Example 2: Find the next five terms of the sequence 1, 3, 6, 10, 15, ·····.
    Solution: The differences are not constant.
        The next five terms are continued by adding 6, 7, 8, 9, and 10.
        The next five terms are 21, 28, 36, 45, and 55.

$$1 \ , \ 3 \ , \ 6 \ , \ 10 \ , \ 15 \ , \ \mathbf{21} \ , \ \mathbf{28} \ , \ \mathbf{36} \ , \ \mathbf{45} \ , \ \mathbf{55}$$
$$\quad 2 \quad 3 \quad 4 \quad 5 \quad \mathbf{6} \quad \mathbf{7} \quad \mathbf{8} \quad \mathbf{9} \quad \mathbf{10} \qquad \text{(differences)}$$

Example 3: Each of the five teams plays each of the other teams just once. How many games will there be ? How many games will there be for 8 teams ?
    Solution:

Number of teams	1	2	3	4	5	6	7	8
Number of games	0	1	3	6	10	15	21	28

There will be 10 games for 5 teams.
There will be 28 games for 8 teams.

Finding the sum of the first *n* terms using Summation Notation " Σ "

The sum (S) of the first *n* terms of a sequence is represented by:

$$S = \sum_{k=1}^{n} a_k = a_1 + a_2 + a_3 + a_4 + \cdots + a_n$$

The symbol $\sum$ is simply to represent " to sum ", or to " add up ". It is read **"sigma"**.

The expression $\sum_{k=1}^{n} a_k$ is to represent to add the terms of the sequence from $k = 1$ to $k = n$.

Example 4: $\sum_{k=1}^{5} 2k = 2 + 4 + 6 + 8 + 10 = 30$.

Example 5: $\sum_{k=1}^{3} (k^2 - 3k + 4) = \sum_{k=1}^{3} k^2 - \sum_{k=1}^{3} 3k + \sum_{k=1}^{3} 4 = (1^2 + 2^2 + 3^2) - (3 + 6 + 9) + (4 + 4 + 4)$

$$= 14 - 18 + 12 = 8.$$

We may consider an infinite sequence as a function that is one-to-one correspondence.

In an infinite sequence, the domain is the set of positive integers and the range is the set of real numbers.
For example, the sequence $1, \frac{1}{2}, \frac{1}{3}, \frac{1}{4}, \cdots, \frac{1}{n}, \cdots$ can be represented as a function $f(n) = \frac{1}{n}$ for every positive integer n.
A sequence is a function. Therefore, we can graph a sequence on the coordinate plane. All points on the graph whose x-coordinates are positive integers only.

If a sequence in which each term is defined by a pattern or rule that involves the first few terms preceding it, the rule is called a **Recursive Formula**. It is called a **Recursive Sequence**. When the pattern of a sequence is known, we can represent the sequence by the formula for the **general** or n**th** term by placing braces without writing the entire sequence.

The sequence $\{a_n\} = \{2n\}$ represents the sequence: $2, 4, 6, 8, 10, \cdots, 2n, \cdots$.

The sequence $\{b_n\} = \begin{cases} n & if\ n\ is\ even \\ n^2 & if\ n\ is\ odd \end{cases}$ has the terms: $a_1 = 1, a_2 = 2, a_3 = 9, a_4 = 4, \cdots$.

Examples

1: Write the first five terms of the recursive sequence: $a_1 = 1, a_2 = 1, a_{n+2} = a_n + a_{n+1}$
Solution:
$$a_1 = 1, a_2 = 1, a_3 = 2, a_4 = 3, a_5 = 5.$$
$$1, 1, 2, 3, 5, 8, 13, 21, \cdots$$
It is called **Fibonacci Sequence**.

2. Write the first six terms of the recursive sequence:
$$a_1 = 1, a_2 = 2, a_{n+2} = 2a_n + a_{n+1}$$
Solution:
$$a_1 = 1, a_2 = 2, a_3 = 4, a_4 = 8$$
$$a_5 = 16, a_6 = 32. \text{ Ans.}$$

3. Write a formula for the nth term:
$$1, 4, 9, 16, 25, \cdots$$
Solution:
$$1^2, 2^2, 3^2, 4^2, 5^2, \cdots$$
$$\therefore a_n = n^2. \text{ Ans.}$$

4. Write a formula for the nth term:
$$2, -4, 6, -8, 10, \cdots$$
Solution:
$$(-1)^{1+1}(2 \cdot 1), (-1)^{2+1}(2 \cdot 2), (-1)^{3+1}(2 \cdot 3),$$
$$\cdots \quad \therefore a_n = (-1)^{n+1} \cdot 2n. \text{ Ans.}$$

5. Write the first four terms of the sequence:
$$\{t_n\} = \left\{\left(\tfrac{1}{2}\right)^n\right\}$$
Solution:
$$t_1 = \left(\tfrac{1}{2}\right)^1, t_2 = \left(\tfrac{1}{2}\right)^2, t_3 = \left(\tfrac{1}{2}\right)^3, t_4 = \left(\tfrac{1}{2}\right)^4$$
$$\therefore t_1 = \tfrac{1}{2}, t_2 = \tfrac{1}{4}, t_3 = \tfrac{1}{8}, t_4 = \tfrac{1}{16}. \text{ Ans.}$$

6. Write the first four terms of the sequence:
$$\{a_n\} = \left\{2^{n-1}\right\}$$
Solution:
$$a_1 = 2^{1-1}, a_2 = 2^{2-1}, a_3 = 2^{3-1}, a_4 = 2^{4-1}$$
$$\therefore a_1 = 1, a_2 = 2, a_3 = 4, a_4 = 8. \text{ Ans.}$$

EXERCISES

Find the next five terms of each sequence.

1. 2, 4, 7, 11, 16, ·····   **2.** 1, 2, 5, 10, 17, ·····   **3.** 0, 2, 6, 12, 20, ·····

4. 2, 3, 6, 11, 18, ·····   **5.** 1, 1, 2, 3, 5, 8, ·····   **6.** 1, 5, 6, 11, 17, ·····

7. 1, 2, 4, 8, 16, ·····   **8.** 3, 4, 7, 11, 18, ·····   **9.** 2, 3, 5, 8, 13, ·····

10. 1, 4, 9, 16, 25, ·····   **11.** 9, 10, 13, 18, 25, ·····   **12.** 0, 0.5, 2, 4.5, 8, ·····

13. 4, 5.5, 8, 11.5, 16, ·····   **14.** −9, −8, −6, −3, 1, ·····   **15.** −9, −10, −12, −15, −19, ·····

16. 5, −10, 15, −20, 25, ·····   **17.** $1, \frac{1}{2}, 3, \frac{1}{4}, 5,$ ·····   **18.** $\frac{1}{2}, \frac{2}{3}, \frac{3}{4}, \frac{4}{5}, \frac{5}{6},$ ·····

19. $1, -\frac{1}{2}, 3, -\frac{1}{4}, 5,$ ·····   **20.** $\frac{1}{2}, -\frac{1}{4}, \frac{1}{6}, -\frac{1}{8}, \frac{1}{10}$

Find the sum of each sequence.

21. $\displaystyle\sum_{k=1}^{5} 4$ **22.** $\displaystyle\sum_{k=1}^{50} 6$ **23.** $\displaystyle\sum_{k=1}^{10} k$ **24.** $\displaystyle\sum_{k=1}^{10} (-k)$ **25.** $\displaystyle\sum_{k=1}^{4} (k^2 + 1)$

26. $\displaystyle\sum_{k=0}^{3} (k^3 + 2)$ **27.** $\displaystyle\sum_{k=2}^{5} (-1)^k 2^k$ **28.** $\displaystyle\sum_{k=2}^{5} (-1)^k k^2$ **29.** $\displaystyle\sum_{k=0}^{4} (-1)^k \frac{1}{k+1}$ **30.** $\displaystyle\sum_{y=0}^{5} (x + y)$

Write the first five terms of each recursive sequence.

31. $a_1 = 5$, $a_{n+1} = a_n + 2$ **32.** $a_1 = 5$, $a_{n+1} = a_n - 2$

33. $a_1 = -3$, $a_{n+1} = a_n + n$ **34.** $a_1 = x$, $a_{n+1} = a_n + d$

35. $a_1 = 1$, $a_2 = 2$, $a_{n+2} = a_n a_{n+1}$ **36.** $a_1 = 1$, $a_2 = 2$, $a_{n+2} = na_n + a_{n+1}$

Write each sum using summation notation.

37. $1 + 2 + 3 + 4 + \cdots\cdots + 30$ **38.** $1^2 + 2^2 + 3^2 + 4^2 + \cdots\cdots + 10^2$

39. $1 + \frac{1}{2} + \frac{1}{3} + \frac{1}{4} + \cdots\cdots + \frac{1}{10}$ **40.** $1 - \frac{1}{2} + \frac{1}{4} - \frac{1}{8} + \cdots\cdots - \frac{1}{128}$

41. $\frac{1}{1 \cdot 3} + \frac{1}{2 \cdot 4} + \frac{1}{3 \cdot 5} + \cdots\cdots + \frac{1}{10 \cdot 12}$ **42.** $a + (a + d) + (a + 2d) + [a + (n - 1)d]$

Write the first five terms of each sequence.

43. $\left\{ n^2 - 1 \right\}$ **44.** $\left\{ \dfrac{n}{2^n} \right\}$ **45.** $\left\{ \dfrac{n}{n+3} \right\}$ **46.** $\left\{ (-1)^n n^2 \right\}$ **47.** $\left\{ (-1)^{n+1} \dfrac{1}{n(n+1)} \right\}$

48. The bus schedule at a station is at these times: 8:00, 8:05, 8:15, 8:30. If this pattern continues, what is the time of the next bus ?

49. How many games have to be played to determine a final winner from 16 teams in a single-elimination competition ?

50. There are 9 students in the party. Each student shakes hands exactly once with each of the others. How many handshakes take place ?

51. What is the greatest number of pieces we can get from a round pizza with 1, 2, 3, 4 or 5 cuts ?

52. Evaluate $\displaystyle\sum_{k=1}^{9} \log_{10} \frac{k+1}{k}$.   **53.** Evaluate $\displaystyle\sum_{k=2}^{11} \log_{10} \frac{k-1}{k}$.

10-2 Arithmetic Sequences

Arithmetic Sequence: A sequence which has common different between any two successive terms.

Arithmetic Mean (Average): A single term between two terms of an arithmetic sequences. If p, m, q is an arithmetic sequence, then $m = \frac{p+q}{2}$.

The following formula is used for finding the nth term of an arithmetic sequence:

Formula: The **general** or n**th** term of an arithmetic sequence is given by the formula:

$$a_1, a_1 + d, a_1 + 2d, a_1 + 3d, \cdots\cdots \quad \textbf{Formula:} \ a_n = a_1 + (n-1)d$$

Where $a_1 \to 1^{st}$ term , $n \to$ position of a_n, $d \to$ common difference.

Examples

1. Find the 10th term of $25, 33, 41, \ldots, a_{10}$.

Solution:
$$d = 33 - 25 = 8$$
$$a_{10} = a_1 + (n-1)d = 25 + (10-1)8 = 97 .$$
Ans.

2. Find the arithmetic mean between 17 and 53.

Solution: $m = \frac{p+q}{2} = \frac{17+53}{2} = 35$. Ans.

(Hint: $17, \mathbf{35}, 53, 71, \cdots\cdots$)

3. Insert four arithmetic means between 14 and 54.

Solution: $14, ?, ?, ?, ?, 54$.
$$a_1 = 14, \quad n = 6, \quad a_6 = 54$$
$$a_6 = a_1 + (6-1)d$$
$$54 = 14 + (6-1)d \quad \therefore \ d = 8$$
Ans: $14, 22, 30, 38, 46, 54$.

4. Find a_{20} of the arithmetic sequence if $a_5 = 36$ and $a_{11} = 48$.

Solution: $a_{11} = a_5 + (7-1)d$
$$48 = 36 + 6d \quad \therefore \ d = 2$$
$$a_{11} = a_1 + (11-1)d$$
$$48 = a_1 + 10 \cdot 2 \quad \therefore \ a_1 = 28$$
$$a_{20} = a_1 + (20-1)d = 28 + 19 \cdot 2 = 66. \ \text{Ans.}$$

5. How many multiples of 6 are there between 100 and 1,000 ?

Solution: $102, 108, 114, \ldots, 996$.
$$a_n = a_1 + (n-1)d$$
$$996 = 102 + (n-1)6$$
$$894 = 6n - 6 \quad \therefore \ n = 150$$
Ans: 150 multiples of 6

6. Find a formula for the nth term of the arithmetic sequence whose first term is 3 and the common difference is 5.

Solution:
$$a_1 = 3, \quad d = 5$$
$$a_n = a_1 + (n-1)d$$
$$a_n = 3 + (n-1)5$$
$$\therefore \ a_n = 5n - 2. \ \text{Ans.}$$

7. Find a formula for the nth term of the arithmetic sequence if $a_1 = 5$ and $a_7 = 25$.

Solution: $a_7 = a_1 + (7-1)d$
$$25 = 5 + (7-1)d \quad \therefore \ d = \frac{10}{3}$$
$$a_n = 5 + (n-1) \cdot \frac{10}{3}$$
$$\therefore \ a_n = \frac{10}{3}n + \frac{5}{3}. \ \text{Ans.}$$

EXERCISES

Find the next five terms of each arithmetic sequence.

1. $4, 7, 10, 13, 16, \cdots$
2. $5, 8, 11, 14, 17\cdots$
3. $2, 6, 10, 14, 18, \cdots$
4. $14, 22, 30, 38, 46, \cdots$
5. $21, 19, 17, 15, 13, \cdots$
6. $29, 26, 23, 20, 17, \cdots$
7. $1, 5, 9, 13, 17, \cdots$
8. $56, 52, 48, 44, 40, \cdots$
9. $3, 8, 13, 18, 23, \cdots$
10. $16, 23, 30, 37, 44, \cdots$
11. $1.5, 2.0, 2.5, 3.0, 3.5, \cdots$
12. $1.5, 3, 4.5, 6, 7.5, \cdots$
13. $18, 15, 12, 9, 6, \cdots$
14. $-3, -7, -11, -15, -19, \cdots$
15. $-21, -18, -15, -12, -9 \cdots$
16. $\{2n - 4\}$
17. $\{\ln 5^n\}$
18. $\{e^{\ln(n-1)}\}$

Find the nth of the arithmetic sequence.

19. $a_1 = 2$; $d = 5$
20. $a_1 = 2$; $d = -5$
21. $a_1 = -2$; $d = -5$
22. $a_1 = 2$; $d = -\frac{1}{2}$
23. $a_1 = 5$; $d = \frac{10}{3}$
24. $a_1 = 10$; $d = 2\pi$
25. $a_1 = \sqrt{5}$; $d = \sqrt{5}$
26. $a_1 = 3$; $a_5 = 23$
27. $a_5 = 36$; $a_{11} = 48$

State whether each sequence is an arithmetic sequence.

28. $6, 9, 12, 15, 18, \cdots$
29. $7, 9, 12, 16, 21, \cdots$
30. $\frac{9}{4}, \frac{3}{2}, \frac{3}{4}, 0, -\frac{3}{4}, \cdots$
31. $\frac{2}{3}, 1, \frac{4}{3}, \frac{2}{3}, \frac{5}{6}, \cdots$
32. $\ln 1, \ln 2, \ln 3, \ln 4, \ln 5 \cdots$
33. $\ln 1, \ln 2, \ln 4, \ln 8, \ln 16, \cdots$
34. $\{a_n\} = \{2n - 4\}$

Find the indicated term in each arithmetic sequence.

35. 12th term of $5, 3, 1, -1, -3, \cdots$
36. 20th term of $\sqrt{3}, 3\sqrt{3}, 5\sqrt{3}, 7\sqrt{3}, 9\sqrt{3}, \cdots$
37. 30th term of $a, a+c, a+2c, a+3c, \cdots$
38. 25th term of $\ln 1, \ln 2, \ln 4, \ln 8, \cdots$

39. Find x of the arithmetic sequence $x+5, 2x-1, 5x+1, \cdots$.
40. Find x of the arithmetic sequence $3x, 4x+5, 7x-3, \cdots$.
41. Find m such that $-3, m, -11$ is an arithmetic sequence.
42. Insert three arithmetic means between 1.5 and 7.5.
43. Is the sequence $1, \frac{1}{2}, \frac{1}{3}, \frac{1}{4}, \frac{1}{5}$ an arithmetic sequence ?
44. John plans to jog for 10 minutes the first day and increases 5 minutes every day. At which day will John be able to jog for 1 hour ?
45. If the sequence a, b, c and d is an arithmetic sequence, is the sequence a^2, b^2, c^2 and d^2 an arithmetic sequence ?
46. If the sequence a, b, c and d is an arithmetic sequence, is the sequence $2a, 2b, 2c$ and $2d$ an arithmetic sequence ?
47. An object is dropped from a plane. The object falls 16 feet during the first second. It falls 48 feet during the second second. It falls 80 feet during the third second. It falls 112 feet during the fourth second. How many feet does it fall during the tenth second ?

10-3 Arithmetic Series

Arithmetic Series: The sum of a finite arithmetic sequence with n terms or the sum of the first n terms of an infinite arithmetic geometric.

$$1+3+5+7+9+11+\cdots\cdots+(2n-1) \text{ is an arithmetic series.}$$

Formula: The sum of the first **n terms** of an arithmetic series is given by:

$$S_n = \frac{n}{2}(a_1 + a_n) \text{ Where } a_1 \to 1^{st} \text{ term}, \quad a_n \to \text{last term}, \quad n \to n \text{ terms}$$

Proof:
$$S_n = a_1 + a_2 + a_3 + \cdots + a_{n-2} + a_{n-1} + a_n$$
$$= a_1 + (a_1 + d) + (a_1 + 2d) + (a_1 + 3d) + \cdots\cdots + [a_1 + (n-1)d]$$

Write the expression in reverse order:
$$S_n = a_n + a_{n-1} + a_{n-2} + \cdots + a_3 + a_2 + a_1$$
$$= a_n + (a_n - d) + (a_n - 2d) + (a_n - 3d) + \cdots + [a_n - (n-1)d]$$

Add the above two expressions of S_n:
$$2S_n = (a_1 + a_n) + (a_1 + a_n) + (a_1 + a_n) + \cdots + (a_1 + a_n) = n(a_1 + a_n)$$

$$\therefore S_n = \frac{n}{2}(a_1 + a_n)$$

To memorize this formula, we may consider it as the area of a trapezoid.

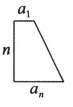

Examples

1. Find the sum of the arithmetic sequence:
$$1 + 2 + 3 + 4 + \cdots\cdots + 500$$
Solution: $a_1 = 1$, $a_{500} = 500$, $n = 500$
$$S_{500} = \frac{n}{2}(a_1 + a_{500}) = \frac{500}{2}(1 + 500)$$
$$= 250(501) = 125,250 \text{. Ans.}$$

2. Evaluate the arithmetic series:
$$1 + 2 + 3 + 4 + \cdots\cdots + 1000$$
Solution: $a_1 = 1$, $a_{1000} = 1000$, $n = 1000$
$$S_{1000} = \frac{1000}{2}(1 + 1000) = 500,500 \text{. Ans.}$$

3. Find the sum of the arithmetic sequence:
$$1 + 3 + 5 + 7 + \cdots\cdots + (2n - 1)$$
Solution: $a_1 = 1$, $a_n = 2n - 1$
$$S_n = \frac{n}{2}[1 + (2n - 1)] = n^2 \text{. Ans.}$$

4. Evaluate the series $\sum_{k=1}^{30}(-5k + 10)$.
Solution: $5, 0, -5, -10, \cdots\cdots$
$$a_1 = 5, \quad d = -5$$
$$a_{30} = 1 + (30 - 1)(-5) = -144$$
$$S_{30} = \frac{30}{2}[5 + (-144)] = -2,085 \text{. Ans.}$$

EXERCISES

Find the indicated sum of the arithmetic sequence (Evaluate each series).

1. $1+2+3+\cdots\cdots+50$

2. $2+4+6+\cdots\cdots+50$

3. $1+3+5+\cdots\cdots+49$

4. $2+4+6+\cdots\cdots+2n$

5. $-3-1+1+3+\cdots\cdots+55$

6. $-3-1+1+3+\cdots\cdots+(2n-5)$

7. $1+5+9+\cdots\cdots+157$

8. $1+5+9+\cdots\cdots+(4n-3)$

9. $2.1+4.5+6.9+\cdots\cdots+59.7$

10. $2.1+4.5+6.9+\cdots\cdots+(2.4n-0.3)$

11. $2.1-0.3-2.7-\cdots\cdots-62.7$

12. $2.1-0.3-2.7-\cdots\cdots-(2.4n-4.5)$

13. $2-4-10-\cdots\cdots-a_{20}$

14. $2-4-10-16-\cdots\cdots-(6n-8)$

15. $\dfrac{1}{2}+\dfrac{1}{4}+0-\dfrac{1}{4}+\cdots\cdots+a_{20}$

16. $\dfrac{1}{2}+\dfrac{1}{4}+0-\dfrac{1}{4}+\cdots\cdots+[-\dfrac{1}{4}(n-3)]$

17. $-\dfrac{1}{2}+\dfrac{1}{4}+1+\dfrac{7}{4}+\cdots\cdots+a_{20}$

18. $-\dfrac{1}{2}+\dfrac{1}{4}+1+\dfrac{7}{4}+\cdots\cdots+\dfrac{1}{4}(3n-5)$

19. $\displaystyle\sum_{k=1}^{30} 5k$

20. $\displaystyle\sum_{k=1}^{n} 5k$

21. $\displaystyle\sum_{k=1}^{12} (9k-6)$

22. $\displaystyle\sum_{k=1}^{n} (9k-6)$

23. $\displaystyle\sum_{k=1}^{300} k$

24. $\displaystyle\sum_{k=1}^{n} k$

25. $\displaystyle\sum_{k=50}^{200} k$

26. $\displaystyle\sum_{k=50}^{n} k$

27. $\displaystyle\sum_{k=1}^{30} (\tfrac{3}{4}k-\tfrac{1}{2})$

28. $\displaystyle\sum_{k=1}^{n} (\tfrac{3}{4}k-\tfrac{1}{2})$

29. $\displaystyle\sum_{n=0}^{100} (\tfrac{1}{2}-\tfrac{3}{16}n)$

30. $\displaystyle\sum_{n=50}^{100} n - \sum_{n=1}^{49} n$

31. If $\displaystyle\sum_{k=1}^{12}[a+(k-1)d]=630$ and $\displaystyle\sum_{k=1}^{20}[a+(k-1)d]=1770$, find a and d.

32. A theater has 50 rows of seats. There are 25 seats in the first row, 31 seats in the second row, 37 seats in the third row, and so on. How many seats are there in the theater ?

33. An object is dropped from a plane. The object falls 16 feet during the first second. It falls 48 feet during the second second. It falls 80 feet during the third second. It falls 112 feet during the fourth second. How many feet will it fall in 15 seconds ?

34. An object is dropped from a plane. The object falls 4.9 meters during the first second. During the successive second, it falls 9.8 meters more than in the preceding second.

      a. How many meters does it fall during the tenth second ?

      b. How many meters will the object fall in 15 seconds ?

10-4 Geometric Sequences and Series

Geometric Sequence: A sequence in which the ratios of consecutive terms are the same.

$1, 3, 9, 27, 81, 243 \cdots$ is a geometric sequence. It has a common ratio $r = 3$.

Geometric Mean: If p, m, q is a geometric sequence, then $m = \pm\sqrt{pq}$.

Formula: The general or nth term of a geometric sequence is given by the formula:

$$a_1, \; a_1 r, \; a_1 r^2, \; a_1 r^3, \; \cdots \qquad \textbf{formula:} \; a_n = a_1 r^{n-1}$$

Where $a_1 \to 1^{\text{st}}$ term, $n \to$ position of a_n, $r \to$ common ratio.

Example: If $a_1 = 1$ and $r = 3$, we have $a_4 = a_1 r^{4-1} = 1 \cdot 3^3 = 27$.

Geometric Series: The summation of a geometric sequence (either finite or infinite).

Formula: The sum of a finite geometric series with n terms is given by:

$$S_n = a_1 + a_1 r + a_1 r^2 + a_1 r^3 + \cdots\cdots + a_1 r^{n-1} = \frac{a_1(1 - r^n)}{1 - r} , \; r \neq 1$$

where $a_1 \to 1\underline{\text{st}}$ term ; $n \to$ The n terms ; $r \to$ common ratio.

(**Hint:** Find n from $a_n = a_1 r^{n-1}$ if n is not given.)

Proof:

$$S_n = a_1 + a_1 r + a_1 r^2 + a_1 r^3 + \cdots\cdots + a_1 r^{n-2} + a_1 r^{n-1}$$

Multiply each side by r:

$$r S_n = a_1 r + a_1 r^2 + a_1 r^3 + a_1 r^4 + \cdots\cdots + a_1 r^{n-1} + a_1 r^n$$

Subtract the above two equations:

$$S_n - r S_n = a_1 - a_1 r^n$$

$$S_n(1 - r) = a_1(1 - r^n) \qquad \therefore S_n = \frac{a_1(1 - r^n)}{1 - r} , \; r \neq 1.$$

When we use the above formula, the index must begin at $k = 1$.
If not, we need to adjust the formula (see example 2 below).

Examples:

1. $\displaystyle\sum_{k=1}^{12}\left(\frac{1}{2}\right)^k = \frac{1}{2} + \frac{1}{4} + \frac{1}{8} + \cdots\cdots + \left(\frac{1}{2}\right)^{12} = \frac{\frac{1}{2}[1 - (\frac{1}{2})^{12}]}{1 - \frac{1}{2}} = 1 - \left(\frac{1}{2}\right)^{12}.$

2. $\displaystyle\sum_{k=0}^{12}\left(\frac{1}{2}\right)^k = \left(\frac{1}{2}\right)^0 + \sum_{k=1}^{12}\left(\frac{1}{2}\right)^{12} = 1 + \left[1 - \left(\frac{1}{2}\right)^{12}\right] = 2 - \left(\frac{1}{2}\right)^{12}.$

Or: $\displaystyle\sum_{k=0}^{12}\left(\frac{1}{2}\right)^k = 1 + \frac{1}{2} + \frac{1}{4} + \frac{1}{8} + \cdots\cdots + \left(\frac{1}{2}\right)^{12} = \frac{1 \cdot [1 - (\frac{1}{2})^{13}]}{1 - \frac{1}{2}} = 2 \cdot \left[1 - \left(\frac{1}{2}\right)^{13}\right] = 2 - \left(\frac{1}{2}\right)^{12}.$

Examples

3. Find the 10th term of the geometric sequence
$$2, 6, 18, \cdots .$$

Solution:

The common ratio $r = \dfrac{6}{2} = 3$

$$a_{10} = a_1 r^{10-1} = 2 \cdot 3^9 = 39366 . \text{ Ans.}$$

4. Find the 8th term of the geometric sequence
$$2, \tfrac{2}{3}, \tfrac{2}{9}, \tfrac{2}{27}, \cdots .$$

Solution:

The common ratio $r = \dfrac{\tfrac{2}{3}}{2} = \dfrac{2}{3} \cdot \dfrac{1}{2} = \dfrac{1}{3}$

$$a_8 = a_1 r^{8-1} = 2 \cdot \left(\dfrac{1}{3}\right)^7 = \dfrac{2}{3^7} = \dfrac{2}{2187}$$
$$= 0.00091 . \text{ Ans.}$$

5. Find the nth term of the geometric sequence
$$2, 6, 18, \cdots .$$

Solution:
$$a_1 = 2, \quad r = 3$$
$$a_n = a_1 r^{n-1} = 2 \cdot 3^{n-1} . \text{ Ans.}$$

6. Find the nth term of the geometric sequence
$$2, \tfrac{2}{3}, \tfrac{2}{9}, \tfrac{2}{27}, \cdots .$$

Solution:
$$a_1 = 2, \quad r = \tfrac{1}{3}$$
$$a_n = a_1 r^{n-1} = 2 \cdot (\tfrac{1}{3})^{n-1} . \text{ Ans.}$$

7. Find the geometric mean of -4 and -9.
Solution:
$$m = \pm\sqrt{pq} = \pm\sqrt{-4 \cdot -9} = \pm 6 . \text{ Ans.}$$
Hint: $-4, 6, -9, \cdots .$ or $-4, -6, -9, \cdots .$

8. Insert three geometric means between 4 and 324.

Solution: $4, ?, ?, ?, 324$
$$a_1 = 4. \quad a_5 = 324, \quad n = 5$$
$$a_5 = a_1 r^{5-1}, \quad 324 = 4 \cdot r^4$$
$$81 = r^4 \quad \therefore r = \pm 3$$
Ans: $4, 12, 36, 108, 324$
Or: $4, -12, 36, -108, 324$

9. Find the sum of the geometric sequence:
$$1 + 3 + 9 + 27 + \cdots \text{ for the first 8 terms.}$$
Solution: $a_1 = 1, n = 8, r = 3$

$$S_8 = \dfrac{a_1(1-r^n)}{1-r} = \dfrac{1 \cdot (1-3^8)}{1-3} = \dfrac{1-6561}{-2}$$
$$= \dfrac{-6560}{-2} = 3,280 . \text{ Ans.}$$

10. Find the sum of the geometric sequence:
$$a_1 = 2, \quad r = 3, \quad n = 100$$
Solution:

$$S_{100} = \dfrac{a_1(1-r^{100})}{1-r} = \dfrac{2 \cdot (1-3^{100})}{1-3}$$
$$= -1 + 3^{100} . \text{ Ans.}$$

11. Evaluate the geometric series $\displaystyle\sum_{k=1}^{50} 5(\tfrac{1}{2})^k$.

Solution:
$$a_1 = 5(\tfrac{1}{2}), \quad a_2 = 5(\tfrac{1}{2})^2, \cdots, a_{50} = 5(\tfrac{1}{2})^{50}$$
$$\sum_{k=1}^{50} 5(\tfrac{1}{2})^k = \dfrac{\tfrac{5}{2} \cdot [1-(\tfrac{1}{2})^{50}]}{1-\tfrac{1}{2}} = 5(1 - \dfrac{1}{2^{50}}) . \text{ Ans.}$$

12. Evaluate the geometric series $\displaystyle\sum_{k=1}^{50} 5(\tfrac{1}{2})^{k-1}$.

Solution:
$$a_1 = 5, \quad a_2 = 5(\tfrac{1}{2}), \cdots, a_{50} = 5(\tfrac{1}{2})^{49}$$
$$\sum_{k=1}^{50} 5(\tfrac{1}{2})^{k-1} = \dfrac{5[1-(\tfrac{1}{2})^{50}]}{1-\tfrac{1}{2}} = 10(1 - \dfrac{1}{2^{50}}) . \text{ Ans.}$$

Infinite Geometric Series

Infinite Geometric Series: The geometric series having its terms increases without limit (without bound).

$$a_1 + a_1 r + a_1 r^2 + a_1 r^3 + \cdots\cdots + a_1 r^{n-1} + \cdots\cdots \quad \text{as } n \to \infty.$$

We can use the formula for the sum of a finite geometric series $S_n = \sum_{k=1}^{n} a_1 r^{k-1} = \dfrac{a_1(1 - r^n)}{1-r}$ to find the formula for the sum of an infinite geometric series.

1) If $|r| \geq 1$, then $|r^n|$ increases without limit as $n \to \infty$. The sum increases without limit as n increases without limit (the series has no sum).

2) Formula: If $|r| < 1$, then $|r^n|$ approaches 0 as $n \to \infty$. The sum is a finite number given by:

$$S_\infty = \sum_{k=1}^{\infty} a_1 r^{k-1} = a_1 + a_1 r^2 + a_1 r^3 + \cdots\cdots + a_1 r^{n-1} + \cdots\cdots = \frac{a_1}{1-r} \cdot \quad |r| < 1$$

Examples

1. Find the sum of the infinite geometric series: $9 - 6 + 4 - \frac{8}{3} + \cdots\cdots$.

Solution: $r = \frac{-6}{9} = -\frac{2}{3}$

$|r| = \left|-\frac{2}{3}\right| = \frac{2}{3} < 1$

The series has a sum (finite number).

$$S_\infty = \frac{a_1}{1-r} = \frac{9}{1-(-\frac{2}{3})} = \frac{9}{\frac{5}{3}} = \frac{27}{5} \cdot \text{ Ans}$$

2. Find the sum of the infinite geometric series: $1 - \sqrt{2} + 2 - 2\sqrt{2} + 4 - 4\sqrt{2} + \cdots$

Solution: $r = \frac{-\sqrt{2}}{1} = -\sqrt{2}$

$|r| = \left|-\sqrt{2}\right| = \sqrt{2} = 1.414 > 1.$

The series has no sum.

The sum increases without limit as n increases without limit. Ans.

3. Prove $0.\overline{5} = \frac{5}{9}$.

Proof:

$0.\overline{5} = 0.555\cdots = 0.5 + 0.05 + 0.005 + \cdots$

$a_1 = 0.5, \quad r = \frac{0.05}{0.5} = 0.1 < 1$

$$\therefore 0.\overline{5} = \frac{a_1}{1-r} = \frac{0.5}{1-0.1} = \frac{0.5}{0.9} = \frac{5}{9} \cdot \text{ Ans.}$$

4. Find the sum of the infinite geometric series: $1 + \frac{1}{2} + \frac{1}{4} + \frac{1}{8} + \frac{1}{16} + \cdots\cdots$.

Solution:

$r = \frac{1}{2} \div 1 = \frac{1}{2} < 1$

The series has a sum (finite number).

$$S_\infty = \frac{a_1}{1-r} = \frac{1}{1-\frac{1}{2}} = 2 . \text{ Ans.}$$

5. Prove $0.\overline{ab} = \frac{ab}{99}$.

Proof:

$0.\overline{ab} = 0.ab + 0.00ab + 0.0000ab + \cdots$

$a_1 = 0.ab, \quad r = \dfrac{0.00ab}{0.ab} = 0.01 < 1$

$$\therefore 0.\overline{ab} = \frac{a_1}{1-r} = \frac{0.ab}{1-0.01} = \frac{0.ab}{0.99} = \frac{ab}{99} \cdot \text{ Ans.}$$

Check: $\frac{ab}{99} = (ab)\frac{1}{99} = (ab)(0.010101\cdots)$

$= (ab)(0.01 + 0.0001 + \cdots)$

$= 0.ab + 0.00ab + \cdots\cdots = 0.\overline{ab}.$

EXERCISES

Find the indicated sum of the geometric sequence (Evaluate each series).

1. $1+2+4+8+\cdots\cdots+512$

2. $1+3+9+27+\cdots\cdots+2187$

3. $2-4+8-16+\cdots\cdots+a_{15}$

4. $1-3+9-27+\cdots\cdots+a_{12}$

5. $0.3+(0.3)^2+(0.3)^3+\cdots\cdots+(0.3)^{12}$

6. $2(0.3)+2(0.3)^2+2(0.3)^3+\cdots\cdots+2(0.3)^{12}$

7. $6+6(\frac{1}{2})+6(\frac{1}{2})^2+\cdots\cdots+a_6$

8. $6+6(\frac{1}{2})+6(\frac{1}{2})^2+\cdots\cdots+a_{100}$

9. $\dfrac{7}{100}+\dfrac{7}{1000}+\dfrac{7}{10000}+\cdots\cdots+a_8$

10. $\sqrt{2}+1+\dfrac{1}{\sqrt{2}}+\cdots\cdots+a_{10}$

11. $1-\dfrac{1}{2}+\dfrac{1}{4}+\cdots\cdots+a_n$

12. $\dfrac{3}{2}+\dfrac{9}{2}+\dfrac{27}{2}+\cdots\cdots+a_n$

13. $\displaystyle\sum_{k=1}^{12}2^{k-1}$

14. $\displaystyle\sum_{k=1}^{12}(-2)^{k-1}$

15. $\displaystyle\sum_{k=0}^{12}(-2)^{k-1}$

16. $\displaystyle\sum_{k=1}^{12}(-2)^{k-2}$

17. $\displaystyle\sum_{k=1}^{5}(\tfrac{2}{3})^{k-2}$

18. $\displaystyle\sum_{k=1}^{n}(\tfrac{1}{2})^{k}$

19. $\displaystyle\sum_{k=1}^{n}6\cdot3^{k-1}$

20. $\displaystyle\sum_{k=1}^{8}3(\tfrac{1}{10})^{k+1}$

Find the sum of each infinite geometric series.

21. $16+8+4+2+\cdots\cdots$

22. $1-\tfrac{1}{2}+\tfrac{1}{4}-\tfrac{1}{8}+\cdots\cdots$

23. $1+0.1+0.01+0.001+\cdots\cdots$

24. $2-\tfrac{1}{2}+\tfrac{1}{8}-\tfrac{1}{32}+\cdots\cdots$

25. $27+3+\tfrac{1}{3}+\tfrac{1}{27}+\cdots\cdots$

26. $3-1+\tfrac{1}{3}-\tfrac{1}{9}+\cdots\cdots$

27. $\displaystyle\sum_{k=1}^{\infty}(\tfrac{1}{2})^{k}$

28. $\displaystyle\sum_{k=0}^{\infty}(\tfrac{1}{2})^{k}$

29. $\displaystyle\sum_{k=1}^{\infty}(\tfrac{5}{4})^{k-1}$

30. $\displaystyle\sum_{k=1}^{\infty}5(10)^{-k-1}$

31. Find the sum of the geometric series from $a_1=12$ to $a_5=48$.

32. Find the sum of the geometric series $a^{n-1}+a^{n-1}b+a^{n-2}b^2+\cdots\cdots$ with n terms and $a\neq b$.

33. If a,b,c is a geometric sequence, the sum is 19, and the product is 216, find a,b,c.

34. If a,b,c is an arithmetic sequence and also a geometric sequence, find $a:b:c$.

35. If a,b,c is an arithmetic sequence and c,a,b is a geometric sequence, find $a:b:c$.

36. Write the repeating decimal $0.121212\cdots\cdots$ in fractional form.

37. Write the repeating decimal $0.12222\cdots\cdots$ in fractional form.

38. A ball is dropped from a tower of 40 feet height and bounces straight up and down. Each time it strikes the ground and bounces up exactly one-half of the previous height.

    a) How far will the ball have traveled after it reaches the top of sixth bounce ?

    b) What total distance does the ball travel before it stops bouncing ?

39. John decides to deposit $3,000 annually in his IRA account that pays 10% interest compounded annually for the next 20 years. Find the sum (A) of all deposits made plus all interest paid (it is called the amount A of the **annuity**) at the end of 20 years.

40. If P represents the deposit in dollars annually at i percent interest compounded annually, prove that the amount A of the annuity at the end of n years is $A=P(1+i)\dfrac{(1+i)^n-1}{i}$.

10-5 Finding the nth Term of a Sequence

Linear Sequence: A sequence which has a common difference between any two successive terms.

Geometric Sequence: A sequence which has a common ratio between any two successive terms.

Quadratic Sequence: A sequence which can be factored with two linear factors.

Rules of finding the nth term of a sequence:

   1. The nth term of a linear sequence is given by the formula: (See Section 10-2.)

   Formula: $a_n = a_1 + (n-1)d$

   a_1: 1st term. n: position of a_n. d: common difference.

   2. The nth term of a geometric sequence is given by the formula: (See Section 10-4.)

   Formula: $a_n = a_1 r^{n-1}$

   a_1: 1st term. n: position of a_n. r: common ratio

   3. The nth term of each of the following sequences are derived from the above formulas.

$1, 2, 3, 4, 5, \cdots, n$	$1, 3, 5, 7, 9, \cdots, 2n-1$
$2, 3, 4, 5, 6, \cdots, n+1$	$3, 5, 7, 9, 11, \cdots, 2n+1$
$4, 5, 6, 7, 8, \cdots, n+3$	$2, 5, 8, 11, 14, \cdots, 3n-1$
$2, 4, 6, 8, 10, \cdots, 2n$	$1, 2, 4, 8, 16, 32, \cdots, 2^{n-1}$
$4, 6, 8, 10, 12, \cdots, 2n+2$	$2, 4, 8, 16, 32, 64, \cdots, 2^n$

   4. If a sequence is not linear, we factor it with two linear factors.
   5. If a sequence is neither linear nor quadratic, we double it and factor it.
   6. If it is not easy to factor a sequence with two linear factors, we use the theoretical method. (See the Examples 4, 5, 6 in this section)

Examples

   1. Find the nth term of the sequence: $2, 6, 12, 20, 30, 42, \cdots$.
   Solution:

$$2, \quad 6, \quad 12, \quad 20, \quad 30, \quad 42, \cdots.$$
$$\text{Factors} \quad 1 \cdot 2, \; 2 \cdot 3, \; 3 \cdot 4, \; 4 \cdot 5, \; 5 \cdot 6, \; 6 \cdot 7, \cdots$$

We have: $1, 2, 3, 4, 5, 6, \cdots, n$
$2, 3, 4, 5, 6, 7, \cdots, n+1$

$$\therefore \; a_n = n(n+1). \text{ Ans.}$$

Examples

2. Find the nth term of the sequence: 3, 6, 12, 24, 48, 96, $\cdots$.

Solution:

$$3, \quad 6, \quad 12, \quad 24, \quad 48, \quad 96, \cdots$$

Factors $3 \cdot 1, \quad 3 \cdot 2, \quad 3 \cdot 4, \quad 3 \cdot 8, \quad 3 \cdot 16, \quad 3 \cdot 32, \cdots$

We have: $1, 2, 4, 8, 16, 32, \cdots, 2^{n-1}, \cdots$

$$\therefore a_n = 3(2^{n-1}). \text{ Ans.}$$

3. Find the nth term of the sequence: 1, 3, 6, 10, 15, 21, $\cdots$.

Solution:

$$1, \quad 3, \quad 6, \quad 10, \quad 15, \quad 21, \cdots$$

Double $2, \quad 6, \quad 12, \quad 20, \quad 30, \quad 42, \cdots$

Factors $1 \cdot 2, \quad 2 \cdot 3, \quad 3 \cdot 4, \quad 4 \cdot 5, \quad 5 \cdot 6, \quad 6 \cdot 7, \cdots$

We have:

$$1, 2, 3, 4, 5, 6, \cdots, n$$
$$2, 3, 4, 5, 6, 7, \cdots, n+1$$

The nth term of the doubled sequence is $n(n+1)$.

$$\therefore a_n = \frac{n(n+1)}{2}. \text{ Ans.}$$

4. Find the nth term of the sequence: 6, 25, 56, 99, 154, 221, $\cdots$.

Solution:

Let b_n = A sequence which has all the differences between any two successive terms of the original sequence

$$a_n : 6, 25, 56, 99, 154, 221, \cdots$$
$$b_n : \quad 19, 31, 43, 55, \quad 67, \cdots$$

$$a_2 = a_1 + b_1 ; \quad a_3 = a_1 + b_1 + b_2 ; \quad \cdots \quad \therefore a_n = a_1 + (b_1 + b_2 + \cdots + b_{n-1}) = a_1 + \sum_{k=1}^{n-1} b_k$$

$$b_1 = 19, \ d = 12, \quad b_n = 19 + (n-1) \cdot 12 = 12n + 7$$
$$b_{n-1} = 12(n-1) + 7 = 12n - 5$$
$$a_n = 6 + [(19 + 31 + 43 + 55 + 67 + \cdots + (12n - 5)]$$

Find the sum of the $(n-1)$ terms of the above arithmetic series:

$$a_n = 6 + [\tfrac{n-1}{2}(19 + 12n - 5)] = 6 + \frac{(n-1)(14 + 12n)}{2} = 6 + \frac{12n^2 + 2n - 14}{2}$$

$$= 6 + 6n^2 + n - 7 = 6n^2 + n - 1 = (3n - 1)(2n + 1)$$

$$\therefore a_n = (3n - 1)(2n + 1). \text{ Ans.}$$

Examples

5. Find the nth term of the sequence: $2, 8, 26, 80, 242, \cdots$.

Solution:

Let b_n = A sequence which has all the differences between any two successive terms of the original sequence

$$a_n: 2, 8, 26, 80, 242, \cdots$$
$$b_n: 6, 18, 54, 162, \cdots$$

$$a_2 = a_1 + b_1 \;;\quad a_3 = a_1 + b_1 + b_2 \;;\quad \cdots \quad \therefore a_n = a_1 + (b_1 + b_2 + \cdots + b_{n-1}) = a_1 + \sum_{k=1}^{n-1} b_k$$

$$b_1 = 6, \; r = 3, \; b_n = b_1 r^{n-1} = 6 \cdot 3^{n-1}$$
$$b_{n-1} = 6 \cdot 3^{n-2} = 2 \cdot 3 \cdot 3^{n-2} = 2 \cdot 3^{n-1}$$

$$a_n = 2 + (6 + 18 + 54 + 162 + \cdots + 2 \cdot 3^{n-1})$$

Find the sum of the $(n-1)$ terms of the above geometric series:

$$a_n = 2 + \frac{6(1 - 3^{n-1})}{1 - 3} = 2 - 3(1 - 3^{n-1}) = 2 - 3 + 3^n = 3^n - 1$$

$$\therefore a_n = 3^n - 1. \text{ Ans.}$$

6. Find the nth term of the sequence: $2, 10, 30, 68, 130, 222, \cdots$.

Solution:

$$a_n: 2, 10, 30, 130, 222, \cdots$$
$$b_n: 8, 20, 38, 92, \cdots$$
$$c_n: 12, 18, 24, \cdots, (6n+6) \qquad \text{Hint: } c_n = 12 + (n-1) \cdot 6 = 6n + 6$$

$$b_n = b_1 + \sum_{k=1}^{n-1} c_k = b_1 + \sum_{k=1}^{n-1} (6k + 6) = 8 + (12 + 18 + 24 + \cdots + 6n)$$
$$= 8 + \tfrac{n-1}{2}(12 + 6n) = 8 + (n-1)(3n+6) = 3n^2 + 3n + 2$$

$$a_n = a_1 + \sum_{k=1}^{n-1} b_k = a_1 + \sum_{k=1}^{n-1} (3k^2 + 3k + 2) = 2 + 3\sum_{k=1}^{n-1} k^2 + 3\sum_{k=1}^{n-1} k + \sum_{k=1}^{n-1} 2$$

(Hint: $\sum_{k=1}^{n} k^2 = 1^2 + 2^2 + 3^2 + \cdots + n^2 = \dfrac{n(n+1)(2n+1)}{6}$ on page 367)

$$a_n = 2 + 3 \cdot \frac{(n-1) \cdot n \cdot [2(n-1)+1]}{6} + 3 \cdot \frac{(n-1)n}{2} + 2(n-1) = n^3 + n$$

$$\therefore a_n = n^3 + n. \text{ Ans.}$$

EXERCISES

Find an expression for the nth terms of each sequence.

1. $-1, 0, 1, 2, 3, \ldots$

2. $1, 4, 7, 10, 13, \ldots$

3. $-2, 1, 4, 7, 10, \ldots$

4. $3, 8, 13, 18, 23, \ldots$

5. $4, 8, 16, 32, 64, \ldots$

6. $\frac{1}{2}, \frac{1}{4}, \frac{1}{8}, \frac{1}{16}, \frac{1}{32}, \ldots$

7. $-\frac{1}{2}, \frac{1}{4}, -\frac{1}{8}, \frac{1}{16}, -\frac{1}{32}, \ldots$

8. $2, -1, \frac{1}{2}, -\frac{1}{4}, \frac{1}{8}, \ldots$

9. $3, 9, 27, 81, 243, \ldots$

10. $1, 4, 9, 16, 25, \ldots$

11. $1, 3, 9, 27, 81, \ldots$

12. $3, 8, 15, 24, 35, \ldots$

13. $1, 6, 15, 28, 45, \ldots$

14. $5, 10, 20, 40, 80, \ldots$

15. $2, 6, 18, 54, 162, \ldots$

16. $5, 12, 21, 32, 45, \ldots$

17. $3, 18, 81, 324, \ldots$

18. $8, 16, 32, 64, 128, \ldots$

19. $8, 15, 24, 35, 48, \ldots$

20. $2, 5, 9, 14, 20, \ldots$

21. $3, 7, 12, 18, 25, \ldots$

22. $4, 12, 24, 40, 60, \ldots$

23. $4, 20, 48, 88, 140, \ldots$

24. $1, 2, 4, 7, 11, 16, \ldots$

25. $1, 3, 6, 10, 15, 21, \ldots$

26. $1, 3, 7, 15, 31, 63, \ldots$

27. $-1, 2, 7, 14, 23, 34, \ldots$

28. $1, 4, 13, 40, 121, \ldots$

29. $2, 4, 7, 11, 16, \ldots$

30. $1, 9, 29, 67, 129, \ldots$

31. $1, 3, 8, 18, 35, \ldots$

32. $1, 3, 6, 11, 20, \ldots$

33. $2, 5, 12, 31, 86, \ldots$

34. $\frac{1}{2}, \frac{2}{3}, \frac{3}{4}, \frac{4}{5}, \frac{5}{6}, \ldots$

35. $\frac{1}{3}, \frac{3}{4}, \frac{5}{5}, \frac{7}{6}, \frac{9}{7}, \ldots$

36. $2, \frac{3}{2}, \frac{4}{3}, \frac{5}{4}, \frac{6}{5}, \ldots$

37. $\frac{1}{2 \cdot 3}, \frac{2}{3 \cdot 4}, \frac{3}{4 \cdot 5}, \frac{4}{5 \cdot 6}, \ldots$

38. $-\frac{1}{2}, \frac{2}{3}, -\frac{3}{4}, \frac{4}{5}, -\frac{5}{6}, \ldots$

39. $\frac{3}{4}, \frac{9}{16}, \frac{27}{64}, \frac{81}{256}, \ldots$

40. Find the nth term of the sequence: $1+\sqrt{2}$, $\sqrt{2}+\sqrt{3}$, $\sqrt{3}+2$, $2+\sqrt{5}$, $\ldots$.

41. If a sequence has $a_1 = 2$ and $a_{n+1} = a_n + n$, find a_{10}.

42. If $\frac{1}{a}$ and $\frac{1}{b}$ are the first two terms of an arithmetic sequence, find a_n.

43. Find the sum of the multiples of 12 between 100 and 999.

10-6 Finding the Sums of Special Series

The following formulas are useful to find the sums of many special series.

1. $\displaystyle\sum_{k=1}^{n} k = 1 + 2 + 3 + \cdots + n = \frac{n(n+1)}{2}$

2. $\displaystyle\sum_{k=1}^{n} k^2 = 1^2 + 2^2 + 3^2 + \cdots + n^2 = \frac{n(n+1)(2n+1)}{6}$

3. $\displaystyle\sum_{k=1}^{n} k^3 = 1^3 + 2^3 + 3^3 + \cdots + n^3 = \frac{n^2(n+1)^2}{4}$ **Hint:** $\displaystyle\sum_{k=1}^{n} k^3 = \left(\sum_{k=1}^{n} k\right)^2$

4. $\displaystyle\sum_{k=1}^{n} k^4 = 1^4 + 2^4 + 3^4 + \cdots + n^4 = \frac{n(n+1)(2n+1)(3n^2+3n-1)}{30}$

5. $\displaystyle\sum_{k=1}^{n} (2k-1)^2 = 1^2 + 3^2 + 5^2 + \cdots + (2n-1)^2 = \frac{n(2n+1)(2n-1)}{3}$

6. $\displaystyle\sum_{k=1}^{n} k(k+1) = 1\cdot 2 + 2\cdot 3 + 3\cdot 4 + \cdots + n(n+1) = \frac{n(n+1)(n+2)}{3}$

7. $\displaystyle\sum_{k=1}^{n} k(k+1)(k+2) = 1\cdot 2\cdot 3 + 2\cdot 3\cdot 4 + \cdots + n(n+1)(n+2) = \frac{n(n+1)(n+2)(n+3)}{4}$

8. $\displaystyle\sum_{k=1}^{n} \frac{1}{k(k+1)} = \frac{1}{1\cdot 2} + \frac{1}{2\cdot 3} + \frac{1}{3\cdot 4} + \cdots + \frac{1}{n(n+1)} = \frac{n}{n+1}$

9. $\displaystyle\sum_{k=1}^{n} \frac{1}{k(k+1)(k+2)} = \frac{1}{1\cdot 2\cdot 3} + \frac{1}{2\cdot 3\cdot 4} + \frac{1}{3\cdot 4\cdot 5} + \cdots + \frac{1}{n(n+1)(n+2)} = \frac{n(n+3)}{4(n+1)(n+2)}$

Example 1: Prove $\displaystyle\sum_{k=1}^{n} k^2 = 1^2 + 2^2 + 3^2 + \cdots + n^2 = \frac{n(n+1)(2n+1)}{6}$.

Proof: We use the following expression to prove this formula:

$$(n+1)^3 - n^3 = n^3 + 3n^2 + 3n + 1 - n^3 = 3n^2 + 3n + 1$$

$$2^3 - 1^3 = 3\cdot 1^2 + 3\cdot 1 + 1$$

$$3^3 - 2^3 = 3\cdot 2^3 + 3\cdot 2 + 1$$

$$4^3 - 3^3 = 3\cdot 3^2 + 3\cdot 3 + 1$$

$$\vdots \qquad\qquad \vdots$$

$$(n+1)^3 - n^3 = 3\cdot n^2 + 3\cdot n + 1$$

Add the left and the right sides of the above expressions:

$$(2^3 - 1^3) + (3^3 - 2^3) + (4^3 - 3^3) + \cdots + [(n+1)^3 - n^3]$$

$$= 3(1^2 + 2^2 + 3^2 + \cdots + n^2) + 3(1 + 2 + 3 + \cdots + n) + (1 + 1 + 1 + \cdots + 1)$$

$$(n+1)^3 - 1^3 = 3\sum_{k=1}^{n} k^2 + 3\cdot\frac{n(n+1)}{2} + n \qquad \therefore\ 3\sum_{k=1}^{n} k^2 = n^3 + 3n^2 + 3n - \tfrac{3}{2}n(n+1) - n$$

$$3\sum_{k=1}^{n} k^2 = n^3 + \tfrac{3}{2}n^2 + \tfrac{1}{2}n = \frac{2n^3 + 3n^2 + n}{2} \qquad \therefore\ \sum_{k=1}^{n} k^2 = \frac{n(n+1)(2n+1)}{6}. \text{ Ans.}$$

EXERCISES

1. Prove the formula: $\displaystyle\sum_{k=1}^{n} k^3 = 1^3 + 2^3 + 3^3 + \cdots\cdots + n^3 = \frac{n^2(n+1)^2}{4}$.

   Hint: Use the expression $(n+1)^4 - n^4 = n^4 + 4n^3 + 6n^2 + 4n + 1 - n^4 = 4n^3 + 6n^2 + 4n + 1$

2. Prove the formula: $\displaystyle\sum_{k=1}^{n} k(k+1) = 1 \cdot 2 + 2 \cdot 3 + 3 \cdot 4 + \cdots\cdots + n(n+1) = \frac{n(n+1)(n+2)}{3}$.

   Hint: $\displaystyle\sum_{k=1}^{n} k(k+1) = \sum_{k=1}^{n}(k^2 + k) = \sum_{k=1}^{n} k^2 + \sum_{k=1}^{n} k$

3. Prove the formula:

   $$\sum_{k=1}^{n} k(k+1)(k+2) = 1 \cdot 2 \cdot 3 + 2 \cdot 3 \cdot 4 + \cdots\cdots + n(n+1)(n+2) = \frac{n(n+1)(n+2)(n+3)}{4}.$$

   Hint: $\displaystyle\sum_{k=1}^{n} k(k+1)(k+2) = \sum_{k=1}^{n}(k^3 + 3k^2 + 2k) = \sum_{k=1}^{n} k^3 + 3 \cdot \sum_{k=1}^{n} k^2 + 2 \cdot \sum_{k=1}^{n} k$

4. Prove the formula: $\displaystyle\sum_{k=1}^{n} \frac{1}{k(k+1)} = \frac{1}{1 \cdot 2} + \frac{1}{2 \cdot 3} + \frac{1}{3 \cdot 4} + \cdots\cdots + \frac{1}{n(n+1)} = \frac{n}{n+1}$.

   Hint: Apply the partial fractions (See Section 6~9, Page 211)

   $$\sum_{k=1}^{n} \frac{1}{k(k+1)} = \sum_{k=1}^{n}\left(\frac{1}{k} - \frac{1}{k+1}\right)$$

5. Find the formula of the sum with n terms: $1 + (1+2) + (1+2+3) + \cdots\cdots + (1+2+3+\cdots+n)$

   Hint: $\displaystyle\sum_{k=1}^{n}(1+2+3+\cdots\cdots+k) = \sum_{k=1}^{n}\frac{k(k+1)}{2} = \frac{1}{2}\left[\sum_{k=1}^{n} k^2 + \sum_{k=1}^{n} k\right]$

6. Find the formula of the sum with n terms: $1^2 \cdot 3 + 2^2 \cdot 5 + 3^2 \cdot 7 + \cdots\cdots + n^2(2n+1)$.

   Hint: $\displaystyle\sum_{k=1}^{n} k^2(2k+1) = 2 \cdot \sum_{k=1}^{n} k^3 + \sum_{k=1}^{n} k^2$

7. Find the formula of the sum with n terms: $\dfrac{1}{1 \cdot 2 \cdot 3} + \dfrac{1}{2 \cdot 3 \cdot 4} + \dfrac{1}{3 \cdot 4 \cdot 5} + \dfrac{1}{n(n+1)(n+2)}$.

   Hint: Apply the partial fractions (See Section 6~9, page 211)

   $$\sum_{k=1}^{n} \frac{1}{k(k+1)(k+2)} = \frac{1}{2} \cdot \sum_{k=1}^{n}\left(\frac{1}{k(k+1)} - \frac{1}{(k+1)(k+2)}\right)$$

8. Use the formulas on Page 367 to find the sum of each series:

   a. $1^2 + 2^2 + 3^2 + \cdots\cdots + 10^2$ **b.** $1 \cdot 2 + 2 \cdot 3 + 3 \cdot 4 + \cdots\cdots + 15 \cdot 16$

   c. $\dfrac{1}{1 \cdot 2} + \dfrac{1}{2 \cdot 3} + \dfrac{1}{3 \cdot 4} + \cdots\cdots + \dfrac{1}{18 \cdot 19}$ **d.** $1^2 + 3^2 + 5^2 + \cdots\cdots + 19^2$

10-7 Mathematical Induction

The statement $3n < 80$ is true for several values of n. But, it is not true for all values of n. Therefore, simply testing a statement (equation, rule, or formula) by its pattern for several values of n is not adequate to establish a formal proof of the statement for all values of n.

Inductive reasoning is a reasoning from a pattern of examples to a conclusion. It applies a specific case to a general case.

To prove a mathematical statement, we use the following principal.
The Principal of Mathematical Induction:

Let P_n be a statement about positive integers n. If the following two conditions are satisfied:

    I). P_n is true when $n = 1$.

    II). Assume P_n is true for any positive integer k, and it is also proven true for the next positive integer $k + 1$.

    then P_n is true for all positive integers n.

Positive integers are also called natural numbers.
The Principal of Mathematical Induction allows us to conclude that a statement is true for all positive integers only after both Condition I and Condition II have been satisfied. If the statement is true for $n = 1$, we assume that the statement is true for $n = k$ and try to prove that it is also true for $n = k + 1$.

Examples

    1. Use the Principal of Mathematical Induction to prove that the following statement is true for all positive integers n.

$$1 + 2 + 3 + 4 + \cdots + n = \frac{n(n+1)}{2}.$$

Proof: 1. $S_1 = \frac{1(1+1)}{2} = 1$. It is true when $n = 1$.

      2. Assume that it is true for any positive integer k, we have

$$S_k = 1 + 2 + 3 + 4 + \cdots + k = \frac{k(k+1)}{2}.$$

When $n = k + 1$, we have

$$S_{k+1} = 1 + 2 + 3 + 4 + \cdots + k + (k+1) = \frac{k(k+1)}{2} + (k+1)$$

$$= \frac{k^2 + k + 2k + 2}{2} = \frac{k^2 + 3k + 2}{2} = \frac{(k+1)(k+2)}{2}.$$

It is true for $n = k + 1$.

Therefore, the statement is true for all positive integers n.

Examples

2. Use the Principal of Mathematical Induction to prove that $(n+1)(n+2)(n+3)$ is divisible by 6 for all positive integers n.

Solution:

1. $(1+1)(1+2)(1+3) = 24$. It is divisible by 6 when $n=1$.

2. Assume $(k+1)(k+2)(k+3)$ is divisible by 6 for any positive integer k, then we have (for $n = k+1$)
$$(k+2)(k+3)(k+4) = (k+2)(k+3)[(k+1)+3]$$
$$= (k+1)(k+2)(k+3) + 3(k+2)(k+3).$$

$(k+1)(k+2)(k+3)$ is divisible by 6.

$3(k+2)(k+3)$ is divisible by 6 since either $(k+2)$ or $(k+3)$ must be even.

Therefore, it is divisible by 6 for $n = k+1$.

Thus, it is divisible by 6 for all positive integers n.

3. Use the Principal of Mathematical Induction to prove that the following statement is true for all positive integers n.
$$\text{If } n \geq 2, \text{ then } (1+x)^n > 1 + nx$$

Solution:

1. $(1+x)^2 = 1 + 2x + x^2 > 1 + 2x$. It is true when $n=2$.

2. Assume that it is true for any positive integer k, we have
$$(1+x)^k > 1 + kx \text{ if } k \geq 2$$
When $n = k+1$, we have
$$(1+x)^{k+1} = (1+x)^k(1+x) > (1+kx)(1+x) = 1 + x + kx + kx^2$$
$$= 1 + (k+1)x + kx^2 > 1 + (k+1)x$$

It is true for $n = k+1$.

Therefore, the statement is true for all positive integers n.

4. Use the Principal of Mathematical Induction to prove that $x^n - y^n$ is divisible by $(x-y)$ for all positive integers n whenever $x \neq y$.

Solution:

1. $\dfrac{x^n - y^n}{x-y} = \dfrac{x-y}{x-y} = 1$. It is divisible by $(x-y)$ when $n=1$.

2. Assume that $x^k - y^k$ is divisible by $(x-y)$ for any positive integer k, then we have (for $n = k+1$)
$$x^{k+1} - y^{k+1} = x^{k+1} - xy^k + xy^k - y^{k+1} = x(x^k - y^k) + y^k(x-y).$$

$x(x^k - y^k)$ is divisible by $(x-y)$ and $y^k(x-y)$ is divisible by $(x-y)$.

Therefore, it is divisible by $(x-y)$ for $n = k+1$.

Thus, it is divisible by $(x-y)$ for all positive integers n.

Example

5. Use the Principal of Mathematical Induction to prove that the following statement is true for all positive integers n.

$$\frac{1}{1\cdot 2}+\frac{1}{2\cdot 3}+\frac{1}{3\cdot 4}+\cdots\cdots+\frac{1}{n(n+1)}=\frac{n}{n+1}$$

Proof:

   1. $S_1 = \dfrac{1}{1+1}=\dfrac{1}{2}$. It is true when $n=1$

   2. Assume that it is true for any positive integer k, we have

$$S_k=\frac{1}{1\cdot 2}+\frac{1}{2\cdot 3}+\frac{1}{3\cdot 4}+\cdots\cdots+\frac{1}{k(k+1)}=\frac{k}{k+1}$$

When $n=k+1$, we have

$$S_{k+1}=\frac{1}{1\cdot 2}+\frac{1}{2\cdot 3}+\frac{1}{3\cdot 4}+\cdots\cdots+\frac{1}{k(k+1)}+\frac{1}{(k+1)(k+2)}=\frac{k}{k+1}+\frac{1}{(k+1)(k+2)}$$

$$=\frac{k(k+2)+1}{(k+1)(k+2)}=\frac{k^2+2k+1}{(k+1)(k+2)}=\frac{(k+1)(k+1)}{(k+1)(k+2)}=\frac{k+1}{k+2}$$

It is true for $n=k+1$.

Therefore, the statement is true for all positive integers n.

6. Prove that the statement "n^2-n+23 is a prime number" is true for $n=1$, but is not true for $n=23$.

Proof:

$$n=1:\ 1^2-1+23=23 \text{ is a prime number}$$
$$n=23:\ 23^2-23+23=23^2 \text{ is not a prime number.}$$

7. Prove that the formula $2+4+6+\cdots\cdots+2n=n^2+n+2$ is false for $n=1$, but is true for $n=k+1$ if the formula is true for any positive integer k.

Proof:

$$n=1:\ 1^2+1+2=4\neq 2.\ \text{It is false for } n=1.$$

Assume that it is true for any positive integer k, we have

$$2+4+6+\cdots\cdots+2k=k^2+k+2$$

When $n=k+1$, we have

$$2+4+6+\cdots\cdots+2k+2(k+1)=k^2+k+2+2(k+1)$$

$$=k^2+k+2+2k+2=k^2+3k+4=k^2+2k+1+k+1+2$$

$$=(k+1)^2+(k+1)+2$$

It is true for $n=k+1$.

Hint: The formula does not satisfy both Condition 1 and Condition 2 of Mathematical Induction. It is not true for any positive integer.

The formula $2+4+6+\cdots\cdots+2n=n^2+n+2$ is not correct.

The formula $2+4+6+\cdots\cdots+2n=n^2+n$ is correct.

EXERCISES

Use the Principal of Mathematical Induction to prove that the following statement is true for all positive integers n (natural numbers).

1. $1 + 5 + 9 + \cdots + (4n - 3) = n(2n - 1)$.

2. $1 + 3 + 5 + \cdots + (2n - 1) = n^2$.

3. $1 + 4 + 7 + \cdots + (3n - 2) = \frac{1}{2} n(3n - 1)$.

4. $1^2 + 2^2 + 3^2 + \cdots + n^2 = \frac{1}{6} n(n + 1)(2n + 1)$.

5. $1 + 2 + 2^2 + \cdots + 2^{n-1} = 2^n - 1$.

6. $1 + 2 + 3 + \cdots + n = \dfrac{n(n + 1)}{2}$

7. $1^3 + 2^3 + 3^3 + \cdots + n^3 = \frac{1}{4} n^2 (n + 1)^2$.

8. $2^n > n$.

9. If $n \geq 5$, then $2^n > n^2$.

10. $(n + 2)(n + 3)$ is divisible by 2.

11. $4^n - 1$ is divisible by 3.

12. $3^n - 1$ is divisible by 2.

13. $a - b$ is a factor of $a^n - b^n$.

14. $a^{2n} - b^{2n}$ is divisible by $(a + b)$ if $a \neq -b$

15. $n^2 - n + 2$ is divisible by 2.

16. $2^n > (n + 1)^2$ if $n \geq 6$.

17. $n + 3 < 5n^2$.

18. $1 + \frac{1}{\sqrt{2}} + \frac{1}{\sqrt{3}} + \cdots + \frac{1}{\sqrt{n}} > \sqrt{n}$ if $n \geq 2$.

19. $1 \cdot 2 + 2 \cdot 3 + 3 \cdot 4 + \cdots + n(n + 1) = \dfrac{1}{3} n(n + 1)(n + 2)$

20. $1 \cdot 3 + 2 \cdot 4 + 3 \cdot 5 + \cdots + n(n + 2) = \dfrac{1}{6} n(n + 1)(2n + 7)$.

21. $1^4 + 2^4 + 3^4 + \cdots + n^4 = \dfrac{n(n + 1)(2n + 1)(3n^2 + 3n - 1)}{30}$.

22. $1^5 + 2^5 + 3^5 + \cdots + n^5 = \dfrac{n^2 (n + 1)^2 (2n^2 + 2n - 1)}{12}$.

23. Show that 2 is a factor of $n^2 - n$ for all positive integers n.

24. Show that 3 is a factor of $n^3 + 5n + 6$ for all positive integers n.

25. Prove that the formula $1 + 2 + 3 + 4 + \cdots + n = \frac{1}{2}(n^2 + n + 1)$ is false for $n = 1$, but is true for $n = k + 1$ if the formula is true for any positive integer k.

26. Use Mathematical Induction to prove that the sum of the interior angles in a convex polygon with n sides is $180°(n - 2)$.

   (Hint: A convex polygon of $(n + 1)$ sides consists of a polygon of n sides plus a triangle.)

27. Use Mathematical Induction to prove that the number of diagonals in a convex polygon with n sides is $\frac{1}{2} n(n - 3)$.

   (Hint: A convex polygon of $(n + 1)$ sides consists of the diagonals of n sides plus $(n - 1)$ additional diagonals.)

10-8 Pascal's Triangle and Binomial Theorem

Pascal's Triangle: The expansion of $(a+b)^n$ with different power of n forms a pattern of " **Triangular array** " by the coefficients of its expansion. Each coefficient is the sum of the two numbers above it.

The nth row in Pascal's Triangle are the coefficients of $(a+b)^n$.

$$
\begin{array}{c}
1 \\
1 \quad 1 \\
1 \quad 2 \quad 1 \\
1 \quad 3 \quad 3 \quad 1 \\
1 \quad 4 \quad 6 \quad 4 \quad 1 \\
1 \quad 5 \quad 10 \quad 10 \quad 5 \quad 1 \\
1 \quad 6 \quad 15 \quad 20 \quad 15 \quad 6 \quad 1
\end{array}
$$

$$(a+b)^0 = 1$$
$$(a+b)^1 = a+b$$
$$(a+b)^2 = a^2 + 2ab + b^2$$
$$(a+b)^3 = a^3 + 3a^2b + 3ab^2 + b^3$$
$$(a+b)^4 = a^4 + 4a^3b + 6a^2b^2 + 4ab^3 + b^4$$
$$(a+b)^5 = a^5 + 5a^4b + 10a^3b^2 + 10a^2b^3 + 5ab^4 + b^5$$
$$(a+b)^6 = a^6 + 6a^5b + 15a^4b^2 + 20a^3b^3 + 15a^2b^4 + 6ab^5 + b^6$$

Examples

1. Expand $(x+2)^4$.

   Solution:
   $$(x+2)^4 = x^4 + 4x^3 \cdot 2 + 6x^2 \cdot 2^2 + 4x \cdot 2^3 + 2^4 = x^4 + 8x^3 + 24x^2 + 32x + 16.$$
   Ans,

2. Expand $(x-2)^4$.

   Solution:
   $$(x-2)^4 = x^4 + 4x^3(-2) + 6x^2(-2)^2 + 4x(-2)^3 + (-2)^4$$
   $$= x^4 - 8x^3 + 24x^2 - 32x + 16. \text{ Ans.}$$

3. Expand $(2x-y)^5$.

   Solution:
   $$(2x-y)^5 = (2x)^5 + 5(2x)^4(-y) + 10(2x)^3(-y)^2 + 10(2x)^2(-y)^3 + 5(2x)(-y)^4 + (-y)^5$$
   $$= 32x^5 - 5(16x^4)y + 10(8x^3)y^2 - 10(4x^2)y^3 + 10xy^4 - y^5$$
   $$= 32x^5 - 80x^4y + 80x^3y^2 - 40x^2y^3 + 10xy^4 - y^5. \text{ Ans.}$$

4. Expand $(2n+3)^6$.

   Solution:
   $$(2n+3)^6 = (2n)^6 + 6(2n)^5 \cdot 3 + 15(2n)^4 \cdot 3^2 + 20(2n)^3 \cdot 3^3 + 15(2n)^2 \cdot 3^4 + 6(2n) \cdot 3^5 + 3^6.$$
   $$= 64n^6 + 18(32n^5) + 135(16n^4) + 540(8n^3) + 1215(4n^2) + 1458(2n) + 729$$
   $$= 64n^6 + 576n^5 + 2160n^4 + 4320n^3 + 4860n^2 + 2916n + 729. \text{ Ans.}$$

Binomial Theorem

To find the expansion of $(a+b)^n$, we may use **Pascal's Triangle** or **Binomial Theorem.**

Binomial Theorem is more convenient when n is large or to find a particular term of $(a+b)^n$.

Binomial Theorem (Binomial Formula):

If n is a positive integer, the general expansion of $(a+b)^n$ is:

$$(a+b)^n = a^n + {}_nC_1 a^{n-1}b + {}_nC_2 a^{n-2}b^2 + \cdots + {}_nC_r a^{n-r}b^r + \cdots + b^n \quad \text{where } {}_nC_r = \frac{n!}{(n-r)!r!}$$

Or: $(a+b)^n = a^n + \dfrac{n}{1!}a^{n-1}b + \dfrac{n(n-1)}{2!}a^{n-2}b^2 + \cdots + \dfrac{n!}{(n-r)!r!}a^{n-r}b^r + \cdots + b^n$

The symbol ${}_nC_r$ is the number of combinations. (See Section 11 ~ 2)

${}_nC_r$ is read "the combinations of n taken r at a time" .

The general term ${}_nC_r a^{n-r}b^r$ is the $(r+1)st$ term of the expansion.

The symbol $C(n, r)$ and $\begin{pmatrix} n \\ r \end{pmatrix}$ are often used to denote ${}_nC_r$.

(Factor notation !):   The product of consecutive integers beginning with r ,
                  decreasing and ending with 1. Read as "r factorial ".

$0! = 1, \quad 1! = 1, \quad 2! = 2 \cdot 1 = 2, \quad 3! = 3 \cdot 2 \cdot 1 = 6, \quad 4! = 4 \cdot 3 \cdot 2 \cdot 1 = 24, \quad 5! = 5 \cdot 4 \cdot 3 \cdot 2 \cdot 1 = 120 ,$

$6! = 6 \cdot 5 \cdot 4 \cdot 3 \cdot 2 \cdot 1 = 720, \quad 7! = 7 \cdot 6! = 7 \cdot 6 \cdot 5! = 7 \cdot 6 \cdot 5 \cdot 4! = 7 \cdot 6 \cdot 5 \cdot 4 \cdot 3! = 7 \cdot 6 \cdot 5 \cdot 4 \cdot 3 \cdot 2!$

$$= 7 \cdot 6 \cdot 5 \cdot 4 \cdot 3 \cdot 2 \cdot 1 = 5{,}040 .$$

$n! = n(n-1)! = n(n-1)(n-2)! = n(n-1)(n-2)(n-3)! = \cdots$

$$= n(n-1)(n-2)(n-3)(n-4)(n-5) \cdots 3 \cdot 2 \cdot 1.$$

$$\frac{n!}{(n-2)!} = \frac{n(n-1)\cancel{(n-2)!}}{\cancel{(n-2)!}} = n(n-1). \qquad \frac{5!}{4!} = \frac{5 \cdot \cancel{4!}}{\cancel{4!}} = 5 .$$

$$C(10, 3) = \begin{pmatrix} 10 \\ 3 \end{pmatrix} = {}_{10}C_3 = \frac{10!}{(10-3)! \cdot 3!} = \frac{10 \cdot 9 \cdot 8 \cdot \cancel{7!}}{\cancel{7!} \cdot 3!} = \frac{10 \cdot 9 \cdot 8}{3 \cdot 2 \cdot 1} = 120$$

Examples (See Page 373)

1. $(a+b)^2 = a^2 + \dfrac{2}{1!}ab + b^2 = a^2 + 2ab + b^2$.

2. $(a+b)^3 = a^3 + \dfrac{3}{1!}a^2b + \dfrac{3 \cdot 2}{2!}ab^2 + b^3 = a^3 + 3a^2b + 3ab^2 + b^3$.

3. $(a+b)^4 = a^4 + \dfrac{4}{1!}a^3b + \dfrac{4 \cdot 3}{2!}a^2b^2 + \dfrac{4 \cdot 3 \cdot 2}{3!}ab^3 + b^4 = a^4 + 4a^3b + 6a^2b^2 + 4ab^3 + b^4$.

4. $(a+b)^5 = a^5 + \dfrac{5}{1!}a^4b + \dfrac{5 \cdot 4}{2!}a^3b^2 + \dfrac{5 \cdot 4 \cdot 3}{3!}a^2b^3 + \dfrac{5 \cdot 4 \cdot 3 \cdot 2}{4!}ab^4 + b^5$

$$= a^5 + 5a^4b + 10a^3b^2 + 10a^2b^3 + 5ab^4 + b^5 .$$

Examples

5. Expand $(x+2)^4$.

Solution:

$$(x+2)^4 = x^4 + \frac{4}{1!}x^3 \cdot 2 + \frac{4 \cdot 3}{2!}x^2 \cdot 2^2 + \frac{4 \cdot 3 \cdot 2}{3!}x \cdot 2^3 + 2^4 = x^4 + 8x^3 + 24x^2 + 32x + 16 .$$

Ans.

6. Expand $(2x-y)^5$.

Solution:

$$(2x-y)^5 = (2x)^5 + \frac{5}{1!}(2x)^4(-y) + \frac{5 \cdot 4}{2!}(2x)^3(-y)^2 + \frac{5 \cdot 4 \cdot 3}{3!}(2x)^2(-y)^3$$

$$+ \frac{5 \cdot 4 \cdot 3 \cdot 2}{4!}(2x)(-y)^4 + (-y)^5$$

$$= 32x^5 - 5(16x^4)y + 10(8x^3)y^2 - 10(4x^2)y^3 + 5(2x)y^4 - y^5$$

$$= 32x^5 - 80x^4y + 80x^3y^2 - 40x^2y^3 + 10xy^4 - y^5 . \text{ Ans.}$$

7. Find the 6th term of $(2x-y)^{15}$.

Solution:

The $(r+1)st$ terms of $(a+b)^n$ is $_nC_r a^{n-r}b^r$. For the 6th term, we use $r = 5$.

The 6th term is: $_{15}C_5(2x)^{10}(-y)^5 = \dfrac{15!}{(15-5)! \cdot 5!}(1024x^{10})(-y^5)$

$$= \frac{15 \cdot 14 \cdot 13 \cdot 12 \cdot 11}{5 \cdot 4 \cdot 3 \cdot 2 \cdot 1}(1024x^{10})(-y^5)$$

$$= 3003(1024x^{10})(-y^5) = -3{,}075{,}072x^{10}y^5 . \text{ Ans.}$$

8. Find the 7th term of $(2x-y^2)^{10}$.

Solution:

For the 7th term, we use $r = 6$ for the term $_nC_r a^{n-r}b^r$ of $(a+b)^n$.

The 7th term is: $_{10}C_6(2x)^4(-y^2)^6 = \dfrac{10!}{(10-6)! \cdot 6!}(16x^4)y^{12}$

$$= \frac{10 \cdot 9 \cdot 8 \cdot 7 \cdot 6 \cdot 5}{6 \cdot 5 \cdot 4 \cdot 3 \cdot 2 \cdot 1}(16x^4)y^{12}$$

$$= 210(16x^4)y^{12} = 3{,}360x^4y^{12} . \text{ Ans.}$$

9. Find the term with y^{12} in the expansion of $(x-y^2)^{15}$.

Solution:

$$x^{15}, \ x^{14}(-y^2), \ x^{13}(-y^2)^2, \ x^{12}(-y^2)^3, \ x^{11}(-y^2)^4, \ x^{10}(-y^2)^5, \ x^9(-y^2)^6, \ldots$$

It is the 7th term : $r = 7 - 1 = 6$

$$_{15}C_6(x^9)(-y^2)^6 = \frac{15!}{(15-6)! \cdot 6!}x^9y^{12} = \frac{15 \cdot 14 \cdot 13 \cdot 12 \cdot 11 \cdot 10}{6 \cdot 5 \cdot 4 \cdot 3 \cdot 2 \cdot 1}x^9y^{12} = 5{,}005x^9y^{12}. \text{ Ans.}$$

EXERCISES

Expand each expression using Pascal's Triangle.

1. $(x+1)^4$ **2.** $(x+1)^5$ **3.** $(x-1)^4$ **4.** $(x-1)^5$

5. $(x+1)^7$ **6.** $(x-1)^7$ **7.** $(x-2)^5$ **8.** $(x+2y)^4$

9. $(3x-y)^5$ **10.** $(a+3)^3$ **11.** $(a^2+1)^5$ **12.** $(x^2+5)^4$

Evaluate each expression.

13. $_5C_5$ **14.** $_5C_4$ **15.** $_5C_3$ **16.** $_5C_2$

17. $_5C_1$ **18.** $_{20}C_5$ **19.** $_{100}C_{98}$ **20.** $_{100}C_2$

21. $\binom{10}{2}$ **22.** $\binom{10}{7}$ **23.** $C(50,49)$ **24.** $C(60,2)$

Evaluate each expression using the Binomial Theorem.

25. $(x+1)^3$ **26.** $(x-2)^5$ **27.** $(x+2)^7$ **28.** $(2x-3)^4$

29. $(2x+3)^5$ **30.** $(x-2y)^4$ **31.** $(x^2+y^2)^5$ **32.** $(\frac{1}{x}+y)^5$

33. $(a+2b)^6$ **34.** $(a^2+1)^5$ **35.** $(x^2+y^2)^6$ **36.** $(x^2-y^2)^6$

37. Find the 6<u>th</u> term of $(a+b)^8$.   **38.** Find the 5<u>th</u> term of $(a+2b)^6$.

39. Find the 4<u>th</u> term of $(x-2y)^{10}$.   **40.** Find the 6<u>th</u> term of $(x+2)^9$.

41. Find the 7<u>th</u> term of $(a+2b)^8$.   **42.** Find the 6<u>th</u> term of $(3x+2)^8$.

43. Find the 3<u>rd</u> term of $(3x-2)^9$.   **44.** Find the 5<u>th</u> term of $(\sqrt{x}+\sqrt{2})^6$.

45. Find the 4<u>th</u> term of $(\frac{1}{x}+\sqrt{x})^7$.   **46.** Find the 6<u>th</u> term of $(3a-\frac{b}{2})^{10}$.

47. Find the term with x^4 of $(x+3)^{12}$.   **48.** Find the term with x^3 of $(2x+1)^{12}$.

49. Find the term with x^2y^8 of $(4x-y)^{10}$.   **50.** Find the term with x^4y^6 of $(x^2+y^2)^5$.

51. The probability P of r successes in the n trials of an experiment is given by the formula $P(r)=_nC_r\,p^r(1-p)^{n-r}$, where p is the probability of a success on each trial. Find the probability of obtaining 5 heads on tossing a fair coin eight times.
Hint: Evaluate $P(5) = _8C_5(\frac{1}{2})^5(\frac{1}{2})^3$ in the Binomial Expansion $(\frac{1}{2}+\frac{1}{2})^8$.

CHAPTER 10   EXERCISES

Write the next four terms of each sequence.

1. 1, 2, 4, 7, 11, ····

2. −2, −4, −7, −11, −16, ·····

3. $\frac{1}{2}$, $\frac{3}{4}$, $\frac{6}{7}$, $\frac{10}{11}$, $\frac{15}{16}$, ····

Write the first five terms of each sequence.

4. $a_1 = 1$, $a_2 = 3$, $a_{n+2} = a_n + a_{n+1}$

5. $a_1 = 2$, $a_2 = 3$, $a_{n+2} = 2a_n + a_{n+1}$

6. $\{n^2 + 1\}$

7. $\left\{ \left(-\frac{1}{2}\right)^n \right\}$

8. $\{(-1)^{n+1} \cdot 2n\}$

9. $\left\{ \dfrac{1}{n(n+1)} \right\}$

10. $\left\{ \log \dfrac{n+1}{n} \right\}$

Find the sum of each sequence.

11 $\displaystyle\sum_{k=1}^{10} 5$

12. $\displaystyle\sum_{k=1}^{50} 10$

13. $\displaystyle\sum_{k=1}^{5} k^2$

14. $\displaystyle\sum_{k=0}^{5} (-1)^k \cdot (k^2 + 1)$

Find the specified term of each arithmetic sequence.

15 1, 4, 7, 10, 13, ..., t_{17}

16. 1, 5, 9, 13, 17, ..., t_{12}

17. 7, 5, 3, 1, −1, ..., t_{10}

18. 1, $\frac{1}{2}$, 0, $-\frac{1}{2}$, −1, ..., t_{12}

19. 1, $\frac{3}{2}$, 2, $\frac{5}{2}$, 3, ..., t_8

20. $3a$, $7a$, $11a$, ..., t_{15}

21. 2.1, 3.3, 4.5, 5.7, ..., t_9

22. $\log 2$, $\log 8$, $\log 32$, ..., t_{14}

23. $4\frac{1}{2}$, $3\frac{2}{3}$, $2\frac{5}{6}$, 2, ..., t_{13}

24. $t_2 = 10$, $t_5 = 43$, t_{16}

25. $t_2 = \log 8$, $t_4 = \log 128$, t_7

26. $t_3 = 9$, $t_8 = 29$, t_{40}

Find the arithmetic mean between each pair of numbers.

27. 4, 10

28. 9, 17

29. 3, −1

30. $\frac{3}{2}$, $\frac{5}{2}$

31. $\log 2$, $\log 32$

32. $7a$, $19a$

33. $4\frac{1}{2}$, $2\frac{5}{6}$

34. $-\frac{1}{2}$, $-\frac{3}{2}$

Insert five arithmetic means between the given two numbers.

35. 1 and 16

36. 7 and −5

37. $3a$ and $22a$

38. 1 and $\frac{9}{2}$

39. 2.1 and 9.3

40. $\log 2$ and $\log 1{,}458$

41. $\frac{1}{2}$ and $-\frac{7}{2}$

42. 1 and 4

Find the specified term of each geometric sequence.

43. 1, 2, 4, 8, 16, ..., t_{17}

44. 3, 9, 27, 81, 243, ..., t_{13}

45. $\frac{1}{2}$, $\frac{1}{4}$, $\frac{1}{8}$, $\frac{1}{16}$, ..., t_{13}

46. 9, −3, 1, $-\frac{1}{3}$, $\frac{1}{9}$, ..., t_{12}

47. 1, $-\frac{1}{2}$, $\frac{1}{4}$, $-\frac{1}{8}$, ..., t_{10}

48. $\sqrt{2}$, 1, $\frac{1}{\sqrt{2}}$, $\frac{1}{2}$, ..., t_{15}

49. 1, 2^{-2}, 2^{-4}, 2^{-6}, ..., t_{11}

50. $1\frac{1}{2}$, $4\frac{1}{2}$, $13\frac{1}{2}$, $40\frac{1}{2}$, ..., t_{12}

51. $a = 2$, $r = 3$, t_8

52. $\log 2$, $2(\log 2)^2$, $4(\log 2)^3$, ..., t_{10}

53. $a = \sqrt{2}$, $t_4 = \frac{1}{2}$, t_9

-----Continued-----

Find the geometric mean between each pair of numbers.

54. 4, 16   **55.** 9, 81   **56.** $-3, -\frac{1}{3}$ **57.** $\frac{1}{8}, \frac{1}{32}$

58. $\log 2, 4(\log 2)^2$ **59.** $1\frac{1}{2}, 13\frac{1}{2}$ **60.** $2^{-6}, 2^{-10}$ **61.** $\frac{3}{2}, \frac{2}{3}$

Find four geometric means between the given two numbers.

62. 1 and 32   **63.** 9 and $-\frac{1}{27}$ **64.** 4, $\frac{1}{8}$ **65.** $\sqrt{2}$ and $\frac{1}{4}$

66. 1 and 2^{-10} **67.** -27 and $-\frac{1}{27}$ **68.** x^2y^{-1} and x^7y^{-6} **69.** a^x and a^{11x}

70. Write a formula for the *n*th term: $2, 2+\sqrt{2}, 2+2\sqrt{2}, 2+3\sqrt{2}, \dots$
71. Write a formula for the *n*th term: $2, 2\sqrt{2}, 4, 4\sqrt{2}, 8, \dots$
72. How many multiples of 7 are there between 100 and 1,000 ?
73. How many multiples of 3 are there between 200 and 1,500 ?
74. How many multiples of 8 are there between 150 and 5,000 ?

Find the sum of each series.

75. $1 + 4 + 7 + 10 + \dots + t_{18}$.

76. $3a + 7a + 11a + \dots + t_{15}$.

77. $1 + \frac{1}{2} + 0 - \frac{1}{2} - 1 + \dots + t_{17}$.

78. $\sum_{n=1}^{30}(n+3)$.

79. $1 + 2 + 4 + 8 + \dots + t_8$.

80. $\frac{1}{2} + \frac{1}{4} + \frac{1}{8} + \frac{1}{16} + \dots + t_9$.

81. $9 - 3 + 1 - \frac{1}{3} + \dots + t_{12}$.

82. $t_1 = 1\frac{1}{2}, r = 3, n = 13$.

Find the sum of each infinite geometric series.

83. $9 - 3 + 1 - \frac{1}{3} + \dots$

84. $2 - \sqrt{2} + 1 - \frac{\sqrt{2}}{2} + \dots$

85. $2 - 2\sqrt{2} + 4 - 4\sqrt{2} + \dots$

86. $3 + \frac{3}{2} + \frac{3}{4} + \frac{3}{8} + \dots$

87. $\sum_{n=1}^{\infty}(-\frac{1}{2})^{n-1}$

88. $\sum_{n=1}^{\infty}(-\sqrt{2})^{n-1}$

Simplify: **89.** $\frac{10!}{6!}$.   **90.** $\frac{9!0!}{4!7!}$.   **91.** $\frac{(n+1)!}{(n-2)!(n-1)}$.

Prove: **92.** $0.\overline{13} = \frac{13}{99}$.   **93.** $0.1\overline{3} = \frac{12}{90}$.   **94.** $0.01\overline{3} = \frac{12}{900}$.

Find an expression for the *n*th term of each sequence.

95. $1, 6, 15, 28, 45, \dots$ **96.** $6, 20, 42, 72, 110, \dots$ **97.** $-3, 8, 25, 48, 77, \dots$

98. $21, 40, 63, 90, 121$ **99.** $2, 8, 24, 64, 160, \dots$ **100.** $2, 3, 5, 9, 17, 33, \dots$

101. $1, 5, 12, 22, 35, 51, \dots$ **102.** $5, 7, 11, 19, 35, \dots$ **103.** $15, 49, 99, 165, 247, \dots$

104. $4, 18, 48, 100, 180, \dots$ **105.** $2, 4, 7, 12, 21, \dots$

-----Continued-----

Prove the formula of each series.

106. $1 + 3 + 5 + \cdots + (2n - 1) = n^2$

107. $\dfrac{1}{1 \cdot 3} + \dfrac{1}{2 \cdot 4} + \dfrac{1}{3 \cdot 5} + \cdots + \dfrac{1}{n(n+2)} = \dfrac{n(3n+5)}{4(n+1)(n+2)}$

108. $1 + (1 + 2) + (1 + 2 + 3) + \cdots + (1 + 2 + 3 + \cdots + n) = \dfrac{n(n+1)(n+2)}{6}$

Use Mathematical Induction to prove the formula of each series.

109. $1 + 3 + 5 + \cdots + (2n - 1) = n^2$

110. $\dfrac{1}{1 \cdot 3} + \dfrac{1}{2 \cdot 4} + \dfrac{1}{3 \cdot 5} + \cdots + \dfrac{1}{n(n+2)} = \dfrac{n(3n+5)}{4(n+1)(n+2)}$

111. $1 + (1 + 2) + (1 + 2 + 3) + \cdots + (1 + 2 + 3 + \cdots + n) = \dfrac{n(n+1)(n+2)}{6}$

112. Expand $(x - 3)^4$ by Pascal's Triangle and Binomial Theorem.

113. Expand $(x^2 - 3)^4$ by Pascal's Triangle and Binomial Theorem.

114. Expand $(x - 2y)^6$ by Pascal's Triangle and Binomial Theorem.

115. Expand $(2x - y^2)^5$ by Pascal's Triangle and Binomial Theorem.

116. Expand $(2x - y)^7$ by Pascal's Triangle and Binomial Theorem.

117. Find 6th term of $(x + 2y)^9$.

118. Find 7th term of $(2x - y)^{10}$.

119. Find 11th term of $(x - 2y)^{10}$.

120. Find the term with x^6 in the expansion of $(2x + 3)^{10}$.

121. Find the term with x^{20} in the expansion of $(x^2 + y)^{15}$.

122. Find the term with $x^5 y^3$ in the expansion of $(2x + 3y)^8$.

123. Find the sum of multiples of 4 and 6 between 100 and 999.

124. If a, b, c are the lengths of three sides of a right triangle, c is the hypotenuse. a, b, c form an arithmetic sequence. Find $a : b : c$.

125. The probability (P) of r successes in the n trials of an experiment is given by the formula $P(r) = {}_n C_r \, p^r (1 - p)^{n-r}$, where p is the probability of a success on each trial. The probability of a basketball player getting a hit on the basket on free throws is $\frac{3}{5}$. Find the probability of the player of getting 5 hits on the basket during the next 8 free throws.

Hint: Evaluate $ {}_8 C_5 (\frac{3}{5})^5 (\frac{2}{5})^3 $ in the Binomial Expansion $(\frac{3}{5} + \frac{2}{5})^8$.

Space for Taking Notes

Answers

CHAPTER 10 – 1   Sequences and Number Patterns   page 354

1. 22, 29, 37, 46, 56   **2.** 26, 37, 50, 65, 82   **3.** 30, 42, 56, 72, 90   **4.** 27, 38, 51, 66, 83   **5.** 13, 21, 34, 55, 89

6. 28, 45, 73, 118, 191   **7.** 32, 64, 128, 256, 512   **8.** 29, 47, 76, 123, 199   **9.** 21, 34, 55, 89, 144

10. 36, 49, 64, 81, 100   **11.** 34, 45, 58, 73, 90   **12.** 12.5, 18, 24.5, 32, 40.5   **13.** 21.5, 28, 35.5, 44, 53.5

14. 6, 12, 19, 27, 36   **15.** $-24, -30, -37, -45, -54$ **16.** $-30, 35, -40, 45, -50$ **17.** $\frac{1}{6}, 7, \frac{1}{8}, 9, \frac{1}{10}$ **18.** $\frac{6}{7}, \frac{7}{8}, \frac{8}{9}, \frac{9}{10}, \frac{10}{11}$

19. $-\frac{1}{6}, 7, -\frac{1}{8}, 9, -\frac{1}{10}$ **20.** $-\frac{1}{12}, \frac{1}{14}, -\frac{1}{16}, \frac{1}{18}, -\frac{1}{20}$ **21.** 20   **22.** 300   **23.** 55   **24.** -55 **25.** 34   **26.** 44   **27.** -20

28. -14 **29.** $\frac{47}{60}$ **30.** $6x + 15$ **31.** 5, 7, 9, 11, 13   **32.** 5, 3, 1, -1, -3 **33.** $-3, -2, 0, 3, 7$

34. $x, x+d, x+2d, x+3d, x+4d$ **35.** 1, 2, 2, 4, 8   **36.** 1, 2, 3, 7, 16   **37.** $\sum_{k=1}^{30} k$ **38.** $\sum_{k=1}^{10} k^2$ **39.** $\sum_{k=1}^{10} \frac{1}{k}$ **40.** $\sum_{k=1}^{8} (-\frac{1}{2})^{k-1}$

41. $\sum_{k=1}^{10} \frac{1}{k(k+2)}$ **42.** $\sum_{k=1}^{n} [a + (k-1)d]$ **43.** 0, 3, 8, 15, 24   **44.** $\frac{1}{2}, \frac{1}{2}, \frac{3}{8}, \frac{1}{4}, \frac{5}{32}$ **45.** $\frac{1}{4}, \frac{2}{5}, \frac{1}{2}, \frac{4}{7}, \frac{5}{8}$ **46.** $-1, 4, -9, 16, -25$

47. $\frac{1}{2}, -\frac{1}{6}, \frac{1}{12}, -\frac{1}{20}, \frac{1}{30}$ **48.** 8:50   **49.** 15 games (Hint: $8+4+2+1=15$)

50. 36 handshakes (Hint: 0, 1, 3, 6, 10, 15, 21, 28, 36)   **51.** 2, 4, 7, 11, 16 pieces   **52.** 1   **53.** $-\log_{10} 11$

CHAPTER 10 – 2   Arithmetic Sequences   page 356

1. 19, 22, 25, 28, 31   **2.** 20, 23, 26, 29, 32   **3.** 22, 26, 30, 34, 38   **4.** 54, 62, 70, 78, 86   **5.** 11, 9, 7, 5, 3

6. 14, 11, 8, 5, 2   **7.** 21, 25, 29, 33, 37   **8.** 36, 32, 28, 24, 20   **9.** 28, 33, 38, 43, 48   **10.** 51, 58, 65, 72, 79

11. 4.0, 4.5, 5.0, 5.5, 6.0   **12.** 9.0, 10.5, 12, 13.5, 15   **13.** 3, 0, -3, -6, -9 **14.** $-23, -27, -31, -35, -39$

15. $-6, -3, 0, 3, 6$ **16.** $-2, 0, 2, 4, 6$ **17.** $\ln 5, 2\ln 5, 3\ln 5, 4\ln 5, 5\ln 5$ **18.** 0, 1, 2, 3, 4   **19.** $a_n = 5n - 3$

20. $a_n = -5n + 7$ **21.** $a_n = -5n + 3$ **22.** $a_n = -\frac{1}{2}n + 2\frac{1}{2}$ **23.** $a_n = \frac{10}{3}n + \frac{5}{3}$ **24.** $a_n = 2n\pi - 2\pi + 10$ **25.** $a_n = \sqrt{5}\, n$

26. $a_n = 5n - 2$ **27.** $a_n = 2n + 26$ **28.** Yes   **29.** No   **30.** Yes   **31.** No   **32.** No   **33.** Yes   **34.** Yes   **35.** $a_{12} = -17$

36. $a_{20} = 39\sqrt{3}$ **37.** $a_{30} = a + 29c$ **38.** $a_{25} = \ln 2^{24}$ **39.** $x = -4$ **40.** $x = 6.5$ **41.** $m = -7$ **42.** 1.5, 3, 4.5, 6, 7.5   **43.** No

44. 11 days   **45.** No   **46.** Yes   **47.** $S_{10} = 304$ feet (In science: $S = 16\,t^2$, $S_{10} = 16 \cdot 10^2 - 16 \cdot 9^2 = 16(100 - 81) = 304$)

CHAPTER 10 – 3   Arithmetic Series   page 358

1. $S_{50} = 1{,}275$ **2.** $S_{25} = 650$ **3.** $S_{25} = 625$ **4.** $S_n = n(n+1)$ **5.** $S_{30} = 780$ **6.** $S_n = n(n-4)$ **7.** $S_{40} = 3{,}160$

8. $S_n = n(2n-1)$ **9.** $S_{25} = 772.5$ **10.** $S_n = 0.3n(4n+3)$ **11.** $S_{28} = -848.4$ **12.** $S_n = -3n(0.4n - 1.1)$

13. $S_{20} = -1{,}100$ **14.** $S_n = n(-3n+5)$ **15.** $a_{20} = -37\frac{1}{2}$ **16.** $S_n = -\frac{1}{8}n(n-5)$ **17.** $S_{20} = 132\frac{1}{2}$ **18.** $S_n = \frac{1}{8}n(3n-7)$

19. $S_{30} = 2{,}325$ **20.** $S_n = \frac{5}{2}n(n+1)$ **21.** $S_{12} = 630$ **22.** $S_n = \frac{3}{2}n(3n-1)$ **23.** $S_{300} = 45{,}150$ **24.** $S_n = \frac{n}{2}(1+n)$

25. $S_{200} = 18{,}875$ **26.** $S_n - S_{49} = \frac{n}{2}(n+1) - 1225$ **27.** $S_{30} = 333\frac{3}{4}$ **28.** $S_n = \frac{n}{8}(3n-1)$ **29.** $S_{100} = -896\frac{3}{8}$

30. 2,600   **31.** $a = 3$, $d = 9$ **32.** 8,600 seats   **33.** $S_{15} = 3{,}600$ feet (Hint: In science $S_{15} = 16t^2 = 16(15)^2 = 3{,}600$ feet)

34. a. $a_{10} = 93.1$ meters   b. $S_{15} = 1102.5$ meters (Hint: In science $S_{15} = 4.9t^2 = 4.9(15)^2 = 1102.5$ meters)

CHAPTER 10 – 4   Geometric Sequences and Series   page 362

1. $S_{10} = 1{,}023$ **2.** $S_8 = 3{,}280$ **3.** $S_{15} = 21{,}846$ **4.** $S_{12} = -132{,}860$ **5.** $S_{12} = 0.4285712$ **6.** $S_{12} = 0.8571424$

7. $S_8 = 11\frac{13}{16}$ **8.** $S_{100} = 12(1 - 2^{-100}) \approx 12$ **9.** $S_8 = 0.077777777$ **10.** $S_{10} = \frac{31}{16}(\sqrt{2} + 1)$ **11.** $S_n = \frac{2}{3}[1 - (-\frac{1}{2})^n]$

12. $S_n = \frac{3}{4}(3^n - 1)$ **13.** $S_{12} = 4{,}095$ **14.** $S_{12} = -1{,}365$ **15.** $a_0 + S_{12} = -1365\frac{1}{2}$ **16.** $S_{12} = 682\frac{1}{2}$ **17.** $S_5 = 3\frac{49}{54}$

18. $S_n = 1 - (\frac{1}{2})^n$ **19.** $S_n = -3(1 - 3^n)$ **20.** $S_8 = 0.033333333$ **21.** $S_\infty = 32$ **22.** $S_\infty = \frac{2}{3}$ **23.** $S_\infty = 1\frac{1}{9}$

24. $S_\infty = 1\frac{3}{5}$ **25.** $S_\infty = 30\frac{3}{8}$ **26.** $S_\infty = 2\frac{1}{4}$ **27.** $S_\infty = 1$ **28.** $S_\infty = 2$ **29.** No sum (Hint: $r = \frac{5}{4} > 1$)   **30.** $S_\infty = \frac{1}{18}$

31. Two answers: $S_5 = 84 + 36\sqrt{2}$ when $r = \sqrt{2}$ and $S_5 = 84 - 36\sqrt{2}$ when $r = -\sqrt{2}$ **32.** $S_n = (a^n - b^n)/(a-b)$

33. 4, 6, 9   **34.** $a:b:c = 1:1:1$ **35.** $a:b:c = 1:1:1$ or $a:b:c = -2:1:4$ **36.** $S_\infty = \frac{4}{33}$ **37.** $S_\infty = \frac{11}{90}$

38. For the first bounce it goes 40 feet down and 20 feet up, a total of 60 feet and so on: 60, 30, 15, $\frac{15}{2}, \frac{15}{4}, \frac{15}{8}, \ldots$

    a) $S_6 = 118\frac{1}{8}$ feet   b) $S_\infty = 120$ feet

39. $A = 3000(1 + 0.1)^{20} + 3000(1 + 0.1)^{19} + \cdots + 3000(1 + 0.1)$

    $= 3000(1 + 0.1) + 3000(1 + 0.1)^2 + \cdots + 3000(1 + 0.1)^{20}$

    $A = \dfrac{a_1(1 - r^{20})}{1 - r} = \dfrac{3000(1 + 0.1)[1 - (1 + 0.1)^{20}]}{1 - (1 + 0.1)} = \$189{,}007.50$

40. $A = P(1 + i)^n + p(1 + i)^{n-1} + \cdots + P(1 + i)$

    $= P[(1 + i) + (1 + i)^2 + \cdots + (1 + i)^n]$

    $A = P \cdot \dfrac{(1 + i)[1 - (1 + i)^n]}{1 - (1 + i)} = P(1 + i)\dfrac{(1 + i)^n - 1}{i}$

CHAPTER 10 – 5   Finding the nth Term of a Sequence   page 366

1. $a_n = n - 2$ **2.** $a_n = 3n - 2$ **3.** $a_n = 3n - 5$ **4.** $a_n = 5n - 2$ **5.** $a_n = 2^{n+1}$ **6.** $a_n = (\frac{1}{2})^n$ **7.** $a_n = (-\frac{1}{2})^n$

8. $a_n = (-1)^{n-1}(\frac{1}{2})^{n-2}$ **9.** $a_n = 3^n$ **10.** $a_n = n^2$ **11.** $a_n = 3^{n-1}$ **12.** $a_n = n(n+2)$ **13.** $a_n = n(2n-1)$ **14.** $a_n = 5(2^{n-1})$

15. $a_n = 2 \cdot 3^{n-1}$ **16.** $a_n = n(n+4)$ **17.** $a_n = n \cdot 3^n$ **18.** $a_n = 2^{n+2}$ **19.** $a_n = (n+1)(n+3)$ **20.** $a_n = \frac{1}{2}n(n+3)$

21. $\frac{1}{2}n(n+5)$ **22.** $a_n = 2n(n+1)$ **23.** $a_n = 2n(3n-1)$ **24.** $a_n = \frac{1}{2}(n^2 - n + 2)$ **25.** $a_n = \frac{1}{2}n(n+1)$ **26.** $a_n = 2^n - 1$

27. $a_n = n^2 - 2$ **28.** $a_n = \frac{1}{2}(3^n - 1)$ **29.** $a_n = \frac{1}{2}(n^2 + n + 2)$ **30.** $a_n = n^3 + n - 1$ (see Example 6, page 365)

31. $a_n = \frac{1}{6}n(2n^2 - 3n + 7)$ (see Example 6, page 365)   **32.** $a_n = 2^{n-1} + n - 1$ (see example 6, page 365)

33. $a_n = 3^{n-1} + n$ (see Example 6, page 365)   **34.** $a_n = \frac{n}{n+1}$ **35.** $a_n = \frac{2n-1}{n+2}$ **36.** $a_n = \frac{n+1}{n}$ **37.** $a_n = \frac{n}{(n+1)(n+2)}$

38. $a_n = (-1)^n(\frac{n}{n+1})$ **39.** $a_n = (\frac{3}{4})^n$ **40.** $a_n = \sqrt{n} + \sqrt{n+1}$ **41.** $a_{10} = 47$ **42.** $a_n = \frac{b+(n-1)(a-b)}{ab}$ **43.** 41,400

CHAPTER 10 – 6   Finding the Sums of Special Series   page 368

1 ~ 7 proofs are left to the students   **8. a.** $S_{10} = 385$ **b.** $S_{15} = 1,360$ **c.** $S_{18} = \frac{18}{19}$ **d.** $S_{10} = 1,330$

CHAPTER 10 – 7   Mathematical Induction   page 372

1. (I) $n = 1$; $4 \cdot 1 - 3 = 1$ and $1(2 \cdot 1 - 1) = 1$. It is true when $n = 1$.

(II) Assume that it is true for $n = k$, we have $1 + 5 + 9 + \cdots + (4k - 3) = k(2k - 1)$.

Then for $n = k + 1$;
$$1 + 5 + 9 + \cdots + (4k - 3) + [4(k+1) - 3] = k(2k-1) + (4k+1) = 2k^2 - k + 4k + 1 = 2k^2 + 3k + 1$$
$$= (k+1)(2k+1) = (k+1)[2(k+1) - 1]$$

It is true for $n = k + 1$. Therefore, it is true for all positive integers of n.

2. (I) $n = 1$; $2 \cdot 1 - 1 = 1$ and $1^2 = 1$. It is true when $n = 1$.

(II) Assume that it is true for $n = k$, we have $1 + 3 + 5 + \cdots + (2k - 1) = k^2$.

Then for $n = k + 1$; $1 + 3 + 5 + \cdots + (2k - 1) + [2(k+1) - 1] = k^2 + (2k + 1) = (k+1)^2$

It is true for $n = k + 1$. Therefore, it is true for all positive integers of n.

3 ~ 5 Proofs are left to the students   **6.** See Example 1, page 369   **7 ~ 18** Proofs are left to the students

19. (I) $n = 1$; $1(1 + 1) = 2$ and $\frac{1}{3} \cdot 1 \cdot (1 + 1) \cdot (1 + 2) = 2$. It is true when $n = 1$.

(II) Assume that it is true for $n = k$, we have $1 \cdot 2 + 2 \cdot 3 + 3 \cdot 4 + \cdots + k(k+1) = \frac{1}{3}k(k+1)(k+2)$.

Then for $n = k + 1$;
$$1 \cdot 2 + 2 \cdot 3 + 3 \cdot 4 + \cdots + k(k+1) + (k+1)(k+2) = \frac{1}{3}k(k+1)(k+2) + (k+1)(k+2)$$
$$= \frac{1}{3}(k+1)(k+2)(k+3)$$

It is true for $n = k + 1$. Therefore, it is true for all positive integers of n.

20 ~ 24 Proofs are left to the students   **25.** Follow Example 7, page 371

26. (I) $n = 3$; $180°(3 - 2) = 180°$. It is true when $n = 3$.

$n = 4$; $180°(4 - 2) = 360°$. $n = 5$; $180°(5 - 2) = 540°$

The sum of the interior angles increases by $180°$ when sides increase by 1.

(II) Assume that it is true for $n = k$, we have $180°(k - 2)$.

Then for $n = k + 1$; $180°(k - 2) + 180° = 180°(k - 1) = 180°[(k+1) - 2]$

It is true for $n = k + 1$. Therefore, it is true for all positive integers of n.

27. (I) $n = 4$; $\frac{1}{2} \cdot 4 \cdot (4 - 3) = 2$ diagonals. It is true when $n = 4$.

$n = 5$; $\frac{1}{2} \cdot 5 \cdot (5 - 3) = 5$ diagonals. $n = 6$; $\frac{1}{2} \cdot 6 \cdot (6 - 3) = 9$ diagonals.

A convex polygon with $(n + 1)$ sides consists of all the diagonals of n sides plus $(n - 1)$ diagonals.

(II) Assume that it is true for $n = k$, we have $\frac{1}{2}k(k - 3)$ diagonals.

Then for $n = k + 1$; $\frac{1}{2}k(k - 3) + (k - 1) = \frac{1}{2}k^2 - \frac{3}{2}k + k - 1 = \frac{1}{2}k^2 - \frac{1}{2}k - 1$
$$= \frac{1}{2}(k^2 - k - 2) = \frac{1}{2}(k+1)(k-2)$$

It is true for $n = k + 1$. Therefore, It is true for all positive integers of n.

Answers

CHAPTER 10 – 8   Pascal's Triangle and Binomial Theorem   page 376

1. $(x+1)^4 = x^4 + 4x^3 + 6x^2 + 4x + 1$ **2.** $(x+1)^5 = x^5 + 5x^4 + 10x^3 + 10x^2 + 5x + 1$

3. $(x-1)^4 = x^4 - 4x^3 + 6x^2 - 4x + 1$ **4.** $(x-1)^5 = x^5 - 5x^4 + 10x^3 - 10x^2 + 5x - 1$

5. $(x+1)^7 = x^7 + 7x^6 + 21x^5 + 35x^4 + 35x^3 + 21x^2 + 7x + 1$

6. $(x-1)^7 = x^7 - 7x^6 + 21x^5 - 35x^4 + 35x^3 - 21x^2 + 7x - 1$

7. $(x-2)^5 = x^5 - 10x^4 + 40x^3 - 80x^2 + 80x - 32$ **8.** $(x+2y)^4 = x^4 + 8x^3y + 24x^2y^2 + 32xy^3 + 16y^4$

9. $(3x-y)^5 = 243x^5 - 405x^4y + 270x^3y^2 - 90x^2y^3 + 15xy^4 - y^5$ **10.** $(a+3)^3 = a^3 + 9a^2 + 27a + 27$

11. $(a^2+1)^5 = a^{10} + 5a^8 + 10a^6 + 10a^4 + 5a^2 + 1$ **12.** $(x^2+5)^4 = x^8 + 20x^6 + 150x^4 + 500x^2 + 625$

13. 1 **14.** 5 **15.** 10 **16.** 10 **17.** 5 **18.** 15,504 **19.** 4,950 **20.** 4,950 **21.** 45 **22.** 120 **23.** 50 **24.** 1,770

25. $(x+1)^3 = x^3 + 3x^2 + 3x + 1$ **26.** $(x-2)^5 = x^5 - 10x^4 + 40x^3 - 80x^2 + 80x - 32$

27. $(x+2)^7 = x^7 + 14x^6 + 84x^5 + 280x^4 + 560x^3 + 672x^2 + 448x + 128$

28. $(2x-3)^4 = 16x^4 - 96x^3 + 216x^2 - 216x + 81$ **29.** $(2x+3)^5 = 32x^5 + 240x^4 + 720x^3 + 1080x^2 + 810x + 243$

30. $(x-2y)^4 = x^4 - 8x^3y + 24x^2y^2 - 32xy^3 + 16y^4$

31. $(x^2+y^2)^5 = x^{10} + 5x^8y^2 + 10x^6y^4 + 10x^4y^6 + 5x^2y^8 + y^{10}$

32. $\left(\dfrac{1}{x}+5\right)^5 = \dfrac{1}{x^5} + \dfrac{5y}{x^4} + \dfrac{10y^2}{x^3} + \dfrac{10y^3}{x^2} + \dfrac{5y^4}{x} + y^5$

33. $(a+2b)^6 = a^6 + 12a^5b + 60a^4b^2 + 160a^3b^3 + 240a^2b^4 + 192ab^5 + 64b^6$

34. $(a^2+1)^5 = a^{10} + 5a^8 + 10a^6 + 10a^4 + 5a^2 + 1$

35. $(x^2+y^2)^6 = x^{12} + 6x^{10}y^2 + 15x^8y^4 + 20x^6y^6 + 15x^4y^8 + 6x^2y^{10} + y^{12}$

36. $(x^2-y^2)^6 = x^{12} - 6x^{10}y^2 + 15x^8y^4 - 20x^6y^6 + 15x^4y^8 - 6x^2y^{10} + y^{12}$

37. $56a^3b^5$ **38.** $240a^2b^4$ **39.** $-960x^7y^3$ **40.** $4032x^4$ **41.** $1792a^2b^6$ **42.** $48384x^3$ **43.** $314928x^7$ **44.** $60x$

45. $35x^{-5/2}$ **46.** $-\dfrac{15309}{8}a^5b^5$ **47.** $3247695x^4$ **48.** $1760x^3$ **49.** $720x^2y^8$ **50.** $10x^4y^6$ **51.** 0.219

CHAPTER 10 EXERCISES   page 377 ~ 379

1. 16, 22, 29, 37 **2.** −22, −29, −37, −46 **3.** $\frac{21}{22}, \frac{28}{29}, \frac{36}{37}, \frac{45}{46}$ **4.** 1, 3, 4, 7, 11 **5.** 2, 3, 7, 13, 27 **6.** 2, 5, 10, 17, 26

7. $-\frac{1}{2}, \frac{1}{4}, -\frac{1}{8}, \frac{1}{16}, -\frac{1}{32}$ **8.** 2, −4, 6, −8, 10 **9.** $\dfrac{1}{1\cdot2}, \dfrac{1}{2\cdot3}, \dfrac{1}{3\cdot4}, \dfrac{1}{4\cdot5}, \dfrac{1}{5\cdot6}$ **10.** $\ln 2, \ln\frac{3}{2}, \ln\frac{4}{3}, \ln\frac{5}{4}, \ln\frac{6}{5}$

11. $S_{10} = 50$ **12.** $S_{50} = 500$ **13.** $S_5 = 55$ **14.** −15 **15.** 49 **16.** 45 **17.** −11 **18.** $-\frac{9}{2}$ **19.** $\frac{9}{2}$ **20.** $59a$

21. $t_9 = 11.7$ **22.** $27\log 2$ **23.** $t_{13} = -5\frac{1}{2}$ **24.** $t_{16} = 164$ **25.** $t_7 = \log 8{,}192$ **26.** $t_{40} = 157$ **27.** $m=7$ **28.** $m=13$

29. $m=1$ **30.** $m=2$ **31.** $m=\log 8$ **32.** $m=13a$ **33.** $m=3\frac{2}{3}$ **34.** $m=-1$ **35.** $1, 3\frac{1}{2}, 6, 8\frac{1}{2}, 11, 13\frac{1}{2}, 16$

36. 7, 5, 3, 1, −1, −3, −5 **37.** $3a, 6\frac{1}{6}a, 9\frac{1}{3}a, 12\frac{1}{2}a, 15\frac{2}{3}a, 18\frac{5}{6}a, 22a$ **38.** $1, \frac{19}{12}, \frac{13}{6}, \frac{11}{4}, \frac{10}{3}, \frac{47}{12}, \frac{9}{2}$

39. 2.1, 3.3, 4.5, 5.7, 6.9, 8.1, 9.3   **40.** $\log 2, \log 6, \log 18, \log 54, \log 162, \log 486, \log 1{,}458$

41. $\frac{1}{2}, -\frac{1}{6}, -\frac{5}{6}, -\frac{3}{2}, -\frac{13}{6}, -\frac{17}{6}, -\frac{7}{2}$ **42.** $1, \frac{3}{2}, 2, \frac{5}{2}, 3, \frac{7}{2}, 4$ **43.** $t_{17} = 2^{16}$ **44.** $t_{13} = 3^{13}$ **45.** $t_{13} = \frac{1}{2^{13}}$ **46.** $t_{12} = -\frac{1}{3^9}$

47. $t_{10} = -\frac{1}{2^9}$ **48.** $t_{15} = \frac{\sqrt{2}}{128}$ **49.** $t_{11} = 2^{-20}$ **50.** $t_{12} = \frac{3^{12}}{2}$ **51.** $t_8 = 4{,}374$ **52.** $t_{10} = 2^9(\log 2)^{10}$ **53.** $t_9 = \frac{\sqrt{2}}{16}$

54. $m = 8$ or -8 **55.** $m = 27$ or -27 **56.** $m = 1$ or -1 **57.** $m = \frac{1}{16}$ or $-\frac{1}{16}$ **58.** $m = 2(\log 2)^{\frac{3}{2}}$ or $-2(\log 2)^{\frac{3}{2}}$

59. $m = 4\frac{1}{2}$ or $-4\frac{1}{2}$ **60.** $m = 2^{-8}$ or -2^{-8} **61.** $m = 1$ or -1 **62.** 1, 2, 4, 8, 16, 32 **63.** 9, −3, 1, $-\frac{1}{3}$, $\frac{1}{9}$, $-\frac{1}{27}$

64. 4, 2, 1, $\frac{1}{2}$, $\frac{1}{4}$, $\frac{1}{8}$ **65.** $\sqrt{2}, 1, \frac{\sqrt{2}}{2}, \frac{1}{2}, \frac{\sqrt{2}}{4}, \frac{1}{4}$ **66.** $1, 2^{-2}, 2^{-4}, 2^{-6}, 2^{-8}, 2^{-10}$

67. $-27, -(27)^{\frac{3}{5}}, -(27)^{\frac{1}{5}}, -(27)^{-\frac{1}{5}}, -(27)^{-\frac{3}{5}}, -\frac{1}{27}$ **68.** $x^2y^{-1}, x^3y^{-2}, x^4y^{-3}, x^5y^{-4}, x^6y^{-5}, x^7y^{-6}$.

69. $a^x, a^{3x}, a^{5x}, a^{7x}, a^{9x}, a^{11x}$ **70.** $t_n = 2 + (n-1)\sqrt{2}$ **71.** $t_n = 2(\sqrt{2})^{n-1}$ **72.** 128 multiples of 7

73. 434 multiples of 3 **74.** 607 multiples of 8 **75.** $S_{18} = 477$ **76.** $S_{15} = 465a$ **77.** $S_{17} = -51$

78. $S_{30} = 555$ **79.** $S_8 = 255$ **80.** $S_9 = 1 - (\frac{1}{2})^9$ **81.** $S_{12} = \frac{27}{4}[1 - (\frac{1}{3})^{12}]$ **82.** $S_{13} = \frac{3}{4}(3^{13} - 1)$ **83.** $S_\infty = 6\frac{3}{4}$

84. $S_\infty = 4 - 2\sqrt{2}$ **85.** No sum (unlimited sum) **86.** $S_\infty = 6$ **87.** $S_\infty = \frac{2}{3}$ **88.** No sum (unlimited sum)

89. 5,040 **90.** 3 **91.** $(n+1)n$ **92 ~ 94** Proofs are left to the students.

CHAPTER 10 EXERXISES page 377 ~ 379 (Continued)

95. $a_n = n(2n-1)$ **96.** $a_n = 2n(2n+1)$ **97.** $a_n = (n+2)(3n-4)$ **98.** $a_n = (n+6)(2n+1)$ **99.** $n(2^n)$

100. $a_n = 2^{n-1}+1$ **101.** $a_n = \frac{n(3n-1)}{2}$ **102.** $2^n + 3$ **103.** $a_n = (2n+3)(4n-1)$

104. $a_n = n^3 + 2n^2 + n$ (See Example 6, page 378)   **105.** $a_n = a^{n-1} + n$ (See Example 6, page 378)

106 ~ 111 Proofs are left to the students.   **112.** $(x-3)^4 = x^4 - 12x^3 + 54x^2 - 108x + 81$

113. $(x^2 - 3)^4 = x^8 - 12x^6 + 54x^4 - 108x^2 + 81$

114. $(x-2y)^6 = x^6 - 12x^5 y + 60x^4 y^2 - 160x^3 y^3 + 240x^2 y^4 - 192xy^5 + 64y^6$

115. $(2x - y^2)^5 = 32x^5 - 80x^4 y^2 + 80x^3 y^4 - 40x^2 y^6 + 10xy^8 - y^{10}$

116. $(2x - 7)^7 = 128x^7 - 448x^6 y + 672x^5 y^2 - 560x^4 y^3 + 280x^3 y^4 - 84x^2 y^5 + 14xy^6 - y^4$

117. $4032x^4 y^5$ **118.** $3360x^4 y^6$ **119.** $1024y^{10}$ **120.** $1088640x^6$ **121.** $3003x^{20} y^5$ **122.** $48348x^5 y^3$

123. $S = 41,400$ **124.** $a:b:c = 3:4:5$ Hint: $b = \frac{1}{2}(a+b)$ and $c^2 = a^2 + b^2$)   **125.** 0.279

A mother asks her 6-year old son about his new eyeglasses.
Mother: Can you see very well with your new eyeglasses
          when you are in the class ?
   Son: Yes. I can see very clear all of my teachers' writing
          on the blackboard.
Mother: Will you get better grades ?
   Son: No. I still have same brain.

A Boy said to his dentist.
    Boy: Why are you charging me $300 ?
          Last time, you charged me only $100.
Dentist: Two patients in the waiting room ran
          away when they heard you screaming.

<u>Finding the Specific Term of a Sequence by a Graphing Calculator</u>

Example 1: Finding the 10th term of the sequence: $1, 3, 5, 7, 9, \cdots$.
        Solution:

     Method 1

         The formula for the nth term is $a_n = 1 + (n-1)2$

         Enter this formula as $y = 1 + (x-1)2$:

            Press: **Y= 1** $\to$ **+** $\to$ **(** $\to$ $\boxed{\textbf{X,T,}\theta\textbf{,}n}$ $\to$ **−** $\to$ **1** $\to$ **)** $\to$ **2**

         Set up the table to represent x and y:

            Press: **2nd** $\to$ **TBLSET** $\to$ **TblStart = 1** $\to$ **ENTER** $\to$ Δ **Tbl = 1** $\to$ **ENTER**

         To view the table:

            Press: **2nd** $\to$ **TABLE**

         Move the cursor down to the desired term

         The screen shows: $Y_1 = 19$ at $X = 10$

X	Y₁
4	7
5	9
6	11
7	13
8	15
9	17
10	19

X=10

     Method 2

         The formula for the nth term is $a_n = 1 + (n-1)2$

         Open the "seq (" command

            Press: **2nd** $\to$ **LIST** $\to$ **OPS** $\to$ **5:seq (** $\to$ **ENTER**

         Enter the formula as $y = 1 + (N-1)2$

            Press: **seq (** $\to$ **1** $\to$ **+** $\to$ **(** $\to$ **ALPHA** $\to$ **N** $\to$ **−** $\to$ **1** $\to$ **)** $\to$ **2** $\to$ **,**

         Enter N with starting value 1, ending value 10, term increment 1

            Press: **ALPHA** $\to$ **N** $\to$ **,** $\to$ **1** $\to$ **,** $\to$ **10** $\to$ **,** $\to$ **1** $\to$ **)** $\to$ **ENTER**

         Move the cursor to the right to see the 10th term is 19

         The screen shows:

```
seq(1+(N-1)2,N,1
,10,1)
...11 13 15 17 19}
```

Example 2: Finding the 8th term of the sequence: $2, 6, 18, 54, \cdots$
        Solution:

         The formula for the nth term is $a_n = 2 \cdot 3^{n-1}$

         Open the "seq (" command

            Press: **2nd** $\to$ **LIST** $\to$ **OPS** $\to$ **5:seq (** $\to$ **ENTER**

         Enter the formula as $Y = 2 \times 3^{N-1}$

            Press: **seq (** $\to$ **2** $\to$ **×** $\to$ **3** $\to$ **∧** $\to$ **(** $\to$ **ALPHA** $\to$ **N** $\to$ **−** $\to$ **1** $\to$ **)** $\to$ **,**

         Enter N with starting value 1, ending value 8, term increment 1

            Press: **ALPHA** $\to$ **N** $\to$ **,** $\to$ **1** $\to$ **,** $\to$ **8** $\to$ **,** $\to$ **1** $\to$ **)** $\to$ **ENTER**

         Move the cursor to the right to see the 8th term is 4374.

         The screen shows:

```
seq(2*3^(N-1),N,
1,8,1)
... 486 1458 4374}
```

Graphing a Sequence
by a Graphing Calculator

Example: Graphing the sequence: 1, 3, 5, 7, 9, ⋯⋯ .

Solution:

The formula for the nth term is $a_n = 2n - 1$

Highlight the "Seq" and "Dot" in MODE command:

Press: **MODE** → **Seq** → **ENTER** → **Dot** → **ENTER**

Enter the formula as $u(n) = 2n - 1$:

Press: **Y₁=** → $u(n) =$ → **2** → $\boxed{\text{X,T,θ,}n}$ → **–** → **1**

Adjust the viewing window, press: **WINDOW**

nMin = 0	Xmax = 0	Ymin = –5
nMax =10	Xmin = 10	Ymax = 25

Graph the sequence, press: **GRAPH**

Find the value of a specific term, Press: **TRACE**

Move the cursor to the right or left

The screen shows:

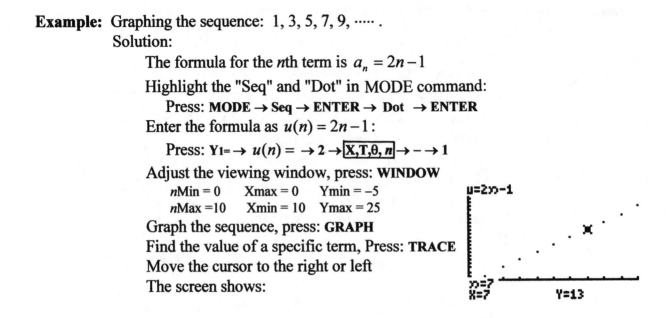

Finding the Sum of a Series
by a Graphing Calculator

Example: Finding the sum of the series: $\displaystyle\sum_{n=1}^{20}(2n-1) = 1 + 3 + 5 + 7 + 9 + \cdots\cdots + a_{20}$.

Solution:

Enter the "sum" and the "sequence" functions:

Press: **2nd** → **LIST** → **MATH** → **5:sum(** → **ENTER**

Press: **2nd** → **LIST** → **OPS** → **5:seq(** → **ENTER**

Enter the expression of the series, variable n, first term 1, last term 20, step size 1.

Press: **2** → **ALPHA** → **N** → **–** → **1** → **ALPHA**
→ **N** →**,**→ **1** →**,** → **20** → **,** → **1** → **)**

Find the sum, press: **ENTER**

The screen shows: (The sum = 400)

```
sum(seq(2N-1,N,1
,20,1)
              400
```

Finding the Specific Term of a Recursive Sequence by a Graphing Calculator

Example: Finding the 6th term of the recursive sequence: $a_1 = 1$, $a_n = 2\,a_{n-1}$.

Solution:

The previous term a_{n-1} is represented by $u(n-1)$.

Highlight the "Seq" and "Dot" in MODE command

Press: **MODE** → **Seq** → **ENTER** → **Dot** → **ENTER**

Enter $a_n = 2\,a_{n-1}$ using $u(n) = 2 \times u(n-1)$ and

Press: **Y1=** → $u(n)$ = → **2** → **×** → **2nd** → **u** → **(** → **X,T,θ,n** → **–** → **1** → **)**

Enter an initial value $a_1 = 1$ using $u(n\text{Min}) = 1$

Press: $u(n\text{Min}) = 1$

Adjust the viewing window, press: **WINDOW**

$n\text{Min} = 1$	PlotStart = 1 (first term)	Xmin = 1	Ymin = –15
$n\text{Max} = 10$	PlotStep = 1 (increment of term)	Xmax = 10	Ymax = 50

Graph the sequence, press: **GRAPH**

Find the value of a specific term, Press: **TRACE**

Move the cursor to the right or left | Examine the terms with a table
The screen shows: | Press: **2nd** → **TBLSET** → **TblStart = 2**
                                                          | → **ΔTbl =1** → **2nd** → **TABLE**
                                                          | The screen shows:

u=2*u(n-1)

n=6
X=6   Y=32

n	$u(n)$
2	2
3	4
4	8
5	16
6	32
7	64
8	128

$n=6$

Evaluate a Combination by a Graphing Calculator

Example: Evaluate $_{65}C_{15}$.

Solution:

Press: **65** → **MATH** → **PRB**
→ **3: nCr** → **15** → **ENTER**
The screen shows:

```
65 nCr 15
   2.073746998ε14
```

($_{65}C_{15} = 2.073746998 \times 10^{14}$)

Evaluate a factorial by a Graphing Calculator

Example: Evaluate 7!.

Solution:

Press: **7** → **MATH** → **PRB**
→ **4: !** → **ENTER** → **ENTER**
The screen shows:

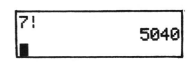

```
7!
          5040
```

<u>**Space for Taking Notes**</u>

CHAPTER ELEVEN

Probability and Statistics

11-1 Basic Counting Principles

1) Principle of Multiplication

If we select one from *m* ways, then we select another one from *n* ways; then all the possible selections are:

$$m \times n \text{ different ways}$$

2) Principle of Addition

If we select one from *m* ways, then **mutually exclusive**, we select another one from *n* ways; then all the possible selections are:

$$m + n \text{ different ways}$$

Examples

1. How many routes can be selected from City A to City D ?

Solution:

$$3 \times 2 \times 5 = 30 \text{ routes. Ans.}$$

2. How many different ways can be selected from A to D ?

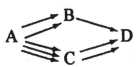

Solution:

$$A \rightarrow B \rightarrow D : \ 2 \times 1 = 2$$
$$A \rightarrow C \rightarrow D : \ 3 \times 2 = 6$$
$$A \longrightarrow D : \ 2 + 6 = 8 \text{ ways.}$$
$$\text{Ans.}$$

3. How many different selections are possible to assign 4-letter codes from 26 alphabets ?

Solution:

$$26 \times 26 \times 26 \times 26 = 456,976 \text{ codes.}$$
$$\text{Ans.}$$

4. How many different ways can a 10-question true-false test be answered if each question must be answered ?

Solution:

Each question has two selections:

$$2 \times 2 \times 2 \times 2 \times 2 \times 2 \times 2 \times 2 \times 2 \times 2 = 1,024$$
$$\text{Or: } 2^{10} = 1,024 \text{ ways. Ans.}$$

5. How many license plates of 6 symbols can be made using letters for the first 2 symbols and digits for the remaining symbols ?

Solution:

$$26 \times 26 \times 10 \times 10 \times 10 \times 10$$
$$= 6,760,000 \text{ license plates. Ans.}$$

6. How many positive even integers less than 1,000 can be made using the digits 5, 6, 7 and 8 ?

Solution: 1-digit even integers: $\qquad 2$
$\qquad$ 2-digit even integers: $\quad 4 \times 2 = \ 8$
$\qquad$ 3-digit even integers: $\underline{4 \times 4 \times 2 = 32}$
$\qquad\qquad\qquad\qquad$ Total $\quad 42$

Ans: Total 42 even integers can be made.

EXERCISES

1. There are 4 routes from City A to City B, 3 routes from City B to City C, and 6 routes from City C to City D. How many different routes can be selected from City A to City D ?

2. There are 3 routes from City A to City B, and 4 routes from City B to City D. Mutually exclusive, there are 5 routes from City A to City C, and 2 routes from City C to City D. How many different routes can be selected from City A to City D ?

3. How many different selections are possible to assign 5-letter codes from 26 alphabets ?

4. How many different selections are possible to assign 5-letter codes from 26 alphabets, each code with no letter repeated ?

5. How many different ways can a 8-question true-false test be answered if each question must be answered ?

6. How many license plates of 7 symbols can be made, using digit for the first symbol, letters for next three symbols, and digits for the remaining symbols ?

7. How many telephone numbers of 7 digits can be assigned within each area code by the AT & T Company if each telephone number cannot begin with 0 or 1 ?

8. How many positive odd integers less than 10,000 can be made using the digits 3, 4, 5, 6, 7 ?

9. A club consists of 30 boys and 25 girls. They want to select a girl as president, a boy as vice president, and a treasurer who may be of either sex. How many choices are possible ?

10. How many four-digit positive integers can be formed if zero is not allowed in the ones place and repetitions of digits are not allowed ?

11. How many 5-digit passwords can be formed for an office security system, using digits from 1 to 9 if the digits can be used more than once in any password ?

12. Five cards are drawn at random from a 52-card bridge deck. How many possibilities are there for this hand if three are black and two are red ?

13. Event 1 can occur in 6 ways, Event 2 can occur in 5 ways, and Event 3 can occur in 4 ways. How many different ways all of the events can occur ?

14. How many different ways can a multiple-choice test with 10 questions be answered if each of the question has 4 possible answers and all questions must be answered ?

15. Eight different books are to be arranged on the shelf of a bookstore. How many arrangements are possible ?

16. In a computer store, you want to buy one of the four monitors, one of the three keyboards, and one of the six computers. If all of the choices are compatible, how many possible choices you have ?

17. Four couples have reserved seats in one row for a concert. How many different ways can they be seated if two members of each couple must sit together ?

18. There are 10 empty seats in a row in a conference rooms. 5 teachers want to sit together in the row. How many different ways can they be seated ?

19. How many positive factors of 540 are there ?

20. How many 3-digit integers are multiples of 5 ?

21. How many 4-digit positive integers can be made using the digits 0, 1, 2, 3, 4, 5 and repetitions of digits are not allowed ?

11-2 Permutations and Combinations

Permutation: An arrangement of elements without repetitions in a definite order.

The number of permutations of the letters a, b, c taken 3 letters at a time:

abc, acb, bac, bca, cab, cba $\rightarrow$ total 6 permutations. $3! = 3 \cdot 2 \cdot 1 = 6$

The number of permutations of the letters a, b, c, d taken 3 letters at a time:

abc, acb, bac, bca, cab, cba

bcd, bdc, cbd, cdb, dbc, dcb

cda, cad, dca, dac, acd, adc

dab, dba, adb, abd, bda, bad $\rightarrow$ total 24 permutations. $\dfrac{4!}{(4-3)!} = \dfrac{4 \cdot 3 \cdot 2 \cdot 1!}{1!} = 24$

The number of permutations of *n* different elements taken *n* elements at a time is :

$$_nP_n = n \cdot (n-1) \cdot (n-2) \cdot (n-3) \cdots 3 \cdot 2 \cdot 1$$

Formula: $_nP_n = n!$ **Example:** $_{10}P_{10} = \dfrac{10!}{(10-10)!} = \dfrac{10!}{1!} = 10! = 3{,}628{,}800$

The number of permutations of *n* different elements taken *r* elements at a time is:

$$_nP_r = n \cdot (n-1) \cdot (n-2) \cdot (n-3) \cdots (n-r+1)$$

Formula: $_nP_r = \dfrac{n!}{(n-r)!}$ **Example:** $_{18}P_2 = \dfrac{18!}{(18-2)!} = \dfrac{18 \cdot 17 \cdot \cancel{16!}}{\cancel{16!}} = 306$

The symbols $P(n, \text{r})$ and P_r^n are often used to denote $_nP_r$.

Examples

1. Find the number of permutations of the letters a, b, c taken 3 letters at a time ?
    Solution:
    $_3P_3 = 3! = 3 \cdot 2 \cdot 1 = 6$ permutations.
    Ans.

2. How many permutations can be made from the letters a, b, c, d taken 4 letters at a time ?
    Solution:

    $_4P_4 = 4! = 4 \cdot 3 \cdot 2 \cdot 1 = 24$ permutations.
    Ans.

3. How many permutations can be made from the letters a, b, c, d taken 3 letters at a time ?
    Solution:

    $_4P_3 = \dfrac{4!}{(4-3)!} = \dfrac{4 \cdot 3 \cdot 2 \cdot \cancel{1}}{\cancel{1!}} = \dfrac{24}{1} = 24.$

    Ans: 24 permutations.

4. How many arrangements can be selected on a shelf from 10 different books taken 10 books for each arrangement ?
    Solution:
    $_{10}P_{10} = 10! = 3{,}628{,}800$ arrangements.
    Ans.

5. How many arrangements can be selected on a shelf from 10 different books taken 5 books for each arrangement ?
    Solution:

    $_{10}P_5 = \dfrac{10!}{(10-5)!} = \dfrac{10 \cdot 9 \cdot 8 \cdot 7 \cdot 6 \cdot \cancel{5!}}{\cancel{5!}}$

    $= 30{,}240$ arrangements. Ans.

6. There are 10 empty seats in a row. 5 students want to sit in the row. How many different ways can they be seated ?
    Solution:

    $_{10}P_5 = \dfrac{10!}{(10-5)!} = \dfrac{10 \cdot 9 \cdot 8 \cdot 7 \cdot 6 \cdot \cancel{5!}}{\cancel{5!}}$

    $= 30{,}240$ ways. Ans.

Combination: An arrangement of elements without repetitions and regardless of the order.

The number of combinations of the letters a, b, c taken 3 letters at a time:
$$\text{abc} \rightarrow 1 \text{ combination.}$$
The number of combinations of the letters a, b, c, d, take 3 letters at a time:
$$\text{abc, bcd, cda, dab} \rightarrow 4 \text{ combinations.}$$

Note that the only difference between a permutation and a combination is that we consider "abc, acb, bac, bca, cab, cba" ($3! = 6$) objects as one combination (regardless of the order). Therefore, we divide the formula of $_nP_r$ by $r!$ to obtain the formula of $_nC_r$.

The number of combinations of n different elements taken r elements at a time:

Formula: $_nC_n = 1$ **Example:** $_{10}C_{10} = \dfrac{10!}{(10-10)! \cdot 10!} = \dfrac{1}{0!} = \dfrac{1}{1} = 1$

The number of combinations of n different elements taken r elements at a time:

Formula: $_nC_r = \dfrac{n!}{(n-r)! \cdot r!}$; $_nC_r = {_nC_{n-r}}$

Examples: $_{18}C_2 = \dfrac{18!}{(18-2)! \cdot 2!} = \dfrac{18!}{16! \cdot 2} = \dfrac{18 \cdot 17 \cdot 16!}{16! \cdot 2} = 153$

$_{18}C_{16} = \dfrac{18!}{(18-16)! \cdot 16!} = \dfrac{18!}{2! \cdot 16!} = \dfrac{18 \cdot 17 \cdot 16!}{2! \cdot 16!} = 153$

The symbols $C(n, r)$, C_r^n and $\binom{n}{r}$ are often used to denote $_nC_r$.

Note: If both n and r are too large to compute $_nP_r$ or $_nC_r$, we use a graphing calculator to approximate the values. (See page 387)

Examples

1. Find the number of combinations of the letters a, b, c taken 3 letters at a time ?
Solution:
$$_3C_3 = \dfrac{3!}{(3-3)! \cdot 3!} = \dfrac{1}{0!} = \dfrac{1}{1} = 1 \text{ combination.}$$
Ans.

2. How many combinations can be made from the letters a, b, c, d taken 4 letters at a time ?
Solution: $_4C_4 = 1$ combination. Ans.

3. How many combinations can be made from the letters a, b, c, d taken 3 letters at a time ?
Solution: $_4C_3 = \dfrac{4!}{(4-3)! \cdot 3!} = \dfrac{24}{1! \cdot 6} = 4$. Ans.

4. There are 20 boys in a class. How many different committees of 3 boys can be formed ?
$$_{20}C_3 = \dfrac{20!}{(20-3)! \cdot 3!} = \dfrac{20 \cdot 19 \cdot 18 \cdot 17!}{17! \cdot 6} = 1140 .$$
Ans.

5. There are 20 boys and 15 girls in a class. A committee to be formed consisting of 3 boys and 3 girls. How many different committees are possible ?
Solution:
$$_{20}C_3 \cdot {_{15}C_3} = \dfrac{20!}{17! \cdot 3!} \cdot \dfrac{15!}{12! \cdot 3!}$$
$$= \dfrac{20 \cdot 19 \cdot 18}{6} \cdot \dfrac{15 \cdot 14 \cdot 13}{6}$$
$$= 1140 \times 455 = 518,700 . \text{ Ans.}$$

Permutations with Repeating Selections

We have learned to find the number of permutations of n different elements taken "n" or "r" elements at a time without repetitions (each element can be taken only once) . Now, we will find the number of permutations if each element can be taken with repetitions.

The number of permutations of the letters a, b, c taken 3 letters at a time, each letter can be taken repeatedly:

aaa, aab, aac, aba, abb, abc, aca, acb, acc

baa, bab, bac, bba, bbb, bbc, bca, bcb, bcc

caa, cab, cac, cba, cbb, cbc, cca, ccb, ccc $\rightarrow$ 27 permutations. $3^3 = 27$

The number of permutations of n different elements taken r elements at a time and each element can be taken repeatedly:

Formula: $P = n \cdot n \cdot n \cdot n \cdots = n^r$

Examples

1. Find the number of permutations of the letters a, b, c taken 3 letters at a time, each letter can be taken repeatedly ?

Solution:

$P = n^r = 3^3 = 27$ permutations. Ans.

2. How many permutations can be made from the letters a, b, c, d taken 4 letters at a time, each letter can be taken repeatedly ?

Solution:

$P = n^r = 4^4 = 256$ permutations Ans.

3. How many permutations can be made from the letters a, b, c, d taken 3 letters at a time, each letter can be taken repeatedly ?

Solution:

$P = n^r = 4^3 = 64$ permutations. Ans.

4. How many permutations can be formed from the letters a, b, c, d, e taken 4 letters at a time, each letter can be taken repeatedly ?

Solution:

$P = n^r = 5^4 = 625$ permutations. Ans.

5. How many 4-digit integers can be formed from the digits 1, 2, 3, 4, 5 and repetitions of digits are allowed ?

Solution:

$P = 5^4 = 625$. Ans.

6. How many different ways can be chosen to drop 4 letters into 6 different mail boxes ?

Solution: Each letter has 6 choices.

$P = 6 \cdot 6 \cdot 6 \cdot 6 = 6^4 = 1{,}296$ ways. Ans.

Hint: 4^6 is incorrect. (4^6 include one same letter to be dropped into 6 boxes.)

7. How many 4-digit integers can be made from the digits 0, 1, 2, 3, 4, 5 and repetitions of digits are allowed ?

Solution.

$P = 5 \cdot 6^3 = 1{,}080$ integers. Ans.

(Hint: It cannot begin with 0.)

8. How many 3-digit integers which are larger than 220 can be made using the digits 0,1,2 3,4 and repetitions of digits are allowed ?

Solution:

(2) (2) (1,2,3,4) $\rightarrow 1 \times 1 \times 4 = 4$

(2) (3,4,) (0,1,2,3,4,) $\rightarrow 1 \times 2 \times 5 = 10$

(3,4,) (0,1,2,3,4) (0,1,2,3,4) $\rightarrow 2 \times 5 \times 5 = 50$

$4 + 10 + 50 = 64$. Ans.

Permutations with Elements Alike

Now, we learn how to find the number of permutations of n elements that are not all different.

The number of permutations of n elements taken n at a time, with n_1 elements alike, n_2 elements alike, n_3 elements alike, ...is:

$$\textbf{Formula: } P = \frac{n!}{n_1! \cdot n_2! \cdots n_k!}$$ It is also called **distinquishable permutations**.

Examples

1. How many permutations can be formed from the letters in ACADIA, taken 6 at a time ?
Solution:

$$P = \frac{6!}{3!} = \frac{720}{6} = 120 \text{ permutations. Ans.}$$

2. How many permutations can be formed from the letters in ACACIA, taken 6 at at a time ?
Solution:

$$P = \frac{6!}{3! \cdot 2!} = \frac{720}{6 \cdot 2} = 60 \text{ permutation. Ans.}$$

3. How many five-digit numerals can be made from 3, 6, 9, and 7 if 3, 6, and 9 can be used once, 7 can be used twice ?
Solution: (3, 6, 9, 7, 7)

$$P = \frac{5!}{2!} = \frac{120}{2} = 60 \text{ numerals. Ans.}$$

4. How many different ways to line up 10 balls if 6 are red, 3 are white, and 1 is black ?
Solution:

$$P = \frac{10!}{6! \cdot 3! \cdot 1!} = \frac{10 \cdot 9 \cdot 8 \cdot 7 \cdot \cancel{6!}}{\cancel{6!} 6 \cdot 1} = 840 . \text{ Ans.}$$

5. How many five-digit numerals can be made from 6 and 2 if 6 can be used three times, 2 can be used twice ?
Solution: (6, 6, 6, 2, 2)

$$P = \frac{5!}{3! \cdot 2!} = \frac{120}{6 \cdot 2} = 10 \text{ numerals. Ans.}$$

6. How many seven-digit numerals can be made from 2, 5, and 8 if we use 2 and 5 twice, 8 three times ?
Solution: (2, 2, 5, 5, 8, 8, 8)

$$P = \frac{7!}{2! \cdot 2! \cdot 3!} = \frac{5040}{2 \cdot 2 \cdot 6} = 210 \text{ numerals. Ans.}$$

7. How many seven-digit numerals can be made from 2, 5, and 8 if we use 2 and 5 three times, 8 once ?
Solution: (2, 2, 2, 5, 5, 5, 8)

$$P = \frac{7!}{3! \cdot 3!} = \frac{5040}{6 \cdot 6} = 140 \text{ numerals. Ans.}$$

8. How many short-cut routes can be selected from A to B ?

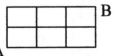

Solution: $(\rightarrow \rightarrow \rightarrow \uparrow \uparrow)$

$$P = \frac{5!}{3! \cdot 2!} = \frac{5 \cdot 4 \cdot \cancel{3!}}{\cancel{3!} 2} = 10 \text{ routes. Ans.}$$

Circular and Necklace Permutations

We have learned how to find the number of permutations of n objects placed in a line.

    The number of permutations of 3 objects taken 3 at a time is $3! = 3 \cdot 2 \cdot 1 = 6$.

    The number of permutations of 6 objects taken 6 at a time is $6! = 6 \cdot 5 \cdot 4 \cdot 3 \cdot 2 \cdot 1 = 720$.

However, when the 3 letters A, B, C are arranged in a circle, once an arrangement ABC is chosen, we cannot distinguish the three arrangements ABC, BCA and CAB by rotating the circle.

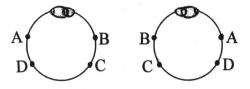

Therefore, 3 permutations in a line form 1 permutation in a circle in this example.

The total number of permutations of 3 objects taken 3 at a time **in a circle** is: $\dfrac{3!}{3} = \dfrac{6}{3} = 2$

The total number of permutations of 6 objects taken 6 at a time **in a circle** is: $\dfrac{6!}{6} = \dfrac{720}{6} = 120$

    Look at these: $\dfrac{3!}{3} = \dfrac{\not{3} \cdot 2!}{\not{3}} = 2!$; $\dfrac{6!}{6} = \dfrac{\not{6} \cdot 5!}{\not{6}} = 5!$; $\dfrac{n!}{n} = \dfrac{\not{n} \cdot (n-1)!}{\not{n}} = (n-1)!$

Circular Permutations: The number of permutations of n distinct objects taken n at a time **in a circle** is:

Formulas: 1. $_nP_r$ (in a circle) $= \dfrac{_nP_r}{r}$ **2.** $_nP_n$ (in a circle) $= (n-1)!$

When we turn over a necklace, one circular permutation is the same one with opposite direction. Therefore, two circular permutations form one permutation in a necklace.

Necklace Permutations: The number of permutations of n beads **in a necklace** is one-half of the number of circular permutations.

Formula: $_nP_n$ (a necklace) $= \dfrac{(n-1)!}{2}$

Examples:

1. Find the number of different arrangements of 10 people seated in a round table.

    Solution:

$$(10-1)! = 9! = 362{,}880 . \text{ Ans.}$$

2. Find the number of necklace strings that can be made from 10 beads of different colors.

    Solution:

$$\frac{(10-1)!}{2} = \frac{9!}{2} = 181{,}440 . \text{ Ans.}$$

The Odds of Lottery Tickets

1. How many combinations of Lottery can be selected that we select 5 numbers from 1 to 49, then a Powerball number from 1 to 42 ?
Solution:

$$_{49}C_5 \cdot _{42}C_1 = \frac{49!}{(49-5)!\cdot 5!} \cdot 42 = \frac{49 \cdot 48 \cdot 47 \cdot 46 \cdot 45 \cdot \cancel{44!}}{\cancel{44!} \cdot 120} \cdot 42 = 1{,}906{,}884 \cdot (42) = 80{,}089{,}128.$$

2. How many combinations of Multistates Powerball Lottery can be selected that we select 5 numbers from 1 to 55, then a Powerball number from 1 to 42 ?
Solution:

$$_{55}C_5 \cdot _{42}C_1 = \frac{55!}{(55-5)!\cdot 5!} \cdot 42 = \frac{55 \cdot 54 \cdot 53 \cdot 52 \cdot 51 \cdot \cancel{50!}}{\cancel{50!} \cdot 120} \cdot 42 = 3{,}478{,}761 \cdot (42) = 146{,}107{,}962.$$

3. How many combinations of California SuperLotto Plus lottery can be selected that we select 5 numbers from 1 to 47, then a Mega number from 1 to 27 ?
Solution:

$$_{47}C_5 \cdot _{27}C_1 = \frac{47!}{(47-5)!\cdot 5!} \cdot 27 = \frac{47 \cdot 46 \cdot 45 \cdot 44 \cdot 43 \cdot \cancel{42!}}{\cancel{42!} \cdot 120} \cdot 27 = 1{,}533{,}839 \cdot (27) = 41{,}416{,}353.$$

4. Find the odds of the prize matching-numbers of Multistates MegaMillions Lottery that we select 5 numbers from 1 to 56, then a Mega number from 1 to 46.
Solution:

a) All 5 of 5 and Mega : $\dfrac{_5C_5}{_{56}C_5} \cdot \dfrac{1}{46} = \dfrac{1}{3819816} \cdot \dfrac{1}{46} = \dfrac{1}{175{,}711{,}536}.$

b) All 5 of 5 and no Mega: $\dfrac{_5C_5}{_{56}C_5} \cdot \dfrac{45}{46} = \dfrac{1}{3819816} \cdot \dfrac{45}{46} = \dfrac{45}{175711536} = \dfrac{1}{3{,}904{,}701}.$

c) Any 4 of 5 and Mega: $\dfrac{_5C_4 \cdot _{51}C_1}{_{56}C_5} \cdot \dfrac{1}{46} = \dfrac{5(51)}{3819816} \cdot \dfrac{1}{46} = \dfrac{255}{175711536} = \dfrac{1}{689{,}065}.$

(Hint: Getting 4 of the 5 matching numbers and 1 of the 51 non-matching numbers.)

d) Any 4 of 5 and no Mega: $\dfrac{_5C_4 \cdot _{51}C_1}{_{56}C_5} \cdot \dfrac{45}{46} = \dfrac{5(51)}{3819816} \cdot \dfrac{45}{46} = \dfrac{11475}{175711536} = \dfrac{1}{15{,}313}.$

e) Any 3 of 5 and Mega: $\dfrac{_5C_3 \cdot _{51}C_2}{_{56}C_5} \cdot \dfrac{1}{46} = \dfrac{10(1275)}{3819816} \cdot \dfrac{1}{46} = \dfrac{12750}{175711536} = \dfrac{1}{13{,}781}.$

f) Any 3 of 5 and no Mega: $\dfrac{_5C_3 \cdot _{51}C_2}{_{56}C_5} \cdot \dfrac{45}{46} = \dfrac{10(1275)}{3819816} \cdot \dfrac{45}{46} = \dfrac{573750}{175711536} = \dfrac{1}{306}.$

g) Any 2 of 5 and Mega: $\dfrac{_5C_2 \cdot _{51}C_3}{_{56}C_5} \cdot \dfrac{1}{46} = \dfrac{10(20825)}{3819816} \cdot \dfrac{1}{46} = \dfrac{208250}{175711536} = \dfrac{1}{844}.$

h) Any 1 of 5 and Mega: $\dfrac{_5C_1 \cdot _{51}C_4}{_{56}C_5} \cdot \dfrac{1}{46} = \dfrac{5(249900)}{3819816} \cdot \dfrac{1}{46} = \dfrac{1249500}{175711536} = \dfrac{1}{141}.$

i) None of 5, only Mega: $\dfrac{_{51}C_5}{_{56}C_5} \cdot \dfrac{1}{46} = \dfrac{2349060}{3819816} \cdot \dfrac{1}{46} = \dfrac{2349060}{175711536} = \dfrac{1}{75}.$

EXERCISES

Evaluate each expression. If both n and r are too large to compute $_nP_r$ and $_nC_r$, use a graphing calculator to approximate the values. (See page 387)

1. $_8P_1$ **2.** $_8P_7$ **3.** $_7P_4$ **4.** $_{30}P_2$ **5.** P_2^{100} **6.** P_4^{99} **7.** $P(20,17)$ **8.** $_{50}P_{35}$

9. $_8C_1$ **10.** $_8C_7$ **11.** $_7C_4$ **12.** $\binom{30}{2}$ **13.** C_2^{100} **14.** C_4^{99} **15.** $C(20,17)$ **16.** $_{50}C_{35}$

17. Find the number of permutations of the letters a, b, c, d, e, f :
   a. taken 6 letters at a time.   b. taken 4 letters at a time.   c. taken 1 letter at a time.

18. Find the number of combinations of the letters a, b, c, d, e. f:
   a. taken 6 letters at a time.   b. taken 4 letters at a time.   c. taken 1 letter at a time.

19. How many arrangements can be selected on a shelf if you have 20 books taken 4 books for each arrangement ?

20. There are 20 students in a class. A committee consists of 4 students. How many different committees can be formed ?

21. How many permutations can be made from the letters a, b, c, d, e, f taken 4 letters at a time, each letter can be taken repeatedly ?

22. A basketball team consists of 3 guards, 2 centers, and 4 forwards. How many different ways can the coach select a starting team of 2 guards, 1 center, and 2 forwards ?

23. How many different ways can be used to distribute among 6 kids if each kid is allowed to choose one fruit from 5 different kinds of fruits ?

24. How many 3-digit integers which are larger than 310 can be made using the digits 0, 1, 2, 3, 4, 5 and repetitions of digits are allowed ?

25. How many 3-digit integers which are larger than 220 can be made using the digits 0, 1, 2, 3, 4, 5 and repetitions of digits are allowed ?

26. A bag contains 5 red balls and 4 white balls. Three balls are drawn at random from the bag. How many ways can the three balls be drawn if all 3 balls are red ?

27. A bag contains 5 red balls and 4 white balls. Three balls are drawn at random from the bag. How many ways can the three ball be drawn if 2 balls are red and 1 ball is white ?

28. Find the total number of subsets of a set that has 5 elements ?

29. How many odd integers are there from 300 to 700 ?

30. How many different ways can be formed using all the letter in the word CONGEST ?

31. How many distinguishable ways can be written using all the letters of ALGEBRA ?

32. How many distinguishable ways can the letters in MISSISSIPPI be written ?

33. How many different ways can the letters of the word REPETITION be arranged ?

34. How many four-digit numerals can be made from 2, 4, and 6 if 2 and 4 can be used twice, 6 can be used once ?

35. How many different ways to arrange 10 flags if 5 are red, 3 are white, and 2 are yellow ?

-----**Continued**-----

36. How many short-cut routes can be selected from A to B ?

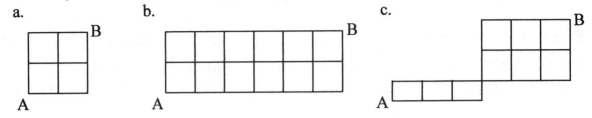

a.   b.   c.

37. How many different ways to arrange 10 flags of different colors in a line ?

38. How many different ways to arrange 10 flags of different color around a track field ?

39. How many different ways to arrange 10 flags of different colors taken 6 flags at a time ?

40. How many different ways to arrange 10 flags of different colors taken 6 flags at a time around a track field ?

41. How many different arrangements can be made for 8 students seated in a row ?

42. How many different arrangements can be made for 8 students seated in a round table ?

43. How many different arrangements can be made for 8 students taken 4 students seated in a row ?

44. How many different arrangements can be made for 8 students taken 4 students seated in a round table.

45. Find the number of permutations of the letters a, b, c, c, c taken 5 at a time in a line.

46. Find the number of permutations of the letters a, b, c, c, c taken 5 at a time in a circle.

47. Find the number of necklace strings that can be made from 8 beads of different colors ?

48. How many different arrangements are possible for 6 different keys placed on a line ?

49. How many different arrangements are possible for 6 different keys placed on a circle ?

50. How many different arrangements are possible for 6 different keys placed on a key ring ?

51. How many different arrangements are possible for 6 keys placed on a key ring if three of the keys are alike ?

52. Find the odd of picking 4 out of 5 prize matching-number of Fantasy Lottery that we select 5 numbers from 1 to 39 ?

53. In Multistates Powerball Lottery, find the odd of matching 4 out of 5 prize matching-number and the Powerball matching-number that we select 5 from 1 to 55 and a Powerball number from 1 to 42 ?

54. Evaluate $S = {}_{30}C_0 + {}_{30}C_1 + {}_{30}C_2 + \cdots\cdots + {}_{30}C_{30}$.

Solve for *n*.

55. ${}_nP_2 = 30$

56. ${}_nP_3 = 210$

57. ${}_nP_2 = 8 \cdot {}_nP_1$

58. ${}_nP_5 = 21 \cdot {}_{n-2}P_4$

59. $11 \cdot {}_nP_2 = {}_{n+8}P_2$

60. ${}_nC_2 = 15$

61. ${}_nC_3 = 35$

62. ${}_nC_2 = 4 \cdot {}_nC_1$

63. ${}_nC_4 = 7 \cdot {}_{n-2}C_4$

64. $11 \cdot {}_nC_2 = {}_{n+8}C_2$

65. ${}_8P_n = 8!$

66. $3 \cdot {}_nC_3 = 10 \cdot {}_{n-2}C_2$

On graduation day, a high school graduate said to his math teacher.
Student: Thank you very much for giving me a "D" in your class.
For your kindness, do you want me to do anything for you ?
Teacher: Yes. Please do me a favor. Just don't tell anyone that I was
Your math teacher.

11-3 Sets and Venn Diagrams

Set: Any particular group of elements or members. We use braces, { }, to name a set.

U: The **universal** set. It is the set consisting of all the elements.

A∩B: The **intersection** of two sets. The set consists of the elements that are common to both sets A and B.

A∪B: The **union** of two sets. The set consists of all members of both sets A and B.

ϕ: The **empty** set, or null set. The set consists of no member.

The set {0} contains exactly one member "0". The set {0} is not an empty set.

∈: It indicates "**it is a member of**".   **∉**: It indicates " **it is not a member of** ".

A ⊂ B : Set A is a **subset** of set B. Every member in set A is also a member of set B.

Every set is a subset of itself, $A \subset A$. The empty set is a subset of every set.

A ⊄ B : Set A is **not a subset** of set B.

A' : The **complement** of set A. A' is the set consisting of all the elements not in A.

$A \cup A' = U, \quad A \cap A' = \phi$

We can specify a set of numbers by writing a description within the braces.

{The even numbers between 1 and 9} = {2, 4, 6, 8}

We can specify a set of numbers by graphing the numbers on a number line.

{The real numbers greater than or equal to −3 and less than 2}

$$-3 \le x < 2$$

Venn Diagram: A diagram used to indicate the relations between sets.

1. $n(A \cup B) = n(A) + n(B) - n(A \cap B)$ **2.** $n(A \cup B \cup C) = n(A) + n(B) + n(C) - n(A \cap B) - n(B \cap C)$
$$- n(A \cap C) + n(A \cap B \cap C)$$

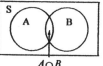

Examples

1. In a high school class, there are 25 students who went to see movie A, 18 students went to see move B. These figures include 9 students went to see both movies. How many students are in this class ?
   Solution:

   25 + 18 − 9 = 34 students. Ans.

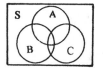

2. In a high school class, there are 17 students in the basketball team, 18 in the volleyball team, 20 in the baseball team. Of these, there are 9 students in both the basketball and baseball teams, 6 in both volleyball and baseball teams, and 5 in both volleyball and basketball teams. These figures include 2 students who are in all three teams. How many students are in these class ?
   Solution:

   17 + 18 + 20 − 5 − 6 − 9 + 2 = 37 students. Ans.

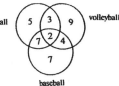

EXERCISES

Specify each of the following sets, referring to the Venn Diagram.

1. $A \cap B$ **2.** $A \cup B$ **3.** $B \cap C$ **4.** $B \cup C$

5. $(A \cup B) \cap C$ **6.** $(A \cap B) \cap C$ **7.** $A \cap \phi$ **8.** $A \cup \phi$

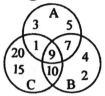

Insert a symbol of sets to make a true statement.

9. 5 ___ $\{0, 5, 8, 11\}$ **10.** 4 ___ $\{0, 5, 8, 11\}$ **11.** $\{5, 8\}$ ___ $\{0, 5, 8, 11\}$

12. $\{5\}$ ___ $\{0, 5, 8, 11\}$ **13.** $\{0, 5, 8, 11\}$ ___ $\{5, 11, 8, 0\}$ **14.** $\{3, 9\}$ ___ $\{0, 5, 8, 11\}$

15. ϕ ___ $\{0, 5, 8, 11\}$ **16.** $\{0, 5, 8, 11)$ ___ $\{5, 8, 0, 12\}$ **17.** ϕ ___ $\{0\}$

Specify each set of numbers.

18. {the even numbers between 1 and 10}   **19.** {the odd numbers between 6 and 16}

20. {the positive integers less than 4}   **21.** {the positive integers less than -1}

22. {the real number that are neither positive nor negative}

23. {the whole numbers less than 6}

24. {the real numbers less than 1 and the real numbers greater than 5}

25. {the real numbers greater than or equal to -4 and less than 2 }

26. List all the subsets of $\{1, 2, 3\}$.

27. What is the total number of subsets of $\{1, 2, 3, 4)$?

28. What is the total number of subsets of $\{1, 2, 3, 4. 5\}$?

29. If a set contains n numbers, what is the total number of subsets of n ?

30. Draw a Venn Diagram to illustrate $A \supset B$.

31. Draw a Venn Diagram to illustrate $C \subset B$ and $B \subset A$.

32. What is the relation between M an N if $M \subset N$ and $N \subset M$?

33. There are 34 students in a class, 25 students went to see movie A, 18 students went to see movie B. These figures include the students who went to see both movie A and movie B. How many students went to see both movie A and movie B ?

34. There are 56 students in a class, 24 were registered in Algebra, 32 students were registered in biology, and 8 were in both courses. How many were registered in neither course ?

35. In a survey of 50 investors in the stock market, 17 owned shares in company A, 18 owned shares in company B, 20 owned shares in company C, 5 owned shares in both company A and B, 6 owned shares in both company B and C, 9 owned shares in both company A and C, 2 owned shares in all three companies.

       a. How many owned only company A shares ?

       b. How many owned shares in both company A and B , but no company C ?

       c. How many did not own shares in any of the three companies ?

36. Write the set consisting of the possible outcomes from tossing two coins. Use H for "heads" and T for "tails".

37. One-third of all Matts are Patts. Half of all Patts are Natts. No Matt is a Natt. All Natts are Patts. There are 25 Natts and 45 Matts. How many Patts are neither Matts nor Natts ?

11-4 Sample Spaces and Events

Random Experiment: It is a chance experiment repeated under independent and similar conditions. An experiment is random when every outcome has an equal chance to be occurred.

Although we never know what the outcome will be in a random experiment, but we know what the possible outcomes are.

Sample Space: The set of all possible outcomes (elements) in a random experiment.

Event: Any subset of the sample space in a random experiment.

Sample(sample point): Each member (element) in the sample space.

Simple Event: An event which has only a single member (element).

The number of simple events in a sample space is denoted by $|S|$.

The sample space for tossing a fair (unbiased) coin is {H, T} and the sample space consists of two simple events (two outcomes) {H} and {T}.

The sample space for tossing two fair (unbiased) coins is {(H, H), (H, T), (T, H), (T, T) } and the sample space consists of four simple events (four outcomes) {H, H}, {T, H}, {T, T}.

The sample space for rolling a fair (unbiased) die is {1, 2, 3, 4, 5, 6} and the sample space consists of six simple events (six outcomes) {1}, {2}, {3}, {4}, {5}, {6}.

Examples

1. For rolling two fair dice, specify:
   a. the sample space for the experiment.
   b. how many simple events are there ?
   c. the event that the sum of the numbers turned up on the two dice is 6.
   d. the event that the sum of the numbers turned up on the two dice is less than 6.

   Solution:
   a. Sample space
   $S = \{(1, 1), (1, 2), (1, 3), (1, 4), (1, 5), (1, 6),$
   $(2, 1), (2, 2), (2, 3), (2, 4), (2, 5), (2, 6),$
   $(3, 1), (3, 2), (3, 3), (3, 4), (3, 5), (3, 6),$
   $(4, 1), (4, 2), (4, 3), (4, 4), (4, 5), (4, 6),$
   $(5, 1), (5, 2), (5, 3), (5, 4), (5, 5), (5, 6),$
   $(6, 1), (6, 2), (6, 3), (6, 4), (6, 5), (6, 6)\}.$

   b. $|S| = 6 \times 6 = 36$ simple events.

   c. $A = \{(1, 5), (2, 4), (3, 3), (4, 2), (5, 1)\}.$

   d. $A = \{(1, 1), (1, 2), (1, 3), (1, 4), (2, 1), (2, 2),$
   $(2, 3), (3, 1), (3, 2), (4, 1)\}.$

2. For tossing three fair coins, specify:
   a. the sample space for the experiment.
   b. the event that only two coins turned up heads.
   c. the event that at least two coins turned up heads.

   Solution:
   a. Sample space
   $S = \{(H, H, H), (H, H, T), (H, T, H), (T, H, H),$
   $(T, H, T), (T, T, H), (H, T, T), (T, T, T)\}.$

   b. $A = \{(H, H, T), (H, T, H), (T, H, H)\}.$

   c. $A = \{(H, H, H), (H, H, T), (H, T, H), (T, H, H)\}.$

3. For rolling three fair dice, how many simple events are in the sample space ?

   Solution:
   $|S| = 6 \times 6 \times 6 = 216$ simple events.   Ans

EXERCISES

Write the sample space (S) for the given random experiment.

1. Tossing one fair coin, then a fair die.
2. Tossing one fair die, then one fair coin.
3. Tossing two fair coins, then a fair die.
4. Tossing two fair dice, then one fair coin.
5. Rolling a fair dice twice, the sum of the numbers turned up on the dice.
6. Rolling a fair dice three times, the sum of the numbers turned up on the dice.
7. Choosing three different shoes a, b, and c, according to your preference.
8. A bag contains two red, two white and one black balls. Two balls are drawn.
9. A box contains ten bulbs. Four are defective. The numbers of bulbs needed to be checked to find all four defectives if all defective bulbs must be checked to be found.
10. There are five different books a, b, c, d and e on the shelf. Two books are selected.

Write the event (A) for the given random experiment.

11. Rolling a fair die, the event the number turned up on the die is greater than 2.
12. Rolling a number cube twice, the event that the sum of the rolls is 7.
13. A fair coin is tossed four times. The event that all four tosses are tails.
14. A fair coin is tossed twice. The event that at least one coin turned up tail.
15. Rolling two fair dice, the event that the numbers turned up on the two dice are equal.
16. Rolling two fair dice, the event that the sum of the numbers turned up on the two dice is greater than 10.
17. Rolling two fair dice, the event that the sum of the numbers turned up on the two dice is greater than or equal to 10.
18. Rolling a fair coin three times, The event that all rolls are the same.
19. A bag contains two red, two white and one black balls. Two balls are drawn. The event that at least one is white.
20. There are five different books a, b, c, d, and e on the shelf. Two books are selected. The event that book "a" must be selected.

21. For rolling one fair die once, how many simple events are in the sample space ?
22. For rolling one fair die twice, how many simple events are in the sample space ?
23. For rolling one fair die twice, how many events in the sample space ?
24. For tossing one fair die and one fair coin together, write the sample space (S) and the event (A) that the coin turned up head and the die turned an even number.
25. For tossing one fair die and one fair coin together, how many simple events are in the sample space.
26. For tossing one fair die and one fair coin together, how many events are in the sample space ?
27. For tossing one fair die and two fair coins, how many simple events are in the sample space ?
28. For tossing one fair die and two fair coins, how many events are in the sample space ?

11-5 Probability

In mathematics, we want to find the possibilities of uncertainty. For example, we want to know the chance of winning a state lottery, or the chance to have a life of more than 90 years.

There are countless of uncertainty in our everyday life. To measure the uncertainties, we use the term "the probability". For example, the probability of tossing a fair coin turned up head is one-half, 0.5, or 50%.

The main use for the theory of probability is to make decision by performing a survey or experiment that will allow us to study, predict, and make conclusions from data.

In the study of probability, the various results in a random experiment are called the **outcomes**. The collection of all possible outcomes is called the **sample space**.

Any subset of the sample space is called an **event**.

There are two types of probability, theoretical probability and experimental probability.

Theoretical Probability: It is simply called the **probability**.

If the sample space S of an experiment has $n(S)$ equally likely outcomes, and the event A has $n(A)$ equally likely outcomes, then the probability of event A is

$$P(A) = \frac{n(A)}{n(S)} \quad ; \quad 0 \le P(A) \le 1$$

The probability of tossing a fair coin turned up head is $P(H) = \frac{1}{2}$.

The probability of tossing a fair coin turned up tail is $P(T) = \frac{1}{2}$.

The probability of rolling a die turned up an even number is $P(E) = \frac{3}{6} = \frac{1}{2}$.

The probability of an event must be between 0 and 1, or between 0% to 100%.

The statement $P(A) = 0$ means that event A cannot occur (0%).

The statement $P(A) = 1$ means that event A must occur (100%).

If p is the probability of an event, $1 - p$ is the probability of an event not occurring.

Experimental Probability: In real-life situations, sometimes it is not possible to find the theoretical probability of an event. Therefore, we find the experimental probability by performing an experiment or a survey. We can use experimental probability to predict future occurrences of events.

If a sufficient number of trials is conducted n times, and an event A occurs e of these times, then the experimental probability that the event A will occur in another trial is given by:

$$P(A) = \frac{e}{n}$$

A school with 2,000 students has 40 students who are left-handed this year. Then the probability of a student chosen at random will be left-handed is $\frac{40}{2000} = \frac{1}{50}$.

If there is no other information available, we can best estimate that the probability of a student chosen at random next year will be left-handed is $\frac{1}{50}$, 0.02, or 2%.

A **tree diagram** is helpful in listing all the possible outcomes (the sample space).
The sample space for tossing two coins is shown below. Use H for heads and T for tails.
There are four possible outcomes (four simple events) {(H, H), (H, T), (T, H), (T, T)}.

The probability of two heads is $\frac{1}{4}$.

The probability of two tails is $\frac{1}{4}$.

The probability of "one is heads and one is tails" is $\frac{1}{4}+\frac{1}{4}=\frac{1}{2}$.

The probability of "at least one is heads" is $\frac{3}{4}$.

Independent Events: In an experiment, A and B are two independent events if the occurrence
of one event does not affect the occurrence of the other.

Dependent Events: In an experiment, A and B are two dependent events if the occurrence
of one event affect the occurrence of the other.

Probability of Independent Events:

If A and B are two independent events, then the probability that both A and B occur is:
$$P(A \cap B) = P(A) \times P(B). \text{ Or, } P(A \text{ and } B) = P(A) \times P(B).$$

Example 1. A bag contains 5 red balls and 4 white balls. Two balls are drawn at random from
the bag. The first ball drawn is put back into the bag before the second ball is
drawn. What is the probability that the two balls drawn are both red ?
Solution: $P(\text{red}) = \frac{5}{9}$; $P(\text{red and red}) = \frac{5}{9} \times \frac{5}{9} = \frac{25}{81}$. Ans.

Example 2. A and B are two independent events. $P(A) = 0.3$, $P(B) = 0.9$. Find $P(A \text{ and } B)$.
Solution: $P(A \text{ and } B) = P(A) \cdot P(B) = 0.3(0.9) = 0.27.$ Ans.

Probability of Dependent Events:

If A and B are two dependent events, then the probability that both A and B occur is:
$$P(A \cap B) = P(A) \times P(B|A). \text{ Or, } P(A \text{ and } B) = P(A) \times P(B|A).$$

$P(B|A)$ is called the **conditional probability** of B given that A has occurred.

Example 3. A bag contains 5 red balls and 4 white balls. Two balls are drawn at random from
the bag. The first ball drawn is not put back into the bag before the second ball is
drawn. What is the probability that the two balls drawn are both red ?
Solution:
$$P(\text{red}) = \frac{5}{9} \text{ ; } P(\text{red} | \text{red}) = \frac{4}{8} \text{ ; } P(\text{red and red}) = \frac{5}{9} \times \frac{4}{8} = \frac{20}{72} = \frac{5}{18}. \text{ Ans.}$$

Example 4. A and B are two dependent events. $P(A) = 0.3$, $P(B/A) = 0.6$. Find $P(A \text{ and } B)$.
Solution: $P(A \text{ and } B) = P(A) \cdot P(B|A) = 0.3(0.6) = 0.18.$ Ans.

Example 5. Tell whether A and B are dependent or independent events.
$$P(A) = 0.24, \quad P(B) = 0.45, \quad P(A \cap B) = 0.036$$
Solution: $P(A) \cdot P(B) = 0.24(0.45) = 0.036.$

$P(A) \cdot P(B) = P(A \cap B).$ They are independent. Ans.

Mutually Exclusive Events

Mutually exclusive events: Two events which have no elements in common.

The probability of the **union** of any two events having elements in common in a sample space is: (not mutually exclusive events)

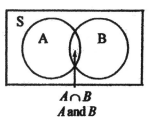

Formula:

$$P(A \cup B) = P(A) + P(B) - P(A \cap B)$$
$$\text{OR: } P(A \text{ or } B) = P(A) + P(B) - P(A \text{ and } B)$$

The probability of the **union of** two **mutually exclusive** events in a sample space is: (For mutually exclusive events, $P(A \cap B) = 0$)

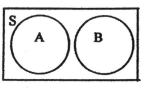

Formula:

$$P(A \cup B) = P(A) + P(B)$$
$$\text{OR: } P(A \text{ or } B) = P(A) + P(B)$$

Examples

1. A number is drawn at random from {1, 2, 3, 4, 5, 6}. What is the probability that either a number larger than 3 or an even number will occur ?

Solution:

$$S = \{1, 2, 3, 4, 5, 6\}$$
$$A = \{4, 5, 6\} \quad \therefore P(A) = \tfrac{1}{2}$$
$$B = \{2, 4, 6\} \quad \therefore P(B) = \tfrac{1}{2}$$
$$A \cap B = \{4, 6\} \quad \therefore P(A \cap B) = \tfrac{2}{6} = \tfrac{1}{3}$$
$$A \cup B = \{2, 4, 5, 6\}$$

$$P(A \cup B) = P(A) + P(B) - P(A \cap B) = \frac{1}{2} + \frac{1}{2} - \frac{1}{3} = \frac{2}{3}. \text{ Ans.}$$

2. A bag contains 5 red, 4 white and 3 black balls. One ball is drawn at random. What is the probability that it is a red or a white ball ?

Solution: (A and B are mutually exclusive events.)

Let A → the event that the ball is red.   B → the event that the ball is white.

$$P(A \cup B) = P(A) + P(B) = \frac{5}{12} + \frac{4}{12} = \frac{9}{12} = \frac{3}{4}. \text{ Ans.}$$

3. A bag contains 5 red balls and 4 white balls. Two balls are drawn together at random from the bag. What is the probability that the two balls drawn are of the same color ?

Solution: (A and B are mutually exclusive events.)

Let A → the event that two balls are red.   B → the event that two balls are white.

$$P(A) = \frac{{}_5C_2}{{}_9C_2} = \frac{10}{36} \quad ; \quad P(B) = \frac{{}_4C_2}{{}_9C_2} = \frac{6}{36}$$

$$P(A \cup B) = P(A) + P(B) = \frac{10}{36} + \frac{6}{36} = \frac{16}{36} = \frac{4}{9}. \text{ Ans.}$$

EXERCISES

One card is drawn at random from a 52-card bridge deck. Find the probability of each event.

1. It is a 10. **2.** It is a club. **3.** It is the ace of heart. **4.** It is a black heart.

5. It is a 4, 5, or 6. **6.** It is a red card. **7.** It is a 5 or 6. **8.** It is an even number.

9. It is not a spade. **10.** It is an ace or a face card.

One card is drawn from number cards 1, 2, 3, ..., 50. Find the probability of the card drawn turned up:

11. an even number. **12.** a multiple of 5. **13.** a multiple of 7

14. a multiple of both 5 and 7. **15.** a multiple of 5 or 7.

A die is tossed. Find the probability of each event.

16. It is a 3. **17.** It is a 7. **18.** It is an even number. **19.** It is an odd number.

20. It is greater than 4. **21.** It is greater than or equal to 2. **22.** It is less than 5.

23. It is between 1 and 5.

What is the probability that after a spin the arrow will stop on:

24. region 7 ? **25.** region 1 ? **26.** region 5 or 1 ?

27. an odd-numbered region ? **28.** a region with a number less than 6 ?

29. A box contains 12 red balls, 10 white balls, and 8 green balls. A ball is drawn at random. What is the probability that the ball drawn is red ?

30. John throws a dart and hit the bull's eye 12 times out of 50 throws.

   a. What is the experimental probability that he will hit the bull's eye in the next throw ?

   b. What is the experimental probability that he will not hit the bull's eye in the next throw ?

31. John tosses a fair coin 20 times and 12 times turned up heads.

   a. What is the experimental probability that it will turn up head in the next toss ?

   b. What is theoretical probability that it will turn up head in the next toss ?

32. In a basketball game, John makes the basket on 6 free throws out of 15.

   a. What is the experimental probability that he will make a basket on the next free throw ?

   b. How many times would you expect John to make the basket on the next 5 free throws ?

   c. If John actually makes the baskets 4 times out of next 5 free throws, what is the experimental probability that John will make a basket on the next free throw ?

33. Draw a tree diagram to show the results of tossing three coins.

34. Three coins are tossed. How many outcomes are possible ?

35. Four coins are tossed. How many outcomes are possible ?

36. Two dice are rolled. How many outcomes are possible ?

37. Three dice are rolled. How many outcomes are possible ?

38. Two coins are tossed. Find the probability that both are heads.

-----Continued-----

39. Two coins are tossed. Find the probability that one comes up head and one comes up tail.

40. Three coins are tossed. Find the probability that at least one comes up heads.

41. In the basketball game, the probability that you will make the basket on free throws is $\frac{3}{5}$. How many times can you expect to make the basket if you makes 125 free throws ?

42. In the basketball game, the probability that you will make the basket on free throws is 0.25. How many of 120 free throws would probable make the basket ?

43. If you take a 5-question true-false test by just guessing at each answer. Find the probability that your score will be 100% ?

44. In a survey of city mayor election, 34% of voters preferred Candidate A, 20% preferred Candidate B, and 46% preferred Candidate C. Based on this survey, how many of 3,900 voters would probably prefer:

    a. Candidate A   **b.** Candidate B   **c.** Candidate C

45. The probability of a person being colorblind is $\frac{1}{40}$. There are 2,400 students in a high school. How many students would probably be colorblind ?

46. The probability of a person being left-handed is $\frac{1}{30}$. There are 2,400 students in a high school. How many students would probably be left-handed ?

47. The probability of a person being colorblind is $\frac{1}{40}$, and the probability of a person being left-handed is $\frac{1}{30}$. There are 2,400 students in a high school. How many students would probably be both colorblind and left-handed ?

48. Spin two spinners shown below.

    a. What is the probability of getting a 7 and a D.

    b. What is the probability of getting a 4 or a 5, followed by A ?

    c. What is the probability of getting a 2, followed by a A or B ?

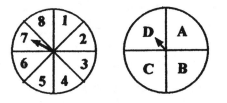

49. Two cards are drawn at random from a 52-card bridge deck with the first card replaced before the second card is drawn. Find the probability of each event.

    a. Both cards are red.   **b.** Both cards are clubs.   **c.** Both cards are the ace of hearts.

    d. The first card is a red card and the second card is a black card.

50. Two cards are drawn at random from a 52-card bridge deck without replacement. Find the probability of each event.

    a. Both cards are red.   **b.** Both cards are clubs.   **c.** Both cards are the ace of hearts.

    d. The first card is a red card and the second card is a black card.

51. You roll two fair dice. If the total of the numbers equals 10 or more, you win. What is the probability that you will win ?

52. Scientists tag 100 salmons in a river. Later, they caught 50 salmons. Of these, 2 have tags. Estimate how many salmons are in the river ?

-----Continued-----

53. One card is drawn at random from a 52-card bridge deck. Find the probability that it is a 10 or a face card ?

54. One card is drawn at random from a 52-card bridge deck. Find the probability that it is a 10 or a red card ?

55. One card is drawn at random from a 52-card bridge deck. Find the probability that it is a 10 and a red card ?

56. Two fair dice are tossed. Find the probability that the sum of either 5 or 7.

57. One card is drawn at random from a 52-card bridge deck. Find the probability that it is an ace or a heart.

58. One card is drawn at random from a 52-card bridge deck. Find the probability that it is an ace and a heart.

59. Five cards are drawn from a 52-card bridge deck. Find the probability that exactly 2 of the cards are hearts.

60. Five cards are drawn from a 52-card bridge deck. Find the probability that at least 2 of the cards are hearts.

61. 13 cards are drawn from a 52-card bridge deck. Find the probability that exactly 2 of the cards are diamonds.

62. 13 cards are drawn from a 52-card bridge deck. Find the probability that at least 2 of the cards are diamonds.

63. There are 20 bulbs in a box and 5 are defective. You select 4 bulbs from the box. What is the probability that all four are defective ?

64. There are 20 bulbs in a box and 5 are defective. You select 4 bulbs from the box. What is the probability that exactly 3 are defective ?

65. There are 20 bulbs in a box and 5 are defective. You select 4 bulbs from the box. What is the probability that at least 3 are defective ?

66. Five fair coins are tossed. What is the probability of getting exactly 3 heads and 2 tails ?

67. Five fair coins are tossed. What is the probability of getting exactly 4 heads and 1 tails ?

68. Five fair coins are tossed. What is the probability of getting at least 3 heads ?

69. There are 8 boys and 12 girls in a class. A committee of 4 students is to be chosen at random. Find the probability that the committee will consist of 2 girls and 2 boys.

70. A company sells a certain product that averages one defective for every 50 it sells. Find the probability that an order of 20 units will have one or more defectives.

71. An electrical circuit is shown below. The current will flow from A to E either switch B is closed or if both switches C and D are closed. The switches operate independently of one another. The probabilities that the current will flow each switch when it is closed are $P(B) = \frac{1}{2}$, $P(C) = \frac{3}{5}$ and $P(D) = \frac{3}{10}$. Find the probability that the current will flow from A to E.

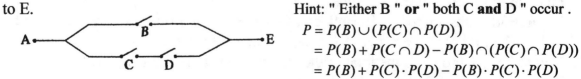

Hint: " Either B " **or** " both C **and** D " occur .
$$P = P(B) \cup (P(C) \cap P(D))$$
$$= P(B) + P(C \cap D) - P(B) \cap (P(C) \cap P(D))$$
$$= P(B) + P(C) \cdot P(D) - P(B) \cdot P(C) \cdot P(D)$$

72. There are 20 students in a class. Find the probability that everyone has a different birthday.

73. How many students must be in a class before the probability of at least two of them having the same birthday is: a) greater than $\frac{1}{2}$.   b) less than $\frac{1}{2}$.

11-6 Histogram and Normal Distribution

In statistics, we use a table called **frequency distribution** to show information about the frequency of occurrences of statistical data.

Histogram is a statistical graph to describe a frequency distribution.

In a histogram, data are grouped into convenient intervals. A boundary datum in a histogram is included in the interval to its left.

If the midpoints at the tops of the bars are connected, a smooth **"bell-shaped"** curve called **"normal curve"** can result from the plotting of a larger collection of data (see page 411). The area under the normal curve represents the probability of a **normal distributions**. In a normal distribution, the area under the curve and above the x-axis is 1. Therefore, the sum of the probabilities of all possible outcomes is 1 (or 100%).

frequency distribution

Test scores	numbers of students
45 ~ 55	3
55 ~ 65	13
65 ~ 75	21
75 ~ 85	27
85 ~ 95	15
95 ~ 100	7

Histogram

Mean: The arithmetic average of the data. It is denoted by M, $\overline{x}$, or μ.

Range: The difference between the highest and the lowest value of the data.

Mode: The most frequent number in the data.

Median: The middle number in the data when the data are arranged in increasing order.

Deviation: The difference of each datum from the mean.

Variance: The deviation or dispersion of the data. It shows how the data are scattered about the mean. It is denoted by v.

The variance of a frequency distribution is computed by squaring each deviation from the mean, adding these squares, and dividing their sum by the number of the entries.

$$\text{Variance } (v) = \frac{(x_1 - M)^2 + (x_2 - M)^2 + \cdots\cdots + (x_n - M)^2}{n} = \frac{\sum(x_i - M)^2}{n}$$

$x_i \rightarrow$ The ith entry of the data ; $i = 1, 2, 3, 4, \cdots\cdots, n$

$M \rightarrow$ The mean

Standard Deviation: The principal **square root** of the variance. It is denoted by σ.

Standard Deviation $(\sigma) = \sqrt{v}$, where v is the variance.

Examples

1. The frequency distribution of the scores of 5 games of a high school sport team is shown below. Find the mean, range, mode, median, variance, and standard deviation.

Data (scores)	Frequency
8	1
15	2
31	1
41	1

Solution:

$$\text{Mean } (M) = \frac{8 + 15(2) + 31 + 41}{5} = 22 \quad ; \quad \text{Range} = 41 - 8 = 33 \quad ; \quad \text{Mode} = 15$$

$$8, 15, \mathbf{15}, 31, 41 \; \rightarrow \; \text{Median} = 15$$

$$\text{Variance } (v) = \frac{(8-22)^2 + (15-22)^2 + (15-22)^2 + (31-22)^2 + (41-22)^2}{5} = 147.20$$

$$\text{Standard deviation } (\sigma) = \sqrt{147.20} = 12.13$$

2. The frequency distribution of the heights of a group of kids is shown below. Find the mean and the standard deviation.

Data (heights)	Frequency
90 *cm*	3
92	1
96	2
97	4
98	1
104	2

Solution:

x_i	f	$x_i \cdot f$	$x_i - M$	$(x_i - M)^2$	$(x_i - M)^2 \cdot f$
90	3	270	−6	36	108
92	1	92	−4	16	16
96	2	192	0	0	0
97	4	388	1	1	4
98	1	98	2	4	4
104	2	208	8	64	128
Sums	13	1,248			260

$$\text{Mean } (M) = \frac{1,248}{13} = 96 \quad ; \quad \text{Standard deviation } (\sigma) = \sqrt{\frac{260}{13}} = 4.472$$

Normal Distribution

A smooth bell-shaped curve that shows a frequency distribution from the plotting of a large collection of data. The probability of a given region is the area of that region under the curve.

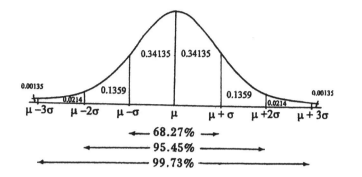

$$y = \frac{1}{\sigma\sqrt{2\pi}} e^{-(x-\mu)^2/2\sigma^2}$$

Standard deviation: σ

Mean: μ

$e = 2.71828\cdots\cdots$

Area = 1

Properties of the normal distribution:

1) The total area under the curve and above the x-axis is equal to 1.
2) The probability of the region $a \le x \le b$ is the area bounded by the curve.
3) It is symmetric with respect to the vertical line above the mean.
4) The probability of the region within 1, 2, or 3 standard deviations of the mean are: $P(\mu \pm \sigma) = 68.27\%$; $P(\mu \pm 2\sigma) = 95.45\%$; $P(\mu \pm 3\sigma) = 99.73\%$.

Standard Normal Distribution

A normal distribution with a mean equal to 0 and a standard deviation equal to 1.
It is symmetric with respect to the y- axis.

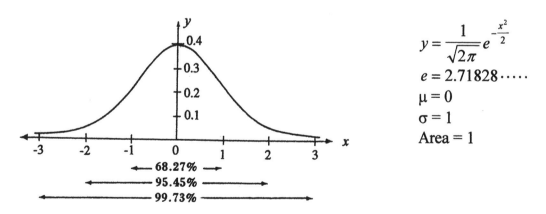

$$y = \frac{1}{\sqrt{2\pi}} e^{-\frac{x^2}{2}}$$

$e = 2.71828\cdots\cdots$

$\mu = 0$

$\sigma = 1$

Area = 1

A table for measuring the area under the curve of standard normal distribution has been made for computation. Table shows the area from $-\infty$ to x. (See Page 413 and Page 414)

$P(-\infty < x < -3.40) = 0.0003 = 0.03\%$; $P(-\infty < x < -3.49) = 0.0002 = 0.02\%$.

$P(-\infty < x < 0.00) = 0.5000 = 50\%$; $P(-\infty < x < 3.49) = 0.9998 = 99.98\%$.

$P(0 < x < 2.20) = P(-\infty < x < 2.20) - 0.50 = 0.9861 - 0.50 = 0.4861 = 48.61\%$.

Applications on Standard Normal Curve, see Problem 17 & 18, page 416.

Examples

1. Find the variance of the standard normal distribution.

Solution:

The standard deviation $\sigma = 1$,

$$\sigma = \sqrt{\text{var} iance} \qquad \therefore \text{Variance} = \sigma^2 = 1^2 = 1. \quad \text{Ans.}$$

2. Assume the following frequency distribution is a normal distribution. Tell whether it is a standard normal distribution.

x_i	$\dfrac{-9}{\sqrt{30}}$	$\dfrac{-6}{\sqrt{30}}$	$\dfrac{-3}{\sqrt{30}}$	$\dfrac{3}{\sqrt{30}}$	$\dfrac{6}{\sqrt{30}}$	$\dfrac{9}{\sqrt{30}}$
f	1	2	3	3	2	1

Solution :

x_i^2	$\dfrac{81}{30}$	$\dfrac{36}{30}$	$\dfrac{9}{30}$	$\dfrac{9}{30}$	$\dfrac{36}{30}$	$\dfrac{81}{30}$
$\sum(x_i^2 \cdot f_i)$						

$$\sum(x_i^2 \cdot f_i) \quad \frac{81}{30} + \frac{72}{30} + \frac{27}{30} + \frac{27}{30} + \frac{72}{30} + \frac{81}{30} = \frac{360}{30} = 12.$$

$$\mu = 0 \ , \ n = 12 \ , \quad \text{Variance} = \frac{\sum(x_i^2 \cdot f_i)}{n} = \frac{12}{12} = 1.$$

$$\therefore \sigma = \sqrt{\text{var} iance} = \sqrt{1} = 1. \qquad \text{Ans: It is a standard normal distribution.}$$

3. The mean of the heights of the students in a high school was found to be 172 cm with a standard deviation of 7 cm. Find the percentage of the students in each of the following category. Assuming a normal distribution of heights.

a) Height > 172 cm.   b) Height < 172 cm.   c) Height > 179 cm.   d) Height < 179 cm.
e) Height > 186 cm.   f) Height < 186 cm.   g) Height > 193 cm.

Solution:

a) 50%   b) 50%   c) $0.13590 + 0.0214 + 0.00135 = 0.15865 = 15.87\%$
d) $0.5 + 0.34135 = 0.84135 = 84.14\%.$ e) $0.0214 + 0.00135 = 0.02275 = 2.28\%.$
f) $1 - 0.0214 - 0.00135 = 0.97725 = 97.73\%.$ g) $0.00135 = 0.14\%.$

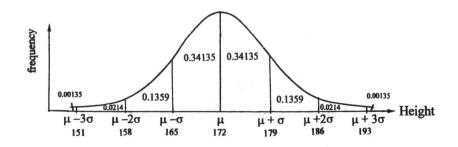

Area under the Standard Normal Curve
(Area from $-\infty$ to x)

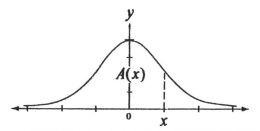

x	0.00	0.01	0.02	0.03	0.04	0.05	0.06	0.07	0.08	0.09
− 3.4	0.0003	0.0003	0.0003	0.0003	0.0003	0.0003	0.0003	0.0003	0.0003	0.0002
− 3.3	0.0005	0.0005	0.0005	0.0004	0.0004	0.0004	0.0004	0.0004	0.0004	0.0003
− 3.2	0.0007	0.0007	0.0006	0.0006	0.0006	0.0006	0.0006	0.0005	0.0005	0.0005
− 3.1	0.0010	0.0009	0.0009	0.0009	0.0008	0.0008	0.0008	0.0008	0.0007	0.0007
− 3.0	0.0013	0.0013	0.0013	0.0012	0.0012	0.0011	0.0011	0.0011	0.0010	0.0010
− 2.9	0.0019	0.0018	0.0017	0.0017	0.0016	0.0016	0.0015	0.0015	0.0014	0.0014
− 2.8	0.0026	0.0025	0.0024	0.0023	0.0023	0.0022	0.0021	0.0021	0.0020	0.0019
− 2.7	0.0035	0.0034	0.0033	0.0032	0.0031	0.0030	0.0029	0.0028	0.0027	0.0026
− 2.6	0.0047	0.0045	0.0044	0.0043	0.0041	0.0040	0.0039	0.0038	0.0037	0.0036
− 2.5	0.0062	0.0060	0.0059	0.0057	0.0055	0.0054	0.0052	0.0051	0.0049	0.0048
− 2.4	0.0082	0.0080	0.0078	0.0075	0.0073	0.0071	0.0069	0.0068	0.0066	0.0064
− 2.3	0.0107	0.0104	0.0102	0.0099	0.0096	0.0094	0.0091	0.0089	0.0087	0.0084
− 2.2	0.0139	0.0136	0.0132	0.0129	0.0125	0.0122	0.0119	0.0116	0.0113	0.0110
− 2.1	0.0179	0.0174	0.0170	0.0166	0.0162	0.0158	0.0154	0.0150	0.0146	0.0143
− 2.0	0.0228	0.0222	0.0217	0.0212	0.0207	0.0202	0.0197	0.0192	0.0188	0.0183
− 1.9	0.0287	0.0281	0.0274	0.0268	0.0262	0.0256	0.0250	0.0244	0.0239	0.0233
− 1.8	0.0359	0.0352	0.0344	0.0336	0.0329	0.0322	0.0314	0.0307	0.0301	0.0294
− 1.7	0.0446	0.0436	0.0427	0.0418	0.0409	0.0401	0.0392	0.0384	0.0375	0.0367
− 1.6	0.0548	0.0537	0.0526	0.0516	0.0505	0.0495	0.0485	0.0475	0.0465	0.0455
− 1.5	0.0668	0.0655	0.0643	0.0630	0.0618	0.0606	0.0594	0.0582	0.0571	0.0559
− 1.4	0.0808	0.0793	0.0778	0.0764	0.0749	0.0735	0.0722	0.0708	0.0694	0.0681
− 1.3	0.0968	0.0951	0.0934	0.0918	0.0901	0.0885	0.0869	0.0853	0.0838	0.0823
− 1.2	0.1151	0.1131	0.1112	0.1093	0.1075	0.1056	0.1038	0.1020	0.1003	0.0985
− 1.1	0.1357	0.1335	0.1314	0.1292	0.1271	0.1251	0.1230	0.1210	0.1190	0.1170
− 1.0	0.1587	0.1562	0.1539	0.1515	0.1492	0.1469	0.1446	0.1423	0.1401	0.1379
− 0.9	0.1841	0.1814	0.1788	0.1762	0.1736	0.1711	0.1685	0.1660	0.1635	0.1611
− 0.8	0.2119	0.2090	0.2061	0.2033	0.2005	0.1977	0.1949	0.1922	0.1894	0.1867
− 0.7	0.2420	0.2389	0.2358	0.2327	0.2296	0.2266	0.2236	0.2206	0.2177	0.2148
− 0.6	0.2743	0.2709	0.2676	0.2643	0.2611	0.2578	0.2546	0.2514	0.2483	0.2451
− 0.5	0.3085	0.3050	0.3015	0.2981	0.2946	0.2912	0.2877	0.2843	0.2810	0.2776
− 0.4	0.3446	0.3409	0.3372	0.3336	0.3300	0.3264	0.3228	0.3192	0.3156	0.3121
− 0.3	0.3821	0.3783	0.3745	0.3707	0.3669	0.3632	0.3594	0.3557	0.3520	0.3483
− 0.2	0.4207	0.4168	0.4129	0.4090	0.4052	0.4013	0.3974	0.3936	0.3897	0.3859
− 0.1	0.4602	0.4562	0.4522	0.4483	0.4443	0.4404	0.4364	0.4325	0.4286	0.4247
− 0.0	0.5000	0.4960	0.4920	0.4880	0.4840	0.4801	0.4761	0.4721	0.4681	0.4641

Source: National Bureau of Standard, Washington D.C.

Area under the Standard Normal Curve
(Continued)

x	0.00	0.01	0.02	0.03	0.04	0.05	0.06	0.07	0.08	0.09
0.0	0.5000	0.5040	0.5080	0.5120	0.5160	0.5199	0.5239	0.5279	0.5319	0.5359
0.1	0.5398	0.5438	0.5478	0.5517	0.5557	0.5596	0.5636	0.5675	0.5714	0.5753
0.2	0.5793	0.5832	0.5871	0.5910	0.5948	0.5987	0.6026	0.6064	0.6103	0.6141
0.3	0.6179	0.6217	0.6255	0.6293	0.6331	0.6368	0.6406	0.6443	0.6480	0.6517
0.4	0.6554	0.6591	0.6628	0.6664	0.6700	0.6736	0.6772	0.6808	0.6844	0.6879
0.5	0.6915	0.6950	0.6985	0.7019	0.7054	0.7088	0.7123	0.7157	0.7190	0.7224
0.6	0.7257	0.7291	0.7324	0.7357	0.7389	0.7422	0.7454	0.7486	0.7517	0.7549
0.7	0.7580	0.7611	0.7642	0.7673	0.7704	0.7734	0.7764	0.7794	0.7823	0.7852
0.8	0.7881	0.7910	0.7939	0.7967	0.7995	0.8023	0.8051	0.8078	0.8106	0.8133
0.9	0.8159	0.8186	0.8212	0.8238	0.8264	0.8289	0.8315	0.8340	0.8365	0.8389
1.0	0.8413	0.8438	0.8461	0.8485	0.8508	0.8531	0.8554	0.8577	0.8599	0.8621
1.1	0.8643	0.8665	0.8686	0.8708	0.8729	0.8749	0.8770	0.8790	0.8810	0.8830
1.2	0.8849	0.8869	0.8888	0.8907	0.8925	0.8944	0.8962	0.8980	0.8997	0.9015
1.3	0.9032	0.9049	0.9066	0.9082	0.9099	0.9115	0.9131	0.9147	0.9162	0.9177
1.4	0.9192	0.9207	0.9222	0.9236	0.9251	0.9265	0.9278	0.9292	0.9306	0.9319
1.5	0.9332	0.9345	0.9357	0.9370	0.9382	0.9394	0.9406	0.9418	0.9429	0.9441
1.6	0.9452	0.9463	0.9474	0.9484	0.9495	0.9505	0.9515	0.9525	0.9535	0.9545
1.7	0.9554	0.9564	0.9573	0.9582	0.9591	0.9599	0.9608	0.9616	0.9625	0.9633
1.8	0.9641	0.9649	0.9656	0.9664	0.9671	0.9678	0.9686	0.9693	0.9699	0.9706
1.9	0.9713	0.9719	0.9726	0.9732	0.9738	0.9744	0.9750	0.9756	0.9761	0.9767
2.0	0.9772	0.9778	0.9783	0.9788	0.9793	0.9798	0.9803	0.9808	0.9812	0.9817
2.1	0.9821	0.9826	0.9830	0.9834	0.9838	0.9842	0.9846	0.9850	0.9854	0.9857
2.2	0.9861	0.9864	0.9868	0.9871	0.9875	0.9878	0.9881	0.9884	0.9887	0.9890
2.3	0.9893	0.9896	0.9898	0.9901	0.9904	0.9906	0.9909	0.9911	0.9913	0.9916
2.4	0.9918	0.9920	0.9922	0.9925	0.9927	0.9929	0.9931	0.9932	0.9934	0.9936
2.5	0.9938	0.9940	0.9941	0.9943	0.9945	0.9946	0.9948	0.9949	0.9951	0.9952
2.6	0.9953	0.9955	0.9956	0.9957	0.9959	0.9960	0.9961	0.9962	0.9963	0.9964
2.7	0.9965	0.9966	0.9967	0.9968	0.9969	0.9970	0.9971	0.9972	0.9973	0.9974
2.8	0.9974	0.9975	0.9976	0.9977	0.9977	0.9978	0.9979	0.9979	0.9980	0.9981
2.9	0.9981	0.9982	0.9982	0.9983	0.9984	0.9984	0.9985	0.9985	0.9986	0.9986
3.0	0.9987	0.9987	0.9987	0.9988	0.9988	0.9989	0.9989	0.9989	0.9990	0.9990
3.1	0.9990	0.9991	0.9991	0.9991	0.9992	0.9992	0.9992	0.9992	0.9993	0.9993
3.2	0.9993	0.9993	0.9994	0.9994	0.9994	0.9994	0.9994	0.9995	0.9995	0.9995
3.3	0.9995	0.9995	0.9995	0.9996	0.9996	0.9996	0.9996	0.9996	0.9996	0.9997
3.4	0.9997	0.9997	0.9997	0.9997	0.9997	0.9997	0.9997	0.9997	0.9997	0.9998

Source: National Bureau of Standard, Washington D. C.

Roger was absent yesterday.
Teacher: Roger, why were you absent ?
Roger: My tooth was aching.
I went to see my dentist.
Teacher: Is your tooth still aching ?
Roger: I don't know. My dentist has it.

EXERCISES

Find the mean (μ), variance (v) and standard deviation (σ) of the data.

1. 2, 3, 4, 5, 6   **2.** 5, 6, 7, 8, 9, 6, 6   **3.** 10, 8, 29, 6, 20, 17

4. 7, 8, 9, 12 18, 18, 21   **5.** 9, 9, 12, 18, 22, 24, 32   **6.** 10, 20, 18, 10, 10, 20, 15, 20

7. The test scores of 50 students are listed below.

```
12  25  41  74  52  70  32  49  54  63
26  44  47  33  55  16  42  24  89  34
44  79  58  21  38  54  39  57  42  51
52  59  29  35  69  43  56  63  30  44
37  40  45  52  51  18  61  75  88  31
```

     a. Make a frequency distribution   b. Complete the following histogram

Test scores	Tally	Frequency
10 ~ 20		
20 ~ 30		
30 ~ 40	𝐓𝐇𝐋 ////	9
40 ~ 50		
50 ~ 60		
60 ~ 70		
70 ~ 80		
80 ~ 90		

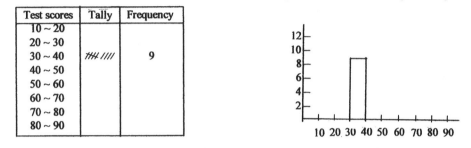

8. The time spent by shoppers in the mall is normally distributed as shown in the graph.

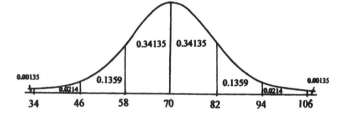

     a. What percent of the shoppers in the mall will spend between 58 and 82 minutes ?
     b. What is the probability that a shopper will spend more than 58 minutes ?
     c. What is the probability that a shopper will spend less than 58 minutes ?
     d. What percent of the shoppers in the mall will spent between 46 and 94 minutes ?

9. In the experiment, the probability of operating life of a new brand battery is normally distributed as shown in the graph.

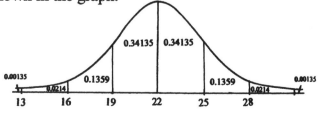

     a. What is the probability that a battery will last more than 25 hours ?
     b. What is the probability that a battery will last less than 25 hours ?   **---Continued---**

10. The scores on a math standardized test gave a normal distribution as shown in the graph.

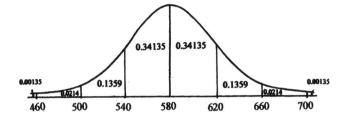

a. What is the probability that a student's score is more than 500 ?
b. What is the probability that a student's score is less than 620 ?

11. The sales of a math book each month are normally distributed as shown in the graph.

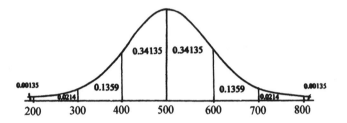

a. What is the probability that a month will sell more than 600 copies in a month ?
b. What is the probability that a month will sell less than 300 copies in a month ?

12. A survey shows that the time spent by shoppers in the mall is normally distributed with a mean of 50 minutes and a standard deviation of 14 minutes.

a. What is the probability that a randomly chosen shopper will spend between 36 and 64 minutes in the mall ?
b. What is the probability that three randomly chosen shoppers will all spend between 36 and 64 minutes in the mall.

13. The scores on a mathematics test in a high school were normally distributed with a mean of 482 and a standard deviation of 94.

a. What is the probability that a random chosen student who took the test scored more than 576 ?
b. What is the probability that two random chosen students who took the test both scored between 388 and 482 ?

14. A normal distribution has a mean of 32 and a standard deviation of 4. Find the probability that a randomly selected x-value is between 24 and 40.

15. A normal distribution has a mean of 72 and a standard deviation of 5. Find the probability that four randomly selected x-value are all between 62 and 77 ?

16. A company makes 100,000 tires a year. The life of a tire is normally distributed with a mean of 32,000 miles and a standard deviation of 3,000 miles. About how many tires will last between 35,000 and 38,000 miles ?

17. Find the percentage of the area under the standard normal curve under $0 < x < 1$.

18. In Problem 16, about how many tires will last less than 40,000 miles ?
(Hint: Problems 17 & 18, see Standard Normal Curve on page 413 ~ 414)

11-7 Binomial Distribution

Binomial Experiment: It is an experiment which has n independent trials and only two possible outcomes "success and failure". The probability of success of each trial is the same.

Binomial Probability: In a binomial experiment, the probability of x successes in n trials is:

Formula: $P(x) =_n C_x P^x (1-P)^{n-x}$ where P is the probability of success on each trial and $(1-P)$ is the probability of failure.

Binomial Distribution: The probabilities of all possible numbers of successes in a binomial experiment.

Suppose we throw a fair coin 10 times. The probability of each throw turned up head is $\frac{1}{2}$.

The probability that none of all 10 throws turned up head is $\left(\frac{1}{2}\right)^{10} = 0.00098$.

The probability that first throw is head and all the other 9 throws are tails is $\left(\frac{1}{2}\right)^1 \left(\frac{1}{2}\right)^9 = 0.00098$. However, the number of combinations of 1 head and 9 tails in 10 throws are $_{10}C_1 = 10$. Therefore, the probability of the event that 1 head and 9 tails in 10 throws is:

$$P(1) =_{10} C_1 \left(\tfrac{1}{2}\right)^1 \left(\tfrac{1}{2}\right)^9 = 10 \cdot (0.00098) = 0.0098$$

Similarly, The probability of the event that 2 heads and 8 tails is:

$$P(2) =_{10} C_2 \left(\tfrac{1}{2}\right)^2 \left(\tfrac{1}{2}\right)^8 = 45 \cdot (0.00098) = 0.0441$$

The probability that the event having each x-value ($x = 0, 1, 2, 3, \cdots, 10$) of throwing a fair coin 10 times can be calculated from:

$$P(x) =_{10} C_x \left(\tfrac{1}{2}\right)^x \left(\tfrac{1}{2}\right)^{10-x} \qquad \text{Therefore: For } n \text{ throws: } P(x) =_n C_x \left(\tfrac{1}{2}\right)^x \left(\tfrac{1}{2}\right)^{n-x}$$

The result of the above probabilities and its histogram of the binomial distribution are shown below:

x	$P(x)$
0	0.00098
1	0.0098
2	0.0441
3	0.1176
4	0.2058
5	0.2470
6	0.2058
7	0.1176
8	0.0441
9	0.0098
10	0.00098

Binomial Distribution

The distribution of the above example is **symmetric** because the probability of success is 0.5. Otherwise, the distribution is **skewed**. When n is large and p is close to 0.5, a binomial distribution is close to a normal distribution.

Examples

1. The probability that John makes the basket on a free throw during the regular basketball game is $\frac{2}{5}$. In the next game, what is the probability that John can make exactly 3 of 4 free throws ?

Solution: Let $P = \frac{2}{5}$ be the probability of success (making the basket)

$1 - P = \frac{3}{5}$ be the probability of failure (missing the basket)

The probability of getting exactly 3 out of 4 free throws is:

$$P(x = 3) = {_4}C_3 (\tfrac{2}{5})^3 (1 - \tfrac{2}{5})^{4-3} = 4 \cdot (\tfrac{2}{5})^3 (\tfrac{3}{5}) = 4 \cdot (0.064) \cdot (0.6) = 0.1536 \,. \text{ Ans.}$$

2. A company makes light bulbs. The defective rate is 0.004 (4 out of 1000). If you buy 600 bulbs from this company, what is the probability that none of the 600 bulbs is defective ?

Solution:

Let $P = 0.996$ be the probability of success (not defective)

$1 - P = 0.004$ be the probability of failure (defective)

The probability of getting exactly 600 non-defective bulbs out of 600 bulbs is:

$$P(x = 600) = {_{600}}C_{600} (0.996)^{600} (0.004)^0 = 1 \cdot (0.996)^{600} \cdot 1 = 0.0903 \,. \text{ Ans.}$$

The mean (μ) and standard deviation (σ) of a binomial distribution

We can prove that the mean and standard deviation of a binomial distribution are given by the following formulas:

(The proofs can be found in the book "Mathematics in Management", written by Rong Yang , 1987)

$$\mu = np \quad \text{and} \quad \sigma = \sqrt{np(1-p)}$$

Suppose we throw a fair coins 20 times and want to compute the probability that the event having at most 12 throws turned up heads. We would have to use the binomial probability formula to compute the following:

$$P(x \le 12) = \sum_{x=0}^{12} {_{20}}C_x (\tfrac{1}{2})^x (\tfrac{1}{2})^{20-x} = {_{20}}C_0 (\tfrac{1}{2})^0 (\tfrac{1}{2})^{20} + {_{20}}C_1 (\tfrac{1}{2})^1 (\tfrac{1}{2})^{19} + \cdots\cdots + {_{20}}C_{12} (\tfrac{1}{2})^{12} (\tfrac{1}{2})^8$$

It is tedious to compute the above binomial expression. However, we may be able to use a normal distribution to find the approximation of a binomial distribution if n is larger and p is close to 0.5.

Example: A fair coin is tossed 20 times. Find the probability of the event having at most 12 throws turned up heads.

Solution:

$$\mu = np = 20 \cdot (\tfrac{1}{2}) = 10 \quad \text{and} \quad \sigma = \sqrt{np(1-p)} = \sqrt{20 \cdot \tfrac{1}{2} \cdot (1 - \tfrac{1}{2})} = 2.236$$

Using the Table of Area under the Standard Normal Curve on page 414

$$Z = \frac{12-10}{2.236} = 0.894 \quad \therefore P(x \le 12) = P(-\infty \le Z \le 0.894) = 0.8143 \,. \text{ Ans.}$$

If we use the binomial probability formula, we have the actual answer $P(x \le 12) = 0.8684$.

A table for binomial probabilities has been made for computation in the college math book "Mathematics In Management", written by Rong Yang, 1987.

EXERCISES

Evaluate each of the following binomial expressions.

1. $_{12}C_2(\frac{1}{2})^2(\frac{1}{2})^{10}$ **2.** $_{12}C_2(\frac{1}{3})^2(\frac{2}{3})^{10}$ **3.** $_{12}C_2(0.45)^2(0.55)^{10}$ **4.** $_{12}C_2(0.55)^2(0.45)^{10}$

5. $_{20}C_{17}(\frac{1}{2})^{17}(\frac{1}{2})^3$ **6.** $_{18}C_5(0.5)^5(0.5)^{13}$ **7.** $_{15}C_4(0.37)^4(0.63)^{11}$ **8.** $_8C_7(\frac{1}{4})^7(\frac{3}{4})$

Find the mean (μ) and standard deviation (σ) of each binomial experiment consisting of n trials with probability p of success on each trial.

9. $n=12,\ p=\frac{1}{2}$ **10.** $n=12,\ p=\frac{1}{3}$ **11.** $n=12,\ p=0.45$ **12.** $n=12,\ p=0.55$

13. $n=20,\ p=\frac{1}{2}$ **14.** $n=18,\ p=0.5$ **15.** $n=15,\ p=0.37$ **16.** $n=8,\ p=\frac{1}{4}$

17. The probability that John makes the basket on a free throw during the regular games is $\frac{2}{5}$. In the coming games,
   a. what is the probability that John can make exactly 4 of 10 free throws ?
   b. what is the probability that John can make at most 4 of 10 free throws ?
   c. what is the probability that John can make at least 4 of 10 free throws ?
   d. What is the probability that John can make at most 50 of 100 free throws ?

18. Six fair coins are tossed. Find the probability of having two throws turned up heads.

19. A bag contains seven white and three black balls. One ball is drawn each time at random. A ball drawn is put back into the bag before the next ball is drawn. A total of five balls are drawn. Find the probability that the five balls having three whites and 2 blacks.

20. A company makes light bulbs. The defective rate is 0.001 (1 out of 1000). If you buy 500 bulbs from this company,
   a. what is the probability that none of the 500 bulbs is defective.
   b. What is the probability that the defective bulbs are at most 3 bulbs.

21. The probability that Roger makes the hit each time at bat in a baseball game is 0.4.
   a. What is the probability that Roger can get at most 3 hits of 5 times at bat.
   a. What is the probability that Roger can get 3 or more hits of 5 times at bat.

22. Rosa guesses at all 10 true-false questions on her math test. Find the probability.
   a. At most 4 are right.   b. At most half are right.
   c. At least half are right.   d. All are right.

23. A fair coin is tossed 100 times. Find the mean and standard deviation of the number of heads.

24. A fair coin is tossed 100 times. Use a normal distribution to find the approximation of the probability of the event having at most 60 heads.

25. An unfair coin is tossed 100 times. The probability of each toss turned up head is $\frac{3}{5}$. Find the mean and standard deviation of the number of heads.

26. An unfair coin is tossed 900 times. The probability of each toss turned up head is $\frac{1}{3}$. Find the mean and standard deviation of the number of heads.

Space for Taking Notes

11-8 Expected Value

Definition of Expected Value:

In a random experiment, the values of the n outcomes are x_1, x_2, x_3, $\cdots\cdots, x_n$ and the corresponding probabilities of the outcomes occurring are p_1, p_2, p_3, $\cdots\cdots, p_n$. The expected value (EV) of the experiment is given by:

$$EV = p_1 x_1 + p_2 x_2 + p_3 x_3 + \cdots\cdots + p_n x_n$$

Examples

1. A fair die is tossed. You win \$30 if the number turned up a 1. What is the expected value of the game ?
    Solution:

    $$EV = \$30 \times \tfrac{1}{6} = \$5 . \text{ Ans}$$

2. A fair die is tossed. You win \$30 if the number turned up a 6 or win \$24 if the number turned up a 5. What is the expected value of the game ?
    Solution: $EV = \$30 \times \tfrac{1}{6} + \$24 \times \tfrac{1}{6} = \$9$. Ans.

3. A fair die is tosses. What is the expected value of all numbers turned up.
    Solution: $EV = 1 \cdot \tfrac{1}{6} + 2 \cdot \tfrac{1}{6} + 3 \cdot \tfrac{1}{6} + 4 \cdot \tfrac{1}{6} + 5 \cdot \tfrac{1}{6} + 6 \cdot \tfrac{1}{6} = 3.5$. Ans.

4. A company offers a lottery of 1,000 tickets numbered from 1 to 1,000. You purchase one ticket for \$5. You must match one of the numbers drawn to win a prize. The prizes are \$500 for one ticket, two tickets of \$300 each, and 10 tickets of \$40 each.

         a. What is the expected value of one ticket ?
         b. Is it a fair game to you ?
    Solution:

         a. $EV = 500 \cdot \tfrac{1}{1000} + 300 \cdot \tfrac{2}{1000} + 40 \cdot \tfrac{10}{1000} = \$1.50$. Ans.

         b. It is an unfair game. You can expect to lose \$5 $-$\$1.50 = \$3.50 on each ticket.

5. Two dice are tossed. Find the expected value of the sum of the two numbers turned up.
    Solution: (See Example 1 on page 401)

Sum	2	3	4	5	6	7	8	9	10	11	12
frequency	1	2	3	4	5	6	5	4	3	2	1
Probability	1/36	2/36	3/36	4/36	5/36	6/36	5/36	4/36	3/36	2/36	1/36

$$EV = 2 \cdot \tfrac{1}{36} + 3 \cdot \tfrac{2}{36} + 4 \cdot \tfrac{3}{36} + 5 \cdot \tfrac{4}{36} + 6 \cdot \tfrac{5}{36} + 7 \cdot \tfrac{6}{36} + 8 \cdot \tfrac{5}{36} + 9 \cdot \tfrac{4}{36} + 10 \cdot \tfrac{3}{36} + 11 \cdot \tfrac{2}{36} + 12 \cdot \tfrac{1}{36}$$

$$= \tfrac{252}{36} = 7 . \text{ Ans}$$

To learn more applications about Expected Value, read the college book "Mathematics in Management, Decision-Making Theory and Game Theory " written by Rong Yang, 1987.

EXERCISES

1. A fair coin is tossed. You win $10 if it is a head. What is the expected value ?

2. A fair coin is tossed. You win $10 if it is a head. You lose $6 if it is a tail. Do you like to play the game ? Explain.

3. A fair die is tossed. You win $8 if the number turned up a 1 or win $5 if the number turned up a 2. Otherwise, you lose $2. Do you like to play the game ? Explain.

4. A company offers a lottery of 500 tickets numbered from 1 to 500. You purchase one ticket for $5. You must match one of the numbers drawn to win a prize. The prizes are 2 tickets of $500 each, 4 tickets of $200 each, 10 tickets of $50 each, and 20 tickets of $20 each. What is the expected value of each ticket ? Do you like to play the game ? Explain.

5. A company offers a lottery of 5,000 tickets. The purchasing price of each ticket is $10. The prizes are 2 tickets of $10,000 each, 5 tickets of $2,000 each, 10 tickets of $500 each, and 50 tickets of $100 each. What is the expected value of each ticket ? Is it a fair game ?

6. A bag contains 3 red balls and 2 white balls. You draw one ball at random from the bag. You win $10 if the ball drawn is red and lose $12 if the ball drawn is white. Find the expected value of each draw.

7. A company offers a fire insurance for houses. The premium of the insurance is $1,000 per year for a $400,000 house. The rate of the houses destroyed on fires is 0.002 (2 out of 1000 houses) annually. What is the expected value of the company's profit for each house ?

8. A bag contains 4 one-dollar bills, 3 five-dollar bills, 5 ten-dollar bills, 2 fifty-dollar bills, and 3 hundred-dollar bills. (Hint: Mean = Expected Value for each bill)
      a. What is the expected value of one bill drawn from the bag at random ?
      b. What is the expected value of two bills drawn from the bag at random ?

9. Two fair dice are tossed. You win $30 if the sum of the two numbers turned up a 6. What is the expected value of each toss ?

10. Two fair coins are tossed. You win $6 if two coins turned up heads and $2 if one coin turned up head. You lose $12 if two coins turned tails. Is the game fair to you ? Explain.

11. Three fair coins are tossed. You win $10 if all three coins turned up heads, $5 if two coins turned up heads, and $2 if one coin turned up head. Considering the game is fair, How much you should lose if all three coins turned up tails ?

12. A company offers a life insurance for young people. The premium of the insurance is $150 per year for $100,000. The death rate of young people from the age 20 to 21 is 0.002. What is the expected value of the profit of this company ?

13. There are 1,000 contestants in a marathon tournament. The prizes are $1000 for the 1st place, $600 for the 2nd place, $300 for the 3rd place, and $100 each of the 4th ~10th place. What is the expected value for each contestant.

14. To win the jackpot of the Multistates MegaMillions Lottery, you must match 5 numbers from 1 to 56, plus a Mega number from 1 to 46. You purchase one ticket for $1. If the jackpot for this coming Saturday is $85,000,000, what is the expected value of your ticket ?

15. To win the jackpot of the Multistates Powerball Lottery, you must match 5 numbers from 1 to 55, plus a Powerball number from 1 to 42. You purchase one ticket for $1. If the jackpot for this coming Saturday is $85,000,000, what is the expected value of your ticket ?

CHAPTER 11 EXERCISES

How many different ways can be selected from city A to city D.

1.   **2.**

$$A \rightleftarrows B \rightarrow C \rightleftarrows D$$

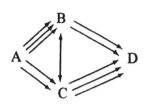

3. How many clubs can be formed selecting one student from each of the three classes 40, 45, 50 students ?

4. How many license plates of 7 symbols can be made using letters for the first 3 symbols and digits for the remaining 4 symbols ?

5. How many positive odd integers less than 1,000 can be made using the digits 4, 5, 6, 7, and 8 ?

Evaluate each factorial expression.

6. $\dfrac{10!}{7!}$ **7.** $\dfrac{12!}{(10-2)!}$ **8.** $\dfrac{35!}{3!(35-3)!}$ **9.** $\dfrac{100!}{98!}$ **10.** $\dfrac{n!}{(n-2)! \cdot n}$

11. How many permutations can be made from the letters a, b, c, d, e, f taken 4 letters at a time ?

12. How many combinations can be made from the letters a, b, c, d, e, f taken 4 letters at a time ?

13. How many different 5-card hands can be formed from 52 bridge cards ?

14. How many clubs can be formed selecting 6 students from 35 students ?

15. Find the odds of California Fantasy Lottery from 1 to 39, win all 5 or any 4, 3, 2 numbers of the 5 winning numbers if we choose 5 numbers at a time ?

16. How many 5-digit positive integers can be formed using the digits 1, 2, 3, 4, 5 if no digit may be used repeatedly ?

17. How many 5-digit positive integers can be formed using the digits 1, 2, 3, 4, 5 if each digit can be used repeatedly ?

18. How many 3-digit positive integers can be formed using the digits from 0 to 9 if each digit can be used repeatedly ?

19. How many permutations can be formed from the letters in SCHOOLHOUSE, taken 11 at a time ?

20. How many permutations can be formed from the letters in LEASEHOLDERS, take 12 at a time ?

21. How many six-digit numerals can be made from 2, 3, 4, 5 if we use 2, 4 twice and 3, 5 once ?

22. How many seven-digit numerals can be formed from 3, 6, 9 if we use 3 four times, 6 twice and 9 once ?

23. How many short-cut routes can be selected from A to B ?

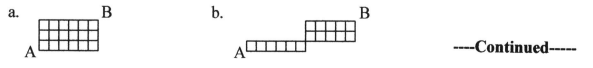

----Continued-----

24. How many different arrangements can be made for 6 students seated in a row ?

25. Find the number of different arrangements of 6 people seated in a round table.

26. Find the number of necklace strings that can be made from 6 beads of different colors.

27. How many different arrangements can be made for 6 students taken 5 at a time in a row ?

28. How many different arrangements can be made for 6 students taken 5 at a time in a round table ?

29. Find the odd of picking 3 out of 5 prize matching-number and no Powerball matching-number if you choose 5 numbers from 1 to 55 and a Powerball number from 1 to 42.

30. Find the odd of picking 3 out of 5 prize matching-number and the Powerball matching-number if you choose 5 numbers from 1 to 55 and a Powerball number from 1 to 42.

31. Specify each of the following sets, referring to the Venn Diagram.
   a. $A \cap B$
   b. $A \cup B$
   c. $B \cap C$
   d. $B \cup C$
   e. $(A \cup B) \cap C$

32. In a high school class, there are 23 students who took Algebra class, 17 students took Biology class. These figure include 5 students took both Algebra and Biology classes. How many students are in this class ?

33. One-half of all Patts are Matts. All Natts are Matts. No Natts is a Patts. One-half of Matts are Patts or Natts. There are 10 Natts and 30 Patts. How many Matts are neither Patts nor Natts ?

34. Two cards are drawn from the letter cards 1, 2, 3, 4, 5. Specify:
a) the sample space   b) the event that the numbers turned up on the two cards total 6.

35. Two bags contain balls. The first bag contains four balls labeled with numbers 1, 3, 5, 7. The second bag contains four balls labeled with numbers 2, 4, 6, 8. One ball is drawn from each bag.
   a) Specify the sample space and the event the numbers turned up on the two balls total 9.
   b) Specify the event that the first number drawn is larger than the second number drawn.

36. Three dice are tossed. Find the probability of the sum turned up: a) 5.   b)10.   c) 15.

37. A bag contains 5 red, 3 white and 2 black balls. Three balls are drawn at random. Find the probability that: a) all three are same colors.   b) two are red and one is white.
                         c) all three are different colors.

38. One card is drawn from number cards 1, 2, 3,.....,50. Find the probability of the card drawn turned up: a) multiple of 5.   b) multiple of 7.   c) multiple of both 5 and 7.

39. A coin is tossed twice. Let A be the event that a head is obtained on the first toss and B the event that a head is obtained on the second toss.
       a) Specify the events A, B, A∩B, A∪B.   b) Find the probability of A, B, A∩B, A∪B.
       c) Are A and B independent events ?

40. A coin is tossed twice. Let A be the event that at least one toss comes tails and B the event two tosses come up different. Tell whether the events are independent ?

-----Continued-----

41. A single die is tossed twice. Let A be the event that the first number turned up an even number and B the event that the second number turned up an odd number.
  a) Specify the events A, B, A∩B, A∪B.   b) Find the probability of A, B, A∩B, A∪B.
  c) Are A and B independent events ?

42. Find the mean, range, mode, median, variance and standard deviation of the data:
  a) 1, 2, 2, 3, 3, 3, 3, 5, 6.   b) 2, 3, 4, 4, 6, 7, 9, 10.   c) 92, 87, 87, 82, 72.

43. The frequency distribution of the test scores(points) of the 50 students is shown as the table. Assuming a normal distribution of points.

Points	Frequency
45	3
55	5
65	14
75	22
85	4
95	2
50 students	

  a. Draw a "bell shaped" distribution curve.
  b. Find the mean.
  c. Find the variance.
  d. Find the standard deviation.
  e. Find the percentage of the test scores in each category:
      points > 70 ; points > 88
      points < 92 ; points < 103.

44. The scores on a math standardized test gave a normal distribution as shown in the graph.

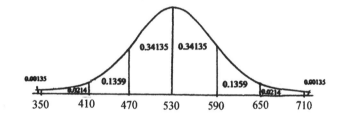

  a. What is the percentage of the students scored between 590 and 650 ?
  b. What is the percentage of the students scored between 410 and 590 ?

45. A company makes screws. The mean of the diameter is 2.20 *mm* and the standard deviation is 0.15 *mm*. It is defective if a screw's diameter is more than 2.50 *mm*. Find the defecive rate of the screws made by the company.

46. 674 boys and girls took the entrance examination of a popular high school. 550 of them with higher scores will receive the admission of the school. The mean of the test scores is 500 and the standard deviation is 100. What is the lowest test score need to receive the admission of the school ?

47. The probability that you make the basket on a free throw is $\frac{3}{4}$. What is the probability that you can make 7 or more of 10 free throws ?

48. A company makes skews. The defective rate is 0.05 (5 out of 100). If you buy 50 screw, what is the probability that the defective screws are at most 2 screws ?

49. a fair coin is tossed 200 times. Find the mean and standard deviation of the number of the heads.

50. An unfair coin is tossed 200 times. The probability of each toss turned up head is $\frac{2}{5}$. Find the mean and standard deviation of the number of the heads.

-----Continued-----

51. An unfair coin is tossed 200 times. The probability of each toss turned up head is $\frac{2}{5}$. Find the probability of the event having at most 90 heads.

52. A fair die is tossed. You win $20 if the number turned up a 1 or win $15 if the number turned up a 2. What is the expected value of the game ?

53. Two fair dice are tossed. You win $20 if two numbers turned up the same. What is the expected value of the game.

54. A company offers a life insurance for people who are 60 and up. The premium of the insurance is $350 per year for $100,000. The death rate of the people from age 60 to 61 is 0.006 annually. What is the expected value of the profit of this company ?

55. To win the jackpot of the Fantasy Lottery, you must match 5 winning numbers from 1 to 39. You purchase one ticket for $1. If the jackpot is 450,000 for this coming Tuesday game, what is the expected value of you ticket ?

56. A young girl sells roses at the airport everyday. She purchases each rose for $1 and sell for $3. A rose is abandoned if it cannot be sold within the day. The girl knows that the probability of selling 30 roses is $\frac{4}{10}$, 40 roses is $\frac{3}{10}$, and 50 roses is $\frac{3}{10}$.

      a. Find the expected value of the profit for each quantity she purchases.

      b. How many roses she should purchase each day to expect the highest profit ?

Hint: We have the following table of profits related to each buy-sell condition.

sell buy	30 (4/10)	40 (3/10)	50 (3/10)
30	$60	60	60
40	50	80	80
50	40	70	100

57. In Decision-Making and Game Theory, it states that the best strategy taken to deal with the opponent is the strategy which guarantees an expected value regardless the strategy taken by the opponent. The following table shows the payoff of two strategies A1 and A2 which Company A can choose to deal with the strategies B1 and B2 taken by Company B. What are the probabilities of the strategies p_1 and p_2 which Company A should choose to guarantee an expected value ? What is the expected value (EV) of Company A ?

A \ B	B1	B2	
A1	$20	−10	p_1
A2	−30	40	$p_2 = 1 - p_1$

Hint: $E(B_1) = 20p_1 - 30(1 - p_1)$

$E(B_2) = -10p_1 + 40(1 - p_1)$

$E(B_1) = E(B_2)$

EV is called **Game Value** in Game Theory.

Solve for *n*

58. $_nC_3 = 120$
    59. $_{n+1}C_3 = 165$
    60. $\frac{17}{5} \cdot {_nP_2} = {_{n+8}P_2}$
    61. $\frac{17}{5} \cdot {_nC_2} = {_{n+8}C_2}$

Answers

CHAPTER 11–1 Basic Counting Principle page 390
1. 72 **2.** 22 **3.** 11,881,376 **4.** 7,893,600 **5.** 256 **6.** 175,760,000 **7.** 8,000,000 **8.** 2,343 **9.** 39,750 **10.** 3,888
11. 59,049 **12.** 10,140,000 **13.** 120 **14.** 1,048,576 **15.** 40,320 **16.** 72 **17.** 384 **18.** 720 **19.** 24 **20.** 180
21. 300

CHAPTER 11–2 Permutations and Combinations page 397 ~ 398
1. 8 **2.** 40,320 **3.** 840 **4.** 870 **5.** 9,900 **6.** 90,345,024 **7.** 390,700,800 **8.** 2.326×10^{52} **9.** 8 **10.** 8 **11.** 35
12. 432 **13.** 4,950 **14.** 3,764,376 **15.** 1,140 **16.** 2.251×10^{12} **17.** a. 720 b. 360 c. 6 **18.** a. 1 b. 15 c. 6
19. 116,280 **20.** 4,845 **21.** 1296 **22.** 36 **23.** 15,626 **24.** 101 **25.** 131 **26.** 10 **27.** 40 **28.** 32 **29.** 200
30. 5,040 **31.** 3,520 **32.** 34,650 **33.** 453,600 **34.** 30 **35.** 2520 **36.** 6 **37.** 3,628,800 **38.** 362,880
39. 151,200 **40.** 25,200 **41.** 40,320 **42.** 5,040 **43.** 1,680 **44.** 420 **45.** 20 **46.** 4 **47.** 2,520 **48.** 720 **49.** 120
50. 60 **51.** 10 **52.** 1/3387 **53.** 1/584432 **54.** 2^{30} **55.** $n=6$ **56.** $n=7$ **57.** $n=9$ **58.** $n=7$ or 15 **59.** $n=4$
60. $n=6$ **61.** $n=7$ **62.** $n=9$ **63.** $n=7$ **64.** $n=4$ **65.** $n=7$ or 8 **66.** $n=5$ or 6

CHAPTER 11–3 Sets and Venn Diagrams page 400
1. $A \cap B = \{7, 9\}$ **2.** $A \cup B = \{1, 2, 3, 4, 5, 7, 9, 10\}$ **3.** $B \cap C = \{9, 10\}$ **4.** $B \cup C = \{1, 2, 4, 7, 9, 10, 15, 20\}$
5. $(A \cup B) \cap C = \{1, 9, 10\}$ **6.** $(A \cap B) \cap C = \{9\}$ **7.** $A \cap \phi = \phi$ **8.** $A \cup \phi = \{1, 3, 5, 7, 9\}$ **9.** $\in$ **10.** $\notin$
11. $\subset$ **12.** $\subset$ **13.** $=$ **14.** $\not\subset$ **15.** $\subset$ **16.** $\neq$ **17.** $\subset$ **18.** $\{2, 4, 6, 8\}$ **19.** $\{7, 9, 11, 13 \ 15\}$ **20.** $\{1, 2, 3\}$ **21.** ϕ
22. $\{0\}$ **23.** $\{0, 1, 2, 3, 4, 5\}$ **24.** ϕ **25.** +++++++++++++++
26. $\{1\}, \{2\}, \{3\}, \{1, 2\}, \{2, 3\}, \{1, 3\}, (1, 2, 3), \phi$
27. 16 subsets **28.** 32 subsets **29.** 2^n **30.** **31.** **32.** M = N

33. 9 students **34.** 16 students **35.** a. 5 investors **37.** 10
b. 3 investors
c. 13 investors

36. {(H, H), (H, T), (T, H), (T, T)}

CHAPTER 11–4 Sample Space and Events page 402
1. S = {(H, 1), (H, 2), (H, 3), (H, 4), (H, 5), (H, 6), (T, 1), (T, 2), (T, 3), (T, 4), (T, 5), (T, 6)}
2. S = {(1, H), (1, T), (2, H), (2, T), (3, H), (3, T), (4, H), (4, T), (5, H), (5, T)}, (6, H), (6, T)}
3. S = {(H,H,1),(H,H,2),(H,H,3),(H,H,4),(H,H,5),(H,H,6),(H,T,1),(H,T,2),(H,T,3),(H,T,4),(H,T,5),(H,T,6)}
4. S = {(1,1,H),(1,2,H),···,(1,6,H),(2,1,H),(2,2,H),···,(2,6,H),(3,1,H),(3,2,H),···,(3,6,H),(4,1,H),(4,2,H),···,(4,6,H),
(5,1,H),(5,2,H),···,(5,6,H),(6,1,H),(6,2,H),···,(6,6,H)} → 36 events
{(1,1,T),(1,2,T),······, (6,6,T)} → 36 events   Total: 72 events
5. S = {2,3,4,5,6,7,8,9,10,11,12}   **6.** S = {3,4,5,·····,18}   **7.** S = {abc,acb,bac,bca,cab,cba}
8. S={(R,R),(R,W),(R,B),(W,W),(W,B)}   **9.** S={4,5,6,7,8,9,10}   **10.** S ={ab,ac,ad,ae,bc,bd,be,cd,ce,de}
11. A = {3,4,5,6} **12.** A = {(1,6),(2,5),(3,4),(4,3),(5,2),(6,1)} **13.** A = {T,T,T,T} **14.** A = {(H,T),(T,H),(T,T)}

15. A = {(1,1),(2,2),(3,3),(4,4),(5,5),(6,6)} **16.** A = {(5,6),(6,5),(6,6)} **17.** A = {(4,6),(5,5),(5,6),(6,4),(6,5),(6,6)}
18. A = {(H,H,H),(T,T,T)} **19.** A = {(R,W),(W,B),(W,W)} **20.** A ={ab, ac, ad, ae} **21.** $|S| = 6$ **22.** $|S| = 36$
23. $|S| = {}_{36}C_0 + {}_{36}C_1 + {}_{36}C_2 + \cdots\cdots + {}_{36}C_{36} = (1+1)^{36} = 2^{36}$
24. S = {(H,1),(H,2),(H,3),(H,4),(H,5),(H,6),(T,1),(T,2),(T,3),(T,4),(T,5),(T,6)}, A = {(H,2),(H,4),(H,6)}
25. $|S| = 12$ **26.** $|S| = {}_{12}C_0 + {}_{12}C_1 + {}_{12}C_2 + \cdots\cdots + {}_{12}C_{12} = (1+1)^{12} = 2^{12}$ **27.** $|S| = 24$ **28.** $|S| = 2^{24}$

CHAPTER 11–5   Probability   page 406 ~ 408

1. $\frac{1}{13}$ **2.** $\frac{1}{4}$ **3.** $\frac{1}{52}$ **4.** 0   **5.** $\frac{3}{13}$ **6.** $\frac{1}{2}$ (Hint: hearts and diamonds)   **7.** $\frac{2}{13}$ **8.** $\frac{5}{13}$ **9.** $\frac{3}{4}$ **10.** $\frac{4}{13}$ (Hint: $\frac{4}{52}+\frac{12}{52}$)   **11.** $\frac{1}{2}$

12. $\frac{1}{5}$ **13.** $\frac{7}{50}$ **14.** $\frac{1}{50}$ **15.** $\frac{17}{50}$ (Hint:$\frac{1}{5}+\frac{7}{50}$)   **16.** $\frac{1}{6}$ **17.** 0   **18.** $\frac{1}{2}$ **19.** $\frac{1}{2}$ **20.** $\frac{1}{2}$ **21.** $\frac{5}{6}$ **22.** $\frac{2}{3}$ **23.** $\frac{1}{2}$ **24.** $\frac{1}{8}$ **25.** $\frac{1}{8}$

26. $\frac{1}{4}$ **27.** $\frac{1}{2}$ **28.** $\frac{5}{8}$ **29.** $\frac{2}{5}$ **30. a.** $\frac{6}{25}$ **b.** $\frac{19}{25}$ **31. a.** $\frac{3}{5}$ **b.** $\frac{1}{2}$ **32. a.** $\frac{2}{5}$ **b.** 2   **c.** $\frac{1}{2}$

33. The tree diagram is left to the students (see example in this section)   **34.** 8   **35.** 16   **36.** 36   **37.** 216   **38.** $\frac{1}{4}$

39. $\frac{1}{2}$ **40.** $\frac{7}{8}$ **41.** 75 times   **42.** 30 times   **43.** $\frac{1}{32}$ **44. a.** 1,326 voters   **b.** 780 voters   **c.** 1,794 voters

45. 60 students   **46.** 80 students   **47.** 2 students   **48. a.** $\frac{1}{32}$ **b.** $\frac{1}{16}$ **c.** $\frac{1}{16}$ **49. a.** $\frac{26}{52}\cdot\frac{26}{52}=\frac{1}{4}$ **b.** $\frac{13}{52}\cdot\frac{13}{52}=\frac{1}{16}$ **c.** $\frac{1}{52}\cdot\frac{1}{52}=\frac{1}{2704}$

  d. $\frac{26}{52}\cdot\frac{26}{52}=\frac{1}{4}$ **50. a.** $\frac{26}{52}\cdot\frac{25}{51}=\frac{25}{102}$ **b.** $\frac{13}{52}\cdot\frac{12}{51}=\frac{1}{17}$ **c.** 0 (there is only one ace of heart.)   **d.** $\frac{26}{52}\cdot\frac{26}{51}=\frac{13}{51}$

51. $\frac{6}{36}=\frac{1}{6}$ (Hint:{(4, 6),(5, 5),(5, 6),(6, 4),(6, 5),(6, 6)}   **52.** 2,500 salmons   **53.** $\frac{4}{13}$ **54.** $\frac{7}{13}$ **55.** $\frac{1}{26}$ **56.** $\frac{11}{36}$ **57.** $\frac{4}{13}$

58. $\frac{1}{52}$ **59.** 0.0624 (Hint: Getting 2 of the 5-hearts and 3 of the 47 non-hearts: $_5C_2\cdot_{47}C_3/_{52}C_5$))

60. 0.0667 (Hint: $P = P(2) + P(3) + P(4) + P(5)$ or $P = 1- P(0) - P(1)$)   **61.** 0.2059 (Hint: $_{13}C_2\cdot_{39}C_{11}/_{52}C_{13}$)

62. 0.9071 (Hint: $P = 1- P(0) - P(1)$)   **63.** 0.00103 (Hint: $_5C_4/_{20}C_4$)   **64.** 0.03096 (Hint: $_5C_3\cdot_{15}C_1/_{20}C_4$)

65. 0.03199   **66.** 0.3125 (Hint: $_5C_3/2^5$)   **67.** 0.15625   **68.** 0.5   **69.** 0.3814 (Hint: $_8C_2\cdot_{12}C_2/_{20}C_4$)

70. 0.3324 (Hint: P $= 1 - (49/50)^{20}$)   **71.** 0.23 (Hint: P = P(B) + P(C) + P(D) − P(B) · P(C) · P(D))

72. 0.5886 (Hint: $P(2) = \dfrac{365}{365}\cdot\dfrac{364}{365}=\dfrac{365\cdot364}{365^2}=\dfrac{365!}{365^2(365-2)!}$, ..., $P(n) = \dfrac{365!}{365^n(365-n)!}$)

73. Use the formula in Problem 72 to calculate the probabilities as shown in the following table:

n	10	20	21	22	23	24
P(n)	0.8831	0.5886	0.5563	0.5243	0.4927	0.4617
1 − P(n)	0.1169	0.4114	0.4437	**0.4757**	**0.5073**	0.5383

a. 23

b. 22

CHAPTER 11 – 6   Histogram and Normal Distribution   page 415 ~ 416

1. $\mu = 4$, $v = 2$, $\sigma = 1.414$ **2.** $\mu = 6\frac{5}{7}$, $v = 1.633$, $\sigma = 1.278$ **3.** $\mu = 15$, $v = 63.33$, $\sigma = 7.958$

4. $\mu = 13\frac{2}{7}$, $v = 27.076$, $\sigma = 5.203$ **5.** $\mu = 18$, $v = 63.714$, $\sigma = 7.982$ **6.** $\mu = 15\frac{3}{8}$, $v = 19.734$, $\sigma = 4.442$

7. The frequency distribution and histogram are left to the students.

8. a. 68.27%   **b.** 0.84135   **c.** 0.15865   **d.** 95.45%   **9. a.** 0.15865   **b.** 0.84135   **10. a.** 0.97725   **b.** 0.84135

11. a. 0.15865   **b.** 0.02275   **12. a.** 0.6827   **b.** 0.3183 (Hint: P = $(0.6827)^3$)   **13. a.** 0.15865   **b.** 0.34135

14. 0.9545   **15.** 0.449 (Hint: P = $(0.8186)^4$)   **16.** 13,590 tires

17. 0.3413 (Hint: See Table on page 414, $p(0 < x <1) = p(-\infty < x <1) - 0.5 = 0.3413$)

18. 99,620 tires (Hint: See Table on page 414, $z = \frac{4000-32000}{3000} = 2.67$. It is under the region within 2.67 standard deviations to the right of y-axis on the standard normal curve. $p(x \le 40000) = p(-\infty \le z \le 2.67) = 0.9962$.)

CHAPTER 11–7   Binomial Distribution   page 419

1. 0.0161   **2.** 0.1272   **3.** 0.0339   **4.** 0.0068   **5.** 0.0011   **6.** 0.0327   **7.** 0.1587   **8.** 0.00037

9. $\mu = 6$, $\sigma = 1.732$ **10.** $\mu = 4$, $\sigma = 1.633$ **11.** $\mu = 5.4$, $\sigma = 1.723$ **12.** $\mu = 6.6$, $\sigma = 1.723$ **13.** $\mu = 10$, $\sigma = 2.236$

14. $\mu = 9$, $\sigma = 2.121$ **15.** $\mu = 5.55$, $\sigma = 1.870$ **16.** $\mu = 2$, $\sigma = 1.723$

17. a. 0.2508 (Hint: $p(x = 4)=_{10}C_4(\frac{2}{5})^4(\frac{3}{5})^6$)   **b.** 0.633 (Hint: $p(x \le 4) = \sum_{x=0}^{4}{}_{10}C_x(\frac{2}{5})^x(\frac{3}{5})^{10-x}$)   **c.** 0.367

  d. 0.9793 (Hint: $\mu = np = 40$, $\sigma = \sqrt{np(1- p)} = 4.9$, $z = \frac{50-40}{4.9} = 2.04$, $p(x \le 50) = p(-\infty \le z \le 2.04 = 0.9793$)

18. $\frac{15}{64}$ **19.** 0.3087   **20. a.** 0.6064   **b.** 0.9983 (Hint: $p(x \ge 497) = \sum_{x=497}^{500}{}_{500}C_x(0.999)^x(0.001)^{500-x}$, see page 414)

21. a. 0.9130   **b.** 0.3174   **22. a.** $\frac{193}{512}$ **b.** $\frac{319}{512}$ **c.** $\frac{193}{1024}$ **d.** $\frac{1}{1024}$ **23.** $\mu = 50$, $\sigma = 5$

24. 0.9772 (Hint: $\mu = np = 50$, $\sigma = \sqrt{np(1- p)} = 5$, $z = \frac{60-50}{5} = 2$, $p(x \le 60) = p(-\infty \le z \le 2) = 0.9772$, page 414)

25. $\mu = 60$, $\sigma = 4.9$ **26.** $\mu = 300$, $\sigma = 14.14$

CHAPTER 11-8 Expected Value page 422

1. $5 **2.** No. It is an unfair game. **3.** Yes. I can expect to win $0.50 each game. **4.** EV = $5.40 ; yes.
5. EV = $8; no. **6.** EV = $10.80; no **7.** $200 **8.** a. EV = $27.59 b. $55.18 **9.** $4.17 **10.** EV = $-0.5; unfair
11. $24 **12.** $50 **13.** $2.60 **14.** $0.48 **15.** $0.58

CHAPTER 11 Exercises page 423 ~ 426

1. 6 **2.** 25 **3.** 90,000 **4.** 175,760,000 **5.** 62 **6.** 720 **7.** 11,880 **8.** 6,545 **9.** 9,900 **10.** $n-1$ **11.** 360 **12.** 15
13. 2,598,960 **14.** 1,623,160 **15.** $\frac{1}{575,757}$, $\frac{1}{3,387}$, $\frac{1}{103}$, $\frac{1}{10}$ **16.** 120 **17.** 3,125 **18.** 900 **19.** 1,663,200
20. 19,958,400 **21.** 180 **22.** 105 **23.** a.84 b.147 **24.** 720 **25.** 120 **26.** 60 **27.** 720 **28.** 144 **29.** $\frac{1}{291}$ **30.** $\frac{1}{11927}$
31. a. $A \cap B = \{9,7\}$ b. $A \cup B = \{1,2,3,4,5,7,9,10\}$ c. $B \cap C = \{9,10\}$ **32.** 35
   d. $B \cup C = \{1,2,4,7,9,10,15,20\}$ e. $(A \cup B) \cap C = \{1,9,10\}$ **33.** 25
34. a. Samplespace:{(1,2),(1,3),(1,4),(1,5),(2,1),(2,3),(2,4),(2,5),(3,1),(3,2),(3,4),(3,5),(4,1),(4,2),(4,3),(4,5),
      (5,1),(5,2),(5,3),(5,4)}
   b. Event: {(1,5),(2,4),(4,2),(5,1)}
35. a. Sample space: {(1,2),(1,4),(1,6),(1,8),(3,2),(3,4),(3,6),(3,8),(5,2),(5,4),(5,6),(5,8),(7,2),(7,4),(7,6),(7,8)}
      Event: {(1,8),(3,6),(5,4),(7,2)}
   b. Event: {(3,2),(5,2),(5,4),(7,2),(7,4),(7,6)}
36. a. $\frac{1}{36}$ b. $\frac{1}{8}$ c. $\frac{5}{108}$ **37.** a) $\frac{11}{120}$ b) $\frac{1}{4}$ c) $\frac{1}{4}$ **38.** a) $\frac{1}{5}$ b) $\frac{7}{50}$ c) $\frac{1}{50}$
39. a. A = {(H,H),(H,T)}, B = {(H,H),(T,H)}, A∩B = {(H,H)}, A∪B = {(H,H),(H,T),(T,H)}
   b. P(A) = $\frac{1}{2}$, P(B) = $\frac{1}{2}$, P(A∩B) = $\frac{1}{4}$, P(A∪) = $\frac{3}{4}$, c) A and B are independent event.
40. A and B are not independent events
41. a. The events A, B, A∩B, A∪B are left to the students
   b. P(A) = $\frac{1}{2}$, P(B) = $\frac{1}{2}$, P(A∩B) = $\frac{1}{4}$, P(A∪B) = $\frac{3}{4}$ c) A and B are independent events.
42. a. mean = $3\frac{1}{9}$, range = 5, mode = 3, median = 3, variance = 2.10, standard deviation = 1.45.
   b. mean = $5\frac{5}{8}$, range = 8, mode = 4, median = 5, variance = 7.23, standard deviation = 2.69.
   c. mean = 84, range = 20, mode = 87, median = 87, variance = 46, standard deviation = 6.78
43. a. The curve is left to the students   b. mean = 70   c. variance = 121   d. standard deviation = 11
   e. P(> 70) = 50%, P(>81) = 15.87%, P(<92) = 97.73%, P(<103) = 99.87%
44. a. 13.59% b. 81.86% **45.** 2.28% (Hint: $z = \frac{2.50-2.2}{0.15} = 2.0$ and use table on page 414)
46. 409 (Hint: Admission rate $P = \frac{550}{674} = 0.8160$, use table on page 414 z = 0.91 when P = 0.8160,
   find x from $z = \frac{500-x}{100} = 0.91$)
47. 0.7759 **48.** 0.5405 **49.** $\mu = 100$, $\sigma = 7.07$ **50.** $\mu = 80$, $\sigma = 6.93$
51. 0.9255 (Hint: $\mu = 80$, $\sigma = 6.93$, $z = \frac{90-80}{6.93} = 1.443$ and use the table on page 414)   **52.** $5.83 **53.** $3.33
54. $250 **55.** $0.78 **56.** 40 roses **57.** $p_1 = \frac{7}{10}$, $p_2 = \frac{3}{10}$ **58.** $n = 10$ **59.** $n = 10$ **60.** $n = 10$ **61.** $n = 10$
                    EV = $5

In a biology class, a student said to his teacher.
Student: I just killed five mosquitoes.
          Three were males and two were females.
Teacher: How did you know which were males
          and which were females.
Student: Three were killed on a beer bottle.
          Two were killed on a mirror.

Finding the Mean, Range, Median, standard deviation and histogram by a Graphing Calculator

Example: Find the mean, median, range, standard deviation, and histogram of the data.

8, 15, 15, 31, 41

Solution:

Enter the data into the list L₁

Press: **STAT → 1:Edit → ENTER**

The screen shows:

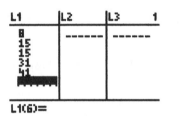

Calculate for the answers:

Press: **STAT → CALC → ENTER**
→ 1-Var Stats → ENTER

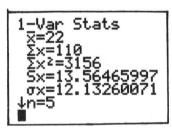

We have: Mean: $\bar{x} = 22$

Standard deviation: $\sigma x = 12.13$

Scrolling down the cursor to find:

Med = 15

minX = 8, maxX = 41

(Range = 41 - 8 = 33)

Plotting the histogram:

Press: **2nd → STATE PLOT → 1:PLOT**
→ ENTER → Plot1 → On → Type:
→ ◫▥▥

Adjust the viewing window:

$$x_{min} = 0 \qquad y_{min} = 0$$
$$x_{max} = 45 \qquad y_{max} = 3$$

To plot the histogram: Press: **GRAPH**

The screen shows:

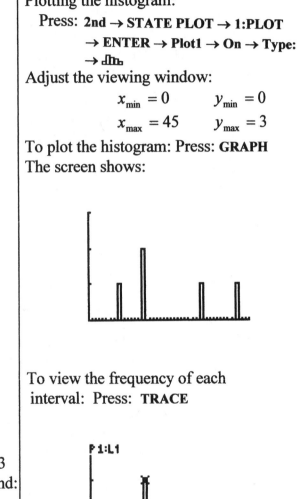

To view the frequency of each interval: Press: **TRACE**

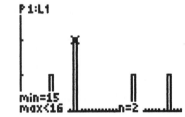

In an English class, the teacher asked the kids to write a letter to his (her) mother regarding what he (she) did today in school. The teacher noticed that John was writing very slowly.

Teacher: John, why do you write so slowly ?

John: My mom could not read fast.

Plotting a Normal Distribution
by a Graphing Calculator

Example: Plotting the histogram and normal distribution of the following data.

12 25 41 74 52 70 32 49 54 63
26 44 47 33 55 16 42 24 89 34
44 79 58 21 38 54 39 57 42 51
52 59 29 35 69 43 56 63 30 44
37 40 45 52 51 18 61 75 88 31

Solution:

Enter the data into the list L1
    Press: **STAT → 1:Edit → ENTER**
    Type the data into the column L1

Plot the histogram:
    Press: **2nd → STAT PLOT → 1:PLOT**
        → ENTER → Plot1 → On → Type:
        → ▫▫▫

Adjust the viewing window:

$$x_{min} = 0, \quad x_{max} = 90, \quad x_{scl} = 10$$
$$y_{min} = 0, \quad y_{max} = 12, \quad y_{scl} = 2$$

Press: **GRAPH**   The screen shows:

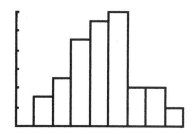

Plot the normal distribution:
The probability density function (pdf) is:

$$f(x) = \frac{1}{\sqrt{2\pi} \cdot \sigma} e^{\frac{-(x-\mu)^2}{2\sigma^2}}$$

Find the mean (μ) and standard deviation (σ):
    Press: **STAT → CALC → ENTER → 1-Var Stats**
        → ENTER
The screen shows: $\bar{x} = 46.86$, $\sigma x = 17.6681$

Compute and plot the normal distribution:
        normalpdf (x, μ, σ)
    Press: **Y= → 2nd → DISTR → 1:normalpdf (**
        → ENTER
    Type: **normalpdf (** X,T,θ,n , 46.86, 17.6681)
    Adjust the window: Press: **WINDOW**

$$x_{min} = 0, \quad x_{max} = 90, \quad x_{scl} = 10$$
$$y_{min} = 0, \quad y_{max} = 0.03, \quad y_{scl} = 0.01$$

(Find y_{max} : Press: **ZOOM → 0:ZoomFit**)

Press: **GRAPH**   The screen shows:

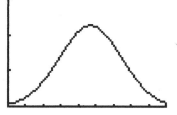

Find and shade the area under the curve:
    Press: **2nd → DISTR → DRAW**
        → 1:ShadeNorm → ENTER
        → ShadeNorm (
Enter **ShadeNorm (** lower, upper, μ, σ)
Type: **ShadeNorm**(10, 30, 46.86, 17.6681)
Press: **ENTER**   The screen shows:

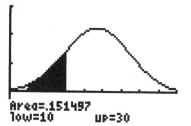

Plotting a Binomial Distribution
by a Graphing Calculator

Example: In a binomial experiment, the probability of success on each trial is $p = 0.4$ and the
number of trials is $n = 6$.
>   1. Evaluate the probability on the trial $x = 5$.
>   2. Prot the binomial distribution.

Solution:
>   The binomial probability density function (pdf) is:
> $$f(x) = {}_nC_x p^x (1-p)^{n-x}, \ x = 0, 1, 2, \ldots, n$$

1. Evaluate the probability at $x = 5$
   binompdf ($n, \ p, x$)
   Press: **2nd → DISTR → 0:binompdf(→ ENTER**
   Type: **binompdf (6, 0.4, 5) → ENTER**
   The screen shows: $f(x) = 0.036864$

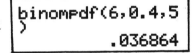

2. Plot the binomial distribution
   Enter the number of trials into the list L₁:
   Press: **STAT → 1:Edit → ENTER**
   The screen shows:

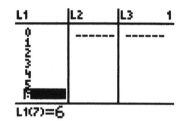

   Calculate and plot the binomial distribution
   binompdf (n, p)
   Press: **2nd → DISTR → 0:binompdf(→ ENTER**
   Type: **binompdf (6, 0.4) → ENTER**
   Store the results (frequencies) into the list L₂:
   Type: **STO → 2nd → L2**
   The screen shows:

   Plot the histogram:
   Press: **2nd → STAT PLOT → 1:Plot → ENTER**
   >   **→ Plot1 → ON → Type: →⊞ → Freq:**
   >   **→ 2nd → L2**
   The screen shows:

   Adjust the viewing window: Press: **WINDOW**
   > $x_{min} = 0, \ x_{max} = 10, \ x_{scl} = 1$
   > $y_{min} = 0, \ y_{max} = 0.5, \ y_{scl} = 0.1$
   Press: **GRAPH**   The screen shows:

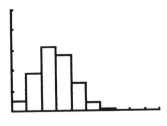

CHAPTER TWELVE

An Introduction to Calculus

12-1 Limits and Continuity

The concept of limits and continuity is very important to the study of calculus.
Suppose we sketch the graph of the following function by direct substitution:

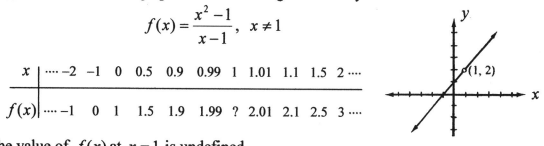

$$f(x) = \frac{x^2 - 1}{x - 1}, \quad x \neq 1$$

x	.... -2	-1	0	0.5	0.9	0.99	1	1.01	1.1	1.5	2 ....
$f(x)$	.... -1	0	1	1.5	1.9	1.99	?	2.01	2.1	2.5	3 ....

The value of $f(x)$ at $x = 1$ is undefined.

Since x cannot equal to 1, we are not sure what the value of $f(x)$ is at $x = 1$.

The graph of $f(x)$ is a straight line that has a hole at the point $(1, 2)$.

The point $(1, 2)$ is not part of the graph.

To find the value of $f(x)$ near the undefined point at $x = 1$, we could use the values of x that approaches 1 from the left, and the values of x from the right. We estimate that the value of $f(x)$ moves close to 2 when x approaches 1 from either the right or the left. We say that " the limit of $f(x)$ is 2 as x approaches 1 ", and we write the result in limit notation: $\lim_{x \to 1} f(x) = 2$

Definition of a limit: A function $f(x)$ is said to approach the limit L as x approaches c. We write: $\lim_{x \to c} f(x) = L$.

Some functions do not have a limit as $x \to c$. If a limit of a function exists, it is unique. In other words, a function cannot have two different limits as $x \to c$.

By reducing the function, we have $f(x) = \frac{x^2 - 1}{x - 1} = \frac{(x+1)(x-1)}{x-1} = x + 1, \; x \neq 1$.

Therefore, for all points other than $x = 1$, $\lim_{x \to 1} \frac{x^2 - 1}{x - 1} = \lim_{x \to 1}(x + 1) = 2$.

We use the following ways to evaluate the limit of $f(x)$ as $x \to c$:

1. Evaluated by direct substitution.
2. If $f(x)$ is undefined at $x = c$, we try to reduce $f(x)$ and find a new function that agrees with $f(x)$ for all x other than c.

433

One-side limits: $\lim\limits_{x \to c^+} f(x) = L$ represents the limit of $f(x)$ from the right. It is the limit of

$f(x)$ as x approaches c from values greater than c.

$\lim\limits_{x \to c^-} f(x) = L$ represents the limit of $f(x)$ from the left. It is the limit of

$f(x)$ as x approaches c from values less than c.

$\lim\limits_{x \to c} f(x) = L$ **exists only and only if** $\lim\limits_{x \to c^+} f(x) = L$ **and** $\lim\limits_{x \to c^-} f(x) = L$.

Infinite Limits: $\lim\limits_{x \to c} f(x) = \infty$ means that $f(x)$ increases without bound as x approaches c.

$\lim\limits_{x \to c} f(x) = -\infty$ means that $f(x)$ decreases without bound as x approaches c.

Since infinite " ∞ " is not a number, we may say that "the limit does not exist".

Limits of the greatest integer functions: If n is an integer, then $\lim\limits_{x \to n}[x]$ does not exist.

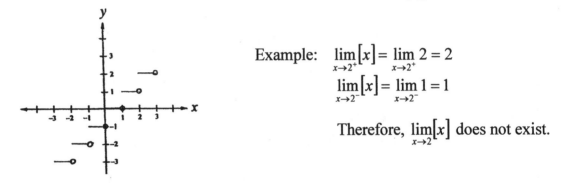

Example: $\lim\limits_{x \to 2^+}[x] = \lim\limits_{x \to 2^+} 2 = 2$

$\lim\limits_{x \to 2^-}[x] = \lim\limits_{x \to 2^-} 1 = 1$

Therefore, $\lim\limits_{x \to 2}[x]$ does not exist.

If the degree of $p(x)$ is greater than the degree of $q(x)$, then $\lim\limits_{x \to \infty} \dfrac{p(x)}{q(x)} = \pm\, \infty$.

If the degree of $p(x)$ is less than the degree of $q(x)$, then $\lim\limits_{x \to \infty} \dfrac{p(x)}{q(x)} = 0$.

To find the limit of a quotient of polynomials, we divide the highest power of x and use the fact that $\lim\limits_{x \to \infty} \frac{1}{x} = 0$.

$$\lim\limits_{x \to \infty} \frac{-x^3}{2 - x^2} = \lim\limits_{x \to \infty} \frac{-1}{\frac{2}{x^3} - \frac{1}{x}} = \frac{-1}{0 - 0} = \frac{-1}{0} = -\infty.$$

The limit does not exist in this example.

> A high school boy is playing the violin.
> His girlfriend sits next to him.
> Boy: Do you like the song I played ?
> Girl: Yes. It makes me miss my father.
> Boy: Was your father a musician ?
> Girl: No. He was a sawmill worker.

Examples

1. $\lim_{x\to 2} 3 = 3$ **2.** $\lim_{x\to 2} x^3 = 2^3 = 8$ **3.** $\lim_{x\to 1}(x^2 + x + 2) = 1^2 + 1 + 2 = 4$ **4.** $\lim_{x\to -2}|x| = 2$

5. $\lim_{x\to 1}\dfrac{x^2 + x - 2}{x - 1} = \lim_{x\to 1}\dfrac{(x+2)(x-1)}{x-1} = \lim_{x\to 1}(x+2) = 1 + 2 = 3$.

6. $\lim_{x\to -4}\dfrac{x^2 + 2x - 8}{x + 4} = \lim_{x\to -4}\dfrac{(x+4)(x-2)}{x+4} = \lim_{x\to -4}(x-2) = -4 - 2 = -6$.

7. $\lim_{x\to 2}\dfrac{x-2}{x^3 - 2x^2 + 3x - 6} = \lim_{x\to 2}\dfrac{x-2}{x^2(x-2) + 3(x-2)} = \lim_{x\to 2}\dfrac{x-2}{(x-2)(x^2+3)} = \lim_{x\to 2}\dfrac{1}{x^2+3} = \dfrac{1}{7}$.

8. $\lim_{x\to 0}\dfrac{x}{x^3} = \lim_{x\to 0}\dfrac{1}{x^2} = \dfrac{1}{0} = \infty$ (undefined). The limit does not exist.

   $f(x)$ increases without bound as x approaches 0 from
   either the right or the left. It has no limit as $x \to 0$.

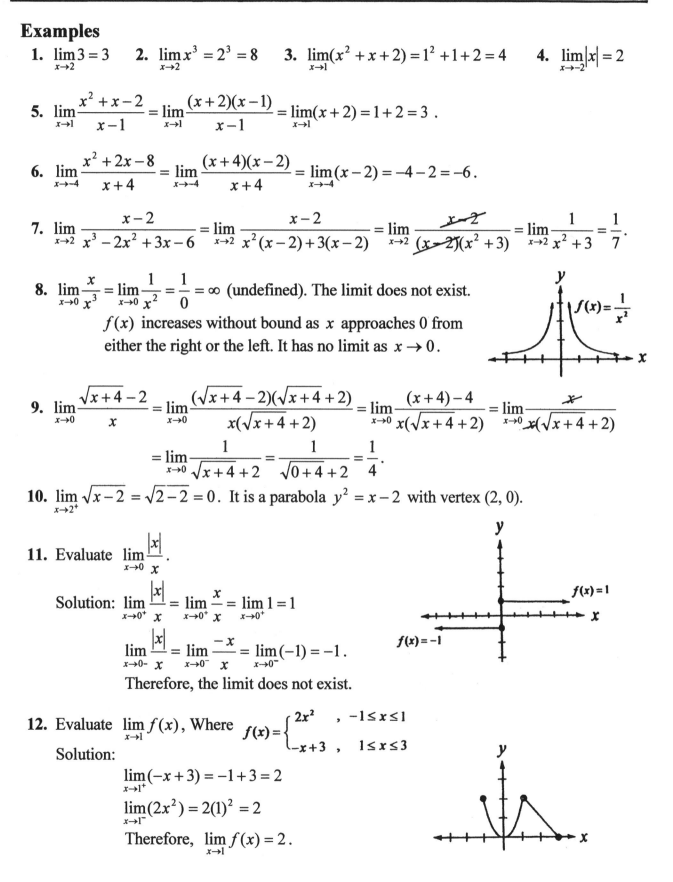

9. $\lim_{x\to 0}\dfrac{\sqrt{x+4} - 2}{x} = \lim_{x\to 0}\dfrac{(\sqrt{x+4} - 2)(\sqrt{x+4} + 2)}{x(\sqrt{x+4} + 2)} = \lim_{x\to 0}\dfrac{(x+4) - 4}{x(\sqrt{x+4} + 2)} = \lim_{x\to 0}\dfrac{x}{x(\sqrt{x+4} + 2)}$

   $= \lim_{x\to 0}\dfrac{1}{\sqrt{x+4} + 2} = \dfrac{1}{\sqrt{0+4} + 2} = \dfrac{1}{4}$.

10. $\lim_{x\to 2^+}\sqrt{x-2} = \sqrt{2-2} = 0$. It is a parabola $y^2 = x - 2$ with vertex $(2, 0)$.

11. Evaluate $\lim_{x\to 0}\dfrac{|x|}{x}$.

   Solution: $\lim_{x\to 0^+}\dfrac{|x|}{x} = \lim_{x\to 0^+}\dfrac{x}{x} = \lim_{x\to 0^+} 1 = 1$

   $\lim_{x\to 0^-}\dfrac{|x|}{x} = \lim_{x\to 0^-}\dfrac{-x}{x} = \lim_{x\to 0^-}(-1) = -1$.

   Therefore, the limit does not exist.

12. Evaluate $\lim_{x\to 1} f(x)$, Where $f(x) = \begin{cases} 2x^2 & , \quad -1 \le x \le 1 \\ -x+3 & , \quad 1 \le x \le 3 \end{cases}$

   Solution:

   $\lim_{x\to 1^+}(-x+3) = -1 + 3 = 2$

   $\lim_{x\to 1^-}(2x^2) = 2(1)^2 = 2$

   Therefore, $\lim_{x\to 1} f(x) = 2$.

Continuity

A function $f(x)$ is said to be continuous at $x = c$ if its graph is unbroken (no holes) at $x = c$. If a function is not continuous at $x = c$, we say that it is discontinuous at $x = c$.

A function $f(x)$ is said to be continuous on the closed interval $[a, b]$ if it is continuous on the interval $a \leq x \leq b$.

A function $f(x)$ that is continuous on the entire real numbers $(-\infty, +\infty)$ is called a continuous function.

Definition of Continuity: A function $f(x)$ is said to be continuous at $x = c$ if the following three conditions are satisfied:

      1. $f(x)$ is defined.   **2.** $\lim\limits_{x \to c} f(x)$ exists.   **3.** $\lim\limits_{x \to c} f(x) = f(c)$.

Continuity on a closed interval: If a function $f(x)$ is continuous on a closed interval

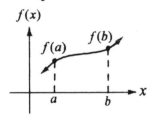

$[a, b]$, we say that it is continuous at every interior point, continuous on the right at a, and continuous on the left at b.

$$\lim\limits_{x \to a^+} f(x) = f(a), \qquad \lim\limits_{x \to b^-} f(x) = f(b).$$

A discontinuity of a function $f(x)$ at $x = c$ is called removable if $f(x)$ can be made continuous by redefining $f(x)$ at $x = c$.

Intermediate Value Theorem: If $f(x)$ is continuous on the closed interval $[a, b]$ and k is any number between $f(a)$ and $f(b)$, then there is at least one number c between a and b such that $f(c) = k$.

The Intermediate Value Theorem states that if a function is continuous on a closed interval, there is no hole in its graph.

If there are two numbers a and b in the function $f(x)$, such that $f(a)$ is negative and $f(b)$ is positive, then the function $f(x)$ must have at least one number c between a and b such that $f(c) = 0$. Its graph must cross the $x - axis$ at $x = c$.
The equation $f(x) = 0$ has a real root at $x = c$.

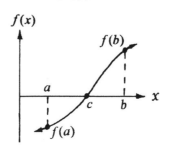

Examples

1. Find the intervals for which the function is continuous.

$$f(x) = x + 1.$$

Solution:

The function is continuous for all real numbers. That is the interval $(-\infty, \infty)$.

2. Find the values for which the function is continuous. Is the discontinuity removable ?

$$f(x) = \frac{1}{x+1}.$$

Solution:

$$\lim_{x \to -1} \frac{1}{x+1} = \frac{1}{-1+1} = \frac{1}{0} \text{ (undefined)}.$$

The function $f(x)$ is continuous for all real numbers except $x = -1$. That is the interval $(-\infty, -1)$ and $(-1, \infty)$. The function is discontinuous at $x = -1$. It is a nonremovable discontinuity.

3. Find the values for which the function is continuous. Is the discontinuity removable ?

$$f(x) = \frac{x^2 - 1}{x - 1}.$$

Solution:

The function $f(x)$ is continuous for all real numbers except $x = 1$. That is the interval $(-\infty, 1)$ and $(1, \infty)$. The function is discontinuous at $x = 1$.

$$\lim_{x \to 1} \frac{x^2 - 1}{x - 1} = \lim_{x \to 1} \frac{(x+1)(x-1)}{x - 1} = \lim_{x \to 1}(x + 1) = 1 + 1 = 2.$$

Therefore, it is a removable discontinuity.

4. Find the value of c in the interval $[-2, 1]$ guaranteed by the Intermediate Value Theorem that the following function has $f(c) = 1$.

$$f(x) = \frac{x^2 - 2x - 3}{x - 3}.$$

Solution:

$f(x)$ is continuous for all real numbers except $x = 3$.

Therefore, it is continuous on $[-2, 1]$.

$$f(-2) = \frac{(-2)^2 - 2(-2) - 3}{-2 - 3} = -1, \quad f(1) = \frac{1^2 - 2(1) - 3}{1 - 3} = 2.$$

$f(-2) < f(c) < f(1)$. Apply the Intermediate Value Theorem:

$$\frac{x^2 - 2x - 3}{x - 3} = 1, \quad x^2 - 2x - 3 = x - 3, \quad x^2 - 3x = 0, \quad x(x - 3) = 0 \quad \therefore x = 0, 3.$$

Therefore, 0 is in the interval $[-2, 1]$. $\therefore c = 0$.

EXERCISES

Find the limit, If it exists. If the function tends to $+\infty$ or $-\infty$, so state.

1. $\lim\limits_{x\to 1} 5$ **2.** $\lim\limits_{x\to 2} 9$ **3.** $\lim\limits_{x\to 1}|x|$ **4.** $\lim\limits_{x\to -1.1}|x|$ **5.** $\lim\limits_{x\to 0}|x|$

6. $\lim\limits_{x\to 0.9}[x]$ **7.** $\lim\limits_{x\to 1.9}[x]$ **8.** $\lim\limits_{x\to 1}[x]$ **9.** $\lim\limits_{x\to 0}[x]$ **10.** $\lim\limits_{x\to -1}[x+1]$

11. $\lim\limits_{x\to 0}(2x-4)$ **12.** $\lim\limits_{x\to -1}(2x-4)$ **13.** $\lim\limits_{x\to 2}(2x^2-4)$ **14.** $\lim\limits_{x\to 2^-}\sqrt{2-x}$

15. $\lim\limits_{x\to 1}\dfrac{2x^2-3x}{x}$ **16.** $\lim\limits_{x\to -1}\dfrac{x^3+x}{x+1}$ **17.** $\lim\limits_{x\to 2}\dfrac{x-2}{x^2-4}$

18. $\lim\limits_{x\to 3^+}\dfrac{x}{\sqrt{x^2-9}}$ **19.** $\lim\limits_{x\to 1}\dfrac{|x-1|}{x-1}$.   **20.** $\lim\limits_{x\to 2}\dfrac{x-2}{3|x-2|}$

21. $\lim\limits_{x\to 9}\dfrac{x-9}{\sqrt{x}-3}$ **22.** $\lim\limits_{x\to 2}\dfrac{\sqrt{x+2}-2}{x-2}$ **23.** $\lim\limits_{x\to\infty}\dfrac{x-4}{x^2+x+2}$

24. $\lim\limits_{x\to\infty}\dfrac{x^3-2x^2+1}{2x^3-2x^2+4}$ **25.** $\lim\limits_{x\to\infty}\dfrac{x^2+x+2}{x-4}$ **26.** $\lim\limits_{x\to -\infty}\dfrac{4x^2}{x+2}$

27. $\lim\limits_{x\to -\infty}\dfrac{2x}{x+1}$ **28.** $\lim\limits_{x\to\infty}\dfrac{x^3+1}{2-x^2}$ **29.** $\lim\limits_{x\to\infty}\dfrac{2x}{\sqrt{x^2-x}}$

30. $\lim\limits_{x\to 0^+}(x^2-\tfrac{1}{x})$ **31.** $\lim\limits_{x\to 0^-}(x^2-\tfrac{1}{x})$ **32.** $\lim\limits_{x\to 0^-}\dfrac{1}{x}$

33. $\lim\limits_{x\to 0^+}\dfrac{1}{x}$ **34.** $\lim\limits_{x\to -\infty}\dfrac{1}{x}$ **35.** $\lim\limits_{x\to 0^-}\dfrac{1}{x^2}$

36. $\lim\limits_{x\to 0^+}\dfrac{1}{x^2}$ **37.** $\lim\limits_{x\to 4^-}\dfrac{x^2}{x^2-16}$ **38.** $\lim\limits_{x\to 4^+}\dfrac{x^2}{x^2-16}$

39. $\lim\limits_{x\to 1^-}\dfrac{1}{x-1}$ **40.** $\lim\limits_{x\to 1^+}\dfrac{1}{x-1}$ **41.** $\lim\limits_{x\to 1^-}\dfrac{1}{(x-1)^2}$

42. $\lim\limits_{x\to 1^+}\dfrac{1}{(x-1)^2}$ **43.** $\lim\limits_{x\to -2^-}\dfrac{1}{x^2-4}$ **44.** $\lim\limits_{x\to -2^+}\dfrac{1}{x^2-4}$

45. Evaluate $\lim\limits_{x\to 2} f(x)$, where $f(x) = \begin{cases} \dfrac{x+1}{2} & , x \le 2 \\[2mm] \dfrac{6-2x}{3} & , x > 2. \end{cases}$

46. Evaluate $\lim\limits_{x\to 3} f(x)$, where $f(x) = \begin{cases} \sqrt{x-2} & , x > 3 \\[2mm] \sqrt{4-x} & , x \le 3. \end{cases}$ -----Continued-----

Find the values for which the function is continuous or discontinuous. Tell whether the discontinuity is removable or nonremovable.

47. $f(x) = 2x - 2$

48. $f(x) = \sqrt{4 - x^2}$

49. $f(x) = |x + 2|$

50. $f(x) = -\dfrac{x^2}{4}$

51. $f(x) = \dfrac{1}{x - 1}$

52. $f(x) = \dfrac{1}{x^2 - 4}$

53. $f(x) = \dfrac{x^2 - 1}{x + 1}$

54. $f(x) = \dfrac{x - 1}{x^2 + 2x - 3}$

55. $f(x) = \begin{cases} 2x + 2, & x \leq 1 \\ 5 - x, & x > 1 \end{cases}$

56. $f(x) = \begin{cases} 2x + 2, & x \leq 2 \\ 5 - x, & x > 2 \end{cases}$

57. $f(x) = \begin{cases} 2 - x, & -2 \leq x \leq 1 \\ \sqrt{x}, & 1 < x \leq 3 \end{cases}$

58. $f(x) = \begin{cases} x + 8, & x \leq 3 \\ x^2 + 2, & x > 3 \end{cases}$

59. $f(x) = \begin{cases} \dfrac{x^2 - 1}{x - 1} & \text{if } x \neq 1 \\ 2 & \text{if } x = 1 \end{cases}$

60. $f(x) = \begin{cases} \dfrac{x - 1}{x^2 + x - 2} & \text{if } x \neq 1, -2 \\ \frac{1}{3} & \text{if } x = 1 \\ 6 & \text{if } x = -2 \end{cases}$

61. Find the value of c in the interval $[2, 4]$ guaranteed by the Intermediate Value Theorem that the following function has $f(c) = 4$.

$$f(x) = \frac{x^2 - 1}{x - 1}.$$

62. Apply the Intermediate Value Theorem to approximate the zero (root) of the following function in the interval $[0, 1]$ by locating the zero in a subinterval of length 0.01.

$$f(x) = x^3 + x^2 - 1.$$

12-2 Vectors

A quantity, such as a velocity or a force, that is given by both its magnitude and direction is represented by a **vector**.

A vector from point A to point B on a coordinate plane is written by an arrow $\overrightarrow{AB}$. We sometimes use a boldface letter to represent a vector. To write a vector, an arrow is placed over the letter.

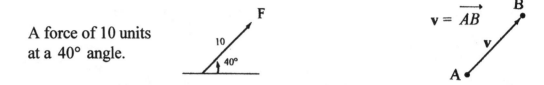

A force of 10 units at a 40° angle.

Point A is the **initial point** and Point B is the **terminal point**. The length of the arrow indicates the **magnitude** (or **norm**) of the vector. The arrow-head of the arrow indicates the **direction** of the vector. The direction of the vector is the angle θ that the vector makes with the positive $x-axis$. The magnitude of a vector **v** is written as $\|v\|$.

Zero Vector 0: A vector whose magnitude is zero. A zero vector has no direction.

Equivalent Vectors: Two vectors that have the same magnitude and direction.

$$\overrightarrow{AB} = \overrightarrow{CD}$$

The **Opposite** of a vector: $-$ **v** is a vector with the same magnitude but the opposite direction as **v**.

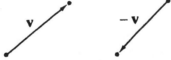

Addition (Resultant) of two vectors: To add two vectors, we place the initial point of second vector at the terminal point of the first vector , connect the initial point of the first vector to the terminal point of the second vector.

Subtraction (difference) of two vectors: To subtract two vectors, we add the opposite of the second vector. $u - v = u + (-v), \quad v - u = v + (-u)$

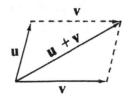
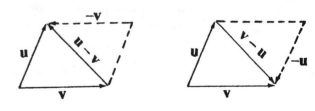

(Form a parallelogram)

The addition of vectors is **commutative** and **associative**:

$$v + u = u + v , \quad u + (v + w) = (u + v) + w$$

In a coordinate plane, a vector **v** with initial point at the origin and terminal point at $p(a, b)$ can be represented by a (the x-component) and b (the y-component) in the form of a **position vector** :

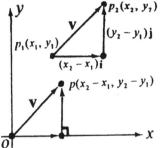

$$\mathbf{v} = \overrightarrow{op} = (a, b) \quad \text{or} \quad \mathbf{v} = a\,\mathbf{i} + b\,\mathbf{j} \quad \text{and} \quad \|\mathbf{v}\| = \sqrt{a^2 + b^2}$$

where **i** is the **unit vector** in the same direction as the positive x-axis, **j** is the **unit vector** in the same direction as the positive y-axis. $\mathbf{i} = (1, 0)$ and $\mathbf{j} = (0, 1)$.

To add two vectors, we add the corresponding components. To subtract two vectors, we subtract the corresponding components.

If any vector **v** whose initial point $p_1(x_1, y_1)$ is not at the origin, and the terminal point is at $p_2(x_2, y_2)$, then we have

$$\overrightarrow{p_1 p_2} = \overrightarrow{op} \quad (\text{they have the same magnitude and direction.})$$

$$\therefore \mathbf{v} = \overrightarrow{p_1 p_2} = (x_2 - x_1)\,\mathbf{i} + (y_2 - y_1)\,\mathbf{j}$$

A vector **u** whose magnitude $\|\mathbf{u}\| = 1$ is called a **unit vector**. To find the **unit vector** having the same direction as a given nonzero vector **v**, we use the following formula:

$$\text{unit vector} = \frac{\mathbf{v}}{\|\mathbf{v}\|}$$

Dot product (inner product) of two vectors: If $\mathbf{v} = a_1\,\mathbf{i} + b_1\,\mathbf{j}$ and $\mathbf{u} = a_2\,\mathbf{i} + b_2\,\mathbf{j}$ are two vectors, then the **dot product** $\mathbf{v} \cdot \mathbf{u} = a_1 a_2 + b_1 b_2$.

The dot product of two vectors are **commutative** and **distributive**:

$$\mathbf{v} \cdot \mathbf{u} = \mathbf{u} \cdot \mathbf{v} \quad ; \quad \mathbf{v} \cdot (\mathbf{u} + \mathbf{w}) = \mathbf{v} \cdot \mathbf{u} + \mathbf{v} \cdot \mathbf{w} .$$

Two vectors **v** and **u** are **perpendicular** (or **orthogonal**) if and only if $\mathbf{v} \cdot \mathbf{u} = 0$.

Two vectors **v** and **u** are **parallel** if there is a nonzero scalar c and $\mathbf{v} = c\,\mathbf{u}$.

Angle between two vectors: If **v** and **u** are two nonzero vectors, the angle θ between **v** and **u** is given by the following formula:

$$\cos\theta = \frac{\mathbf{u} \cdot \mathbf{v}}{\|\mathbf{u}\| \cdot \|\mathbf{v}\|}$$

The vector $\mathbf{v} = 4\mathbf{i} - 2\mathbf{j}$ and $\mathbf{u} = 2\mathbf{i} + 4\mathbf{j}$ are perpendicular since $\mathbf{v} \cdot \mathbf{u} = (4)(2) + (-2)(4) = 0$. Or
$$\cos\theta = 0 \quad \therefore \theta = 90° .$$

The vector $\mathbf{v} = 4\mathbf{i} - 2\mathbf{j}$ and $\mathbf{u} = 2\mathbf{i} - \mathbf{j}$ are parallel since $\mathbf{v} = 2\,\mathbf{u}$. Or

$$\cos\theta = \frac{(4)(2) + (-2)(-1)}{\sqrt{4^2 + (-2)^2} \cdot \sqrt{2^2 + (-1)^2}} = \frac{10}{\sqrt{20} \cdot \sqrt{5}} = \frac{10}{\sqrt{100}} = 1 \quad \therefore \theta = 0° .$$

Examples

1. If $\overrightarrow{AB} = (4, 1)$ and $\overrightarrow{BC} = (1, 2)$, find $\overrightarrow{AB} + \overrightarrow{BC}$.
Solution:

$$\overrightarrow{AB} + \overrightarrow{BC} = (4, 1) + (1, 2) = (4 + 1, 1 + 2) = (5, 3). \text{ Ans.}$$

2. If $\mathbf{v} = (-4, 2)$ and $\mathbf{u} = (3, 4)$, find the **resultant** and the **norm** (magnitude) of $\mathbf{v} + \mathbf{u}$.
Express $\mathbf{v}$ in terms of $\mathbf{i}$ and $\mathbf{j}$.
Solution:

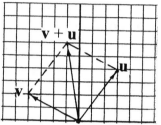

$$\text{Resultant } \mathbf{v} + \mathbf{u} = (-4 + 3, 2 + 4) = (-1, 6)$$
$$\text{Norm } \|\mathbf{v}+\mathbf{u}\| = \sqrt{(-1)^2 + 6^2} = \sqrt{37}$$
$$\mathbf{v} = -4\mathbf{i} + 2\mathbf{j}$$

3. If $\mathbf{u} = (-4, 2)$ and $\mathbf{u} + \mathbf{v} = (-1, 6)$, find $\mathbf{v}$.
Solution:

$$\mathbf{u} + \mathbf{v} = (-4 + v_1, 2 + v_2) = (-1, 6)$$
$$\text{We have } v_1 = 3, \ v_2 = 4 \quad \therefore \ \mathbf{v} = (3, 4). \text{ Ans}$$

4. Find the position vector of $\mathbf{v} = \overrightarrow{p_1 p_2}$ if its initial and terminal points are $p_1(5, 3)$ and
$p_2(-1, -4)$.
Solution: $\mathbf{v} = (-1 - 5)\mathbf{i} + (-4 - 3)\mathbf{j} = -6\mathbf{i} - 7\mathbf{j}$. Ans.

5. If $\mathbf{v} = 3\mathbf{i} + 2\mathbf{j}$ and $\mathbf{u} = 5\mathbf{i} - 4\mathbf{j}$, find $\mathbf{v} + \mathbf{u}$, $\mathbf{v} - \mathbf{u}$, $\mathbf{u} - \mathbf{v}$, $4\mathbf{u}$, $3\mathbf{v} - 4\mathbf{u}$, $\|\mathbf{v}\|$.
Solution:

$$\mathbf{v} + \mathbf{u} = (3\mathbf{i} + 2\mathbf{j}) + (5\mathbf{i} - 4\mathbf{j}) = (3 + 5)\mathbf{i} + (2 - 4)\mathbf{j} = 8\mathbf{i} - 2\mathbf{j}$$
$$\mathbf{v} - \mathbf{u} = \mathbf{v} + (-\mathbf{u}) = (3\mathbf{i} + 2\mathbf{j}) + (-5\mathbf{i} + 4\mathbf{j}) = (3 - 5)\mathbf{i} + (2 + 4)\mathbf{j} = -2\mathbf{i} + 6\mathbf{j}$$
$$\mathbf{u} - \mathbf{v} = \mathbf{u} + (-\mathbf{v}) = (5\mathbf{i} - 4\mathbf{j}) + (-3\mathbf{i} - 2\mathbf{j}) = (5 - 3)\mathbf{i} + (-4 - 2)\mathbf{j} = 2\mathbf{i} - 6\mathbf{j}$$
$$4\mathbf{u} = 4(5\mathbf{i} - 4\mathbf{j}) = 20\mathbf{i} - 16\mathbf{j}$$
$$3\mathbf{v} - 4\mathbf{u} = (9\mathbf{i} + 6\mathbf{j}) + (-20\mathbf{i} + 16\mathbf{j}) = -11\mathbf{i} + 22\mathbf{j}$$
$$\|\mathbf{v}\| = \|3i + 2j\| = \sqrt{3^2 + 2^2} = \sqrt{13}$$

6. Find a **unit vector** having the same direction as $\mathbf{u} = 3\mathbf{i} + 4\mathbf{j}$.
Solution:

$$\|\mathbf{u}\| = \|3i - 4j\| = \sqrt{3^2 + 4^2} = 5 \quad \therefore \text{ unit vector} = \frac{\mathbf{u}}{\|\mathbf{u}\|} = \frac{3i + 4j}{5} = \frac{3}{5}i + \frac{4}{5}j. \text{ Ans.}$$

$$\text{Hint: Its magnitude (norm)} = \sqrt{\left(\tfrac{3}{5}\right)^2 + \left(\tfrac{4}{5}\right)^2} = \sqrt{\tfrac{9}{25} + \tfrac{16}{25}} = 1$$

Examples

7. Express $\mathbf{v} = -4\mathbf{i} + 2\mathbf{j}$ in terms of $\mathbf{a} = (2, 3)$ and $\mathbf{b} = (4, 2)$.

Solution:

$$\mathbf{v} = (-4, 2)$$

$$\mathbf{v} = v_1 \mathbf{a} + v_2 \mathbf{b} = v_1(2, 3) + v_2(4, 2) = (2v_1, 3v_1) + (4v_2, 2v_2) = (2v_1 + 4v_2, 3v_1 + 2v_2)$$

Solve $\begin{cases} 2v_1 + 4v_2 = -4 \\ 3v_1 + 2v_2 = 2 \end{cases}$ we have $v_1 = 2$, $v_2 = -2$ $\therefore \mathbf{v} = 2\mathbf{a} - 2\mathbf{b}$. Ans.

Check : $\mathbf{v} = 2\mathbf{a} - 2\mathbf{b} = 2(2, 3) - 2(4, 2) = (4, 6) - (8, 4) = (-4, 2)$

8. Find the x-component and y-component of a vector $\mathbf{v}$, 10 units long at a direction of $60°$.

Solution:

$x = 10\cos 60° = 10(0.5) = 5$.

$y = 10\sin 60° = 10(0.866) = 8.66$.

$\mathbf{v} = (5, 8.66)$

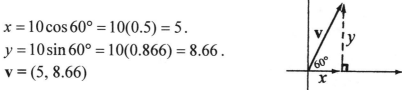

9. If $\mathbf{v} = 3\mathbf{i} + 4\mathbf{j}$ and $\mathbf{u} = 4\mathbf{i} + 3\mathbf{j}$, find the dot product $\mathbf{v} \cdot \mathbf{u}$ and the angle between $\mathbf{v}$ and $\mathbf{u}$.

Solution:

$$\mathbf{v} \cdot \mathbf{u} = 3(4) + 4(3) = 24$$

$$\cos\theta = \frac{\mathbf{v} \cdot \mathbf{u}}{\|\mathbf{v}\| \cdot \|\mathbf{u}\|} = \frac{24}{\sqrt{3^2 + 4^2} \cdot \sqrt{4^2 + 3^2}} = \frac{24}{25} = 0.96 \quad \therefore \theta = 16.26°.$$

10. An object is pulled by a force of 10 units to the south and a force of 6 units at a heading (clock-wise from due north) of $60°$. Find the magnitude and direction of the resultant.

Solution:

The force of 10 units to the south is $\mathbf{f}_1 = -10\mathbf{j}$.

The force of 6 units at a heading of $60°$ is

$$\mathbf{f}_2 = (6\cos 30°)\,\mathbf{i} + (6\sin 30°)\,\mathbf{j} = 3\sqrt{3}\,\mathbf{i} + 3\,\mathbf{j}$$

The resultant $\mathbf{r} = (-10\,\mathbf{j}) + (3\sqrt{3}\,\mathbf{i} + 3\,\mathbf{j}) = 3\sqrt{3}\,\mathbf{i} - 7\,\mathbf{j}$

The magnitude of the resultant is

$$\|\mathbf{r}\| = \sqrt{(3\sqrt{3})^2 + (-7)^2} = \sqrt{76} = 8.72.$$

The direction of the resultant is given by

$$\cos\theta = \frac{\mathbf{f}_2 \cdot \mathbf{r}}{\|\mathbf{f}_2\| \cdot \|\mathbf{r}\|} = \frac{(3\sqrt{3})(3\sqrt{3}) + 3(-7)}{(6)(8.72)} = \frac{6}{52.32} = 0.1147$$

$\therefore \theta = 83.4°$ (it is at a heading of $143.4°$). Ans.

EXERCISES

1. Given $\overrightarrow{AB}$ and $\overrightarrow{BC}$, find the resultant $\overrightarrow{AB} + \overrightarrow{BC}$.

   a. $\overrightarrow{AB} = (3, 1)$, $\overrightarrow{BC} = (2, -3)$ b. $\overrightarrow{AB} = (-4, 2)$, $\overrightarrow{BC} = (-3, -6)$

2. Given $\mathbf{v} = (3, 2)$, $\mathbf{u} = (-2, 4)$, find $\mathbf{v} + \mathbf{u}$, $\mathbf{v} - \mathbf{u}$, $\mathbf{u} - \mathbf{v}$, $\|\mathbf{v}\|$, $\|\mathbf{u}\|$, $\|\mathbf{v} - \mathbf{u}\|$.

3. If $\mathbf{v} = (3, 2)$ and $\mathbf{v} + \mathbf{u} = (1, 6)$, find $\mathbf{u}$.

4. If $\mathbf{u} = (-2, 4)$ and $\mathbf{u} - \mathbf{v} = (-5, 2)$, find $\mathbf{v}$.

5. Find the position vector (in terms of $\mathbf{i}$ an $\mathbf{j}$) of $\mathbf{v}$ having initial point p_1 and terminal point p_2.

   a. $p_1(0, 0)$, $p_2(4, -2)$ b. $p_1(3, 2)$, $p_2(-4, 2)$
   c. $p_1(-3, -2)$, $p_2(6, -1)$ d. $p_1(0, 3)$, $p_2(3, 0)$

6. If $\mathbf{v} = -4\,\mathbf{i} + 2\,\mathbf{j}$ and $\mathbf{u} = 3\,\mathbf{i} + 4\,\mathbf{j}$, find $\mathbf{v} + \mathbf{u}$, $\mathbf{v} - \mathbf{u}$, $\mathbf{u} - \mathbf{v}$, $4\mathbf{v} - 3\mathbf{u}$, $\|\mathbf{v}\|$, $\|\mathbf{u} - \mathbf{v}\|$.

7. Find the unit vector having the same direction as $\mathbf{v}$.
   a. $\mathbf{v} = 3\,\mathbf{i}$ b. $\mathbf{v} = -5\,\mathbf{i}$ c. $\mathbf{v} = -5\,\mathbf{j}$
   d. $\mathbf{v} = \mathbf{i} - \mathbf{j}$ e. $\mathbf{v} = -4\,\mathbf{i} + 3\,\mathbf{j}$

8. Express $\mathbf{v} = -4\,\mathbf{i} + 3\,\mathbf{j}$ in terms of $\mathbf{a} = (2, -1)$ and $\mathbf{b} = (3, -2)$.

9. Find the $x-$ component and $y-$ component of a vector $\mathbf{v}$, 12 units long at a direction of $130°$.

10. Find the magnitude and direction of the vector $\overline{OB} = (-5, 12)$.

11. Find the dot product $\mathbf{v} \cdot \mathbf{u}$ and the angle between $\mathbf{v}$ and $\mathbf{u}$.
   a. $\mathbf{v} = (3, 0)$, $\mathbf{u} = (0, 4)$ b. $\mathbf{v} = (1, -1)$, $\mathbf{u} = (1, 1)$ c. $\mathbf{v} = (3, -4)$, $\mathbf{u} = (4, -3)$
   d. $\mathbf{v} = 3\,\mathbf{i}$, $\mathbf{u} = \mathbf{j}$ e. $\mathbf{v} = 4\,\mathbf{i} - 3\,\mathbf{j}$, $\mathbf{u} = 2\,\mathbf{i} + 5\mathbf{j}$ f. $\mathbf{v} = 2\,\mathbf{i} + 2\,\mathbf{j}$, $\mathbf{u} = \mathbf{i} + 2\,\mathbf{j}$

12. Determine whether or not the following two vectors are parallel, perpendicular, or neither.
   a. $\mathbf{u} = (-4, 2)$ and $\mathbf{v} = (3, 6)$ b. $\mathbf{u} = (-4, 2)$ and $\mathbf{v} = (6, 3)$
   c. $\mathbf{v} = 6\,\mathbf{i} - 2\,\mathbf{j}$ and $\mathbf{u} = 3\,\mathbf{i} - \mathbf{j}$ d. $\mathbf{v} = -2\,\mathbf{i} + \mathbf{j}$ and $\mathbf{u} = -4\,\mathbf{i} + 2\,\mathbf{j}$

13. A jet plane is flying at a constant airspeed of 500 miles per hours due east, and a wind velocity of 80 miles per hours to the northeast. Find the actual speed and direction of the jet plane.

14. If $\mathbf{a}$ and $\mathbf{b}$ are two vectors, and $\|\mathbf{a}\| = \|\mathbf{b}\| = 1$, prove that $-1 \le \mathbf{a} \cdot \mathbf{b} \le 1$.
   (Hint: $\mathbf{a} \cdot \mathbf{b} = \|\mathbf{a}\| \cdot \|\mathbf{b}\| \cos\theta$)

12-3 Slope and Derivative of a Curve

In basic Algebra, we learned how to find the slope of a straight line. The slope of a straight line is used to describe the rate of change when a line rises or falls, $\Delta y / \Delta x$. The slope is the same at every point on a straight line (a linear function). The slope is different from point to point on a curve (a non-linear function).

There are many applications of rates of change in science and our real life. For example, the scientists and astronauts want to calculate the velocity and acceleration of a space shuttle at every point and time on its elliptical orbit.(See page 267 and 305)

In Calculus, we start to find formula for the slope (rates of change) of the **tangent line** from point to point on a curve. The process is called **differentiation**.
In the process of differentiation, we derive the slope formula of the function $y = f(x)$.
The derived function is called the **derivative** of $y = f(x)$ at x.

To find the slope of the tangent line at a point on a curve, we use the following **Newton's Four-Step Process of Differentiation:**

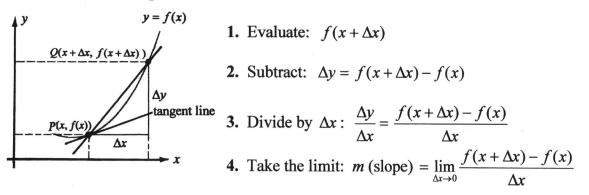

1. Evaluate: $f(x + \Delta x)$

2. Subtract: $\Delta y = f(x + \Delta x) - f(x)$

3. Divide by Δx : $\dfrac{\Delta y}{\Delta x} = \dfrac{f(x + \Delta x) - f(x)}{\Delta x}$

4. Take the limit: $m \text{ (slope)} = \lim_{\Delta x \to 0} \dfrac{f(x + \Delta x) - f(x)}{\Delta x}$

Δx and Δy are **changes** in x and y, or **increments** of x and y. Symbol Δ is read "Delta".

$\Delta x / \Delta y$ is the **average rate of change** of $f(x)$ from point P to point Q. It is the slope of the secant line PQ passing through P (the point of tangent) and any nearby point Q. Therefore, the slope (m) of the secant line PQ is given by:

$$m(\text{ secant line }) = \frac{\Delta y}{\Delta x} = \frac{f(x + \Delta x) - f(x)}{\Delta x}$$

When the point Q moves closer to the point P, we obtain better approximation of the slope of the tangent line at P. When Δx approaches 0 ($\Delta x \to 0$), we can find the exact slope of the tangent line at point P.

The expression in Step 4 is called the derivative of the function $y = f(x)$ at x.

The notations to describe the derivative are y', $f'(x)$, or $\dfrac{dy}{dx}$.

Definition of Derivative of a Function:

The derivative of the function $y = f(x)$ at x is given by:

$$y' = \frac{dy}{dx} = \lim_{\Delta x \to 0} \frac{\Delta y}{\Delta x} = \lim_{\Delta x \to 0} \frac{f(x + \Delta x) - f(x)}{\Delta x} \quad \text{if the limit exists.}$$

The other form to express the derivative of a function:

The derivative of the function $y = f(x)$ at c is given by:

$$y' = \frac{dy}{dx} = \lim_{\Delta x \to 0} \frac{\Delta y}{\Delta x} = \lim_{x \to c} \frac{f(x) - f(c)}{x - c} \quad \text{if the limit exists.}$$

Examples

1. Find the slope of the line $y = 3x - 5$ at the point (2, 1).

Solution:

$$m = \lim_{\Delta x \to 0} \frac{f(x + \Delta x) - f(x)}{\Delta x} = \lim_{\Delta x \to 0} \frac{[3(x + \Delta x) - 5] - (3x - 5)]}{\Delta x} = \lim_{\Delta x \to 0} \frac{3\Delta x}{\Delta x} = \lim_{\Delta x \to 0} 3 = 3. \text{ Ans.}$$

2. Find the slope of the function $y = x^2$ at the point (1, 1) and the point (2,4).

Solution:

$$m = \lim_{\Delta x \to 0} \frac{f(x + \Delta x) - f(x)}{\Delta x} = \lim_{\Delta x \to 0} \frac{(x + \Delta x)^2 - x^2}{\Delta x} = \lim_{\Delta x \to} \frac{x^2 + 2 \cdot x \cdot \Delta x + (\Delta x)^2 - x^2}{\Delta x}$$

$$= \lim_{\Delta x \to 0} \frac{2 \cdot x \cdot \Delta x + (\Delta x)^{2^1}}{\Delta x} = \lim_{\Delta x \to 0} (2x + \Delta x) = 2x$$

Therefore, the slope at the point (1, 1): $m = 2(1) = 2$. At (2, 4): $m = 2(2) = 4$. Ans.

Applying the Four-Step Process of differentiation, we can prove and obtain the following **General Formulas of Differentiation**:

1. If c is a constant, then $\frac{d}{dx}[c] = 0$	**4.** $\frac{d}{dx}[f(x) \pm g(x)] = f'(x) \pm g'(x)$
2. $\frac{d}{dx}[x^n] = nx^{n-1}$	**5.** $\frac{d}{dx}[f(x)g(x)] = f(x)g'(x) + g(x)f'(x)$
3. $\frac{d}{dx}[cf(x)] = cf'(x)$	**6.** $\frac{d}{dx}\left[\frac{f(x)}{g(x)}\right] = \frac{g(x)f'(x) - f(x)g'(x)}{[g(x)]^2}$

Fortunately, we use the above formulas to calculate the derivatives directly and simply without the tedious process by using the limit definition of the derivative.

Examples: 1. $\frac{d}{dx}(2) = 0$ **2.** $\frac{d}{dx}(12) = 0$ **3.** $\frac{d}{dx}(x) = 1x^{1-1} = x^0 = 1$ **4.** $\frac{d}{dx}x^2 = 2x^{2-1} = 2x$

5. $\frac{d}{dx}x^5 = 5x^{5-1} = 5x^4$ **6.** $\frac{d}{dx}(4x^{10}) = 4 \cdot \frac{d}{dx}(x^{10}) = 4 \cdot 10x^{10-1} = 40x^9$

Examples

7. $\dfrac{d}{dx}(x^{-3}) = -3 \cdot x^{-3-1} = -3x^{-4} = -\dfrac{3}{x^4}$

8. $\dfrac{d}{dx}\left(\dfrac{2}{x}\right) = \dfrac{d}{dx}(2x^{-1}) = 2\dfrac{d}{dx}(x^{-1}) = 2(-1)x^{-2} = -\dfrac{2}{x^2}$

9. $\dfrac{d}{dx}(-\dfrac{3}{5x^4}) = -\dfrac{3}{5}\dfrac{d}{dx}(x^{-4}) = -\dfrac{3}{5} \cdot -4x^{-5} = \dfrac{12}{5x^5}$

10. $\dfrac{d}{dx}(3\pi\ x^4) = 3\pi\dfrac{d}{dx}(x^4) = 3\pi \cdot 4x^3 = 12\pi\ x^3$

11. $\dfrac{d}{dx}(\sqrt{x}) = \dfrac{d}{dx}(x^{1/2}) = \dfrac{1}{2}x^{-1/2} = \dfrac{1}{2\sqrt{x}}$

12. $\dfrac{d}{dx}(x^{-2/3}) = -\dfrac{2}{3} \cdot x^{-5/3} = -\dfrac{2}{3x^{5/3}}$

13. $\dfrac{d}{dx}(5x^2 + 3x - 6) = \dfrac{d}{dx}(5x^2) + \dfrac{d}{dx}(3x) + \dfrac{d}{dx}(-6) = 10x + 3 + 0 = 10x + 3$

14. $\dfrac{d}{dx}[(2x^2 - 3x)(4x + 5)] = (2x^2 - 3x)\dfrac{d}{dx}(4x + 5) + (4x + 5)\dfrac{d}{dx}(2x^2 - 3x)$

$$= (2x^2 - 3x)(4) + (4x + 5)(4x - 3) = (8x^2 - 12x) + (16x^2 + 8x - 15)$$

$$= 24x^2 - 4x - 15$$

15. $\dfrac{d}{dx}[(2 + x^{-1})(x - 2)] = (2 + x^{-1})\dfrac{d}{dx}(x - 2) + (x - 2)\dfrac{d}{dx}(2 + x^{-1})$

$$= (2 + x^{-1})(1) + (x - 2)(-x^{-2})$$

$$= 2 + \dfrac{1}{x} - \dfrac{x - 2}{x^2} = \dfrac{2x^2 + x - x + 2}{x^2} = \dfrac{2x^2 + 2}{x^2}$$

16. $\dfrac{d}{dx}\left(\dfrac{x - 1}{x + 1}\right) = \dfrac{(x + 1)\dfrac{d}{dx}(x - 1) - (x - 1)\dfrac{d}{dx}(x + 1)}{(x + 1)^2} = \dfrac{(x + 1)(1) - (x - 1)(1)}{(x + 1)^2} = \dfrac{2}{(x + 1)^2}$

17. $\dfrac{d}{dx}\left(\dfrac{x^2 + 3x - 2}{2x - 1}\right) = \dfrac{(2x - 1)\dfrac{d}{dx}(x^2 + 3x - 2) - (x^2 + 3x - 2)\dfrac{d}{dx}(2x - 1)}{(2x - 1)^2}$

$$= \dfrac{(2x - 1)(2x + 3) - (x^2 + 3x - 2)(2)}{(2x - 1)^2} = \dfrac{4x^2 + 4x - 3 - 2x^2 - 6x + 4}{(2x - 1)^2}$$

$$= \dfrac{2x^2 - 2x + 1}{(2x - 1)^2}$$

18. Find the slope of the line $y = 3x - 5$ at the point (2, 1).

Solution:

$$m = \dfrac{d}{dx}(3x - 5) = 3. \text{ Ans.}$$

19. Find the slope of the function $f(x) = x^3 - 2x$ at the point (1, −1).

Solution:

$$m = \dfrac{d}{dx}(x^3 - 2x) = 3x^2 - 2 = 3(1^2) - 2 = 1. \text{ Ans.}$$

Velocity and Acceleration

Speed is defined as the absolute value of velocity (v). Speed is always nonnegative.
If the distance function for an object moving along a straight line is $s = f(t)$, then the
average velocity (rate of change) is the distance traveled divided by elapsed time ($\Delta s/\Delta t$).

To find the instantaneous velocity (the velocity at exact time t), we take the derivative of s
with respect to t.

> **Velocity:** The instantaneous velocity of an object at time t is $v(t) = \dfrac{ds}{dt}$.

> **Acceleration:** The acceleration of an object at time t is $a(t) = \dfrac{dv}{dt}$.

Example

The distance function of an object which is thrown upward is given by the formula
$s(t) = -4.9t^2 + v_0 t$, where v_0 is the initial velocity in meters per second and t is the time
in seconds. A rocket is fired upward with an initial velocity of 100 meters per second.
1. Find the velocity and acceleration of the rocket after 3 seconds and 12 seconds.
2. How high does the rocket go ?

Solution:

1. Velocity: $v(t) = \dfrac{ds}{dt} = \dfrac{d}{dt}(-4.9t^2 + 100t) = -9.8\,t + 100$

$\therefore v(t = 3) = -9.8(3) + 100 = 70.6 \; m/\sec$ (upward).

$v(t = 12) = -9.8(12) + 100 = -17.6 \; m/\sec$ (downward).

Acceleration: $a(t) = \dfrac{dv}{dt} = \dfrac{d}{dt}(-9.8\,t + 100) = -9.8 \; m/\sec^2$. Ans.

(It is called the **Acceleration due to Gravity**, or $-32.2 \; ft/\sec^2$.)

2. The rocket reaches the highest point when its velocity is zero.
$v(t) = -9.8\,t + 100 = 0$, we have $t = 10.2$.

The highest point the rocket reaches is:
$s(t) = -4.9t^2 + 100\,t$

$s(t = 10.2) = -4.9(10.2)^2 + 100(10.2) = 510.2$ meters. Ans.

The Maximum and Minimum Values of a function

If a function $y = f(x)$ is continuous on an interval $[a, b]$ and has its maximum (or

minimum) at a point x_0 which is in the interval, then $\dfrac{dy}{dx} = f'(x_0) = 0$.

It means that the slope at the maximum (or minimum) point is zero.

EXERCISES

Use the Four-Step limit process to find the slope of each function at the specified point.

1. $y = 4x + 6$, $(2, 5)$

2. $y = -4x + 2$, $(5, 2)$

3. $f(x) = x^2 - 5$, $(-6, 1)$

4. $f(x) = 3x^2 - 2x$, $(-1, 8)$

5. $y = \sqrt{x}$, $(4, -7)$

6. $k(x) = x^3 - x^2 + 1$, $(2, 3)$

7. $g(x) = \dfrac{5}{x}$, $(-3, 4)$

8. $h(x) = \dfrac{1}{x - 3}$, $(5, 3)$

9. $h(x) = \dfrac{3}{x^2}$, $(-\frac{1}{2}, 10)$

Find the derivative of each function.

10. $y = 15$

11. $f(x) = -8$

12. $y = 15x$

13. $f(x) = -8x$

14. $y = 15x^2$

15. $y = 15x^2 + 10x$

16. $h(x) = 15x^3 + 10x^2$

17. $g(x) = -2x^{20} + 3x^{15}$

18. $y = \sqrt{x}$

19. $y = \sqrt{x + 1}$

20. $f(x) = \sqrt{x - 4}$

21. $f(x) = (x + 1)^{2/3}$

22. $h(x) = \dfrac{1}{x + 1}$

23. $g(x) = \dfrac{1}{x - 3}$

24. $k(x) = \dfrac{1}{\sqrt{x}}$

25. $f(x) = (3x^2 + 2x)(2x + 1)$

26. $f(x) = (x^3 - 2x)(2x^2 + 3)$

27. $g(x) = (x^5 - 3x) \cdot \dfrac{1}{x}$

28. $y = \dfrac{x + 1}{x - 1}$

29. $y = \dfrac{x^2 - 1}{x^2 + 1}$

30. $h(x) = \dfrac{2x^2 - 4x + 3}{-3x + 2}$

31. The distance function of a free-falling ball is given by the formula $s(t) = -16t^2 + s_0$, where t is the time in seconds and s_0 is the initial height in feet of the ball. Find the velocity and acceleration of the ball after 3 seconds.

32. The distance function of an object which is thrown upward is given by the formula $s(t) = -16t^2 + v_0 t$, where t is the time in seconds and v_0 is the initial velocity in feet per second. A rocket is fired upward with an initial velocity of 160 feet per second. Find the velocity and acceleration of the rocket after 4 seconds.

33. Find the maximum profit if the equation of profit is $p(x) = -2x^2 + 32x - 38$, where $p(x)$ is the profit and x is the number of products sold. (See Example 7, page 120)
   Hint: The slope of the function at x having the maximum profit is 0.

34. Find the minimum cost if the equation of cost is $c(x) = 2x^2 - 32x + 158$, where $c(x)$ is the cost and x is the number of products made.
   Hint: The slope of the function at x having the minimum cost is 0.

Answers

CHAPTER 12–1 Limits and Continuity page 438

1. 5   **2.** 9   **3.** 1   **4.** 1.1   **5.** 0   **6.** 0   **7.** 1   **8.** The limit does not exist. (see graph on page 434)

9. The limit does not exist. (see graph on page 434)   **10.** The limit does not exist.   **11.** −4   **12.** −6

13. 4   **14.** 0   **15.** −1   **16.** 3   **17.** $\frac{1}{4}$ **18.** $+\infty$ (The limit does not exist.)   **19.** The limit does not exist.

20. The limit does not exist.   **21.** 6   **22.** $\frac{1}{4}$ **23.** 0   **24.** $\frac{1}{2}$ **25.** $+\infty$ (The limit does not exist.)

26. $-\infty$ (The limit does not exist.)   **27.** 2   **28.** $-\infty$ (The limit does not exist.)   **29.** 2

30. $-\infty$ (The limit does not exist.)   **31.** $+\infty$ (The limit does not exist.)   **32.** $-\infty$ (The limit does not exist.)

33. $+\infty$ (The limit does not exist.)   **34.** 0

35 ~ 45 The limit does not exist   **35.** $+\infty$ **36.** $+\infty$ **37.** $-\infty$ **38.** $+\infty$ **39.** $-\infty$ **40.** $+\infty$ **41.** $+\infty$

42. $+\infty$ **43.** $+\infty$ **44.** $-\infty$ **45.** The limit does not exist. (Hint: $\lim\limits_{x\to2^+} f(x)=\frac{2}{3}$, $\lim\limits_{x\to2^-} f(x)=\frac{3}{2}$)

46. 1 (Hint: $\lim\limits_{x\to3^+} f(x)=1$, $\lim\limits_{x\to3^-} f(x)=1$)

47. Continuous for all real numbers   **48.** Continuous on [−2, 2]   **49.** Continuous for all real numbers

50. Continuous for all real numbers

51. Continuous for all real numbers except $x = 1$; Discontinuous at $x =1$ (nonremovable)

52. Continuous for all real numbers except $x = −2$ and 2; Discontinuous at $x = −2$ and 2 (both nonremovable)

53. Continuous for all real numbers except $x = −1$; Discontinuous at $x = −1$ (removable)

54. Continuous for all real numbers except $x = −3$ and 1.
Discontinuous at $x = −3$ (nonremovable) and $x = 1$ (removable).

55. Continuous for all real numbers

56. Continuous for all real numbers except $x = 2$. Discontinuous at $x = 2$ (nonremovable)

57. Continuous on [−2, 3]   **58.** Continuous for all real numbers   **59.** Continuous for all real numbers

60. Continuous for all x except $x = −2$. Discontinuous at $x = −2$ (nonremovable)   **61.** $c = 3$

62. The zero of the function is in the interval [0.75, 0.76].
Hint: Locate the zero in a subinterval of length 0.1 on [0, 1]. We have:
$f(0.7) = −0.167$, $f(0.8) = 0.152$. Therefore, the zero is on [0.7, 0.8].
Locate the zero in a subinterval of length 0.01 on [0.7, 0.8]. we have:
$f(0.75) = −0.0156$, $f(0.76) = 0.0165$. Therefore, the zero is on [0.75, 0.76].

CHAPTER 12–2 Vectors page 444

1. a. (5, −2)   b. (−7, −4)

2. $\mathbf{v} + \mathbf{u} = (1, 6)$, $\mathbf{v} − \mathbf{u} = (5, −2)$, $\mathbf{u} − \mathbf{v} = (−5, 2)$, $\|\mathbf{v}\| = \sqrt{13}$, $\|\mathbf{u}\| = 2\sqrt{5}$, $\|\mathbf{v} − \mathbf{u}\| = \sqrt{29}$

3. $\mathbf{u} = (−2, 4)$ **4.** $\mathbf{v} = (3, 2)$ **5.** a. $\mathbf{v} = 4\,\mathbf{i} − 2\,\mathbf{j}$ b. $\mathbf{v} = −7\,\mathbf{j}$ c. $\mathbf{v} = 9\,\mathbf{i} + \mathbf{j}$ d. $\mathbf{v} = 3\,\mathbf{i} − 3\,\mathbf{j}$

6. $\mathbf{v} + \mathbf{u} = −\mathbf{i} + 6\,\mathbf{j}$, $\mathbf{v} − \mathbf{u} = −7\,\mathbf{i} − 2\,\mathbf{j}$, $\mathbf{u} − \mathbf{v} = 7\,\mathbf{i} + 2\,\mathbf{j}$, $4\mathbf{v} − 3\mathbf{u} = −25\,\mathbf{i} − 4\,\mathbf{j}$, $\|\mathbf{v}\| = 2\sqrt{5}$, $\|\mathbf{u} − \mathbf{v}\| = \sqrt{53}$

7. a. $\mathbf{i}$ b. $−\mathbf{i}$ c. $−\mathbf{j}$ d $\frac{\sqrt{2}}{2}\mathbf{i} − \frac{\sqrt{2}}{2}\mathbf{j}$ e. $−\frac{4}{5}\mathbf{i} + \frac{3}{5}\mathbf{j}$

8. $\mathbf{v} = \mathbf{a} − 2\,\mathbf{b}$ **9.** $x = −7.11$, $y = 7.66$ or $\mathbf{v} = (−7.71, 7.66)$ **10.** 13, 112.6° from the positive $x − axis$

11. a. 0, 90°   b. 0, 90°   c. 24, 16.26°   d. 0, 90°   e. −7, 105°   f. 6, 18.4°.   **14.** $\mathbf{a} \cdot \mathbf{b} = \|\mathbf{a}\| \cdot \|\mathbf{b}\| \cos\theta$

12. a. Perpendicular   b. Neither   c. Parallel.   d. Parallel.   Since $−1 \le \cos\theta \le 1$,

13. 559 miles per hour with a direction about 5.4° north of east (or E 5.4° N).   we have $−1 \le \mathbf{a} \cdot \mathbf{b} \le 1$

CHAPTER 12–3 Slope and Derivative on a curve page 449

1. 4   **2.** −4   **3.** −12   **4.** −8   **5.** $\frac{1}{4}$ **6.** 8   **7.** $−\frac{5}{9}$ **8.** $−\frac{1}{4}$ **9.** 48   **10.** 0   **11.** 0   **12.** 15   **13.** −8   **14.** $30x$

15. $30x + 10$ **16.** $30x^2 + 20x$ **17.** $−40x^{19} + 45x^{14}$ **18.** $\dfrac{1}{2\sqrt{x}}$ **19.** $\dfrac{1}{2\sqrt{x+1}}$ **20.** $\dfrac{1}{2\sqrt{x-4}}$ **21.** $\dfrac{2}{3(x+2)^{1/3}}$

22. $−\dfrac{1}{(x+1)^2}$ **23.** $−\dfrac{1}{(x-3)^2}$ **24.** $−\dfrac{1}{2x^{-5/2}}$ **25.** $18x^2 + 14x + 2$ **26.** $10x^4 − 3x^2 − 6$ **27.** $4x^3 − \dfrac{2}{x}$ **28.** $−\dfrac{2}{(x-1)^2}$

29. $\dfrac{4x}{(x^2+1)^2}$ **30.** $\dfrac{−6x^2 + 8x + 1}{(−3x+2)^2}$ **31.** $v = −96$ ft/sec (downward), $a = −32$ ft/sec^2 (It is called the Acceleration
due to Gravity. It is equal to $−9.8$ m/sec^2.)   **32.** $v = 32$ ft/sec, $a = −32$ ft/sec^2 **33.** \$90   **34.** \$30

INDEX

APPENDIX

Abraham Lincoln was the 16th President of the United States. One day, as he was giving a speech before the crowd, a pro-slave audience wrote " foolish " on a piece of paper and gave it to Lincoln.

Lincoln: Yesterday, I received a letter from a friend who forgot to sign his name.
Today, I just received a letter from a friend who signed his name only.

A teacher is testing a seven-year old boy.
Teacher: What is the answer of 5×6 ?
Boy: 15
Teacher: That is not right.
The answer should be 30.
Boy: At least my answer is half right.

Now, you have completed this book. Are you ready
to study Calculus ?   See you soon in next book !!

The A-Plus Notes Series:

1. A-Plus Notes for Beginning Algebra (Pre-Algebra and Algebra 1)

2. A-Plus Notes for Algebra (Algebra 2 and Pre-Calculus)
   (A Graphing Calculator Approach)

3. A-Plus Notes for Calculus (Coming Soon)

If you are an educator and want to write your own book (all subjects)
to improve the education in schools, please contact us.

" A person's life has an end.
  Honor or dishonor does not last long.
  Only creative writing is eternal. "

Confucius

" 生命有其所終，
  榮辱止於其身，
  未若文章之無窮。"

孔夫子

Simply complete and mail the order form (or a copy of the form).

A-PLUS NOTES FOR ALGEBRA

ORDER FORM

Qty	Paperback	Hardcover	quantity	$ Amount
Regular	$22.50	$31.50		
10 to 50	21.50	30.50		
51 to 100	20.50	29.50		
101 and up	19.50	28.50		

Prices are subject to change without notice.

Please call (310) 542-0029.   Subtotal   _____

$2.50 per copy for S/H & Tax   _____

Total   _____

Ship to:(PLEASE PRINT)

Name _____

School name _____

Address _____

City/State/Zip _____

Daytime Phone _____

Make check payable to **A-Plus Notes Learning Center**.
Mail with form to:

A-Plus Notes Learning Center
2211 Graham Ave.
Redondo Beach, CA 90278

Allow 2 to 3 weeks for delivery. Thank you.